7th International Congress on Combustion Engines (PTNSS 2017)

MATEC Web of Conferences Volume 118 (2017)

Poznan, Poland
27 – 29 June 2017

Editor:

Jacek Pielecha

ISBN: 978-1-5108-4516-9

Printed from e-media with permission by:

Curran Associates, Inc.
57 Morehouse Lane
Red Hook, NY 12571

CURRAN ASSOCIATES INC. proceedings.com

Some format issues inherent in the e-media version may also appear in this print version.

Printed by Curran Associates, Inc. (2017)

For additional information, please contact EDP Sciences – Web of Conferences at the address below.

EDP Sciences – Web of Conferences
17, Avenue du Hoggar
Parc d'Activité de Courtabœuf
BP 112
F-91944 Les Ulis Cedex A
France

Phone: +33 (0) 1 69 18 75 75
Fax: +33 (0) 1 69 28 84 91

contact-edps@webofconferences.org

Additional copies of this publication are available from:

Curran Associates, Inc.
57 Morehouse Lane
Red Hook, NY 12571 USA
Phone: 845-758-0400
Fax: 845-758-2633
Email: curran@proceedings.com
Web: www.proceedings.com

TABLE OF CONTENTS

PTNSS

ISSiT

VII INTERNATIONAL
CONGRESS ON COMBUSTION ENGINES

POLISH SCIENTIFIC SOCIETY
OF COMBUSTION ENGINES

27th-29th June 2017

HONORARY PATRONAGE

POLITECHNIKA POZNAŃSKA – POZNAN UNIVERSITY OF TECHNOLOGY

PAN
POLISH ACADEMY of SCIENCES

WOJEWODA WIELKOPOLSKI

MARSHAL OF THE
WIELKOPOLSKA REGION

POZnan
Honorary Patronage
of the Mayor of Poznań

BOSMAL®

AVL

FCA
POWERTRAIN POLAND

casp system
ADVANCED TECHNOLOGIES

LAVISION
FOCUS ON IMAGING

BINDER
Best conditions for your success

KISTLER
measure. analyze. innovate.

TSI®

VERT

Poznan University of Technology
60-965 Poznan, Piotrowo 3 Street, POLAND

HONORARY PATRONAGE

Rector of Poznan University of Technology – Tomasz Łodygowski
Polish Academy of Sciences Committee of Transport – Wojciech Wawrzyński
Governor of the Wielkopolskie Voivodeship – Zbigniew Hoffmann
Marshal of the Wielkopolska Region – Marek Woźniak
President of the City of Poznan – Jacek Jaśkowiak

HONORARY COMMITTEE OF THE CONFERENCE

Tomasz	**Łodygowski**	Rector of Poznan University of Technology, Poland
Andrzej	**Kowalewicz**	Professor University of Technology and Humanities in Radom, Poland
Hans-Peter	**Lenz**	President Austrian Society of Automotive Engineers, Austria
Helmut	**List**	President AVL List GmbH, Austria
Dariusz	**Michalak**	Deputy CEO of Solaris Bus & Coach S.A.
Jens	**Ochsen**	Plant Manager Volkswagen Poznan
Stefan	**Pischinger**	President FEV Motorentechnik, Germany
Franciszek	**Tomaszewski**	Dean of the Faculty of Machines and Transport, Poznan University of Technology, Poland
Jan A.	**Wajand**	Professor University of Bielsko-Biala, Poland

MEMBERS OF INTERNATIONAL SCIENTIFIC COMMITTEE

(reviewers of papers)

Andrzej **Adamkiewicz**
Andrzej **Augustynowicz**
Yuzo **Aoyagi**
Maciej **Bajerlein**
Piotr **Bielaczyc**
Piotr **Boguś**
Marek **Brzeżański**
Gerardo **Carelli**
Zdzisław **Chłopek**
Bogusław **Cieślikowski**
Jerzy **Cisek**
Tadeu **Cordeiro**
Karol **Cupiał**
Jan **Czerwiński**
Friedrich **Dinkelacker**
Adam **Dużyński**
Jerzy **Ejsmont**
Jacek **Eliasz**
Kurt **Engeljehringer**
Andrzej **Felski**
Hubert **Friedl**
Paweł **Fuć**
Peter **Gehr**
Jerzy **Girtler**
Wojciech **Gis**
Krzysztof **Górski**
Witold **Grzegożek**
Janbernd **Hentschel**
Jacek **Hunicz**
Marek **Idzior**
Antoni **Iskra**
Marianna **Jacyna**
Janusz **Jakóbiec**
Anna **Janicka**
Antoni **Jankowski**
Jarosław **Kałużny**
Andrzej **Kaźmierczak**
Toni **Kinnunen**
Oleh **Klyus**
Tomasz **Knefel**
Tomasz **Kniaziewicz**
Christopher **Kolodziej**
Zbigniew **Korczewski**
Grzegorz **Koszałka**
Mirosław **Kowalski**
Miłosław **Kozak**
Thomas **Krinke**
Stanisław **Kruczyński**
Andrzej **Kulczycki**
Amanda **Lea-Langdon**
Kazimierz **Lejda**
Piotr **Lijewski**
Wincenty **Lotko**
Zbigniew **Lozia**
Sławomir **Luft**
Thomas **Lutz**
Jarosław **Mamala**
Wojciech **Marek**
Jarosław **Markowski**
Andreas **Mayer**
Jerzy **Merkisz**
Władysław **Mitianiec**
Janusz **Mysłowski**
Andrzej **Niewczas**
Marcin **Noga**
Jacek **Nowakowski**

INDUSTRY COMMITTEE

ORGANISING COMMITTEE

Jerzy	**Merkisz**	President of PTNSS Congress
Piotr	**Bielaczyc**	Vice President for Foreign Participants
Zdzisław	**Stelmasiak**	Vice President for Science
Antoni	**Świątek**	Vice President for Industry
Marek	**Idzior**	Vice President for Industry Exhibition
Jacek	**Pielecha**	Secretary for Organisation
Joseph	**Woodburn**	Secretary for Foreign Participants
Dariusz	**Pietras**	Secretary for Science
Dawid	**Gallas**	Secretary for Industry Exhibition
Paweł	**Fuć**	Member
Remigiusz	**Jasiński**	Member
Wojciech	**Karpiuk**	Member
Ireneusz	**Pielecha**	Member
laudiusz	**Waniołka**	Member

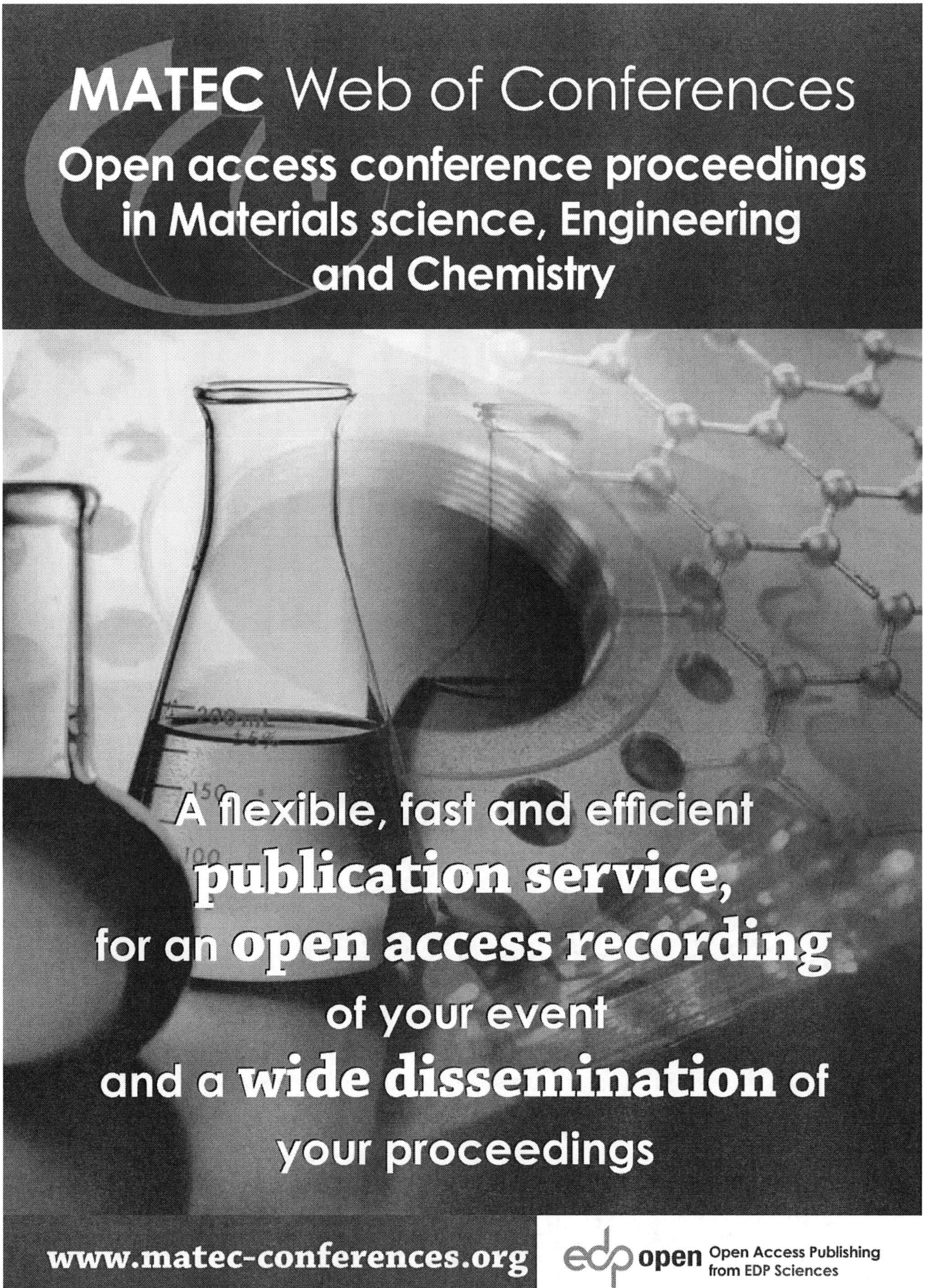
MATEC Web of Conferences
Open access conference proceedings
in Materials science, Engineering
and Chemistry
A flexible, fast and efficient
publication service,
for an open access recording
of your event
and a wide dissemination of
your proceedings
www.matec-conferences.org
edp open
Open Access Publishing
from EDP Sciences

ISSN (electronic edition): 2261-236X

edp sciences

MATEC Web of Conferences

Statement of Peer review

In submitting conference proceedings to MATEC Web of Conferences, I certify to the Publisher that I adhere to the **Policy on Publishing Integrity** of the journal in order to safeguard good scientific practice in publishing.

1. All articles have been subjected to peer review administered by the proceedings editors.
2. Reviews have been conducted by expert referees, who have been requested to provide unbiased and constructive comments aimed, whenever possible, at improving the work.
3. Proceedings editors have taken all reasonable steps to ensure the quality of the materials they publish and their decision to accept or reject a paper for publication has been based only on the merits of the work and the relevance to the journal.

Title, date and place of the conference:

VII International Congress on Combustion Engines, 27-29 June 2017, Poznan, Poland

Proceedings editor(s):

Prof. Jacek Pielecha

Date and editor's signature

30.06.2017

 DOI: 10.1051/matecconf/201711800001

Mathematical modelling malfunctions of marine diesel engine

Marcin Zacharewicz[1], and *Tomasz* Kniaziewicz[1,*]

[1]Polish Naval University, Faculty of Mechanical and Electrical Engineering, ul. Śmidowicza 69, 81-127 Gdynia, Poland

Abstract. The paper describes modelling methods of selected malfunctions in a four-stroke marine diesel engine. The model has been developed for purposes to assess methods of technical conditions selected elements of the engine structure. Modelling malfunctions allow to determine answers to energetic parameters of the faults. The paper is based on earlier publications, in which was carried out studies of the physical model of energetic processes in engine and additionally the mathematical model which was based on the developed physical model.

1 Introduction

The engines which are operated on the board of Polish Navy ships, most of them are supercharged, high speed, four-stroke engines. They are used as a main propulsion and an auxiliary engines both. The auxiliary engines drive electric generators acting together with diesel-generator (ZSE). Most of them they are not equipped with indicator valves, especially auxiliary engines [1, 2]. This is the main reason for using the TBO hourly strategy relative to this engines. This method of operation presents several problems:

- need to shut down an engine from service for maintenance. The combat readiness of the ship is low,
- high operational costs resulting from a replacement of components that could be used in the further. This illustrates the Wöhler curve (Fig. 1).

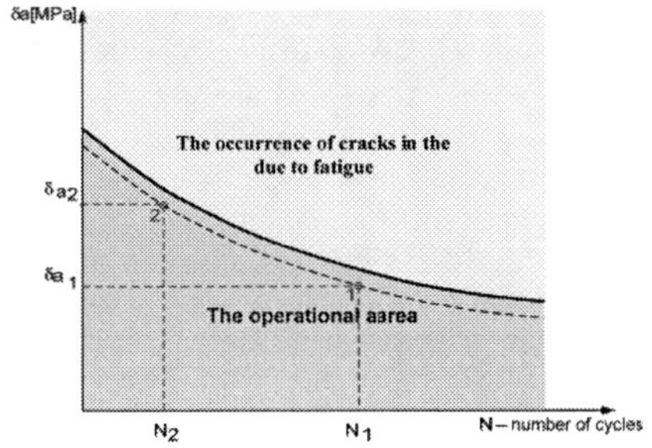

Fig. 1. The Wöhler curve [3].

Wöhler curve is a relationship between tension (periodically variable) and a number of load change cycles that could perform a part of machine until it will be destroyed due to fatigue. Wöhler curve is a statistical curve. An expected number of cycles must be under the curve. What is more, a problem is to estimate a load on an engine. As the result in most cases engine components are changed to fit for further use. These factors make necessary to look for alternative strategies according to exploitation marine engines.

Another operating strategy is the exploitation of a technical condition which is free of mentioned disadvantages [4, 5]. This strategy method involves periodic implementation of diagnostic tests. This allows to determine a technical condition of an engine (the diagnosis). What is more it could be expected changes of a technical condition in a future (prediction). It allows that a part of engine construction could be replaced in the pre-damage moment [6, 7].

One of priorities of research in The Institute of Ships Construction and Operation (IBiEO) Polish Naval Academy (AMW) is development of technical condition of engines based on non-invasive methods. Mentioned methods are based on energetic parameters diesel's marine engines or electric generators. One of the ZSE's energetic parameters is interfacial voltage of synchronous generator which is driven by engine. Analysis of recorded waveform will determine fluctuations of crankshaft engine (shaft generator) angular velocity. Fluctuations of crankshaft engine angular velocity are the result of many factors. The main ones being:

- order of cylinder work (torques generated by gas forces),
- mass reciprocating and rotary movement,
- working cycle disruption as a result of structural engine's elements damage.

Fluctuations are repeated periodically and the period of four-stroke cycle engines is:

$$T = \frac{n}{2} \cdot N_{cyl} \qquad (1)$$

where:
n – the rotational speed of engine's crankshaft [s^{-1}],
N_{cyl} – the engine cylinders number.

* Corresponding author t.kniaziewicz@amw.gdynia.pl

DOI: 10.1051/matecconf/201711800001

Fluctuations are the result from gas forces and mass moments of inertia. The Period of fluctuation rotation speed (four-stroke engine) is:

$$T = \frac{n}{2} \quad (2)$$

It results from working cycle disruption, which is connected with engine's damage.

2 Modelling of marine diesel engine's working cycle

Teams of scientists work on the problem of modeling processes in diesel engines in many countries in the world [8-10]. Most of the research is focused on a design of marine diesel engines and its environmental performance [11-13]. Many commercial programs are usually designed for this purpose. In most cases these programs are based on the finite element FEM and BEM boundary element method. Using them for diagnostic tests is often problematic. This is due to the fact that malfunctions typically are modeled as a change in the geometry of an engine structure. This approach is associated with the commercial models of a need to modify respective mesh elements. It is a very labor and time-consuming process. The source code of commercial programs are unavailable and users have only executable file. User does not have information about calculation methods and what simplifying assumptions have been applied. Therefore, it has been decided to develop a mathematical model of the physical processes in the marine four-stroke diesel engines. For this reason the model must fulfill several conditions, the most important of which are:

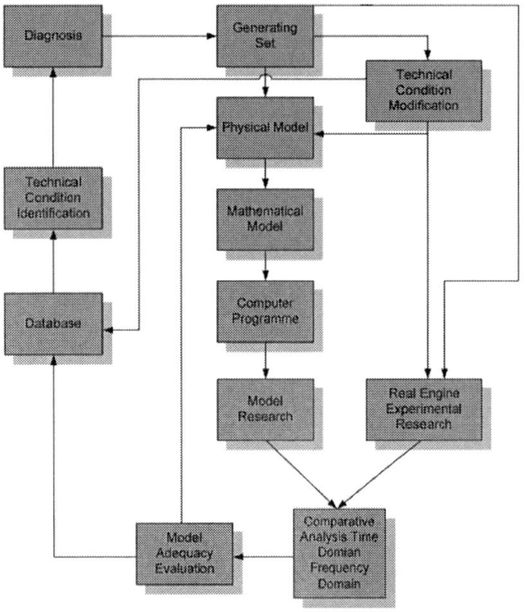

Fig. 2. Schematic implementation of research [14].

- versatility, must make possible of modeling all four stroke engines which are operated in Polish Navy,
- must be able to quickly and easily modify the structure parameters of selected structural engine components (eg. enable of changing the fuel delivery to each cylinder independently),
- performing rapid calculations for efficient and damaged engines both.

The algorithm of research methodology are shown on Fig. 2.

The object of the research is the ZSE – part of a ship's power plant. On a basis of a generating set the physical model of processes occurring in a marine engine is designed [15]. Based on the physical model is developed mathematical model of these processes [16].

3 Modeling fault conditions marine diesel engines

Marine engines is degraded differently in various states of operations (standby, running). This is a result of chemical and physical processes such as erosion, corrosion, friction or semi-dry friction, clogging, fatigue. The consequence of the engine's structure degradation is a change of condition and could be gradual or sudden. The state of condition changes from technical efficiency to the state of unfitness.

The model takes into account the ability to simulate changes in the technical condition of marine diesel engines. This is implemented by modifying the technical structure of engine. Modifications may concern elements comprising all or individual cylinder sections of engines. The modifications concern the structure of engine design (geometric dimensions) as well as the parameters of fuel and air supply. Modeled can be:

- change of fuel dose for each cylinder's sections (modeling of damage to the engine fuel supply),
- change fuel's chemical composition (for the whole engine),
- change timing injection (for each cylinder),
- change active cross-section area of intake and exhaust valves (for each cylinders),
- change of a flow resistance in exhaust channel (for whole cylinders),
- a leak in a piston-rings and cylinder (TPC) (for each of cylinders),
- change active cross-section area of injector holes (for each of cylinders).

The dose calculation of fuel is delivered during the engine cycle which is based on a value of the excess air ratio λ. By the fact that the parameter is easily measurable by exhaust gas analyzer this is the best way to determine the dose of calculation fuel.

The changing of the fuel chemical composition (in the model) is not a result of engine's technical condition. However, it is decided on the possibility of modifying this parameter. The calorific value is changed by the fuel composition, which shows the relation:

$$W = 340 \cdot C + 1017 \cdot H + 63 \cdot N + \\ + 191 \cdot S - 106 - 25 \cdot w \quad (3)$$

where C, H, N, S – are the percentages of the various chemical elements contained in the fuel. And w determines the moisture content of the fuel.

Changing the injection timing may concern the whole engine and its individual cylinders. Change of the injection advance angle for individual cylinders is a result of change injection pomp shaft.

Changing the active cross-section area of the intake valves and exhaust can be modeled for the whole engine and each cylinders. Modelling of active cross-section area is performed by modifying the flow rate μ. Value μ can be between 0 and 1, as shown by the equation:

$$\dot{m} = w \cdot S \cdot \rho \cdot \mu \quad (4)$$

where: w – linear velocity, S – cross-sectional area of the valve, ρ – density factor, μ – flow rate.

Change pressure values anti-outflow of gases from the exhaust channel outlet are modeled for the all cylinders. This is realized by modifying the pressure p_2 at the outlet from the engine cylinders in the equation de Saint-Venant [17]:

$$w = \sqrt{\frac{2 \cdot \kappa_1}{\kappa_1 - 1} \cdot R_1 \cdot T_1 \cdot \left[1 - \left(\frac{p_2}{p_1}\right)^{\frac{\kappa_1 - 1}{\kappa_1}}\right]} \quad (5)$$

In equation 5 parameters marked with the subscript "1" refers tohigher pressure (in-cylinder). Index "2" refers to a value at the output of a cylinder. Equation 5 is true, when the pressure inside a cylinder is higher than a pressure in an outlet channel, and when the linear velocity w is less than the speed of sound a which is determined by the equation [7]:

$$a = \sqrt{\kappa_1 \cdot R_1 \cdot T_1} \quad (6)$$

Modelling of leaks in a piston – ring – cylinder (TPC) is carried out for each cylinder separately. It is realized by specifying a value of a cross-section leaks in a TPC. It is calculated a value of the linear velocity (a leak) for each iteration. A value of the linear velocity w of the thermodynamic medium from a cylinder (a pressure inside the cylinder is higher than in a crankcase pressure) is based on the de Saint-Venant relation (relation 5).

The thermodynamic medium outlet from a cylinder, the value p_1 replies to the pressure inside the cylinder. The pressure p_2 replies to the pressure inside the engine crankcase. Parameters such as the adiabatic exponent κ_1, the individual gas constant R_1 and the temperature T_1 correspond with the parameters inside a cylinder. In case of a calculation all parameters are treated as variable and could be changed in every counting iteration (temperature, pressure, chemical composition). For modeling of engine failures, it is necessary to provide a constant of an indication power value for an engine in the fit state and the out of order.

In the model, this condition is fulfilled by being determined calculations engine's indicated power value for an efficient of engine. In the case of a damaged engine, calculations are repeated and in every iteration – the fuel feed rate (the excess air ratio) for an engine cylinders is modified. If a value of an indication power is less one percent (for damaged engine) then the calculation will be proceeded.

Exemplary results of carried out calculations (for full and partial technical condition of the engine) in the case of modification of fuel injection dose to one of the engine's cylinder are shown in Figures 3 and 4. Figure 3 shows the course of the indicated pressure as a function of the crankshaft's angle. Course "1" is obtained as a result of calculations for the engine in a state of full technical condition. Course "2" is corresponded to a 50% reduction in fuel supply to cylinder number 1.

Analysis of the figure shows that the speed regulator increased the fuel dose in the remaining cylinders to compensate indicate power loss due to reduction in the fuel dose in cylinder 1. For a fully engine-fit engine, the value of the excess air factor λ was 2.2. In the case of a damaged motor the excess air factor λ is 1.9. Indicative power for efficient engine was 405,182 kW and for damaged is 404,997 kW. The indicative power values of the individual cylinders for a fully engine-fit engine and damaged engine are shown in the Figure 4.

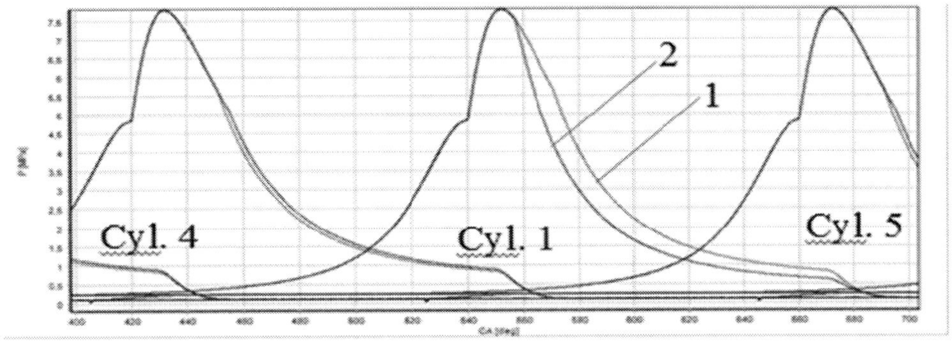

Fig. 3. The part of graph presents a pressure curve of an engine as a function of a crankshaft rotation angle, 1 – efficient engine, 2 – damaged engine.

a)
b)

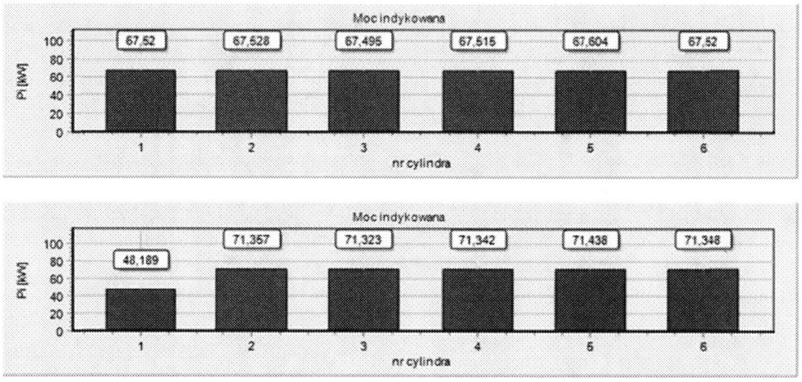

Fig. 4. The graph of power indicated by individual engine cylinders: a) efficient engine, b) damaged engine.

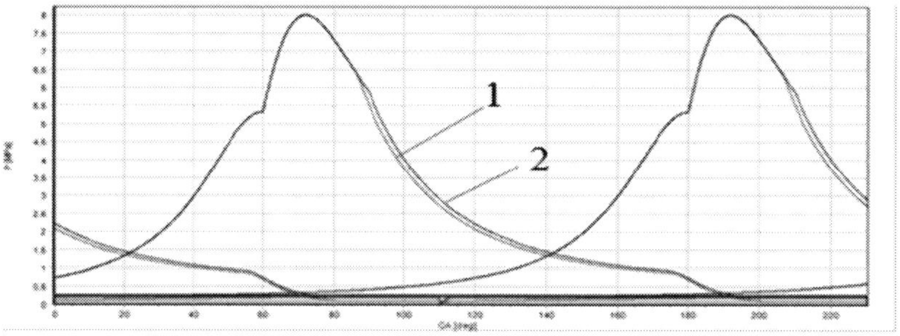

Fig. 5. The graph of pressure as a function of a crankshaft rotation angle, 1 – efficient engine, 2 – damaged engine.

a)

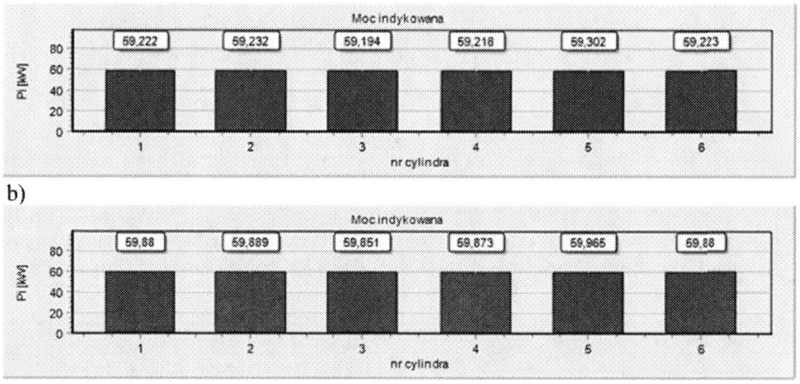

Fig. 6. The graph of power indicated by individual engine cylinders: a) efficient engine, b) damaged engine.

Another example of the research is a modification of an exhaust backpressure value in engine's exhaust duct. Figure 5 shows a process of indicated pressure as a function of the crankshaft rotation angle for an engine in full technical state and for a damaged. The back pressure (in exhaust duct) was doubled than nominal in simulation. As a result of the simulation the dose of fuel injected into the cylinder was increased (the result of the speed controller) and its combustion process was clearly elongated. The increase back pressure was caused an increase in unit fuel consumption from 105.25 to 111.02 g/kWh.

4 Summary

Damage modeling described in the article is one of the stages of method assessment of technical condition development of a marine engine with limited control susceptibility. Ability to simulate energy parameters of an damage engine allows both quantitative and qualitative assessment of their impact on it. The proposed model can also be used for research purposes: assess the energy efficiency of marine engines and assess impact of damage engine on emissions of toxic compounds.

The model is highly sensitive to a technical failures. Simulation of damage by reducing the fuel dose in one of cylinders caused about 30% indicated power drop from a damaged cylinder. Remaining cylinders increased the load by about 6%. The total power indicated by engine cylinders practically did not change due to a simulated technical state. In simulation, the change of flow resistance in an exhaust duct caused about 8% decrease in excess air factor. In addition, an increase back pressure caused an increase about 6%. The indicated power was not changed for this damage.

Currently the mathematical model is being verified by comparing the results obtained during experimental research on Sulzer type 6AL20/24 and Leyland type 6SW400 engines.

References

1. S. Polanowski, R. Pawletko, K. Witkowski, *Key* Engineering *Materials* **588**, 149-156 (2014), DOI: 10.4028/www.scientific.net/KEM.588.149
2. J. Rudnicki, R. Zadrag, Polish Maritime Research **21**, 57-65 (2014), DOI: 10.2478/pomr-2014-0042
3. Z. Korczewski, M. Zacharewicz, Metoda diagnozowania silników okrętów wojennych o ograniczonej możliwości pomiaru ciśnień wewnątrzcylindrowych na podstawie wyników badania procesów gazodynamicznych w układzie turbodoładowania. Projekt badawczy nr 0T00B02129, Gdynia (2008)
4. R. Zadrag, *Solid State Phenomena* **196**, 74-81 (2013), DOI: 0.4028/www.scientific.net/SSP.196.74
5. D. Pietras, Polish Maritime Research **23**, 58-63 (2016), DOI: 10.1515/pomr-2016-0032
6. Z. Korczewski, J. Rudnicki, *Computational Methods in Marine Engineering VI*, 490-501 (2015)
7. C. Specht, J. Rudnicki, Polish Maritime Research **23**, 20-27 (2016), DOI: 10.1515/pomr-2016-0028
8. L. Guzzella, C.H. Onder, *Introduction to Modeling and Control of Internal Combustion Engine Systems* (Springer, 2010)
9. S. Kopeć, A. Witek, Archiwum Technologii Maszyn i Automatyzacji **26**, 2 (2006)
10. Y.Z. Zhang, E.H. Kung, D.C. Haworth, *Proceedings of the Combustion Institute* **30**, 2 (2005)
11. J. Merkisz, J. Pielecha, *Springer Tracts on Transportation and Traffic* **8**, 131-139 (2015), DOI: 10.1007/978-3-319-15928-7_10
12. J. Merkisz, P. Lijewski, P. Fuc, M. Siedlecki, A. Ziolkowski, *IOP Conference Series-Materials Science and Engineering* **148**, UNSP 012077 (2016), DOI: 10.1088/1757-899X/148/1/012077
13. P. Lijewski, J. Merkisz, P. Fuc, A. Ziolkowski, L. Rymaniak, W. Kusiak, European Journal of Forest Research **136**, 153-160 (2017), DOI: 10.1007/s10342-016-1015-2
14. R.H. Cannon, *Dynamika układów fizycznych* (WNT, Warszawa, 1973)
15. A. Cwalina, M. Zacharewicz, Research on energetic processes in a marine diesel engine driving a synchronous generator for diagnostic purposes PART 1 – Physical model of the processes
16. M. Zacharewicz, T. Kniaziewicz, A. Cwalina, *Solid State Phenomena* **236** (2014)
17. W. Pudlik, *Termodynamika* (PG, Gdańsk, 1998)

MATEC Web of Conferences **118**, 00002 (2017) DOI: 10.1051/matecconf/201711800002
VII International Congress on Combustion Engines

Deterioration of the fuel injection parameters as a result of Common Rail injectors deposit formation

Zbigniew Stępień[1,*]

[1]Oil and Gas Institute – National Research Institute, 31-503 Cracow, ul. Lubicz 25A, Poland

Abstract. The article describes external and internal Common Rail injectors deposits formed in dynamometer engine simulation tests. It discussed not only the key reasons and factors influencing injector deposit formation but also the resulting way of fuel preparation and engine test approaches. The effects of external coking deposit as well as internal deposits two most common form types that is carboxylic soaps and organic amides on deterioration of the fuel injection parameters were assessed. The assessments covered both deposits impacts on quantitative and qualitative changes of the injectors diagnostic parameters and as a result on deterioration of the injector performance. Finally the comparisons between characteristic of dosage of one fuel injector before test and characteristics few injectors after engine tests of simulated deposit formation were made.

1 Introduction

Problems arising from build-up of different types of damaging deposits in diesel fuel injectors is well known since many years now and was extensively described in many scientific publications including [1-3]. In the 1980s the build-up of external injector deposits on the pintle of a fuel injector has caused a significant concern of fuel systems producers and introduction of DCA (Deposit Control Additives) to reduce injector fouling occurred [4]. Continuing since 1997 development of HPCR (High Pressure Common Rail) systems which design features are more conductive or susceptible to deposit formation, especially IDID (Internal Diesel Injector Deposits), such as severe high temperature along with high pressure operating conditions as well as precision engineering to operate highly controlled multiple injection strategies [5-7]. This may rise problems in the field due to deposits that have formed on their critical moving parts, such as the needle and control valve and manifested themselves as injector sticking, rough idling, power loss, issues with drivability, high emissions and many more [8-10]. The current industry focus is on internal injector needle deposits. These IDID have been categorized in four main types: higher carboxylate deposit, amide deposit, ester lower carboxylate – cold flow improver deposit and ester lower carboxylate – FAME deposit [10-15]. The mechanism of IDID deposit formation is different from that of external deposits on the injection nozzles, because internal deposits are not exposed to exhaust gas generated during combustion and surface evaporation of fuel. IDID are formed as a result of deteriorated fuel which becomes deposit adheres to the material surface by physical binding or deteriorated fuel which adheres to the material surface by chemical binding [2, 16]. It was concluded that in case of diesel engines with indirect fuel injection the amount and rate of coke formation were widely influenced by the type of injector used, in case of direct fuel injection it is the fuel that plays the key role in formation of the subject coke deposits. Wider use of low sulfur diesel fuels and fuel additives gave way to fuels containing acid components. For example, to a different extent, unsaturated fatty acids are commonly used as lubricant additives. These acids easily react with metallic ions of the fuel contaminants forming carboxylic salts, that, similarly to low molecular mass polar compounds, much worse dissolve in low sulfur diesel fuels of low content of aromatic hydrocarbons compared to old high sulfur content fuels [1-4, 16, 17]. A growing share of FAME (Fatty Acid Methyl Esters) characterized by low stability in the diesel fuels accelerates the process of fuel oxidation and degradation. FAME are a source of weak acids and lead to an increase in the fuel contamination (sodium compounds) – a component of typical catalysts used in the reaction of transestrification [1-4, 16-19]. Hence, due to their low stability, the resistance to oxidation is one of the most important properties of diesel fuels containing FAME. Consequently, in the market today there are a variety of fuel qualities. At the same time DCA additives used in fuels are generally considered to be the most effective way to counteract injector deposit formation. However, currently, widely applied detergents preventing external coke deposits on the injectors are based on PIBSI (Polyisobutylene Succinimide) and in combination with the previously described factors may facilitate the formation of IDID [1-4]. Therefore, the development and manufacture of these type additives is critical to their performance in the field in terms of functioning to control both external and internal deposits but also from a no-charm standpoint [4, 20, 21].

* Corresponding author stepien@inig.pl

This paper reports results generated to develop further knowledge relating to the engine test investigations into external and internal common rail injector deposit formation. The main research problems concerned the possibilities of reproducing different type of injector deposits in engine bench tests that simulated conditions of real operation. The main target of the research was to investigate the problem of injector deposit formation and to try to identify the source of the problem. It was important to study the effects of diesel fuel properties on the common rail injector functioning because the fuel could deteriorate inside an injector at such severe conditions and build-up of damaging deposits. As a criterion for assessing the effects of external and internal deposits reproduced during engine simulation tests were used diagnostic parameters of injector functioning and their characteristics of fuel injection. In addition, visual assessments of the deposits were performed.

2 Methodology

2.1 Reproduction engine test of injector deposit

The engine simulation tests of injector deposits reproduction were performed on a multi-purpose test stand equipped with a modern, widely used HSDI (High Speed Direct Injection) FORD 2.0i 16V Duratorq TDCi diesel engine coupled with an Alpha 160 AF eddy current brake from AVL, with a control module allowing the programming of the engine operation parameters (rpm, load, phase time and ramp time between phases). The basic technical parameters of the FORD 2.0i 16V Duratorq TDCi are given in Table 1.

Table 1. Selected technical parameters of the FORD 2.0i 16V Duratorq TDCi engine.

Engine type	four-stroke, compression ignition
Fuel injection type	direct fuel injection, common rail (Delphi) electronically controlled, cooperating with the Levanta engine control system
Cylinders arrangement	in-line, vertical
No. of cylinders	4
Injection sequence	1-3-4-2
Type of timing gear	DOHC/4 VPC
Cylinder diameter	86.0 mm
Piston stroke	86.0 mm
Displacement	1998 cm^3
Maximum power	130 KM (96 kW) at 3800 rpm
Max. torque	330 Nm at 1800 rpm
Max. speed	4800 rpm
Idle speed	750 ±20 rpm
Compression ratio	18.2
Filling	turbocharged with intercooler and "overboost" function
Valve clearance	hydraulic adjustment
Capacity of lubrication system with filter	6.0 dm^3
Complies with emission standard	Euro IV

The engine tests were performed in a 4-phase, repeatable cycle which reflected the average engine operating conditions in a low-intensity city traffic. The parameters of the 4-phase cycle are given in Table 2.

Table 2. Parameters of the 4-phase engine cycle.

Phase	Time [s]	Engine speed [rpm]	Engine load [Nm]
1	30	1000	~0
2	150	1800	100
3	60	3000	70
4	60	1500	50

The test duration was 100 hours of actual engine operation in 8-hour periods which comprised repetitions of the 4-phase cycle (Table 2) and were separated by 16-hour soak period during which the formed deposits stabilized.

When the test was complete, the injectors were removed from the engine and two of them were sent for full, professional assessment of diagnostics parameters and the other two were left to take photographs of the internal deposits. The necessity for such procedure resulted from the fact that the deposits may be subject to mechanical and chemical damage/changes during the diagnostics and in the final stage they are removed in order to evaluate the parameters and the adjustment possibilities of the "cleaned" injectors. On the other hand, before the diagnostics the injectors must not be dismantled (e.g. for photographing and evaluation of deposits) as this would affect (distort) the diagnostic parameters after the test.

The injector evaluation after the test comprised both the external coke deposits injectors and the IDID. In case of the IDID, the obligatory evaluation was performed on the injector needle and solenoid valve control piston as components with critical impact on the correct injector operation, and in addition most often subject to damage.

The measurements and evaluations of selected diagnostic parameters were performed on the Zapp CRU.2i testing device. The evaluations of each diagnostic parameters are an average value from three measurements. Characteristics injectors fuel delivery were taken – off on the Hartridge CRi-PC test bench.

2.2 Tests Fuels Preparation

In order to accelerate the formation of injector deposits (especially the IDID), the engine was running on the naturally aged, commercially available diesel fuel containing 4.78% (v/v) of FAME – Table 3. Furthermore, an additional admixture of naturally aged FAME – Table 3, was used in some tests fuels.

The results of the to date analysis of sodium soaps (or sodium carboxylate) type IDIDs identified alkenyl succinic acids, like dodecenyl succinic acid (DDSA) and hexadecenyl succinic acid (HDSA), which are commonly used as pipeline corrosion inhibitors in the petroleum industry [22-24]. It was also considered that if higher molecular weight carboxylates and sodium are contained in diesel fuel simultaneously, carboxylate salt deposits are build-up on metal surfaces at temperatures of 130 degress C or higher [8]. Moreover, sodium ions are captured by fatty acids, which are a by-product of FAME blended into diesel fuel. When sodium ions and fatty acids meet, sodium soaps form [25].

Sources of amide deposits as well mechanism for amide lacquer formation is less recognized but it is most likely derived from polyisobutylene succinimides (PIBSI) which are added to diesel fuel as a major component of DCA. Two hypothesis are widely known for the formation of amide type IDID. One hypothesis assumes that deposits are formed through the reaction of PIBSI with fatty carboxylic acid to form an amide [8, 17, 22]. The other assumes that it is the low molecular weight portion of the PIBSI that is responsible for the deposit formation as it is only sparingly soluble in the diesel fuel and thus deposits on the surfaces of the inner injector parts [20, 22].

For this reason, added to the test fuel as 1-litre premixes were mixtures of chemical compounds which according to the available results of tests performed at various laboratories have the greatest impact on the formation of various IDID [9, 18, 20, 26-31]. The said chemical compounds comprised the ingredients present in detergent and lubrication additives, corrosion inhibitors, additives increasing the cetane number and the contaminants which make their way into to the fuel. As a result, three fuels described below were prepared for the simulation engine tests.

In test No. 1, the propensity for the formation of external and internal deposits was tested on the aged, commercially available diesel fuel (physical and chemical properties are given in Table 3) with addition of 0.5 ppm Na^+ + 70 ppm dodecenyl succinic acid (DDSA). This aimed at accelerating the formation of carboxylate deposits, in this case sodium soaps by adding "Na" as metallic contaminant often appears in diesel fuels and dodecenyl succinic acid (DDSA) which is ingredient of corrosion inhibitors also occurs in diesel fuels.

Table 3. Selected physical and chemical properties of fuels used during engine tests.

Tested parameter	Test method	Unit	Diesel fuel	B100
Oxidation stability	PN-EN 15751:2014	h	19.8	4.9
Resistance to oxidation	PN-EN ISO 12205:2011+Ap1:2011	g/m^3	5	32
Oxidation stability	ASTM D 7545-14	minute	39	23
Corrosion properties to steel: • temp. 38°C, test time 5 h •temp. 50°C, test time 3 h	ASTM D 665-14 Procedure A (distilled water)	–	no –	– no
Contaminants	PN-EN 12662:2014 PN-EN 12662:2009	mg/kg	6.0 –	– 182.5
Kinematic viscosity at 40°C	PN-EN ISO 3104:2004	mm^2/s	3.113	4.540
Distribution of contaminants: • 4, 6 and 14 μm • 2 and 5 μm	ISO 11500:2008 and PN-ISO 4406:2005	–	18/16/14 18/16	23/23/23 23/23
Fractional composition: • up to temp. 250°C distils • up to temp. 350°C distils • 95% (v/v) distils up to temp.	PN-EN ISO 3405:2012	% (v/v) % (v/v) °C	29.6 92.6 358.8	–
Density at 15 °C	PN-EN ISO 12185:2002	kg/m^3	838.8	–
Sulfur content	PN-EN ISO 20846:2012	mg/kg	7.7	–
Water content	PN-EN ISO 12937:2005	mg/kg	65	–
Flash point	PN-EN ISO 2719:2007	°C	69	–
Cetane number	PN-EN 5165:2014	–	52.8	–
Cetane index	PN-EN ISO 4264:2010 + A1:2013-07	–	54.1	–
Cold filter plugging point (CFPP)	PN-EN 116:2015	°C	–6	–
Carbon residue from 10% distillation residue	PN-EN ISO 10370:2014	% (m/m)	0.048	–
Ash residue	PN-EN ISO 6245:2008	% (m/m)	< 0.001	–

In test No. 2, the propensity for the formation of external and internal deposits was tested on the aged, commercially available diesel fuel (physical and chemical properties are given in Table 3) with addition of 120 ppm of PIBSI (Polyisobutylene Succinimide) + 100 ppm of hydrogenated dimer of oleic acid + 1ppm of formic acid. This aimed at accelerating the formation of amide deposits by adding PIBSI on which the detergent additives to diesel fuel are based and by adding acidic substances which are ingredients of lubrication additives.

In test No. 3, similarly to test No. 1, the propensity for the formation of external and internal deposits was tested on the aged, commercially available diesel fuel (Table 3) in which the content of biocomponents (4.78% (v/v) FAME) was increased to 10% (v/v) by adding aged B100 (properties given in Table 2). Moreover, 500 ppm of 2 ethyl-hexyl nitrate (2-EHN) + 120 ppm of PIBSI (small molecular mass) were added to the fuel. This aimed at accelerating the formation of amide (polymer-type) deposits by adding PIBSI on which the detergent additives to diesel fuel are based and the formation of ester deposits from aged FAME in tandem with 2 ethyl-hexyl nitrate (2-EHN) added to increase the cetane number.

3 Results

Figure 1 presents the tips of two examples of injectors after test No. 1.

Fig. 1. Photographs of injector tips after engine test No. 1.

On the injector tips were formed quite thick layers unevenly distributed, mat, dry, light grey deposits. The deposits are thicker around the fuel outlet holes. For a few outlet holes their exit cross-sections are limited by the deposits. The cracks and peeling of deposits is present in some areas – Fig. 1. The size and distribution of deposits on the surfaces of evaluated injector tips is similar. Figure 2 shows injector needle of one of injectors after test No. 1.

Fig. 2. Deposit formed at the injector needle after test No. 1.

A light grey coloured, typical carboxylate deposit, was observed the whole surface of the injector needles. Particularly intensive deposits were found on the cylindrical part of needles directly above the conical part – Fig. 2. A layer of uneven, light grey deposit with traces of wipping was also observed on sliding, cylindrical surfaces of the needles. Such deposits may result sticking of the inner injector components or delay of the injector response on the control signal. This can lead in some injectors not injecting at all low rail pressures, thus causing difficulties in restarting and poor running at idle. A similar type and colour IDIDs were observed at the plungers and heads of pistons of the fuel flow control valves – Fig. 3.

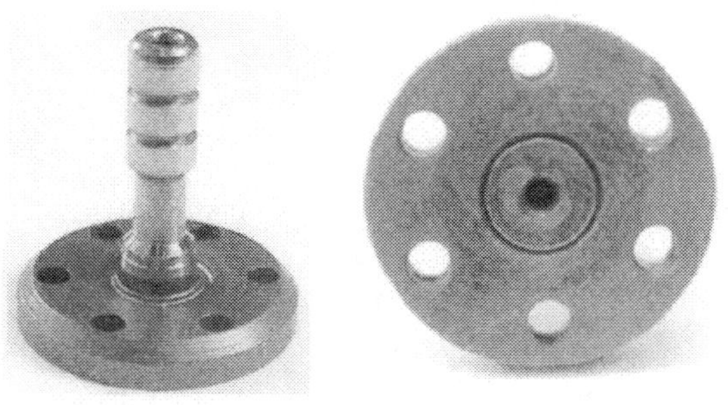

Fig. 3. Deposit formed at the plunger and head of piston of the fuel flow control valve after test No. 1.

The next part of the evaluation comprised a measurement of selected diagnostic parameters of two injectors after test No. 1. The results of the all injector assessments from three carried out tests are summarised in Table 4. The measurements and assessments of these selected diagnostic parameters were performed on the Zapp CRU.2i testing device.

Table 4. Assessment results of selected injector diagnostic parameters.

Assessments after test No. 1		Assessments after test No. 2		Assessments after test No. 3	
Injector No. 1	Injector No. 2	Injector No. 1	Injector No. 2	Injector No. 1	Injector No. 2
Fuel dose size assessment [mm^3/stroke] at injection frequency, pressure, and injector opening time (400 injections/minute, 160 MPa, 400 μs, respectively)					
PASS	FAIL	FAIL	FAIL	FAIL	FAIL
Fuel dose size assessment [mm3/stroke] at injection frequency, pressure, and injector opening time (400 injections/minute, 120 MPa, 412 μs, respectively)					
PASS	PASS	PASS	PASS	PASS	FAIL
Fuel dose size assessment [mm3/stroke] at injection frequency, pressure, and injector opening time (400 injections/minute, 80 MPa, 450 μs, respectively)					
PASS	PASS	PASS	PASS	PASS	FAIL
Fuel dose size assessment [mm3/stroke] at injection frequency, pressure, and injector opening time (1500 injections/minute, 23 MPa, 685 μs, respectively)					
PASS	PASS	PASS	PASS	PASS	PASS
Injector response time assessment [μs] to the electric control pulse at one specific injection frequency, pressure, and injector opening time (1000 injections/minute, 80 MPa, 450 μs, respectively)					
PASS	FAIL	FAIL	PASS	PASS	PASS

The injectors evaluations after test No. 1, presented in Table 4, indicate that the injector No. 2 has failed the requirements for the limit values of dosed fuel at the injection pressure of 160 MPa and the injector opening time of 400 μs as well as for injector response time. As a result, from the point of view of operational parameters, the injector was not fit for further use without maintenance-repair operations comprising washing, adjustment and possibly replacement of some internal parts. The most probable reason of such deterioration of the injector diagnostic parameters were internal deposits formed on the working part of the needle which slides on the injector body, and on the plunger of the fuel flow control piston inside the injector. This changed the injector operation time as a result of a delayed or slowed down response of the needle to the electric control pulse, and even caused a possibility of the needle's hanging in the injector body. Injectors fuel delivery characteristics taken – off on the Hartridge CRi-PC test bench are presented in Fig. 4.

For both characteristics were observed delivery drift especially for low pulse width and delivery restriction in particular for maximum injection pressures (160 MPa). In the case of injector No. 2, the change in injection rate to IDID adhesion manifests itself in injection delayed as shown in Fig. 4. When the injection start is delayed, the fuel injection quantity decrease is observed. This observation are consistent with assessments selected diagnostic parameters and confirm that carboxylate deposits impact on change in injection quality that may have adverse effects on different injection volume quantities among cylinders. For comparison, Fig. 5 presents example of characteristics curves of the new fuel injector before test.

Injector No. 1

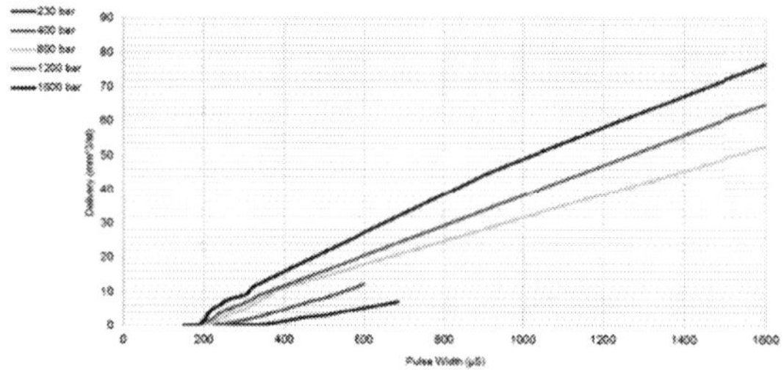

Injector No. 2

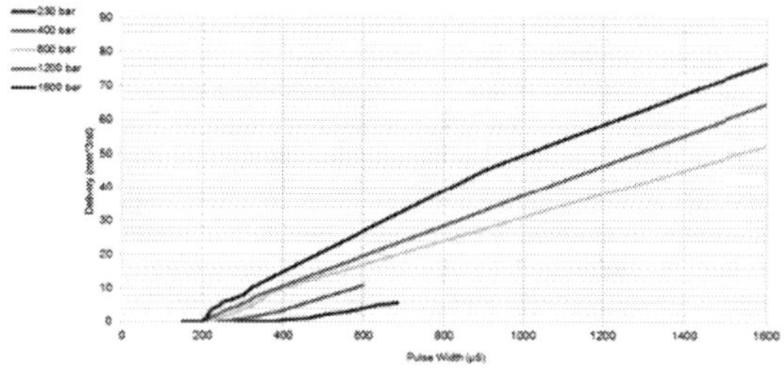

Fig. 4. Comparison of injector fuel delivery characteristics after engine test No. 1.

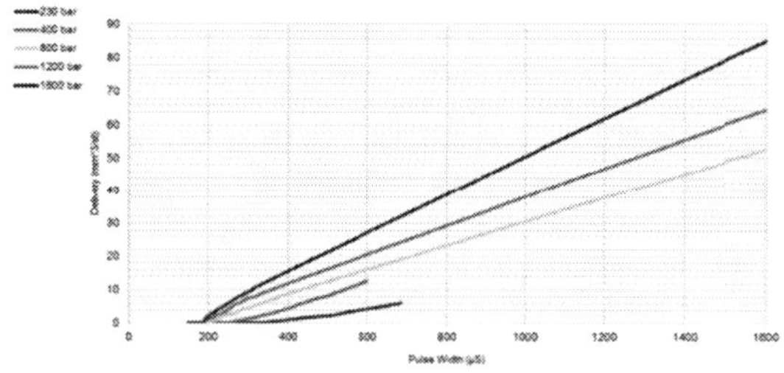

Fig. 5. Fuel delivery characteristics of the new fuel injector before test.

Figure 6 presents the tips of two examples of injectors after test No. 2.

The injector tips have quite thick layers of uneven, mat, dry, dark greyish brown deposits. Deposits are much thicker around the fuel outlet holes. The cracks and peeling of deposits is present in some areas – Fig. 6.

The size and distribution of deposits on the evaluated surfaces of both injectors is similar.

Fig. 6. Photographs of injector tips after engine test No. 2.

Figure 7 shows selected internal parts of one of injectors after test No. 2.

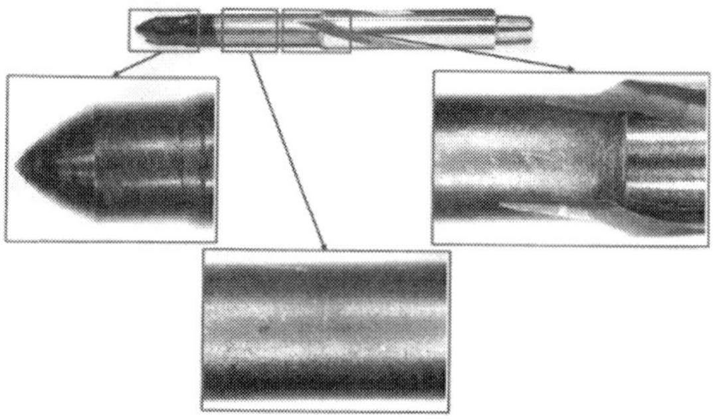

Fig. 7. Deposit formed at the injector needle after test No. 2.

Particularly intensive deposits, covering the surfaces in an irregular manner, were found on the conical ends of injector needles and on the cylindrical parts of needles directly above the conical part – Fig. 7. Colour as well appearance of these varnish-like deposits indicated the amide type deposits. More uniform amide deposits, although of varying thickness, covered the internal surfaces of grooves in the cylindrical guiding part of the needles. A very thin layer of light brown, sticky deposit can be observed around sliding part of the needles. A build-up of light brown deposits was also found on the plungers and heads of pistons of the fuel flow control valves – Fig. 8.

Fig. 8. Deposit formed at the plunger and head of piston of the fuel flow control valve after test No. 2.

These deposits especially those adhered into sliding surfaces of needles and control valves caused sticking or constrain operation of these injector components with critical impact on the correct injector operation.

Further evaluation comprised a measurement of selected diagnostic parameters of two injectors after test No. 2.

Assessment results of selected injector diagnostic parameters carried out using Zapp CRU.2i testing device showed that both injectors did not fulfil the requirements for fuel dose size at the injection pressure of 160 MPa and the injector opening time of 400 μs – Table 4. As a consequence, from the point of view of operational parameters, the injectors were not fit for further use without maintenance-repair operations comprising washing, adjustment and possibly replacement of some internal parts. Apart from that, the injector No. 1 failed requirements regarding injector response time to the electric control pulse at specific injection frequency 1000 injections/minute, measured at pressure 80 MPa, and injector opening time 450 μs respectively – Table 4.

The most probable reason of such deterioration of the injectors' diagnostic parameters were internal sticky, varnish-like deposits formed on sliding part of the injector needles co-acting with injector body, and on the plunger of the fuel flow control piston inside the injector. Observed after the test a few percent engine torque drop was induced by loss of fuel flow, induced likely by the external, nozzle coking deposits build-up around injectors fuel outlet holes and inside fuel channels as well as amide type IDID which changed the injector operation time as a result of a delayed or slowed down response of the needle to the electric control pulse. Injectors fuel delivery characteristics, after test No. 2, taken – off on the Hartridge CRi-PC test bench are presented in Fig. 9.

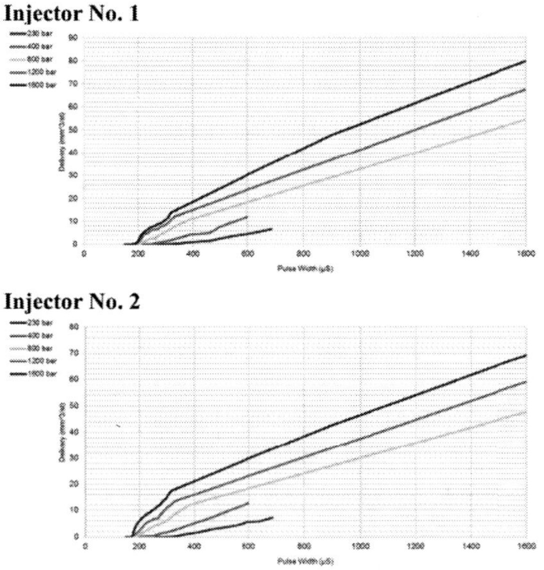

Fig. 9. Comparison of injector fuel delivery characteristics after engine test No. 2.

For both characteristics were observed delivery drift especially for low pulse width – Fig. 9. Delayed injection start is particularly apparent in the case of injector No. 1 what decrease fuel injection quantity especially for maximum injection pressures (160 MPa). The clear differences between both injector characteristics can be observed within pulse width between 200 up to 400 μs. This shows that IDID (in this case amide type) that collected on the nozzle needle guide, metering valves and other parts of the armature group of the fuel injectors could slow the response of the fuel injector or cause sticking of the moving internal parts, impairing the injection timing and quantity and the quality of fuel delivered per injection.

Figure 10 presents the tips of two examples of injectors after test No. 3.

Mat, peeling deposits or varying thickness were formed on the injector tips. Deposits colour locally varies from dark brown to dark grey. The deposits created significant swellings around the fuel outlet holes – Fig. 10. Injector spray-hole deposits have been also observed. So located coking deposits may restrict flow area and therefore led to a flow rate reduction, resulting in loss of torque and power. The size and distribution of deposits on the evaluated surfaces of both injectors is similar.

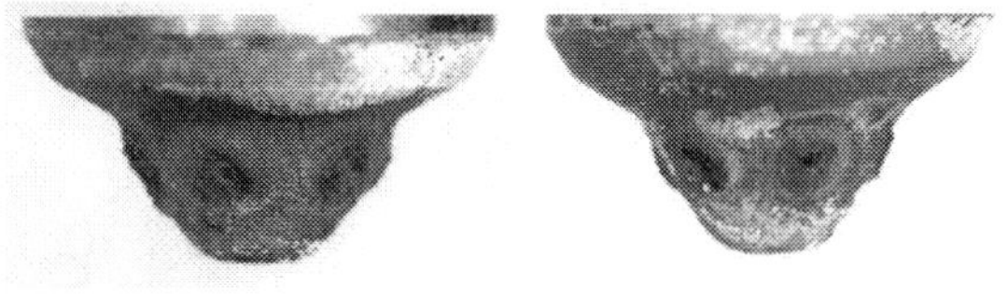

Fig. 10. Photographs of injector tips after engine test No. 3.

Figure 11 shows selected internal parts of injectors after test No. 3.

In the case of injector No. 2 after test No. 3, the needle was removed from its housing with difficulty and was clearly stuck. When both injectors from the test were dismantled, a combinations of amide lacquers as well carboxylate deposits has been observed. Intensive amide (polymer) deposits accompanied by ester deposits were formed on the injectors inside surfaces. On possibilities of the presence of biodiesel oxidation product indicates brownish lacquered appearance some deposits. A particularly large amount of deposits was found on conical ends of injector needles and on the cylindrical part of needles above the conical part. These deposits had an irregular shape and colour changing from yellow through orange, brown to dark brown. – Fig. 11.

A layer of intensive polymer deposits were found on the cylindrical parts of the needle washed with fuel, and in grooves in the cylindrical guiding part of the needles, and also in grooves of plungers of the fuel flow control pistons– Fig. 11, 12. The sliding, cylindrical surfaces of needles and plungers feature areas where the deposits were mechanically rubbed during the injector operation – Fig. 11, 12. On the backside of the fuel flow control valve head were found slight traces of carboxylate deposits – Fig. 12.

The performance of the dismantled injectors including results of selected injector diagnostic parameters carried out in the same manner that is using Zapp CRU.2i testing device. Similarly to test No. 2, showed that both injectors did not fulfil the requirements for fuel dose size at the injection pressure of 160 MPa and the injector opening time of 400 μs – Table 4. Injector No. 2 did not fulfil the requirements also for fuel dose size at the injection pressure of 120 MPa and the injector opening time of 412 μs, and also at the injection pressure of 80 MPa and the injector opening time of 450 μs – Table 4.

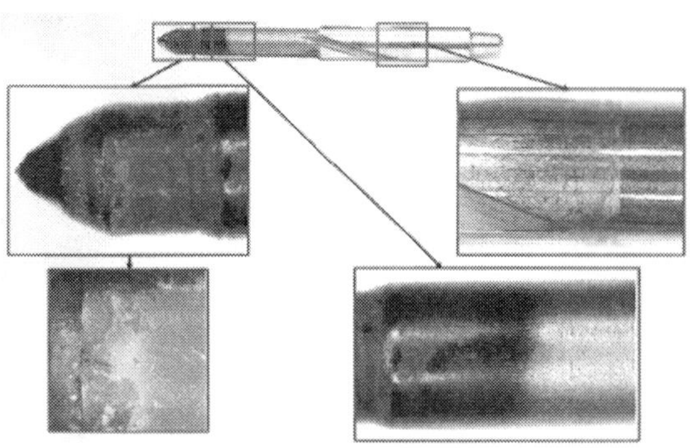

Fig. 11. Deposit formed at the injector needle after test No. 3.

Fig. 12. Deposit formed at the plunger and head of piston of the fuel flow control valve after test No. 3 after Test No. 3.

Similarly to test No. 2, the most probable reason of such deterioration of the injectors' diagnostic parameters were internal deposits formed on the working part of the needle which slides on the injector body, and on the plunger of the fuel flow control piston inside the injector. Observed after the test a few percent engine torque drop was due to a loss of fuel flow, induced likely by the external, external nozzle coking deposits around injectors fuel outlet holes and inside fuel channels as well as combinations of amide lacquers, carboxylate deposits and FAME decomposition deposits. These sticky IDID could have changed the injector operation time as a result of a delayed or slowed down response of the needle to the electric control pulse. Injectors fuel delivery characteristics, after test No. 3, are presented in Fig. 13. The analysis of course of the injectors fuel delivery characteristics fully acknowledges the above observations and hypothesis. For both injectors is visible delayed injection start – Fig. 13. It also notes considerably decrease fuel injection quantity especially for maximum injection pressures (160 MPa), paying particular attention to sudden fall of the fuel delivery characteristic of the injector No. 2, for pulse widths above about 900 μs. Thus the observations are consistent with assessments selected diagnostic parameters and confirm that combinations of amide lacquers, carboxylate deposits and FAME decomposition deposits very significantly influences on change in injection quality that may have adverse effects on different injection volume quantities among cylinders.

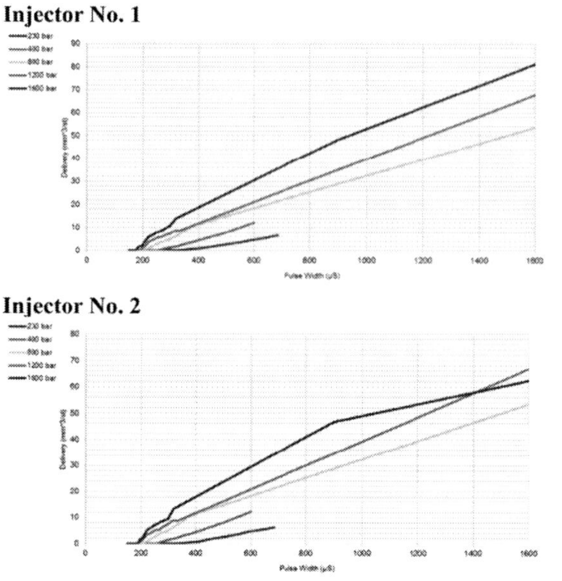

Fig. 13. Comparison of injector fuel delivery characteristics after engine test No. 3.

The analysis have also shown that when IDID forms, some injectors are impacted more than others. These more impacted injectors fail to deliver correct amount of fuel, and the injector better functioning inject more fuel to compensate for this.

4 Conclusions

An experimental investigations were carried out in order to better understand and evaluate impact of different type of injector deposits (both external and internal) on the common rail injector functioning. The research was preceded by reproducing different type of injector deposits in engine bench tests that simulated conditions of real operation. The most interesting aspect of the investigations are given below.

- Modern, advanced diesel fuel injection systems are susceptible and prone to the formation of a variety deposits including both external coking deposits and different IDID. These can occur in different locations, including among other things injector nozzle spray holes and channels, as well inside components of the injector body which threaten orderly functioning fuel injectors.
- The carried out engine tests prove the negative effect of some type both external and internal deposits on the operation and performance of the HPCR system.
- Changes to fuel composition and particularly fuel additives, possible trace metals contaminations and bio-components have in some cases increased potential of reactive species or provided new potential injector deposit precursors.
- External coking deposits in the nozzle orifices and inside fuel channels may have significant effect on engine power loss, which is due to reduction of fuel flow as deposits build up.

- IDID that collected on the nozzle needle guide and inside the nozzle body, metering valves and other parts of the armature group of the fuel injectors could slow the response of the fuel injector or cause sticking of the moving internal parts, impairing the injection timing and quantity and the quality of fuel delivered per injection.
- The change in injection rate due to IDID adhesion manifests itself in injection start and end timing changes which results different fuel injection volume quantities among cylinders and relevant consequences.

References

1. Z. Stępień, Combustion Engines **156**, 1 (2014)
2. Z. Stępień, Combustion Engines **163**, 4 (2015)
3. Z. Stępień, Combustion Engines 165, 2 (2016)
4. G. Żak, L. Ziemiański, Z. Stępień, M. Wojtasik, Nafta-Gaz (2013)
5. I. Pielecha, K. Wislocki, P. Borowski, W. Cieslik, Journal of Thermal Analysis and Calorimetry **122**, 473-485 (2015), DOI: 10.1007/s10973-015-4735-5
6. I. Pielecha, W. Cieslik, Journal of Thermal Analysis and Calorimetry **126**, 815-827 (2016), DOI: 10.1007/s10973-016-5544-1
7. J. Nowakowski, K. Brzozowski, T. Knefel, *Transport Means – Proceedings of the International Conference* 333-336 (2015)
8. J. Ullmann, M. Gedulding, H. Stutzenberger, et al. SAE Technical Paper 2008-01-0926 (2008)
9. R. Quigeley, R. Barbour, E. Fahey, et al., *TAE Fuels 7th Annual Colloquium* (2009)
10. A. Tanaka, K. Yamada, T. Omori, et al., SAE Technical Paper 2013-01-2661 (2013)
11. K. Yamada, S. Bunne, T. Omori, *2015 Powertrains, Fuels and Lubricants Meeting TWS2* (2015)
12. W. Karpiuk, T. Borowczyk, M. Idzior, R. Smolec, *2016 International Conference on Sustainable Energy, Environment and Information Engineering* 16-22 (2016)
13. J. Merkisz, P. Lijewski, P. Fuc, S. Weymann, Eksploatacja i Niezawodnosc – Maintenance and Reliability **15**, 364-368 (2013)
14. W. Karpiuk, R. Smolec, M. Idzior, *2016 International Conference on Sustainable Energy, Environment and Information Engineering* 37-43 (2016)
15. S. Schwab, J. Bennett, S. Dell, J. Galante-Fox et al. SAE Int. J. Fuels Lubricants **201**, 865-878 (2010)
16. Internal diesel injector deposits, Coordinating Research Council, INC. CRC Project No. DP-04-13a, CRC Report No. 665 (2013)
17. J. Ullmann, M. Gedulding H., Stutzenberger, et al., *TAE Fuels 8th Annual Colloquium* (2011)
18. R. Caprotti, N. Bhatti, G. Balfour, SAE Technical Paper 2010-01-2250 (2010)
19. P. Lacey, S. Gail, J Marc, et al. SAE Technical Paper 2012-01-1693 (2012)
20. J. Reid, J. Barker, SAE Technical Paper 2013-01-2682 (2013)
21. A. Suchecki, J. Nowakowski, T. Knefel, *Transport Means – Proceedings of the International Conference* 194-197 (2014)
22. R. Barbour, R. Quigley, A. Panesar, et al., *9th International Colloquium Fuels. Conventional and Future Energy for Automobiles* (2013)
23. S. Schwab, J. Bennett, S. Dell, et al. SAE Technical Paper 2010-01-2242 (2010)
24. M. Almena, O. Esperilla, F. Manzanero, et al. SAE Technical Paper 2012-01-1689 (2012)
25. J. Ullmann, H. Stutzenberger, R. Bosch, *9th International Colloquium Fuels. Conventional and Future Energy for Automobiles* (2013)
26. J. Baker, S. Cook, SAE Technical Paper 2013-01-2687 (2013)
27. J. Baker, C. Snape, D. Scurr, *9th International Colloquium Fuels. Conventional and Future Energy for Automobiles* (2013)
28. P. Lacey, S. Gail, J.M. Kientz, et al. SAE Technical Paper 2011-01-1925 (2011)
29. R. Quigley, R. Barbour, D. Arters, J. Bush, *9th International Colloquium Fuels. Conventional and Future Energy for Automobiles* (2013)
30. H. Shiotani, S. Goto, SAE Technical Paper 2007-01-0073 (2007)
31. Z. Stępień, Nafta-Gaz **10** (2014)

Simulations geometric structures of the stepped profile bearing surface of the piston

Emil Wroblewski[1,*], *Antoni* Iskra[1], and *Maciej* Babiak[1]

[1]Poznan University of Technology, Faculty of Machines and Transport, ul. Piotrowo 3, 60-965 Poznan, Poland

Abstract. The main node piston-pin-piston rings are most responsible for the formation of mechanical losses. It is advisable to reduce friction losses in the piston-cylinder group lead to an increase in the overall efficiency of the engine and thus reduce the fuel consumption. The method to reduce the area covered by the oil film is a modification of the bearing surface of the piston by adjusting the profile. In this paper the results of simulation for the stepped microgeometry piston bearing surface are presented.

1 Introduction

The advancement of modern engines is chiefly aimed at environment protection and low fuel consumption [1-2]. To meet the above-mentioned criteria, engines are designed based on the idea of downsizing characterized by a reduction of their displacement and increasing their effective pressure by the application of a variety of supercharging systems [3]. The consequence of these changes is that engines operate at high thermal and mechanical loads. This is an adverse phenomenon for the piston-crankshaft assembly, particularly for pistons exposed to high temperatures and pressures inside the cylinder [4-6].

Modern engines are also expected to provide the highest overall efficiency [7]. The way to ensure that is to reduce the friction loss in the piston-cylinder pair [8-9]. Constructors aspire to design units, that have preserved its initial state in the longest possible time. In the case of combustion engines can be list a few groups, which consume deteriorating engine parameters and finally, makes it impossible to operation. They are mainly components of the piston-cylinder and crankshaft bearings [10-11].

The diagnostic testing of internal combustion engine can be made by using working processes and methods which take advantage of leftover processes. Working processes give information about general condition of internal combustion engine [12]. Leftover processes give information about condition of particular subassemblies and kinematic couples. This methods are based on analysis of vibrations and noise [13-14].

Over the years combustion engine's pistons changed their geometric proportions to a greater extent, but still their essential role is to close the variable volume combustion chamber. However performing of such main functionality is accompanied by a number of adverse effects, such as [15-16]:

- friction losses,
- forces and moments of inertia transferred to the motor suspension,
- lubricating oil consumption,
- noise emission.

The geometry of the gap between the bearing surface of the piston and the cylinder has large effect on friction losses. The area covered by the oil film contributes to these losses more than the thickness of the oil film. A method to reduce the coverage area of the oil film is to use different bearing surface shapes of the piston-stepped-like or barrel-like. Barrel-like shape is commonly used, but a stepped one was not yet applied for various reasons. The stepped shape can be obtained i.e. by the application of refining coatings of materials with good friction properties [17-19]. One such material is a graphite, commonly used to cover the bearing surface of the piston. Currently the entire bearing surfaces of the piston are coated. Such method produces different boundary friction conditions, but it does not affect the fluid friction formation significantly in the oil film [20-21].

The paper presents the results of simulation research for the reduction of frictional losses and wear of the bearing surface of the piston and the cylinder wall by covering the support surface of the piston with a layer of graphite in the H-shape. A stepped profile obtained this way is characterized by a beneficial tribological properties of the kinematic piston-cylinder system under fluid friction conditions. The solution proposed by the authors hereby extends the scope of tasks performed by the coatings on the bearing surface of the piston.

2 The piston skirt

Piston skirt is to ensure axial displacement of the piston inside the cylinder and a transfer of lateral forces on the cylinder liner.

In reality, however, the cooperation of the piston skirt with the cylinder occurs on a much smaller area,

* Corresponding author: emil.z.wroblewski@doctorate.put.poznan.pl

DOI: 10.1051/matecconf/201711800003

which results in a reduction of stresses between these surfaces. Taking this fact into account, it is easily observable that the piston skirt operates in much worse conditions than those of generously lubricated sliding bearings. The lubrication of the piston skirt is much scanter. Ensuring the required durability of the piston skirt under such conditions is a very difficult task requiring a careful selection of materials and design solutions [22].

The distribution of temperatures on the piston circumference is uneven. The presence of the kingpin hubs and parts connecting the hubs with the piston makes the cross-section of this spot greater than the rest of the circumference. Parts of greater cross-section have lower thermal resistance, which makes the temperature on the piston skirt in these regions much higher than on the plane perpendicular to the kingpin axis. This results in greater thermal deformations towards the kingpin axis, which is why the piston skirt must be oval with a high ovality axis perpendicular to the axis of the kingpin and a low ovality axis alongside the kingpin to ensure that under operating temperatures the piston skirt becomes cylinder-shaped or close to it [6].

The operating force acting on the piston influences not only the friction conditions between the piston and the cylinder liner but also deforms the piston skirt, which is why, when cold, the guiding element should be shaped in such a way as to ensure cooperation with the cylinder liner on the largest possible area upon thermal deformation under operating temperatures and under the influence of the lateral forces.

The above-mentioned factors render the geometry of the piston skirt very complex, particularly in engines of high displacement/power ratio. When the piston reaches its operating temperatures, the play between the piston and the cylinder liner changes. Changes in the play also occur after any change of the operating conditions following the stabilization of the thermal equilibrium [6, 22].

Such play variations are disadvantageous for a variety of reasons. One of them is the difficulty to ensure optimum play in terms of friction and lubrication under a sufficiently wide range of conditions. An increase in the play under lower operating temperatures results in an increase in dynamic forces exerted by the piston on the cylinder, vibration of the cylinder sleeve as well as engine noise and may lead to cavitation damage of cylinder liners in water-cooled engines.

The shape of the piston skirt in the axial cross-section, perpendicularly to the kingpin axis, and in the transverse cross-section should be designed in such a way as to ensure the maximum area of cooperation between the piston skirt and the cylinder liner under operating temperatures and under forces exerted on the piston skirt.

A modification of the geometry of the piston skirt consists in choosing such a shape of the surface that will ensure the continuity of the oil film during the piston reciprocating motion with a possibly low value of friction loss in the piston-cylinder pair.

The adopted microgeometry of the piston skirt is a result of a continuation of former research. The authors assume an application of a layer of graphite of the shape of letter H on the piston skirt. Thus obtained stepped profile is characterized by good tribological properties under liquid friction and, owing to the application of graphite, one can expect good properties under boundary friction.

In order to ascertain the friction loss in the piston – cylinder pair, 3 variants of piston microgeometry have been developed as shown in Fig. 1.

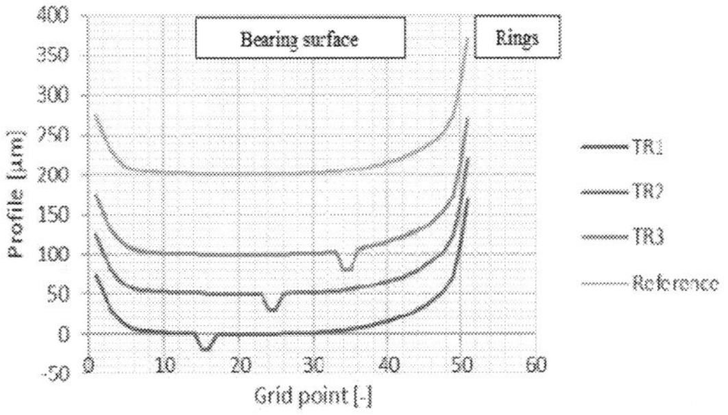

Fig. 1. Profile experimental piston TR1, TR2, TR3.

The horizontal bar of the letter H allows extending the area of high hydrodynamic pressure at the same time obtaining the effect of reduction of the oil film internal friction forces. Because the correctness of the computer model was validated experimentally in earlier research, the results confirm the positive results of the modification of the piston skirt microgeometry of a specified thickness of the H bar, which proves the validity of the replacement of classic barrel-shaped profile of the piston skirt with the stepped one by the application of a layer of graphite of a specified thickness.

3 Simulation research results

Calculations of oil film parameters were carried out for diverse profiles of piston skirt, beginning with barrel shape of standard pistons and ending with a stepped profile [23]. Figure 2 presents the course of oil film thickness with the thickest line, the thickness of oil layer over cylinder bore with the line of medium width and the friction force between piston skirt and bore with the thinnest line.

Most important parameters of engine operation for which the simulation has been carried out like:
om = ω – engine rotational speed,
eta = η – engine oil viscosity,
eps = ε – compression ratio,
hr – piston skit height,
du – piston nominal diameter
are presented over the draft.

At the upper right corner in Fig. 2. there is a schematic draft of piston skirt profile taken for calculations.

The calculations of the parameters of the oil film were performed for different microgeometries of the piston skirt, starting from the barrel-shaped profile of the reference pistons and ending with the stepped piston skirt profile. The thickness of the H bar is variable as well as its height in the bottom or top position against the symmetrical one.

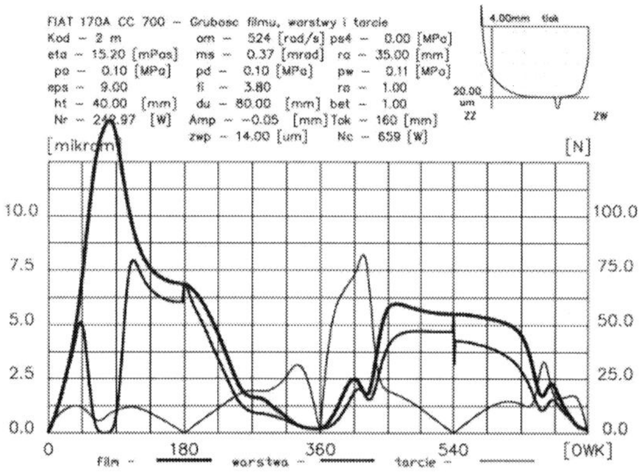

Fig. 2. Course of oil film thickness – the thickest line, thickness of oil layer over cylinder bore – line of medium width and friction force between piston skirt and bore – the thinnest line.

The calculation of oil film parameters were done using different geometry variants beginning from the stepped shape profile, and ending at barrel-like shape one. The analysis was conducted under the following conditions:

- engine revolution speed at 750 rpm (idle) and 5000 rpm (full power),
- initial oil layer thickness at cylinder liner of 2 and 35 μm,
- pressure increase of $\varphi = 2.13$ and 3.81,
- piston rings height: 1.5; 2; 4 mm,
- piston height: 40 mm,
- lubricating oil temperature of 60, 80 and 100 °C,
- lubricating oil type of 10W30.

Please note that the simulation was based on the theory of hydrodynamic lubrication and it does not take into account factors like i.e. tribological properties of materials working under boundary friction conditions. The simulations were aimed at defining the characteristics of a stepped profile under fluid friction conditions that prevail most of the time in the internal combustion engine.

The results of simulations are summarized in Fig. 3–6 obtained values are:

- *Nr* [W] – friction losses for piston bearing and cylinder surfaces,
- *Nc* [W] – total friction losses for piston and piston rings system.

Research shows that all variants have brought the expected reduction in friction losses for the piston-cylinder system in relation to reference pistons.

As it turns out, the most advantageous solution is the symmetrical position of the horizontal *H* bar, as the friction loss decreases by approx. 10% compared to pistons of a standard profile. One could also observe that the reduction in the friction loss is greater for higher engine speeds. This is an important observation because, due to increased friction loss converted into heat when increasing the engine speed, the engine elements warm up faster and the oil viscosity drops, which may lead to boundary friction. This phenomenon may reduce the friction loss assuming that boundary friction in the piston skirt – cylinder liner pair does not occur.

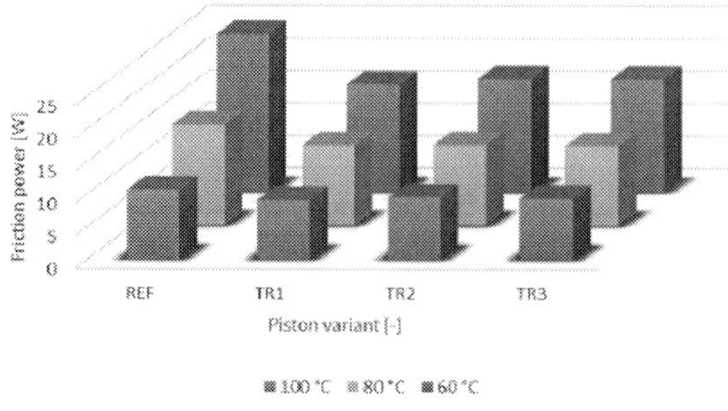

Fig. 3. Friction losses – *Nr* values, all pistons as a function of oil temperature – idle.

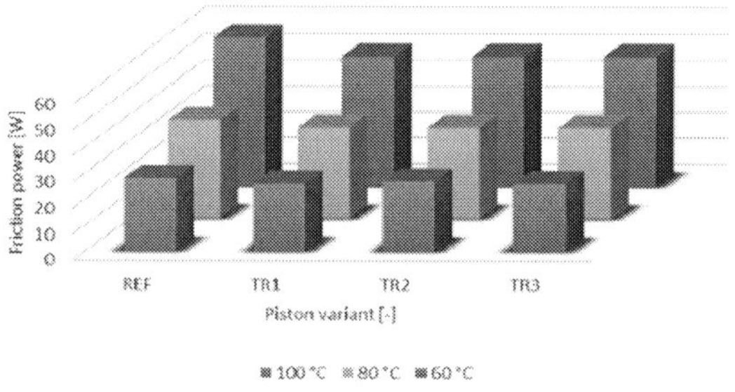

Fig. 4. Friction losses – *Nc* values, all pistons as a function of oil temperature – idle.

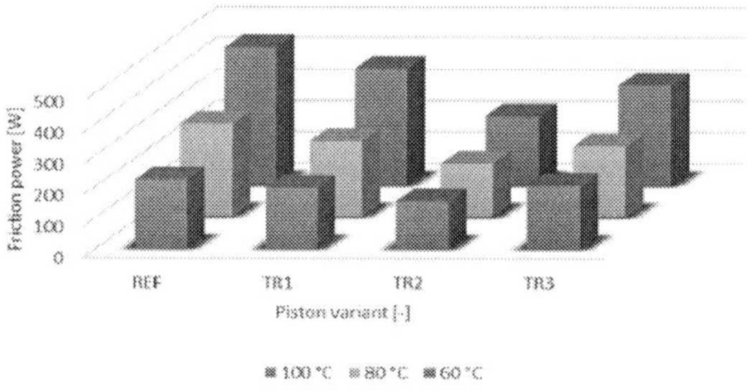

Fig. 5. Friction losses – *Nr* values, all pistons as a function of oil temperature – full power.

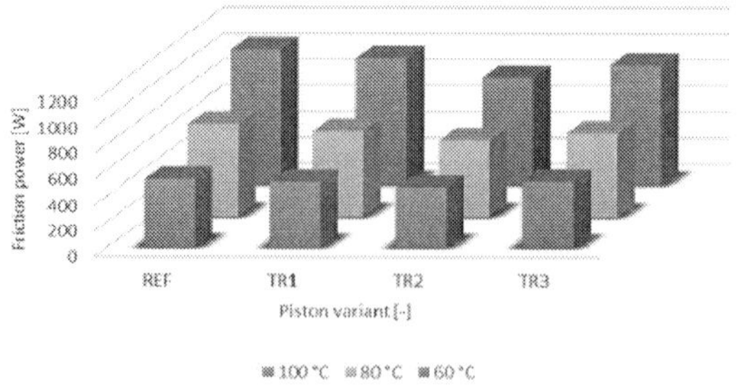

Fig. 6. Friction losses – *Nc* values, all pistons as a function of oil temperature – full power.

DOI: 10.1051/matecconf/201711800003

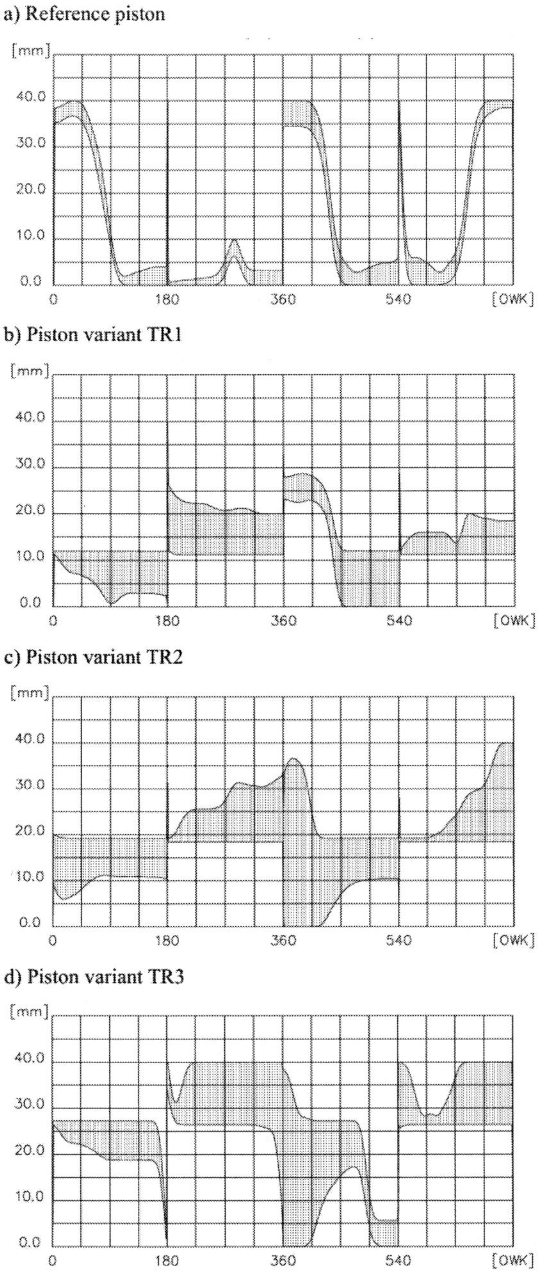

Fig. 7. Oil film height on the piston skirt – idle.

From the equations of hydrodynamic lubrication theory we know that in order to form a hydrodynamic oil film one needs a narrowing oil gap. As results from the performed simulations, the microgeometry of the piston skirt significantly influences the friction loss in the piston – cylinder pair. The obtained results are explained by the Fig. 7–8 presenting the coverage of the piston skirt with oil film.

Comparing the oil films on the piston skirts one can observe that intense oil film results in an increased friction loss. If the differences in the oil film level on the piston skirt are small, even minimum variations are impactful on the friction loss. The oil film thickness is greater for the stepped profile.

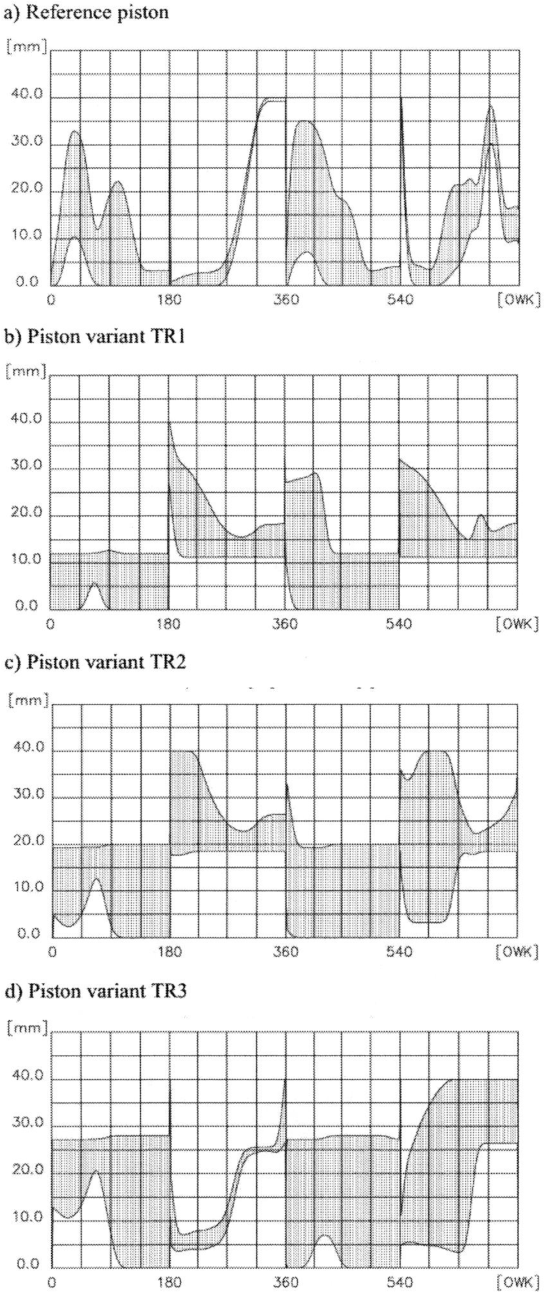

Fig. 8. Oil film height on the piston skirt – full power.

4 Conclusions

The results from the research allow to conclude that:

- friction losses can be reduced by configuring a gap between the piston and the cylinder to provide a continuous oil film separating the cooperating surfaces,

- the results of the simulation showed that the stepped profile reduce friction loss by about 10% in relation to the barrel-like shape,
- surface coating of oil film reduces friction losses but does not eliminate them,
- viscosity forces perform a work that is immediately converted to heat, manifested by an increase in the temperature of the oil film layer forming the oil film,
- assessment of the oil film between the bearing surface of the piston and cylinder wall, must take into account the movement of the oil film layers, both in the direction of movement of the piston caused by the crank mechanism, and in the circumferential direction of the piston,
- slide promoting layers are especially desirable in conditions of permanent shortage of continuous oil film, for example after an engine longer stop and balance of its temperature with the ambient temperature,
- the stepped profile of the bearing surface increases the mechanical efficiency of the combustion engine and reduces fuel consumption,
- a stepped surface can be achieved by the coating of the support surface of the piston using lubricants like i.e. graphite.

The study presented in this article was performed within the statutory research (contract No. 05/52/DSPB/0262).

References

1. M. Blümm, A. Baberg, F. Dörnenburg, D. Leitzmann, MTZ **2** (2016)
2. J. Merkisz, P. Fuc, P. Lijewski, A. Ziolkowski, M. Galant, M. Siedlecki, Journal of Electronic Materials **45**, 4028-4037 (2016), DOI: 10.1007/s11664-016-4543-0
3. J. Merkisz, P. Fuc, P. Lijewski, A. Ziolkowski, K.T. Wojciechowski, Journal of Electronic Materials **44**, 1704-1715 (2015), DOI: 10.1007/s11664-014-3522-6
4. T. Deuss, H. Ehnis, R. Rose, R. Künzel, MTZ **5** (2010)
5. F. Dörnenburg, K. Lades, S. Kenningley, MTZ **4** (2010)
6. A. Iskra, M. Babiak, W. Wroblewski, *IOP Conference Series-Materials Science and Engineering* **148**, UNSP 012068 (2016), DOI: 10.1088/1757-899X/148/1/012068
7. M. Fahr, W. Hanke, C. Klimesch, A. Rehl, MTZ **7-8** (2011)
8. M. Werkmann, A. Bitter, R. Künzel, MTZ **7-8** (2011)
9. G. Koszalka, Eksploatacja i Niezawodnosc – Maintenance and Reliability **3**, 40-44 (2011)
10. A. Baberg, M. Freidhager, H. Mergel, K. Schmidt, MTZ **12** (2012)
11. R. Backhaus, MTZ **12** (2009)
12. M. Orczyk, F. Tomaszewski, B. Czechyra, G.M. Szymanski, *Proceedings of the International Congress on Sound and Vibration* (Athens, 2016)
13. G.M. Szymanski, F. Tomaszewski, *Proceedings of The 17th International Congress on Sound and Vibration* (Cairo, 2010)
14. G.M. Szymanski, M. Josko, F. Tomaszewski, R. Filipiak, Mechanical Systems and Signal Processing **58-59**, 298-307 (2015), DOI: 10.1016/j.ymssp.2014.12.017
15. R. Golloch, VDI Verlag **12** (Düsseldorf, 2001)
16. J. Merkisz, J. Mizera, M. Bajerlein, L. Rymaniak, P. Maj, Experimental and Applied Mechanics **518**, 102-107 (2014), DOI: 10.4028/www.scientific.net/AMM.518.102
17. A. Iskra, M. Babiak, E. Wróblewski, Technika Transportu Szynowego **12**, 644-647 (2015)
18. A. Iskra, M. Babiak, E. Wroblewski, Journal of KONES **22**, 95-101 (2015)
19. A. Iskra, P. Krzymień, E. Wroblewski, Combustion Engines **162**, 192-196 (2015)
20. T. Deuss, H. Ehnis, R. Rose, R. Künzel, MTZ **4** (2011)
21. R. Kuberczyk, J. Berner, M. Bargende, MTZ **1** (2009)
22. W. Kozaczewski, *Konstrukcja grupy tłokowo-cylindrowej silników spalinowych* (Warszawa, 2004)
23. A. Iskra, *Studium konstrukcji i funkcjonalności pierścieni w grupie tłokowo-cylindrowej* (Poznan, 1995)

The impact of microgeometry pistons with a stepped bearing surface for the friction loss of the internal combustion engine

Emil Wroblewski[1,*], *Antoni* Iskra[1], and *Maciej* Babiak[1]

[1]Poznan University of Technology, Faculty of Machines and Transport, ul. Piotrowo 3, 60-965 Poznan, Poland

Abstract. This paper present the results of experimental piston friction losses on stepped bearing surface microgeometry obtained on the test rig. This test rig is equipped with special temperature control system, which provides better stability to temperature than in standard systems. The results of station tests was discussed. Tests was analyzed depending the moment caused by the friction on the oil temperature in the oil sump. Specified conclusions allow to assess the impact of the stepped profile of the pistons bearing surface microgeometry for different values of engine speed and the oil temperature at the friction losses in the main kinematic engine node which is piston-cylinder.

1 Introduction

The bearing surface of the piston, especially on the pressure side, is subjected during operation of internal combustion engine to significant mechanical loads. This situation is associated with the continuous improvement of internal combustion engines and therefore with increasing values of basic indicators of operating an internal combustion engine, such as power and maximum torque [1-2]. To assess technical condition and working conditions of combustion engine, parameters and characteristics of vibration signal generated by combustion engine are more and more often used [3-4]. The growth of these indicators is a direct result of the maximum pressure level increase inside the cylinder during the power stroke and also effective pressure. This fact has its direct impact on the size of the forces acting on the elements of the crank mechanism of piston as well as the strength of the normal forces at the bearing surfaces of the piston [5]. In addition, increasing the mean effective pressure can increase local temperatures in cooperating elements of the piston-crank mechanism, especially those in direct contact with the high temperature exhaust gas, e.g. the bottom of the piston [6-8]. This hard environment where parts of the piston-crank mechanism works, especially the piston, causes the certain problems in providing desired stability and reliability of the piston-crank mechanism and minimizing fuel consumption at the same time [9].

One method of improving the durability of the cylinder liner and the bearing surface of the piston and reduced fuel consumption as well, is the use of appropriate coatings on the side surfaces of the pistons. The coatings are applied to the support surface of almost all modern pistons natural aspirated and supercharged engines [6, 10-11]. Purpose of usage of them is primarily to avoid seizing of the piston in adverse operating conditions as [1]:

- local reduce of the piston slack caused by the deformation resulting from thermal and mechanical cylinder loads,
- reduce the slack of the piston as a result of overheating,
- insufficient oil film thickness at the time of starting the engine,
- worse properties of the oil caused by the heat,
- the state of the engine before the run-in.

The test results clearly confirm the positive role of the standard coating in reduce the possibility of scuffing the piston in the cylinder and reduce the surface wear of the cylinder under the normal engine operating conditions. In real operating conditions the total friction losses depend primarily on the liquid friction conditions so the use of coating does not give direct large benefits in reducing friction [6, 12].

Modification of the bearing surface shape on the piston consists of choosing the shape of the surface which is making the reciprocating motion regard to the cylinder liner to ensure the continuity of the oil film with the smallest possible value of the friction losses at the node between piston and cylinder. Reducing friction losses will contribute to the achievement of greater mechanical efficiency of the internal combustion engine and thus reduce fuel consumption [13-14].

Step-shaped gap can be obtained by the application of the H shaped graphite layers on the supporting surface of the piston. This will give the stepped profile which is characterized by a favorable tribological properties of the cylinder-piston kinematic couple fluid friction conditions, and use of graphite can expected advantageous properties in terms of boundary friction [15-16].

The article presents the results from research model of cooperation between experimental pistons with

* Corresponding author: emil.z.wroblewski@doctorate.put.poznan.pl

DOI: 10.1051/matecconf/201711800004

stepped microgeometry bearing surface and the cylinder wall at simulation bench for measuring friction losses in the conditions of the engine test bench.

2 Developing support surface

Adopted ennobling layer shape on the supporting surface is a continuation of earlier studies [14-17]. Friction losses can be reduced by such shaping of the gap between the piston and the cylinder (figure 1), which will provide continuous generation of the oil film separating the mating surfaces. Oil film reduces friction losses but not eliminate them. This is due to the energy balance. Viscous forces perform a work which immediately is converted into heat, manifested by an increase in temperature of oil layer which form film.

Method for obtaining reduction of friction losses in the piston-cylinder node is the use of a stepped microgeometry profile of the piston bearing surface by applying on a supporting surface of the piston graphite layers H-shaped. Optimize H bar shall be verified experimentally by the method of determining the friction loss in the oil film produced as a result of hydrodynamic phenomena in the layer of oil between the bearing surface of the piston and cylinder.

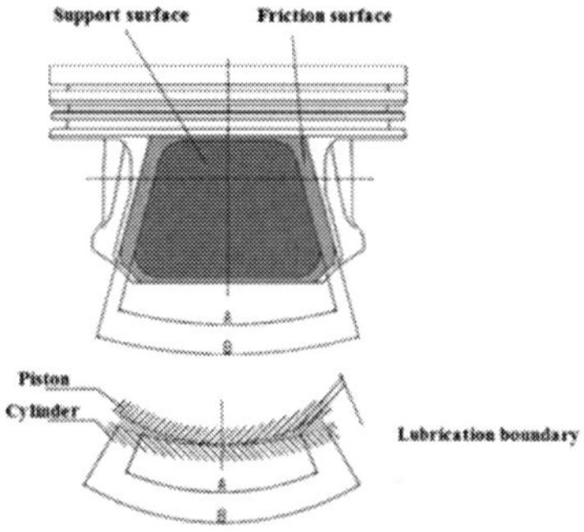

Fig. 1. Contact of the piston with the cylinder wall [6].

There were analyzed three experimental variants of the pistons with different bar position of an H-shape up or down relative to the symmetrical position. Figure 2-4 is shown in the drawings embodiments developed profiles of experimental pistons. Different variants of the pistons are presented in Fig. 5.

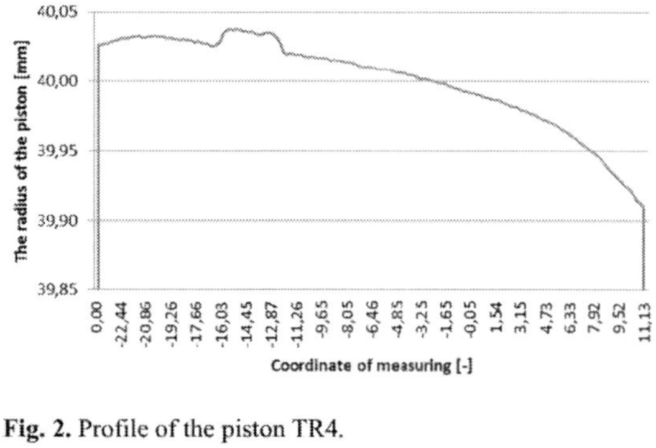

Fig. 2. Profile of the piston TR4.

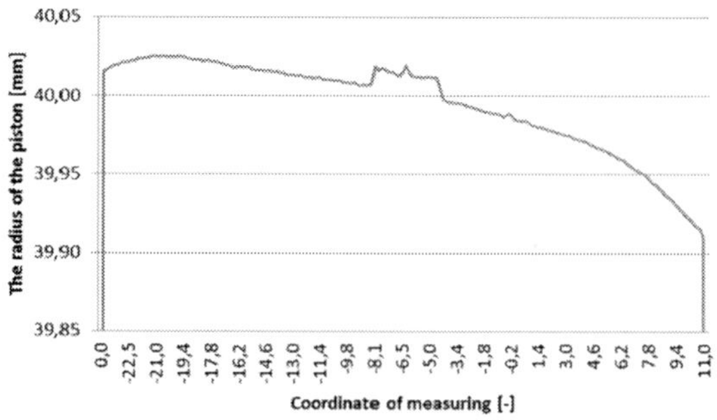

Fig. 3. Profile of the piston TR5.

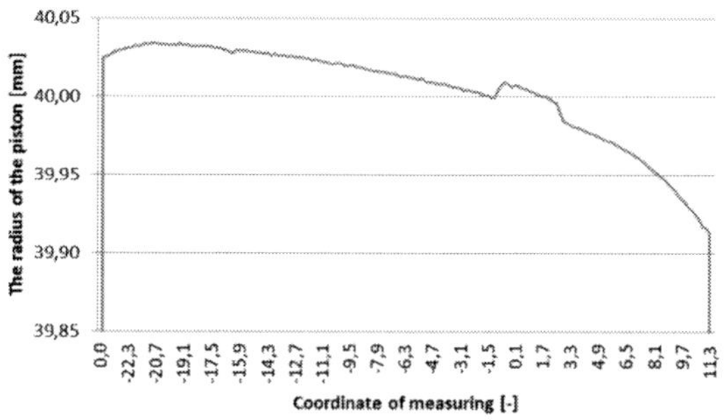

Fig. 4. Profile of the piston TR6.

Fig. 5. View of different piston variants – Reference, TR4, TR5, TR6.

3 Research position

The test bench (Fig. 6) was built with a complete engine block FIAT 170A.046 with the crankshaft, connecting rods, pistons and cylinder head, in order to reproduce as closely as possible the actual working conditions of piston-engine crankshaft in the engine, while ensuring accurate measurement of torque under external drive.

There were built model for test bench to measure friction loss in the piston-cylinder group composed of the modified internal combustion engine driven electric motor from outside. Drive is transmitted through the shaft of the measuring shaft that allow accurate torque measurement with high temporal resolution. In addition, the engine camshaft has been immobilized, leaving all valves in the closed position, the immobilized crankshaft-driven coolant pump and an oil pump outer replacing them, driven by electric motors.

The changes made during each revolution of the crankshaft forces the repeating cycle of compression and expansion of the load confined within the cylinder, wherein the portion of the load due to leakage enters the crankcase. With the valves closed engine loss charge in

DOI: 10.1051/matecconf/201711800004

the cylinder can be supplemented by additional non-return valves mounted instead of spark plugs. The design position allows you to close the valves or their opening to allow free air intake, ambient or bringing it under certain hypertension. Bringing the cylinder at the beginning of the stroke compressed air under increased pressure causes the increase in the maximum cylinder pressure. This is the method to simulate the actual increased engine load.

Fig. 6. The view of engine friction losses test bench.

A very important element of the original system to maintain a constant oil temperature of the heat exchanger oil-cooling liquid engine, the radiator and heater fluid and electrically powered pumps, coolant and oil. The PID controller makes it possible to maintain the desired oil temperature within ±0.2 °C, which ensures repeatability of the measurements. Diagram of stabilizing the oil temperature at the position shown in Fig. 8, and the most important technical data positions in Tab. 1. The torque meter combining combustion engines with an electric motor shown in Fig. 7. A detailed description of the position of the model is in the literature [18-19].

Table 1. Specifications of test bench simulation.

Technical data	Description
Combustion engine	two-cylinder, liquid-cooled, FIAT 170.A046
Electric engine	Maximum continuous power 2,2 kW
Rotation speed	fluently controlled in the range of 500-3000 rpm
Torque measurement	In scope –50…50 Nm, frequency 9,6 kHz
Registration results	production unit HBM Spider 8 type, with integrated power bridge strain gauge measuring shaft
Lubricating oil	Castrol Edge 5W30
Oil temperature	Regulated in the range of 40–110 °C the maximum error adjustment of 0,2 °C
Oil pressure	adjustable, equal to 0,4 MPa
Oil pump	Wing type product of ZF, driven by an electric motor with a power of 0,4 kW
Coolant pump	Grundfos production, the liquid stream equal 5 m^3/h
Fluid heater	Electric, with power adjustable in steps 0, 1, 2 and 4 kW

Fig. 7. Placement of a torque meter combining combustion engines with an electric motor [18].

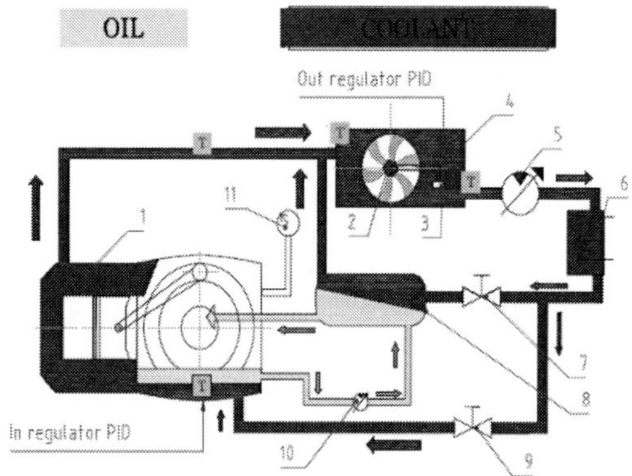

No.	Name
1	Engine Fiat 170A.046
2	Fan radiator fluid
3	Microprocessor control fan speed
4	Cooler
5	Liquid pump with a variable flow rate
6	Electric heating liquids
7	The ball valve of a small flow of liquid
8	The plate heat exchanger of a water / oil
9	The ball valve of a large flow of the liquid
10	Oil pump with variable flow
11	Manometer-pressure oil lubrication system

Fig. 8. Diagram of the oil temperature regulation system used on test bench.

4 Research results

The approach used in construction of test bench give opportunity to measure the torque with high accuracy and measurement frequency. The measured value of the torque due to the friction loss in the piston-crank mechanism is not significantly affected by the action of any other engine mechanism. Torque measurement is affected by the influence of thermodynamic phenomena that occur in the engine and blow-by charge to the crankcase.

The impact of these phenomena cannot be eliminated, but it can be assumed that their progress depends largely on microgeometry or covering coats of piston bearing surface and that allows each other comparison of these variants in the range of generated friction losses.

Measurements of friction losses made for the options of presented experimental microgeometry bearing surface of the experimental stepped piston profile and a surface of the barrel profile pistons serial used in the engine Fiat 170A.046. The measurements were carried out for the following conditions:

- speed 750-3000 rpm,
- oil temperature 50, 80, 110 °C,
- no supplement charge.

For each temperature of the oil was carried out five series of measurements – no refill load, which gave a lower pressure and a force of the compression gas. The results obtained are shown in Fig. 9-11.

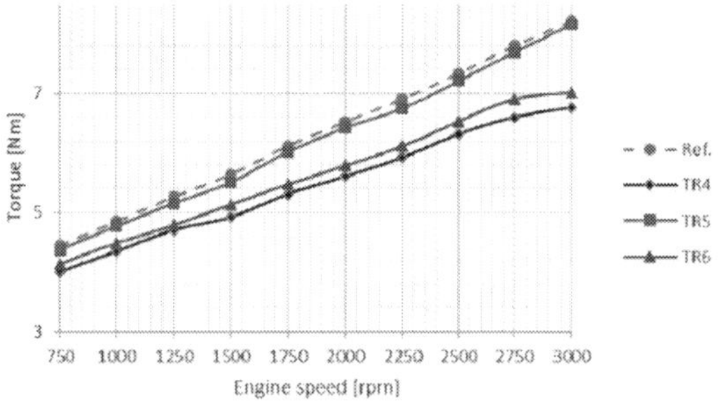

Fig. 9. The results of the measurements for oil temperature 50 °C.

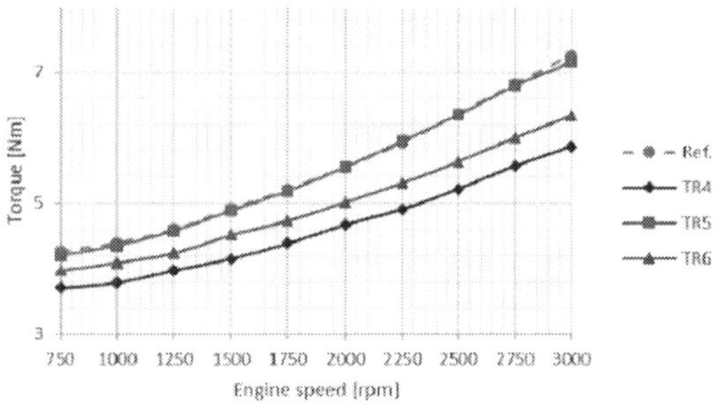

Fig. 10. The results of the measurements for oil temperature 80 °C.

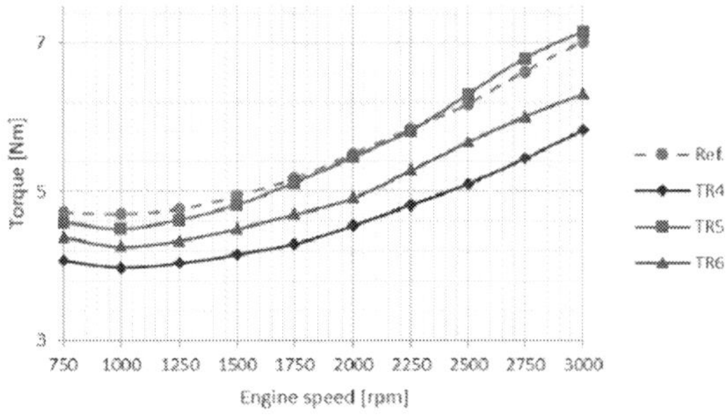

Fig. 11. The results of the measurements for oil temperature 110 °C.

Fig. 12-15 summarizes the results of the percentage reduction in friction losses in all options experimental pistons as a function of rotation speed.

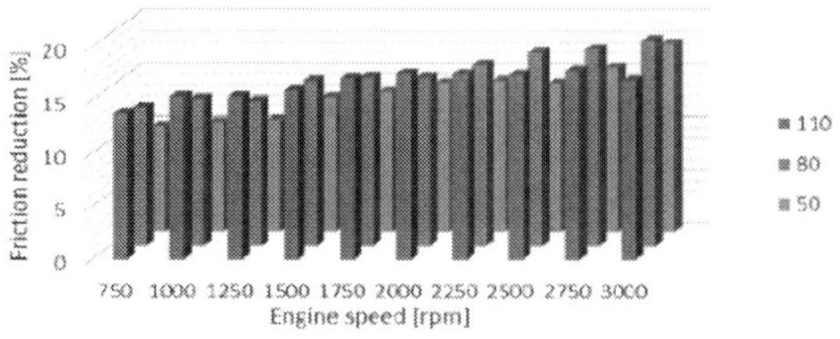

Fig. 12. Reduction of friction losses as a function of temperature and speed variant TR4.

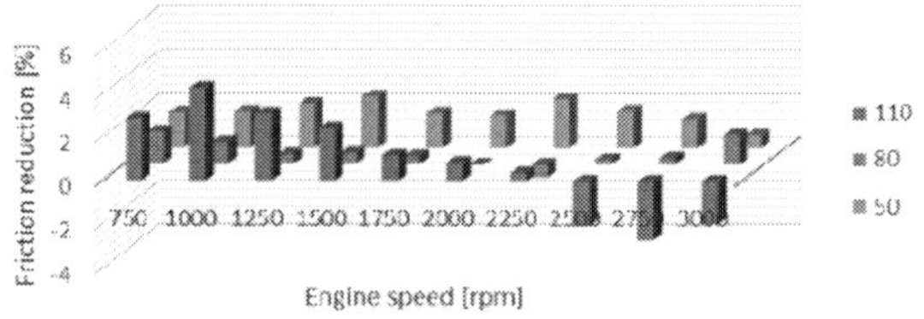

Fig. 13. Reduction of friction losses as a function of temperature and speed variant TR5.

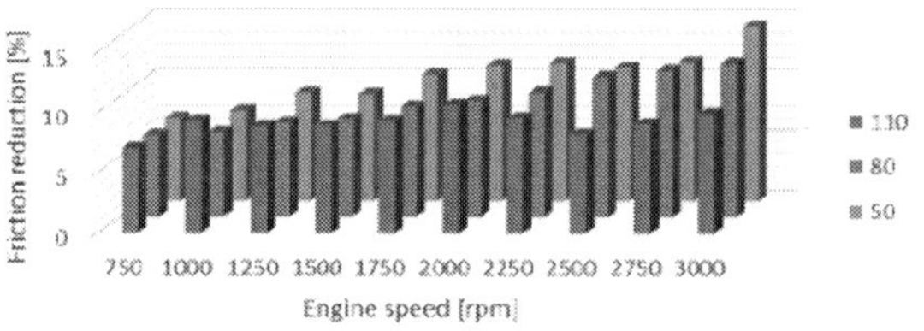

Fig. 14. Reduction of friction losses as a function of temperature and speed variant TR6.

Based on these results it can be seen that all the experimental variants of the piston with a stepped bearing surface microgeometry generate the desired reduction of friction in the piston-cylinder group, relative to the serial pistons. As it was expected from the hydrodynamic lubrication theory, increasing the rotational speed increases the braking torque.

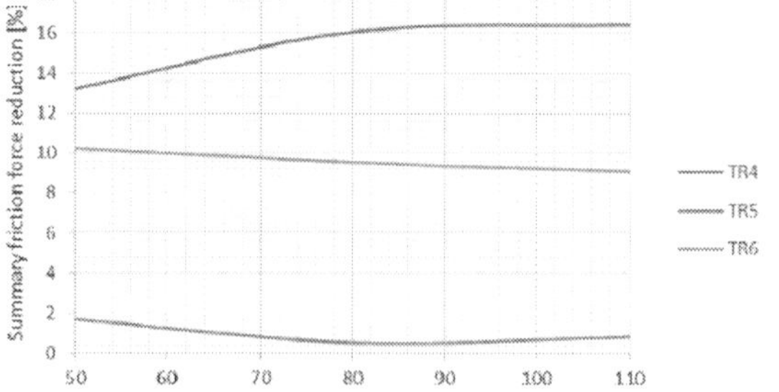

Fig. 15. The total reduction of the friction losses as a function of oil temperature.

The result of measurement, showed the greatest reduction in friction variant TR4 (approx. 15%), TR6 (approx. 10%) and the lowest TR5 (approx. 1%). On the basis of the results it can be also seen that a reduction in friction losses is greater for higher speeds of the engine crankshaft. This is an important observation, because while increasing the engine speed as a result of higher friction losses being converted into heat, it occurs faster heating in cooperating parts of the engine and in the lubricating oil. This leads to conditions of boundary friction. This phenomenon may, however, have a positive effect on the reduction of friction losses in the piston-cylinder group on the assumption that no boundary friction will occur at the interface between the bearing surface of the piston and the cylinder liner.

5 Conclusions

The study used the bench test, in which was measured precisely the total friction loss in the piston-crank mechanism and the other motor mechanisms are fixed or fitted with independent electric drive. Studies in experimental bench model of the motor pistons with a stepped profile bearing surface to estimate the value of the friction losses in the piston-crank mechanism and to draw the following conclusions:

- the use of microgeometry of the bearing surface stepped profile relative to the barrel allows for the reduction of friction losses thus increasing the mechanical efficiency of the engine and reducing fuel consumption,
- the application of the H-shape bearing surface of the piston allows to reduce friction loss average about 10% in relation to the barrel-like shape,
- it was showed the greatest reductions variant of the piston with the bar letter H shifted downwards relative to the position symmetric about 16%, because this variant extended the area, it ensures the generation of a continuous oil film separating the cooperating surfaces,
- the coolant temperature regulates temperature of engine and oil, which affects the friction losses,
- for all oil temperatures along with the increase in rotational speed, experimental pistons showed a marked reduction in friction in relation to the barrel-like shape. This is caused by a higher relative velocity of the moving elements of the engine and as a result of the higher temperature,
- increase speed at relatively constant temperature of the oil increases the loss of mechanical resistance,
- a stepped surface can be achieved by the coating of the support surface of the piston using lubricants like i.e. graphite,
- a layer of lubricants are particularly desirable under lack of continuity of oil film conditions.

The study presented in this article was performed within the statutory research (contract No. 05/52/DSPB/0262).

References

1. M. Blümm, A. Baberg, F. Dörnenburg, MTZ **2** (2016)
2. G.M. Szymanski, F. Tomaszewski, *Proceedings of The 17th International Congress on Sound and Vibration* (Cairo, 2010)
3. M. Orczyk, F. Tomaszewski, B. Czechyra, G.M. Szymanski, *Proceedings of the International Congress on Sound and Vibration* (Athens, 2016)
4. G.M. Szymanski, M. Josko, F. Tomaszewski, R. Filipiak, Mechanical Systems and Signal Processing **58-59**, 298-307 (2015), DOI: 10.1016/j.ymssp.2014.12.017
5. A. Baberg, M. Freidhager, H. Mergel, K. Schmidt, MTZ **12** (2012)
6. M. Fahr, W. Hanke, C. Klimesch, A. Rehl, MTZ **7-8** (2011)
7. J. Merkisz, P. Fuc, P. Lijewski, A. Ziolkowski, K.T. Wojciechowski, Journal of Electronic Materials **44**, 1704-1715 (2015), DOI: 10.1007/s11664-014-3522-6
8. J. Merkisz, P. Fuc, P. Lijewski, A. Ziolkowski, M. Galant, M. Siedlecki, Journal of Electronic Materials **45**, 4028-4037 (2016), DOI: 10.1007/s11664-016-4543-0
9. G. Koszalka, Eksploatacja i Niezawodnosc – Maintenance and Reliability **3**, 40-44 (2011)
10. T. Deuss, H. Ehnis, R. Rose, R. Künzel, MTZ **5** (2010)
11. J. Merkisz, J. Mizera, M. Bajerlein, L. Rymaniak, P. Maj, Experimental and Applied Mechanics **518**, 102-107 (2014), DOI: 10.4028/www.scientific.net/AMM.518.102
12. T. Deuss, H. Ehnis, R. Rose, R. Künzel, MTZ **4** (2011)
13. A. Iskra, P. Krzymień, E. Wróblewski, Combustion Engines **162**, 192-196 (2015)
14. A. Iskra, M. Babiak, E. Wróblewski, Technika Transportu Szynowego **12**, 644-647 (2015)
15. E. Wróblewski, A. Iskra, M. Babiak, Journal of KONES Powertrain and Transport **23**, 431-436 (2016)
16. A. Iskra, M. Babiak, W. Wroblewski, *IOP Conference Series-Materials Science and Engineering* **148**, UNSP 012068 (2016), DOI: 10.1088/1757-899X/148/1/012068
17. A. Iskra, E. Wróblewski, M. Babiak, Journal of Kones Powertrain and Transport **22**, 95-101 (2015)
18. J. Kaluzny, *Wpływ kształtu powierzchni nośnej tłoka na parametry filmu olejowego pokrywającego gładź cylindra* (Poznan, 2004)
19. E. Wróblewski, A. Iskra, M. Babiak, Journal of Mechanical and Transport Engineering **68**, 2 (2016)

The impact of the accuracy of indicator diagrams on the heat release characteristics calculation, used in the diagnosis of marine diesel engine

Kazimierz Witkowski[1,*]

[1]Gdynia Maritime Academy, Mechanical Faculty, ul. Morska 81-87, 81-225 Gdynia, Poland

Abstract. The paper analyzes the possibility to use the electronic type indicators in the diagnosis of marine engines. It has been shown that in-depth analysis of indicator diagrams would be useful – calculation of heat release characteristics. To make this possible, measuring indicated systems should meet a number of important requirements in or-der to ensure that they can be used for the diagnostic purposes. These includes: high precision sensors for the measurement of cylinder pressure, high speed and accuracy of measuring and recording of measured values. These also includes reliable determination of the top dead center piston (TDC). In order to demonstrate the impact of positional error TDC, simulation study was conducted in which indicated diagrams were used, obtained on a medium-speed four-stroke marine diesel engine type A25/30 and the low-speed two-stroke marine diesel engine type RTA76, Sulzer company.

1 Introduction

Nowadays many marine engine-rooms are additionally equipped with portable or stationary electronic indicators. Open (expanded) indicator diagrams are mainly analyzed – diagrams in a coordinate system $p = f(\alpha)$ (cylinder pressure p variation as a function of crank angle α). Indicated parameters are the essential complement to the standard measurement parameters and diagnostic inference can be more reliable [1, 2].

Measuring systems destined for indication should meet a number of important requirements to be met to ensure the possibility of their diagnostic use. These includes: high precision sensors for the measurement of cylinder pressure, high speed and accuracy of measuring and recording of measured values, a reliable determination of the top dead center piston (TDC). These and other important issues cylinder pressure measurement were highlighted in many publications [3-6]. Compliance with these requirements is particularly important when will be determined based on the indicator diagram of heat release characteristics for diagnostic purposes [7-10].

At old-design sea cargo ships with conventional equipment, marine engines diagnostics are still made on the basis of an analysis of changes in the value of the measured parameters of the engine [7, 11]. Diagnostician is the ship's mechanic, who by analyzing changes in the measured parameters relative to their reference values, tries to determine the type of damage. Such action is subjective and usually inefficient, depending on the knowledge and experience of the marine engineer [5, 12, 13].

* Corresponding author wika@am.gdynia.pl

2 Direct and in-depth analysis of indicator diagrams in diagnostics

Increasingly for indicated marine engines, it is being used the electronic type indicator, which includes: the combustion pressure sensors in each cylinder, the angular position sensor crankshaft and microprocessors system used for processing and visualization of the results of the measurements. This indicator is easy to use, characterized by high accuracy and fast developing results. This device gives results indicated as the average of dozen cycles. All essential values are displayed on the screen, and in addition an open indicator diagram is presented. Mean indicated pressure is automatically served beside other parameters, based on our numerical integration.

Figure 1 shows an example of an indicator diagram, obtained using the an electronic indicator. It is marked on the all important characteristic values. In addition to those indicated values in the figure, the mean indicated pressure and indicated power can be counted.

An important feature of electronic indicators is the ability to save the results in memory and play them after some time. This way, you can easily compare the graphs indicated which were performed at different times, which gives a fuller analysis capabilities [14-16]. This is particularly important during comparing the reference indicator diagram drawn in good technical condition of the engine with a graph removed after a longer period of operation; then are visible all deviations from the norm.

Figure 2 shows an example of one coordinate system $p = f(\alpha)$ two graphs – curve 1 and curve 2. The pressure value on the expansion curve (p_{36^o}) graph of measured current (curve 2) is higher than the corresponding value for the standard graph (curve 1). This may be a prove of chronic working process, after-burning of the fuel during the expansion, e.g. due to incorrect fuel atomization. This state can also affect non-leak fuel injector nozzle or very poor fuel quality.

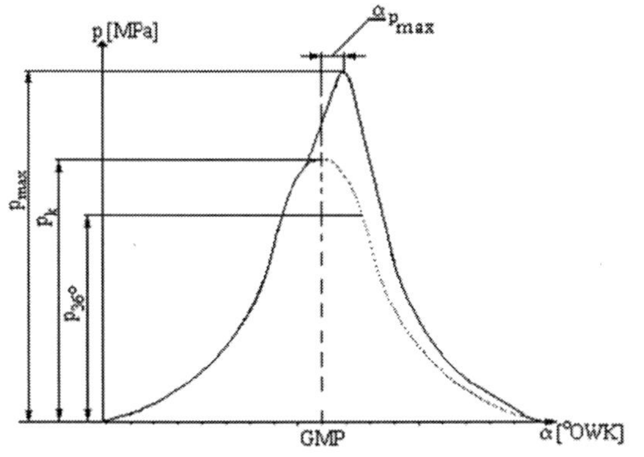

Fig. 1. Indicator diagram obtained from the electronic indicator and its characteristic values: p_{max} – the maximum combustion pressure, p_k – compression pressure, $p_{36}°$ (or p_{exp}) – expansion pressure read 36 degrees of crankshaft rotation after TDC, α_{pmax} – angle after TDC obtain maximum pressure in the cylinder.

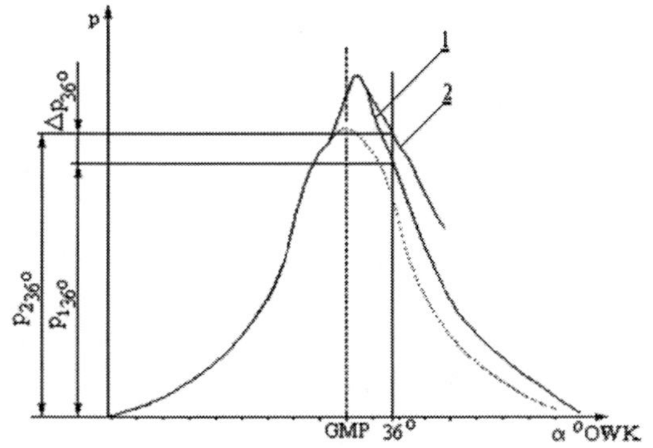

Fig. 2. Exemplary indicator diagram (1) and too high expansion pressure on the diagram (2).

The graph in Figure 3 shows that gain in pressure is too dynamic. Data from the graph allow to calculate the value of $\Delta p2/\Delta\alpha2$. Based on a comparison of the measured $\Delta p2/\Delta\alpha2$ with exemplary $\Delta p1/\Delta\alpha1$, you can quantify the degree of damage. The cause of excessive dynamic rise pressure – $(\Delta p2/\Delta\alpha2) >> (\Delta p1\ /\ \Delta\alpha1)$ – may be premature fuel injection or ignition delay reduction: $\alpha_{zz2} < \alpha_{zz1}$, due to significant thermal overload the elements of the combustion chamber of the engine. A large value of $\Delta p/\Delta\alpha$ causes mechanical engine overload, especially bearings piston-connecting-rod system.

Ongoing monitoring and control of the mean indicated pressure p_i provides a very accurate load current of each cylinder and also of the whole engine. On this basis we can determine whether the load status of each cylinder is the same or if there is too much spread of power between the cylinders [17-19]. In relation to the average of the instantaneous power of the cylinder, these differences should not exceed ±2,5%. With irregular engine regulation, mean indicated pressure pi in one or more cylinders will generally be lower than the average value. There are at least several potential reasons for this: it may be due to defective regulation of the injection pump – too small dose of fuel on account of the piston-cylinder assembly leakage caused by poor technical conditions of injection equipment and air system supercharger.

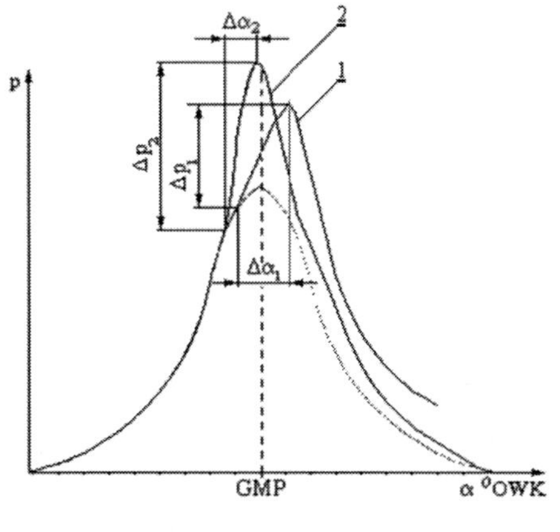

Fig. 3. Exemplary indicator diagram (1) and diagram with high value of $\Delta p/\Delta\alpha$ (2).

The direct analysis of the characteristics feature indicator diagram and changes of indicated parameter values does not always allow on the formulation of correct diagnostic conclusions. Even a few different defects can cause similar changes in the values of indicated parameters and shape indicator diagrams. Therefore this is recommended in-depth analysis of these graphs based on calculated heat release characteristics. Among other things, this is important in the diagnosis combustion process and of damage of injection system elements – injection pumps and injectors. Then, however, it is essential to cylinder pressure measurement accuracy – the accuracy of the indicator diagram. The analysis shows, that on the accuracy of the calculated heat release characteristics has a piston TDC location error on the indicator diagram.

3 Analysis of the influence of selected factors on the accuracy of the indicator diagram

A very significant impact on the accuracy of the determination of the heat release's characteristics, is the correct term for the indicator diagram obtained during the test engine, piston position placed in the top dead center (TDC). The main reasons of difficulty with accurately determine TDC are delays caused by crankshaft torsional vibrations, caused by indicated tunnels and cocks indicated, as well as due to non-uniformity of the crankshaft rotational speed [3, 4, 9,

DOI: 10.1051/matecconf/201711800005

20]. It must be noted that between the kinematic angular position TDC piston and the TDC position shown on the indicator diagram, there is an error that will cause further errors calculated heat release characteristics. In [1, 3] stated that the error of determining TDC the amounting to ±2%, will cause distortion chart calculated heat release rate to about ±12%.

In many cylinder pressure measurement systems, errors occur, which are related to the "flow" of the reference line (ambient pressure or supercharging pressure). This is related primarily to the construction of sensors. It is observed in electronical indicators in which recourse piezo-quartz sensors and strain gauge type sensors. In order to minimize this error is constructed signal amplifiers, using a high impedance input. "Flow" line reference pressure has an impact primarily on heat release rate calculated [4, 21-24].

For most of marine engine, cylinder pressure sensor is mounted on the indicator cock. Engine combustion chamber with an indicator cock were combined with the gas tunnel, the length of which depends on the design of the cylinder head and can rise up to several tens of centimeters. The gas channel and indicator cock are the cause of the measured signal distortions and their delay. In research of four-stroke, medium-speed, marine diesel engines [5], has been shown that delays may be several degrees of crankshaft rotation. Signal distortion characteristic of medium-speed engines, are negligible small in two-stroke slow-speed marine engines. If the sensor is mounted in special tunnels of the cylinder head, these types of measurement problems do not occur. Then however, it can occur the phenomenon of thermal shock sensor, formed by the action of the flame on the front face of the sensor. It leads to a distortion, inter alia, of measured changes of pressure. In study [4] this is shown, that small increase of temperature in piezo-quartz sensors for only about 100 K causes a decrease in the sensitivity of about 3%. Also decreases the resistance self, for about 10%. In marine diesel engine heavy fuel oil feeding, additional distortions in measurement circuit may arise as a result of fouling carbon deposit sensor.

The important thing for the quality of the indicator diagram is averaging the received multiple charts and their smoothing. In electronic indicators used in operational practice (in ship engine room) is averaged at least 16 charts. In the case of piston engines used in inland transport, it can be said that we can use even more than 30 charts [4]. By averaging the appropriate number of execution indicated it is possible to eliminate inter alia, random vibration of high frequency electrical noise. After averaging chart should be further smoothing, to eliminate distortion – get a smooth and continual graphs [3, 4].

Electronic indicator registers the cylinder pressure variation from the relatively low (the beginning of the compression, the beginning of exhalation) less than 1 MPa, up to the maximum even about 20 MPa. The accuracy of measurement of the pressure instantaneous value affects resolution analog-to-digital converters and their accuracy. The error of quantification particularly affects the measurement of low pressures.

4 Study of the impact of the error of the TDC location, on the accuracy calculated heat release characteristics

TDC position errors are a major source errors of determining the mean indicated pressure and thereby greatly influenced by on value of error the maximum value of the net heat release (Q_{max}) and the maximum value heat release rate (q_{max}).

In order to demonstrate the error of the impact positional TDC, simulation study was conducted in which was used indicated diagrams obtained on a medium-speed four-stroke marine engine type A25/30 (under laboratory conditions) and the low-speed two-stroke engine type RTA76, Sulzer company, in the operating of the ship.

Studies in laboratory conditions was carried out on a test bench of the marine engine SULZER type 3AL25/30. This diesel engine was loaded self-excited synchronous generator type GD8-500-50. This generator is loaded the water resistor, equipped with a tower cooler of water (brine). The test stand is located in the laboratories of the Department of Marine Power Plant Gdynia Maritime University and is shown in Figure 4 and 5.

Fig. 4. View of the research stand with the marine diesel engine Sulzer 3AL25/30.

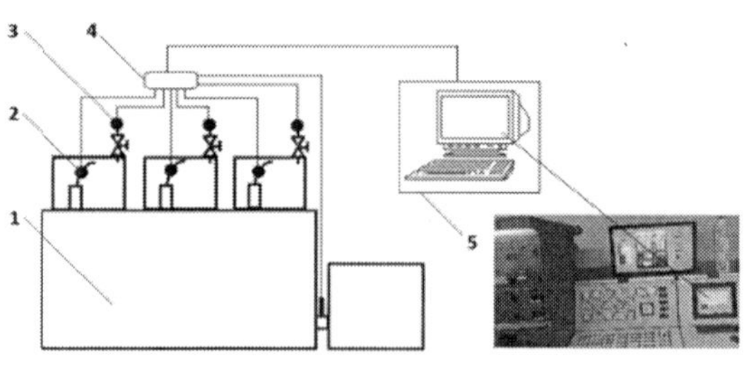

Fig. 5. Scheme of the research stand with the marine diesel engine Sulzer 3AL25/30: 1 – diesel engine Sulzer 3AL25/30, 2 – injection press sensors, 3 – combustion press sensors, 4 – amplifiers and converters A/C, 5 – computer.

This stand also includes the electrical switchgear RK-73, fuel tanks and a central computer system measuring and registration of operating parameters of the engine. An important element of this measuring system is an electronic indicator UNITEST 2008 in which, to measure the pressure in the cylinders, we use sensors from Kistler 6353A24 with transmitters type 6961A250 (element 1 in Figure 6) and sensors to measure the pressure in the injection system of Optrand type Autopsie-S transducers type C922A8 (element 2 in Figure 6).

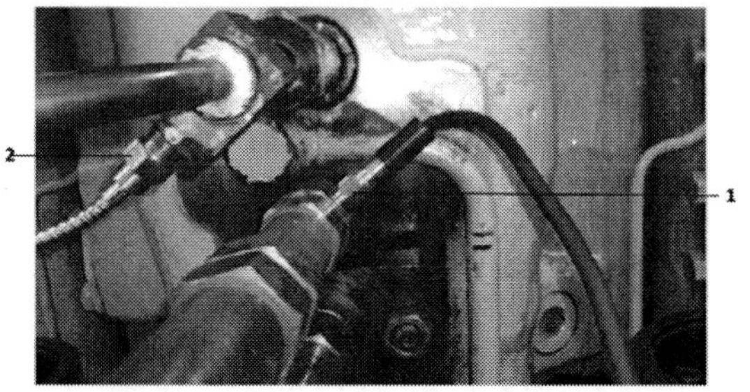

Fig. 6. The research stand with the marine diesel engine Sulzer 3AL25/30 – view on one of cylinder heads of the mounted sensors: 1 – injection press sensor Optrand typu AutoPSI-S, 2 – combustion press sensor Kistler 6353A24.

Basic specifications of the marine engine SULZER type 3AL25/30:

- power cylinder 136 kW,
- rotation speed 750 rpm,
- mean effective pressure 1.575 MPa,
- compression ratio 13,
- specific fuel consumption 204 g/kWh,
- supercharger system pulse system,
- turbocharger BBC VTR160N.

For the engine A25/30 it has been adopted TDC location error amounting to ±1° rotation of the crankshaft and for the low-speed engine RTA76 ±0,5° rotation of the crankshaft.

In Figures 7 and 8 it has been shown the effect of errors in the TDC location of the indicator diagram on the characteristics of q and Q for medium-speed diesel engine (Fig. 5) and low-speed diesel engine (Fig. 6).

Deviations q_{max} are approximately ± 5% for engine A25 with TDC errors on value ±1° rotation of the crankshaft and ±3.5% for the engine RTA76 the TDC with errors ranging ±0.5° rotation of the crankshaft. Q_{max} deviations are approximately ±8% for the A25 engine with TDC errors amounting to ±1° rotation of the crankshaft and ±4.5% for the engine RTA76 with TDC errors amounting to ±0.5° rotation of the crankshaft.

The values of TDC positional errors are intentionally significant. In practice, using the method of determining the TDC based on thermodynamic models, the error should not exceed ±0.3° rotation of the crankshaft in the case of medium-speed engines and ±0.1° rotation of the crankshaft in the case of low-speed engines.

It should be noted the almost negligible impact of error TDC on the first phase (the growth steeply runs) q (Fig. 7, Fig. 8). This is an important observation in the context of the possibility of using of the symptom for diagnostic purposes.

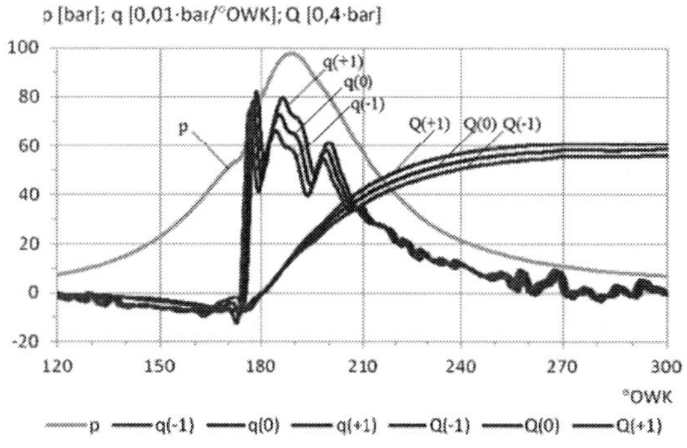

Fig. 7. Effect of TDC position errors on waveforms heat release characteristics *q* i *Q* medium-speed marine diesel engine A25/30. The error value TDC (in degrees rotation of the crankshaft) given in brackets.

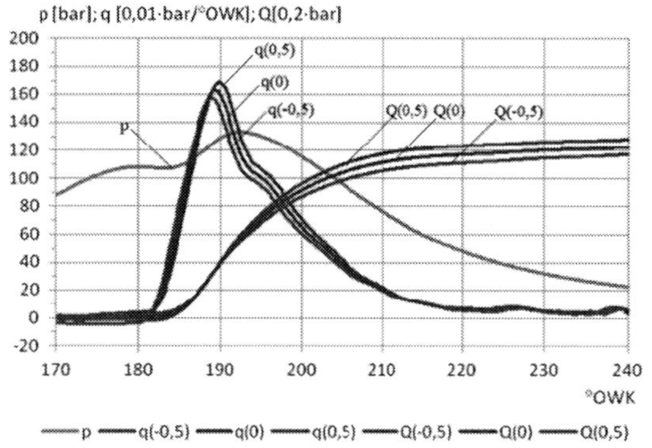

Fig. 8. Effect of TDC position errors on waveforms heat release characteristics *q* i *Q* low-speed marine diesel engine RTA76. The error value TDC (in degrees rotation of the crankshaft) given in brackets.

5 Summary

The analysis of indicated parameters read from the indicator diagram is now widely use on vessels in the diagnosis of marine engines. The direct analysis of the characteristics feature indicator diagram and changes of indicated parameter values does not always allow on the formulation diagnostic correct conclusions.

Even a few different defects can cause similar changes in the values of indicated parameters and shape indicator diagrams. Therefore this is recommended to conduct in-depth analysis of these graphs based on calculated heat release characteristics.

Measuring systems for electronic indication should meet a number of important requirements to guarantee the possibility of their diagnostic use. These include: high precision sensors for the measurement of cylinder pressure, high speed and accuracy of measuring and recording of measured values, a reliable determination of the top dead center piston (TDC).

In many cases, in order to obtain a reliable diagnosis is indicated in-depth analysis of these graphs based on calculated heat release characteristics. Then, however, it is essential to cylinder pressure measurement accuracy – the accuracy of the indicator diagram, including the exact location of TDC in the diagram.

The analysis of the results indicates that the error location TDC to 1% results in a significant deviation of the net heat release maximum value and heat release rate, to a few percent (from 3.5% to 8%).

References

1. S. Polanowski, Journal of Kones (1995)
2. S. Polanowski, Journal of Kones (2000)
3. A. Ambrozik, P. Ambrozik, *Analiza i symulacja wykresu indykatorowego tłokowego silnika spalinowego* (Szczecin, 2001)
4. A. Ambrozik, *Wybrane zagadnienia procesów cieplnych w tłokowych silnikach spalinowych* (Kielce, 2003)
5. S. Polanowski, Zeszyty Naukowe Akademii Marynarki Wojennej (2007)
6. A. Ambrozik, *Analiza cykli pracy czterosuwowych silników spalinowych* (Kielce, 2010)
7. G.M. Szymanski, M. Josko, F. Tomaszewski, R. Filipiak, Mechanical Systems and Signal Processing **58-59**, 298-307 (2015), DOI: 10.1016/j.ymssp.2014.12.017
8. G.M. Szymanski, F. Tomaszewski, *Proceedings of The 17th International Congress on Sound and Vibration* (Cairo, 2010)
9. M. Orczyk, F. Tomaszewski, B. Czechyra, G.M. Szymanski, *Proceedings of the International Congress on Sound and Vibration* (Athens, 2016)
10. T. Rychter, A. Teodorczyk, *Modelowanie matematyczne roboczego cyklu silnika tłokowego* (Warszawa, 1990)
11. J.A. Wajand, *Pomiary szybkozmiennych ciśnień w maszynach tłokowych* (Warszawa, 1974)
12. C. Specht, J. Rudnicki, Polish Maritime Research **23**, 20-27 (2016), DOI: 10.1515/pomr-2016-0028
13. D. Pietras, Polish Maritime Research **23**, 58-63 (2016), DOI: 10.1515/pomr-2016-0032
14. I. Pielecha, W. Cieslik, Journal of Thermal Analysis and Calorimetry **126**, 815-827 (2016), DOI: 10.1007/s10973-016-5544-1
15. I. Pielecha, M. Skowron, W. Bueschke, W. Cieslik, K. Wislocki, *IOP Conference Series-Materials Science and Engineering* **148**, UNSP 012081 (2016), DOI: 10.1088/1757-899X/148/1/012081
16. R. Zadrag, *Solid State Phenomena* **196**, 74-81 (2013), DOI: 0.4028/www.scientific.net/SSP.196.74
17. J. Rudnicki, R. Zadrag, Polish Maritime Research **21**, 57-65 (2014), DOI: 10.2478/pomr-2014-0042
18. I. Pielecha, P. Borowski, J. Czajka, K. Wislocki, J. Kazmierowski, Journal of Thermal Analysis and Calorimetry **119**, 695-703 (2015), DOI: 10.1007/s10973-014-4139-y
19. I. Pielecha, K. Wislocki, P. Borowski, W. Cieslik, Journal of Thermal Analysis and Calorimetry **122**, 473-485 (2015), DOI: 10.1007/s10973-015-4735-5
20. L.J. Kapusta, I. Pielecha, K. Wisłocki, A. Teodorczyk, Journal of Thermal Analysis and Calorimetry **123**, 819-828 (2016), DOI: 10.1007/s10973-015-4927-z
21. M. Skowron, I. Pielecha, K. Wislocki, *IOP Conference Series-Materials Science and Engineering* **148**, UNSP 012085 (2016), DOI: 10.1088/1757-899X/148/1/012085
22. J. Markowski, J. Pielecha, R. Jasinski, T. Kniaziewicz, P. Wirkowski, *E3S Web of Conferences* **10**, UNSP 00140 (2016), DOI: 10.1051/e3sconf/20161000140
23. M. Noga, Applied Sciences – Basel **7**, 295 (2017), DOI: 10.3390/app7030295
24. H. Golas, A. Mazur, J. Gruszka, *ACSR-Advances in Computer Science Research* **57**, 257-264 (2016)

MATEC Web of Conferences **118**, 00006 (2017) DOI: 10.1051/matecconf/201711800006
VII International Congress on Combustion Engines

Optical study of the use of recirculated gases for adiabatization of combustion process in the SIDI engine

Wojciech Cieślik[1,*], *Ireneusz* Pielecha[1], and *Krzysztof* Wislocki[1]

[1]Poznan University of Technology, Faculty of Machines and Transport, ul. Piotrowo 3, 60-965 Poznan, Poland

Abstract. Proper delivery of gaseous components of the charge into the combustion chamber enables controlling of combustion in the aspect of adiabatization of the process. The adiabatization obtained as result of surrounding of combustible mixture by recirculated exhaust gases should contribute to reducing formation of harmful components created during this process. The key issue here is the formation of radicals, which is not sufficiently recognized according fuels surrounded by non-combustible gases. The innovative nature of this work ensues from the experimental confirmation of so defined organizing of combustion process. Currently there are no tests concerning attempts of gas separation in the combustion chamber of engine with external source of ignition. Such separation would contribute to the increase of the adiabatization process while at the same time the combustion rate increases and reduces the combustion duration. This paper presents the next stage of research, which were preceded by simulation and experimental investigations. In the article the results of the impact of the strategy of non-combustible gas injections on combustion ratios for cylinder head with a centrally positioned ignition point have been discussed. The analysis has been based on the photo material for the period from the start of ignition to full coverage of the cylinder by the flame. Authors performed a comparative analysis (against the recorded images) of the thermodynamic indexes of the combustion process obtained from the indicator traces.

1 Introduction

Internal combustion engines still play an important role as a power source for automobiles. Even though there has been a growing trend in the automobile industry to pursue development of hybrid electric vehicles (HEV) or plug-in hybrid electric vehicles (PHEV), innovation of internal combustion engines (ICE) is essential to maximize performance of these types of vehicle. In an era of stringent standards for fuel consumption and emissions of exhaust gases from vehicle drive systems [1, 2], one of the directions of development in the design of these systems is the use of alternative drive systems (hybrid, hydrogen drives) [3-5]. Traditional drive systems using spark-ignition combustion engines and compression-ignition engines are subject to modification [6-9], which can be summed up as actions involving downsizing, downspeeding, direct fuel injection, boosting and rightsizing [4, 10-12]. Possibilities presented by Isenstadt et al. [13] for reducing fuel consumption on the example of Skyactiv-G engine (Fig. 1) indicate the effectiveness of the application of the cooled exhaust gas recirculation system, particularly for medium engine loads.

One particular advantage of using exhaust gas recirculation systems is a relatively small increase in the cost of modifications. For this reason, virtually all modern engines are equipped with exhaust gas recirculation systems.

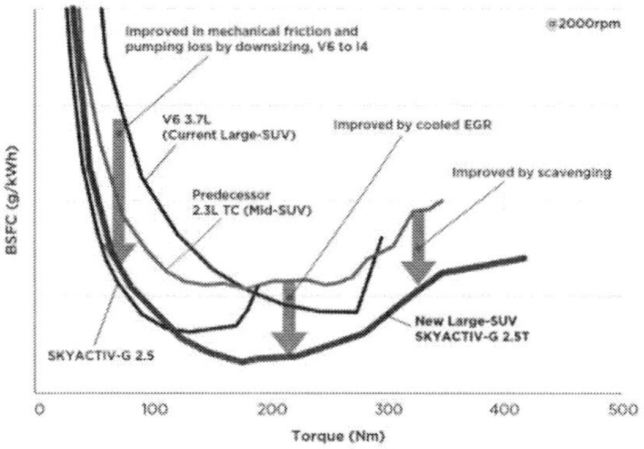

Fig. 1. Reduction in fuel consumption of Skyactiv-G 2.5T [13].

Some ideas of applying of flue gases for reduction in intensity of nitrogen oxides formation was formulated in early 80-th of XX c. However, the level of development of fuelling systems was not high enough for any spectacular progress at this time. Next ideas of this kind were proposed in 1993 by Heitland et al. [14]. Based on the patent application from early 1993 designed for two-stroke engines [15] they suggested to generate the region of air-fuel mixture in the central part of combustion chamber which should be surrounded by exhaust gases from previous cycle. For mixture formation the air-blast atomizer has been proposed, and tangential shaped

* Corresponding author: ireneusz.pielecha@put.poznan.pl

exhaust channels for swirling of exhaust gases remaining around the mixture and isolating combustion process from cylinder walls.

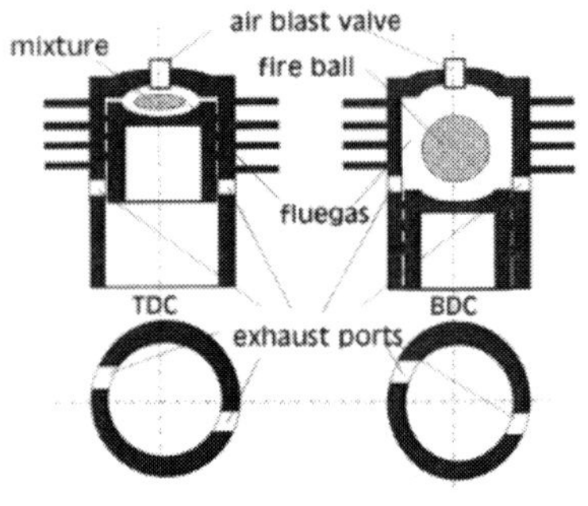

Fig. 2. The conception of mixture-exhaust stratified charge combustion system for opened combustion chamber [15].

For practical realization of this system in open chamber the application of impinging jets has been suggested with electronically controlled fuel and air supplies, or – alternatively – an air assisted fuel injection – Fig. 3. Unfortunately, at that time the electronically driven injectors have not reached enough high level of development and the authors of this idea did not disposed the appropriate equipment for the experimental confirmation of its expected operation. With the present technical investigation possibilities, this task might be newly restarted.

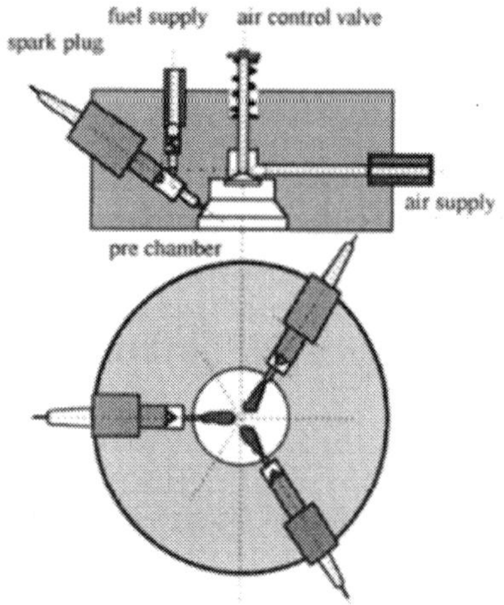

Fig. 3. MESC-type engine with electronically controlled fuel supply and with pneumatic mixture injection [16].

In recent years as such it can be also considered the Turbulent Jet Ignition (TJI), which makes it possible to increase the thermal efficiency of the system up to about 45% [17, 18]. An actual research question should be the recognition of the impact of those prototype combustion systems, which can significantly reduce the emissions.

In order to meet the requirements of the above trends in development involving improving combustion efficiency, it is necessary to identify the essential processes taking place during combustion. One of the methods of diagnostics of the fast-varying processes is the evaluation of the flame initiation and development. The authors of this article undertook the task of such assessment using the performed earlier simulation studies [19] concerning the control of the distribution of combustible and non-combustible gases in the cylinder chamber.

2 Motivation for undertaking the research issues

Earlier publications of the Authors concerning simulation [19] and experimental analysis with one modification of the cylinder head of the rapid compression machine (RCM) [20] have shown the benefits of the proposed solution to use non-combustible gases to control the process. The solution is associated with a reduction in the energy indexes of the process (caused by the reduction of the volume of the combustible mixture). However, the nature of combustion with its benefits in the form of reduced losses of heat exchange with the walls of the operating cylinder justify undertaking the subject. In order to increase the efficiency of the internal combustion engine, it is important to prepare properly the combustible charge in the chamber of the cylinder. The correct conditioning of the combustion process may allow the reduction of fuel consumption.

One way to meet the above demands is to use recirculated exhaust gases to introduce some volume of a non-combustible gas into the cylinder in a strictly defined way, involving the control of the space in which the combustion process will take place. According to the previously mentioned methods of gas distribution inside the cylinder [19, 20], it is possible to obtain several types of swirling: lateral, axial and radial. Radial stratification of the air and exhaust gases is the most suitable choice, from the point of view of the compression stroke, in terms of maintaining the stratification during piston movement. The interior region of the fuel-air mixture is encased in the exhaust gas region. Both of these regions are coaxial in the cylinder, and therefore the compression stroke should not disturb the load stratification (the separation of injected air and exhaust gas). In addition, if the angular momenta of the loads are be kept during the compression stroke then a mass of air and exhaust gas swirling in the same direction will result in a much longer lasting stratification of the mixtures at the end of the compression stroke.

In contrast to the classic solutions supplying the recirculated exhaust gases into the cylinder chamber in combustion engines, the presented system involves supplying non-combustible gases in a strictly defined manner. Presented in the paper [21], as well as in [14-16], possibilities of stratification of a charge with non-combustible gases (separation of air and exhaust gases) allowed the Authors to choose the concept of a combustion system from among several earlier proposals.

Charge stratification can be divided into lateral (starting from the left), axial and radial [21]. Radial

 DOI: 10.1051/matecconf/201711800006

stratification of the discussed gases (air and exhaust gases) from the point of view of the compression stroke is the most suitable choice in terms of preserving stratification during the movement of the piston. An outer layer of the exhaust gases is surrounding the internal area of the air-fuel mixture. Both layers are coaxial in the cylinder of the engine; therefore the compression stroke should not interfere with the separation of the charges of combustible and non-combustible gases. In addition, if during the compression the air momentum and exhaust gas momentum are preserved, then the air and gases swirling in the same direction will result in much longer retention of the mixtures stratification at the end of the compression stroke. Figure 4 shows the implementation of the assumptions using RCM.

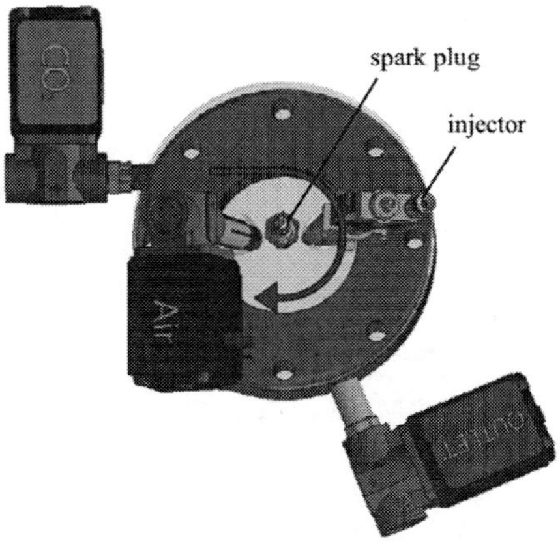

Fig. 4. The implementation scheme of the provide gases to the RCM.

Presented in publication [20] analysis of possibilities of obtaining the combustion process adiabatization in the designed system of lateral excitation of the ignition, indicated obtaining radial stratification, which made it possible to separate air from the exhaust gases in the area of the combustion chamber (in the SIDI engine). Test results indicated restriction in the access of the flames to the cylinder walls. The obtained reduction in the surface area of the flame amounted from 13% to 50% (for three repetitions of the combustion process) compared to combustion of fuel only with the participation of air.

3 Research object

For the tests was used the Rapid Compression Machine (RCM) with the parameters shown in Table 1. Upgrades in relation to the previously presented designs of the combustion chamber [20] involved different location of the spark plug in the cylinder head. In the presented tests the spark plug is placed in the centre of the chamber (Fig. 5) (11). The injector (10) and the inlet valve of air (12) and non-combustible gases (13) are positioned as shown in Fig. 5.

Table 1. Characteristic of RCM.

Quantity	Value
Stroke	89 mm
Bore	80 mm
Volume of combustion chamber	55 cm^3
Method of delivery non-combustible gases	solenoid valves
Compression ratio	11
Type of combustion chamber	hemispherical chamber in the head + chamber in the piston
Optical access	from the bottom of the combustion chamber, quartz glass ϕ50 mm arranged in the piston

Previous studies of the Authors [20] proved the thesis that it was possible to conduct combustion in the presence of recirculated exhaust gases. However, due to positioning of the spark plug in the lateral part of the head, the Authors decided to modify the system, so that now it is in the central part of it. The aim was to increase the concentration of combustible charge in the centre of the combustion chamber, allowing for greater concentration of the flame in the combustion chamber with limited access to the cylinder walls.

4 Research methodology

The tests on combustion process were conducted with the use of two main measurement systems (Fig. 5):

- the system for acquisition of measurements of fast-varying signals (3),
- the system for acquisition high-speed imaging of combustion process (5, 6).

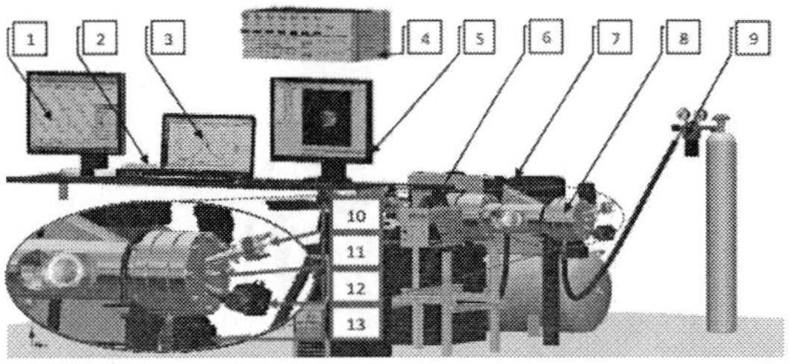

Fig. 5. A 3D design of the test bench (explanations in text).

The other elements necessary to conduct the presented tests were a triggering system – a sequencer (1, 2), a valve control (4), a compressor (7), RCM (8), and a non-combustible gas bottle (9).

The tests were conducted for 3 different sequences of the inlet of non-combustible and combustible gases into the combustion chamber, and for various values of non-combustible gas injection pressure. Gas inlet sequences were as follows:

I. Initial air injection into the chamber through the valve (12), then injection of CO_2 through the valve (13),

II. Initial injection of CO_2 into the chamber through the valve (13), then injection of air through the valve (12),
III. Simultaneous injection of both gases.

The tests results were compared to the reference measurements, where the injection of CO_2 was not used.

5 Thermodynamic indexes of combustion process

Comparison of the characteristics of indicated pressure in the combustion chamber is shown in Fig.6. The characteristics of three sequences of non-combustible gas injection into the combustion chamber were compared to the reference characteristics. For each case three measurement points corresponding to the pressures of CO_2 injection 5, 7, 10 bars are presented. Based on the increase of pressure in the combustion chamber prior to the movement of the piston, the portions of the non-combustible gases in relation to combustible mixtures were determined during the tests, and they amounted to respectively 20 and 32% of the charge volume. These values are variable within a range of ±5% depending on the sequence of the injection of gas into the chamber.

The areas of maximum delay for P_{max} values in relation to the reference point are shown on the diagrams. The biggest delay in occurrence of maximum value of this parameter is observed for the characteristics of sequence II, where for the injection pressure of CO_2 of 10 bars was observed over 25 ms delay in the occurrence of the P_{max} value in comparison to the reference point. Sequence III – simultaneous injection of combustion air and CO_2 – was registered with the maximum delay P_{max} lower by 20% from sequence II. The slightest delay in occurrence of P_{max} value at the CO_2 injection pressure of 10 bars was registered for sequence I (it amounted to 8 ms) which represents 60% of delay compared to sequence II.

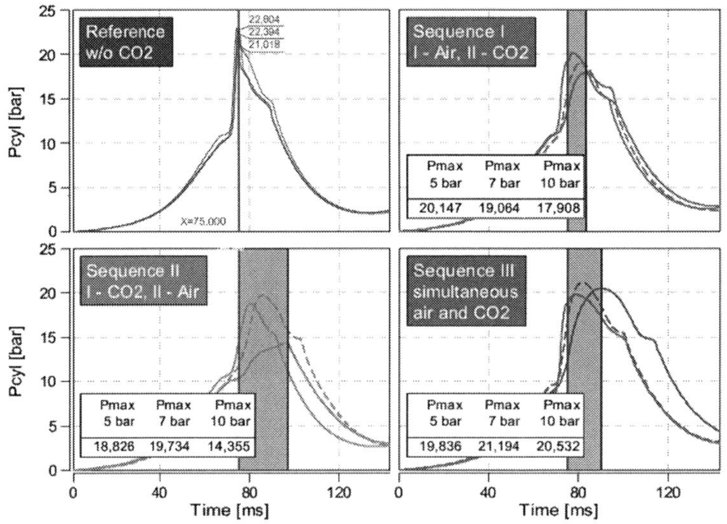

Fig. 6. Indicated pressure characteristics at different sequences of injection of air and non-combustible gases.

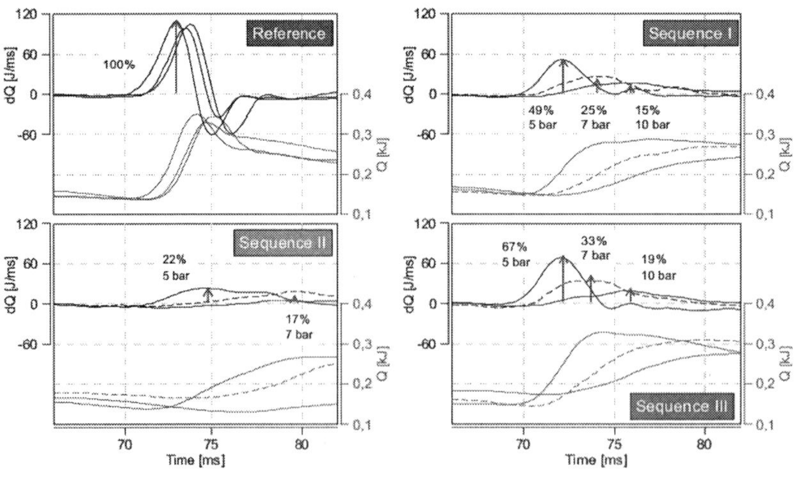

Fig. 7. Characteristics of heat release rate for the analysed sequences of CO_2 injection.

 DOI: 10.1051/matecconf/201711800006

The maximum reported values of the combustion pressure in the RCM chamber were less than for the reference point. The smallest decrease in of P_{max} when compared to the reference point was recorded at the injection pressures of CO_2 of 5 to 10 bars for sequence III. Possible causes of only a several-percent decrease in the value of the P_{max} were an early creation of swirling and combustible areas around the centre of the cylinder. This has allowed achieving combustion without significant deterioration of operating parameters. In the sequences I and II, in which gases are supplied at intervals, charge distribution is differentiated, which affects the process of combustion. For these sequences were recorded decreases in P_{max} pressure by 19% for sequence I and 36% for sequence II, at non-combustible gas injection pressure amounting to 10 bars in relation to the reference characteristics.

An increase in the share of non-combustible gases in the combustion chamber by increasing the injection pressure of CO_2 deteriorates the combustion process in terms of the value of heat release rate (*dQ*), which is shown in Fig. 7.

Comparing to the reference characteristics (100%) it was observed that the heat release rate was reduced for sequence I from 51 to 85%, for sequence II from 78 to 95% and for sequence of III from 33 to 81%. Conditions of changes in heat release rate indicate suppression of the combustion process by addition of non-combustible gases. This process is inevitable in the assumed prototype system of combustion control. It is important, however to recognize the possibility to control the combustion process, depending on the selected sequence of injection of CO_2.

The next stage of the research could therefore be analysed taking into account the change in dose of the injected fuel. Such parameters may allow to obtain the value of the combustion process at the level of the reference characteristics. Comprehensive coverage of more variable parameters, however, affects the size of the measurement matrix extending the time of tests and increasing the range of research material.

6 Optical analysis of the combustion process

Along with recording of the indicated pressure, also the optical recording of the combustion process took place, utilizing for this purpose the optical access in RCM from the bottom of the piston through quartz glass mounted in the piston crown. The frequency of recording was chosen for 5 kHz allowing registration of images with a resolution of 624 × 1024 pixels.

In order to specify the area the flame takes up during combustion in a given system, the recorded data was the subject of digital processing consisting of (show in Fig. 8): (1) defining the viewing window area mask, then (2) choice of analysis area and (3) counting the pixels whose radiation intensity was higher than the background noise. Then, by registering the triggered discharge on the spark plug, it was possible to determine a common time baseline for the indicator and optical research results (synchronization).

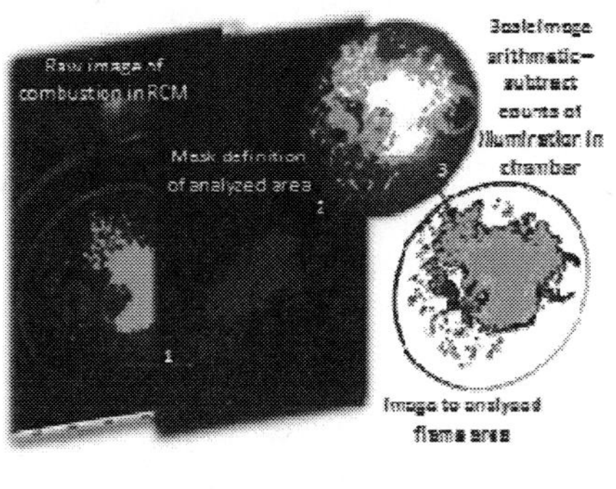

Fig. 8. The course of processing the optical test results.

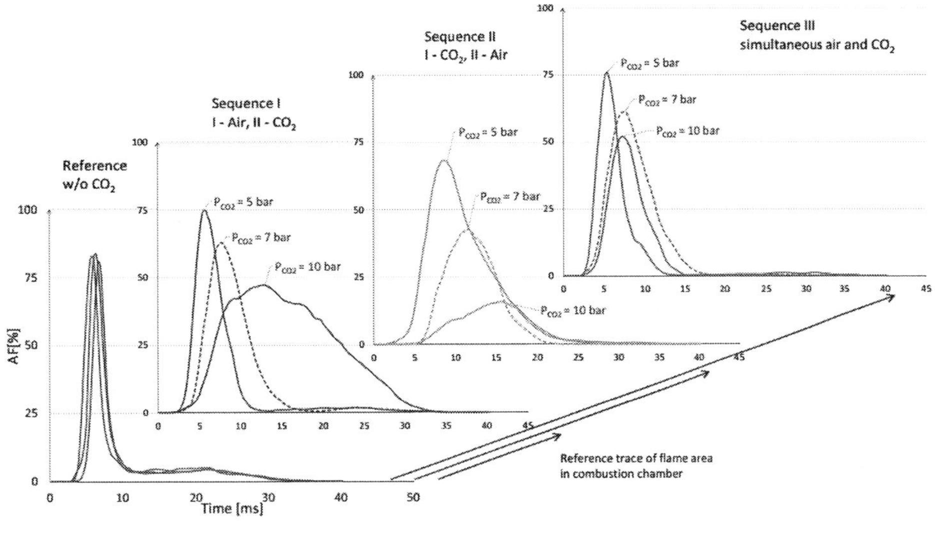

Fig. 9. Analysis of the flame area history in the combustion chamber.

Optical registration of two-dimensional (planar) flame exposure has allowed the determination of the area occupied by the flame in the chamber. For this purpose, a macro of DaVis software was used. Figure 9 shows the results of the calculations of the area of the flame. In consequence (in relation to Fig. 4) the colours and markings used show in succession the areas of reference flame for sequence I, II and III. For the reference measurements, a sharp increase of the flame in the chamber was recorded, on average involving over 80% of the optical window in the piston. The flame in this configuration is maintained for a longer period of time in the optical window, which point to a long combustion process.

The use of non-combustible gas injection into the chamber resulted in significant differences in the recorded image of the combustion process. The presented selected results of tests at 3 different injection pressures of CO_2 show direct influence of the increase of the non-combustible gases dose of on the area occupied by the flame. Combustion at the injection pressure of CO_2 equal to 5 bars results in a slight decrease to the level of 75% of the covered exposure by the flame, the elongation of the combustion process and the lack of previously observed after-burning phase (for sequences I and III).

With increasing pressure of CO_2 injection, the area of the flame decreases to a value of about 50% for a sequences I and III, and to 15% for the sequence II. A decrease of the flame area by 80% (sequence II, P_{CO2} = 10 bar) indicates high by volume concentration of the process of combustion by non-combustible gases. The frame-by-frame analysis indicates the limitation of this process, which is concentration of the flame in the middle of the cylinder. Reduction of the light intensity (luminance), also indicates a reduction in flame intensity.

For the sequences I and II was reported elongation of the combustion process from 50 to 300% relative to the reference point, without taking into account the afterburning phase (flame area below 5% in Fig. 7). In the reference combustion process, however, there was reported reference long-lasting process of afterburning of the flame in the vicinity of the cylinder wall. Such burning is very unfavourable due to the likely formation of soot particles. Non-flammable gas injection tangential to the axis of the cylinder allowed for the introduction of the assumed gas swirling in the chamber in such a way that they surround the resulting flame. Such a way of supplying gases allowed to restrict the spread of flame in the cylinder in a controlled manner, limiting its spread from the centre of the combustion chamber. Analysis of flame areas provided in Fig. 10 shows selected illustrations of flame with interval of 1 ms. In the reference characteristics can be observed persistence of the flame within the cylinder walls. Sequence II is characterized by the greatest suppression of the combustion process and its concentration in the centre of the cylinder. In contrast, sequences I and III are characterised by longer time of spreading of the flame.

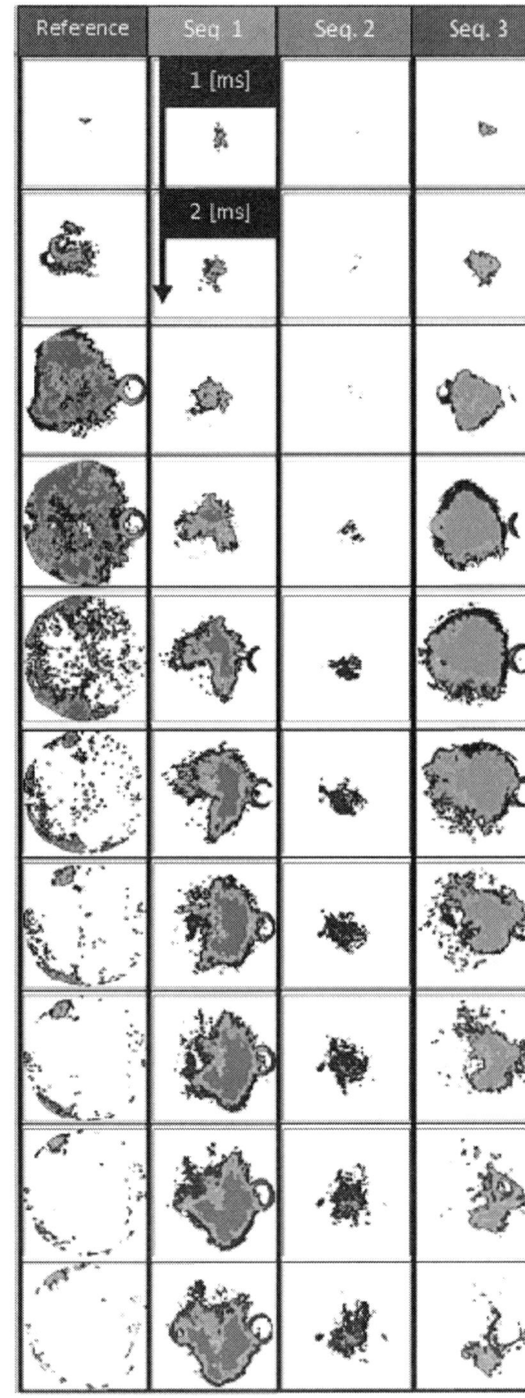

Fig. 10. Comparing the flame area for the presented sequences (P_{CO2} = 10 bar).

The highest flame intensity (for a reference sequence occurring after 6.4 ms) appears in sequence I – after 12.8 ms, in sequence II – after 15.6 ms, and in the third sequence – after 7.4 ms. This means that the maximum flame area instance delay is, respectively, 100, 140 and 15% for non-combustible gases and injection pressure of

10 bars. The diversity of these times results from different supply of the non-combustible gases and different organisation of the combustion process.

Based on the results of experimental studies a conclusion can be presented concerning the benefits of different ways of supplying combustible and non-combustible gases:

- sequence III (that is, the simultaneous supply of gases into the chamber) intensifies the combustion process in the centre of the chamber and at the same time allows its quick conclusion,
- sequence II delivering first non-combustible gases and later air, makes it possible to concentrate combustion in the centre of the combustion chamber, increasing the adiabatization of the process.

It seems that the sequence II is more desired because of the organisation of the combustion process (high concentration of flame around the centre of the combustion chamber), adiabatization of the process (limited heat exchange) and the creation of harmful components (combustion showing characteristics of the low temperature combustion with limited potential for creation of nitrogen oxides).

7 Summary

Thermodynamic analysis of the combustion process of different variants of delivering air and non-combustible gases into the combustion chamber indicates that the smallest deterioration of performance of RCM is observed for sequence III. Also the optical analysis of the combustion process indicates the smallest its elongation of the above-mentioned sequence.

The proposed system makes it possible to introduce for controlling the additional system variables such as: changing the portion of non-combustible gases by extending injection time, advancing of flue-gas injection or division of the injected portions of these gases, and simultaneous control of the process relating to fuel dose quantity and time of injection. The variation of these variables enables in big extent the optimization of the combustion course, especially for sequences I and II (see chapter 4). The above conclusions are justified by the highest concentration of the flame in the centre of the cylinder observed for sequence II.

The presented tests are basic research, showing the validity of thesis that the combustion process can be controlled using non-combustible gases in such a way as to limit the exchange of heat (during combustion) with the cylinder walls. Further analysis of the combustion process will apply to local flame temperature analysis using the two-colour method of optical research. This will make it possible to associate the global indicators of the process with local temperature values in the combustion chamber.

The study presented in this article was performed within the statutory research (contract No. 05/52/DSPB/0246).

References

1. P. Fuc, P. Kurczewski, A. Lewandowska, E. Nowak, J. Selech, A. Ziolkowski, International Journal of Life Cycle Assessment **21**, 1438-1451 (2016), DOI: 10.1007/s11367-016-1104-y
2. M. Sobieszczanski, K. Brzozowski, J. Nowakowski, M. Tekielak, Eksploatacja i Niezawodnosc – Maintenance and Reliability **3**, 53-61 (2008)
3. W. Cieslik, I. Pielecha, A. Szalek, Combustion Engines **161**, 14-27 (2015)
4. M. Crawford, 3 *Emerging Trends in Automotive Engineering* (2013)
5. M. Noga, B. Sendyka, International Journal of Automotive Technology **15**, 1083-1089 (2014), DOI: 10.1007/s12239-014-0112-9
6. W. Smuga, L.J. Kapusta, A. Teodorczyk, Journal of Power Technologies **97**, 1-6 (2017)
7. R. Pyszczek, L.J. Kapusta, A. Teodorczyk, Journal of Power Technologies **97**, 52-60 (2017)
8. M. Noga, Applied Sciences – Basel **7**, 295 (2017), DOI: 10.3390/app7030295
9. G.M. Szymanski, F. Tomaszewski, Mechanical Systems and Signal Processing **68-69**, 479-490 (2016), DOI: 10.1016/j.ymssp.2015.07.015
10. B. Flaig, U. Beyer, M. André, MTZ **71**, 22-27 (2010)
11. Gasoline combustion systems beyond 2020 (2012)
12. I. Hirose, M. Hitomi, MTZ **77**, 28-33 (2016)
13. A. Isendstadt, J. German, M. Dorobantu, et al., International Council on Clean Transportation (2016)
14. H. Heitland, G. Rinne, K. Wislocki, MTZ **2**, 94-101 (1993)
15. K. Wislocki, H. Heitland, Arbeitsverfahren für einen Verbrennungsmotor, Deutsche Patent- und Markenamt, 43 05 468 (1999)
16. G. Hauser, H. Heitland, K. Wislocki, SAE Technical Paper 951962 (1995)
17. W. Attard, E. Toulson, A. Huisjen, et al., SAE Technical Paper 2012-01-0823 (2012)
18. The development of a pre-chamber combustion system for lean combustion of liquid and gaseous fuel (2014)
19. W. Cieslik, I. Pielecha, L. Kapusta, Combustion Engines **162**, 257-263 (2015)
20. W. Cieslik, I. Pielecha, Combustion Engines **168**, 27-31 (2017)
21. H. Wei, T. Zhu, G. Shu, et al., Applied Energy **99**, 534-544 (2012), DOI: 10.1016/j.apenergy.2012.05.011

Comparative studies exhaust emissions of the Euro VI buses with diesel engine and spark-ignition engine CNG fuelled in real traffic conditions

Maciej Gis[1,*]

[1]Motor Transport Institute, ul. Jagiellonska 80, 03-301 Warsaw, Poland

Abstract. You should leave 8 mm of space above the abstract and 10 mm after the abstract. The heading Abstract should be typed in bold 9-point Times. The body of the abstract should be typed in normal 9-point Arial in a single paragraph, immediately following the heading. The text should be set to 1 line spacing. The abstract should be centred across the page, indented 17 mm from the left and right page margins and justified. It should not normally exceed 200 words.

1 Introduction

Air pollution coupled with more and more restrictive requirements concerning the emission of hazardous exhaust gases cause that bus fleets in towns and also in non-urban traffic are changing considerably. In large urban areas the content of hazardous substances usually exceeds standard levels. Kraków is the best example. In October 2016 year once again an alert was announced that admissible PM10 levels exceeded. One of the measurement station recorded 61 $\mu g/m^3$ at 10 am and 127 $\mu g/m^3$ at 1 am. In Nowa Huta a level of 50 $\mu g/m^3$ was observed at 9 am and 63 $\mu g/m^3$ at 1 am. The first-degree alerts were also announced in Warsaw. According to public alerts admissible levels for air quality as regards benzo(a)pyrene were most likely exceeded in the entire Mazowieckie region, and in some towns and communities also for PM_{10} and $PM_{2.5}$ and nitrogen dioxide. To counteract the above motor vehicle companies are launching new technical solutions. One of the possibilities as regards vehicles is to replace traditional city fleets with vehicles powered by alternative systems [1]. Another method of reducing emission of hazardous exhaust gases to atmosphere comprises actions aimed at limiting traffic of passenger cars by opting for public transport, e.g. by intensifying actions promoting and improving the public transportation infrastructure [2].

2 European requirements regarding emission of hazardous substances by buses

The main element that has impact on the construction of buses and allows for their sale are requirements concerning the reduction of exhaust emissions. Because of environmental requirements actions are undertaken to reduce to maximum degree the emission of toxic substances of exhaust gases, in particular the emission of nitrogen oxides – NO_X, hydrocarbons – HC, carbon oxides – CO, PM and PN [3, 4].

Those requirements are laid down in international and European standards on the level of exhaust emissions from Euro V and EEV, i.e. vehicles particularly environmentally friendly – regulations with effect as from 2009 and implemented by way of Directive 2005/55/EC of 28 September 2005 [5]. Starting from 2014 the new Euro VI standard is in force, as set forth in Regulation no. 595/2009 [6] and in the EEV standard. In real conditions of traffic the values of admissible emissions may be exceeded for instance for CO. HC and NO_X 1.5 times [7] (Regulation no. 582/2011 EC [8]).

For conventional SI engines to be able to fulfil the standards of exhaust emissions, SCR (Selective Catalytic Reduction) must be applied. SCR acts by converting hazardous substances in exhaust gases from SI engines in the discharge, in particular nitrogen oxides and their transformation into vapour and neutral nitrogen. The transformation is effected in the presence of non-toxic aqueous urea solution AdBlue® as catalyst of exhaust gases. The electronic system of engine management doses AdBlue® in a separate tank and thereafter under high pressure is injected directly to the reactor, where AdBlue® reacts with ammonia and decomposes into nitrogen and water vapour. Owing to this technology fuel consumption is reduced by approx. 5% [9] and the emission of nitrogen oxides (NO_X) is reduced and also reduced – even more – is the emission of particulate matter. It is estimated that approx. 1,7 dm^3 of AdBlue® is required per 100 km (in case of vehicles with Euro V emission levels) [10, 11].

* Corresponding author: maciej.gis@gmail.com

DOI: 10.1051/matecconf/201711800007

3 Buses fuelled with CNG

Methane is the main constituent of natural gas, with volume ranging from 90 to 98%. The remaining constituents include in particular ethane, propane, butane and nitrogen. Natural gas used typically in a compressed form characterises with low energy density. Under pressure of 17.7 MPa and in a temperature of 21 °C it is equal to 7 kJ/cm^3 (which is 30 kJ/cm^3 for petrol). Due to a much lower density of gas fuels and in consequence lower energy per volume unit the vehicle must be specially adapted to gas fuelling. Effective use of energy contained in gas fuel necessitates changes in the power unit [12, 13].

4 The purpose and scope of the study

The purpose of the studies was to estimate the emission properties and fuel consumption of public transport buses fuelled by CNG and diesel fuel. The test route was selected at random and in two towns A and B along the road between them. CNG-fuelled bus was an articulated bus equipped with an internal combustion engine fulfilling the Euro VI emission standard. The bus production year was 2015 and its engine displacement was equal to 12 816 cm^3, with power of 228 kW. The diesel-fuelled bus was also an articulated bus with engine displacement of 10,518 cm^3 and maximum power of 235 kW, and production year 2015.

The studies were conducted during one day. Fig. 1 presents the tested buses with the installed testing equipment.

Fig. 1. Tested buses with the testing equipment (CNG-fuelled bus and diesel oil bus).

Testing equipment – PEMS used in the studies is presented in Fig. 2. Two mobile analysers Semtech DS were used to measure the content of hazardous compounds in exhaust gases [14]. The equipment was used to measure compounds of harmful exhaust gases, i.e. carbon dioxide, carbon oxide, hydrocarbons and nitrogen oxides and to assess fuel consumption with the carbon balance method. Information in publications regarding the use of mobile exhaust gas analysers coupled with data recorded by on-board diagnostic systems [6, 15], prove that the evaluation of contamination emission in real conditions of traffic with the use of such equipment was reasonable.

Fig. 2. PEMS testing equipment.

5 Research methodology

The exhaust gases emission and fuel consumption had been measured in a range of driving conditions. The emission of hazardous exhaust substances and fuel consumption were compared owing to measurements made with two devices. Both vehicles moved one after another, which allowed for eliminating errors due to driving style or changeable weather conditions. The test route combined elements of urban and country driving (Fig. 3 and 4). Parameters of the test route are represented in Table 1.

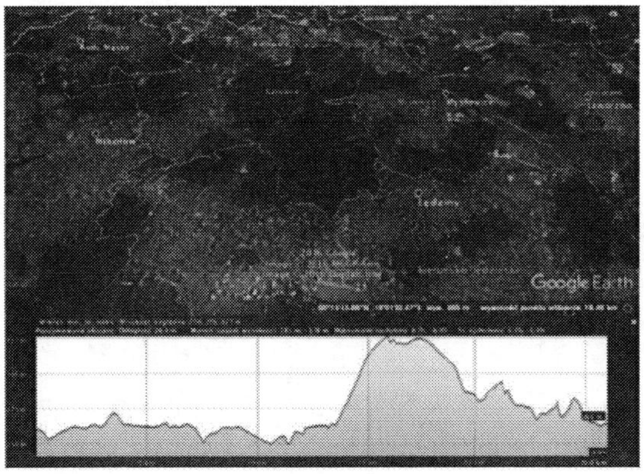

Fig. 3. Characteristics of the test rote.

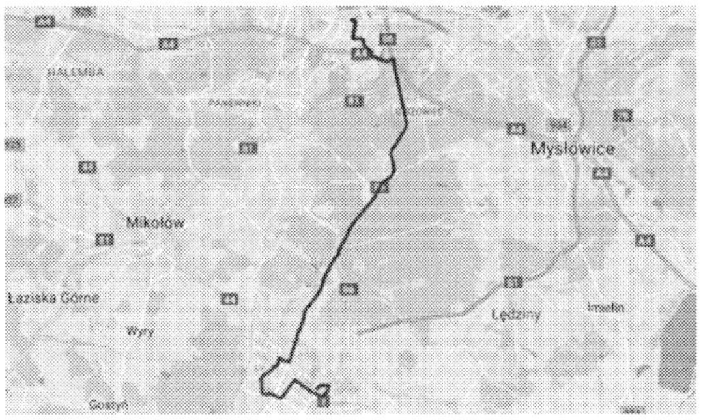

Fig. 4. Test route.

Table 1. Parameters of the test route.

Length	26 km
Change of altitude	min: 183 m, max: 178 m
Max. slope	4.7%
Min. slope	–6.9%
Average positive slope	1.3%
Average negative slope	–1.3%

6 Results and analyses of studies

The results of studies could be evaluated owing to a positive result of the comparison of test routes and their large time consistency. This is because the portion of time of vehicle's driving phases was determined in the test (with values rounded to full units). The compared values for acceleration, fixed velocity, braking and stopover of vehicles are presented below.

Table 2. Values of respective driving phases of tested buses.

	Value	Bus powered by CNG fuel	Bus powered by diesel fuel
Acceleration	%	38	40
Fixed velocity	%	19	9
Braking	%	28	35
Stopover	%	15	17

With the test results, it was possible to determine the content and intensity of flow, depending on the route. Fig. 5 presents the results for CO_2 with reference to the road, for both tested buses.

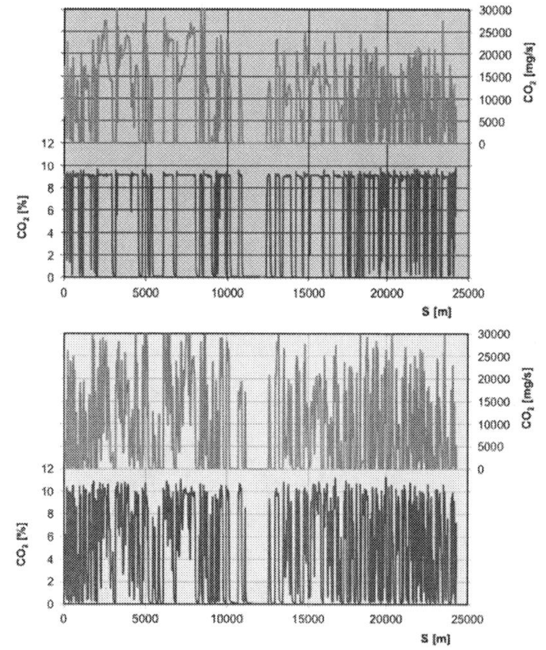

Fig. 5. Concentration and intensity of CO_2 emission with reference to the completed road: ▒ – I bus (CNG-fuelled), ▒ – II bus (diesel fuelled).

Fig. 6 presents temporary and average fuel consumption, also with reference to the route of both vehicles.

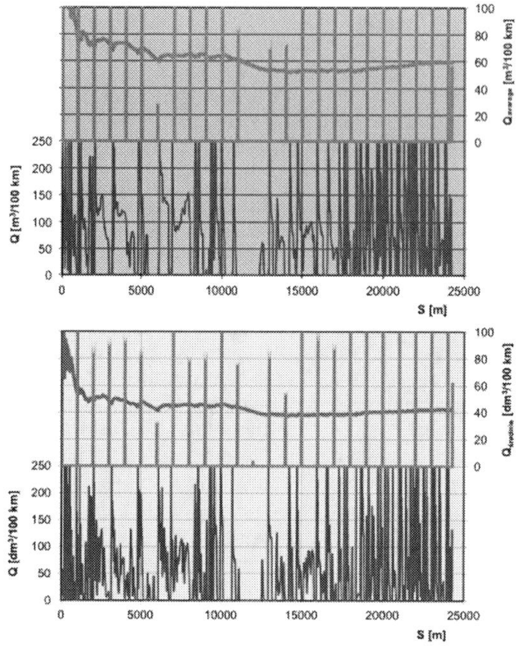

Fig. 6. Characteristics of temporary and average fuel consumption with reference to the completed road: ▒ – I bus (CNG-fuelled), ▒ – II bus (diesel fuelled).

Owing to the tests it was possible to determine time courses for the intensity of emission of hazardous substances of exhaust gases for both buses. Hense, correlations were developed characterising the impact of dynamic properties of public transport buses on the emission of harmful compounds. Dynamic properties of vehicles were included indirectly, using the entire velocity range and the range of mesured acceleration in urban traffic for a matrix of intensity of emissions of respective pollutants. Those figures were averaged in respective velocity and acceleration ranges.

Therefore, characteristics of operation in respective ranges and the matrix of emission of harmful compunds had been determined. According to data presented in Fig. 7 there is a similarity in the portion of operating time for the tested buses. Also, a greater portion of operation is observed in the area of minimum driving speed (stopover) and high driving speeds (10-18 m/s) and zero bus acceleration.

Two-dimensional histogram of intensity of CO_2 emission [mg/s] in the relation velocity – acceleration is presented in Fig. 8. It represented indirectly the fuel consumption during the tests. The bus fuelled with CNG was comparable. Histograms of intensity of CO_2 emission during the tests for averaged velocities are represented in Fig. 9.

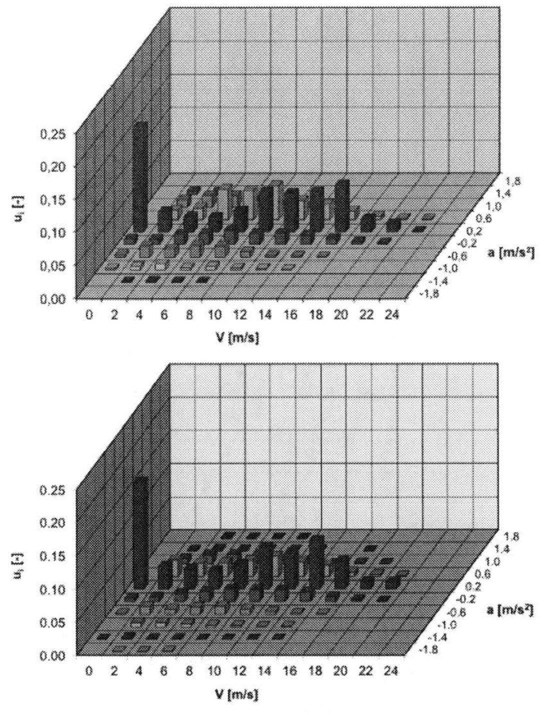

Fig. 7. Two-dimensional histograms of vehicle's operating time in relation velocity – acceleration: ▒ – I bus (CNG-fuelled), ▒ – II bus (diesel fuelled).

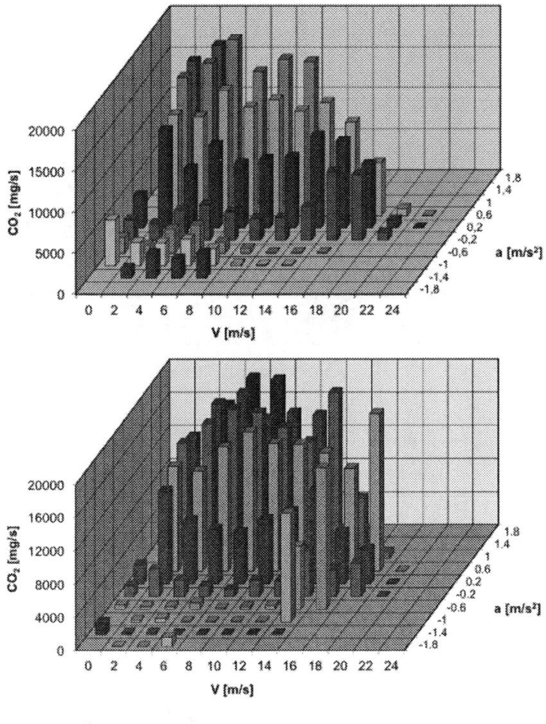

Fig. 8. Two-dimensional histograms of CO_2 emission intensity during the tests in relation velocity – acceleration: ▒ – I bus (CNG-fuelled), ▒ – II bus (diesel fuelled).

In case of the CNG-fuelled bus, attention must be given to NO_X emission, which is lesser than in the diesel-fuelled bus. Fig. 10 presents a two-dimensional histogram of NO_x emission intensity during the tests in the relation velocity – acceleration. The figures shown in Fig. 10 prove the above.

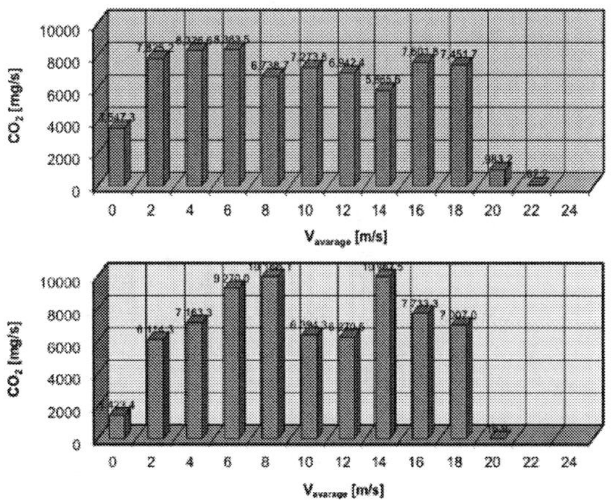

Fig. 9. Histograms of CO_2 emission intensity during the tests for average velocity: ▒ – I bus (CNG-fuelled), ▒ – II bus (diesel fuelled).

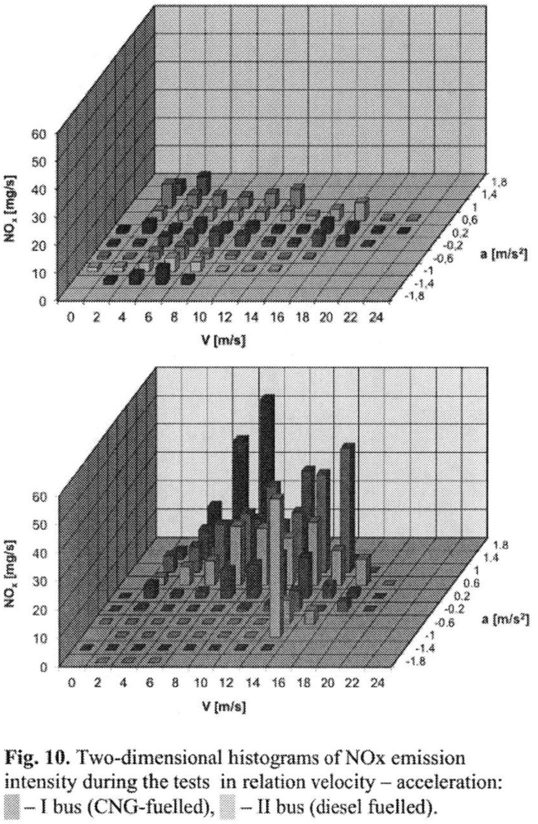

Fig. 10. Two-dimensional histograms of NOx emission intensity during the tests in relation velocity – acceleration: ▒ – I bus (CNG-fuelled), ▒ – II bus (diesel fuelled).

Also, in case of HC emission intensity, the CNG-fuelled bus manifests lower emission than the diesel-fuelled bus. This is reflected in Fig. 11 and corroborated as in Fig. 12.

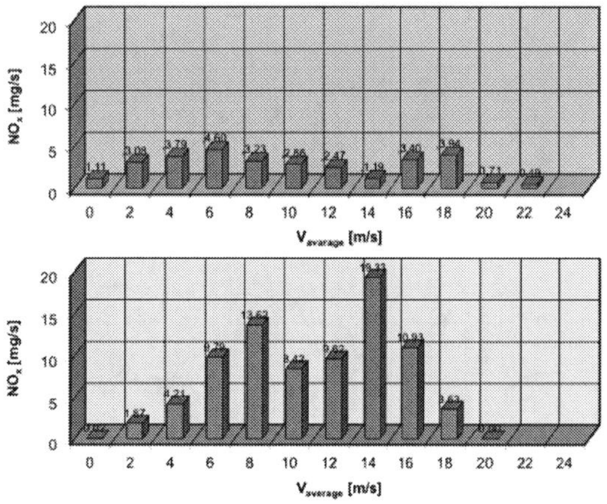

Fig. 11. Histograms of NOx emission intensity during the tests for average velocity: ▒ – I bus (CNG-fuelled), ▒ – II bus (diesel fuelled).

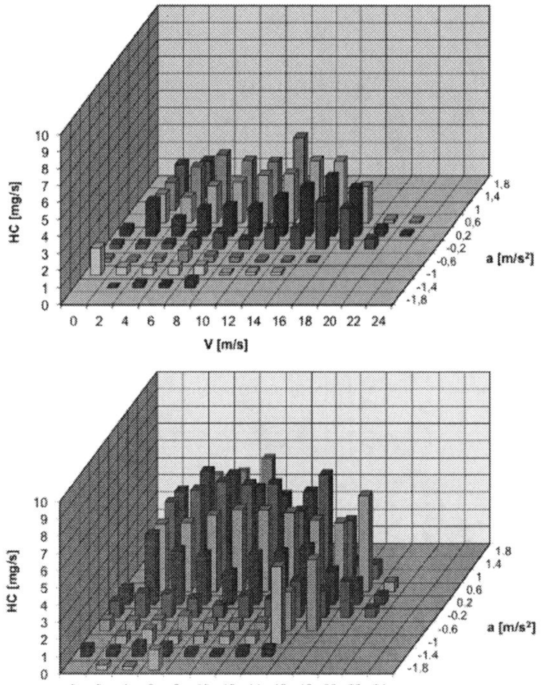

Fig. 12. Two-dimensional histograms of hydrocarbons emission intensity during the tests in relation velocity – acceleration: ▒ – I bus (CNG-fuelled), ▒ – II bus (diesel fuelled).

On Fig. 13 and Fig. 14 are presents the results of fuel consumption and road emission of exhaust gases in the course of tests and a comparison of fuel consumption and exhaust emissions during tests.

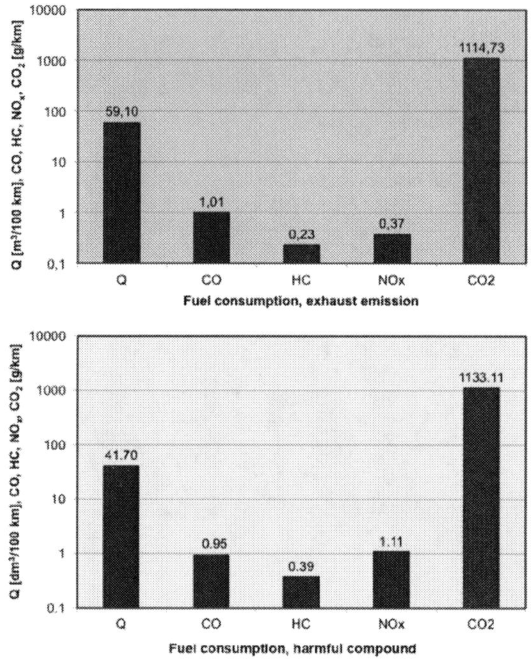

Fig. 13. Results of fuel consumption and road emission of exhaust gases during tests: ▒ – I bus (CNG-fuelled), ▒ – II bus (diesel fuelled).

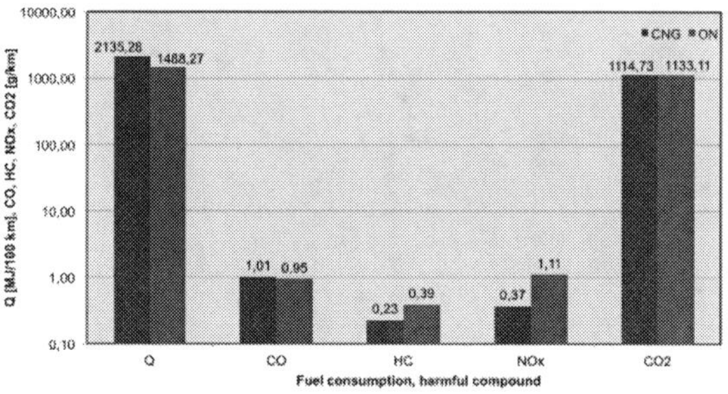

Fig. 14. Comparison of fuel consumption and emission of exhaust gases during tests: █ – I bus (CNG-fuelled), █ – II bus (diesel fuelled).

7 Summary

The tests in real road traffic conditions resulted in the following conclusions:

- consumption (natural gas) 59,1 m^3/100 km (CNG-fuelled bus) and 41.7 dm^3/100 km (diesel-fuelled bus),
- CO emission: 1.01 g/km (CNG-fuelled bus) and 0.95 g/km (diesel-fuelled bus),
- HC emission: 0.23 g/km (CNG-fuelled bus) and 0.39 g/km (diesel-fuelled bus),

- NO_X emission: 0.37 g/km (CNG-fuelled bus) and 1.11 g/km (diesel-fuelled bus),
- CO_2 emission: 1114,73 g/km (CNG-fuelled bus) and 1133,11 g/km (diesel-fuelled bus).

Attention should be given to the fact that despite positive results of using compressed natural gas for fuelling public transport buses, in terms of road emission of – in particular – nitrogen oxides, as proved by tests, also in other aspects the use of methane fuel is reasonable. This applies particularly to compressed bio-methane used instead of CNG, with CO_2 emission in the life cycle is almost null. Because of those aspects the use of vehicles fuelled with methane (CNG, LNG and bio-methane) in bus fleets is reasonable.

References

1. M. Bajerlein, L. Rymaniak, P. Swiatek, A. Ziolkowski, P. Daszkiewicz, M. Dobrzynski, Experimental and Applied Mechanics **518**, 108-113 (2014), DOI: 10.4028/www.scientific.net/AMM.518.108
2. J. Merkisz, M. Bajerlein, P. Daszkiewicz, *Lecture Notes in Information Technology* **13**, 106-113 (2012)
3. P. Lijewski, J. Merkisz, P. Fuc, A. Ziolkowski, L. Rymaniak, W. Kusiak, European Journal of Forest Research **136**, 153-160 (2017), DOI: 10.1007/s10342-016-1015-2
4. J. Merkisz, S. Tarkowski, Eksploatacja i Niezawodnosc – Maintenance and Reliability **2**, 50-58 (2011)
5. Directive 2005/55/EC of the European Parliament and of the Council of 28 September 2005 on the approximation of the laws of the Member States relating to the measures to be taken against the emission of gaseous pollutants from positive – ignition engines fuelled with natural gas or liquefied petroleum gas for use in vehicles. Official Journal of the European Union L 275/1
6. Regulation (EC) No 595/2009 of the European Parliament and of the Council of 18 June 2009 on type-approval of motor vehicles and engines with respect to emissions from heavy duty vehicles (Euro VI) and on access to vehicle repair and maintenance information and amending Regulation (EC) No 715/2007 and Directive 2007/46/EC and repealing Directive 80/1269/EGC, 2005/46/EC and 2005/78/EC. Official Journal of the European Union L188/1
7. K. Brzozowski, J. Nowakowski, Eksploatacja i Niezawodnosc – Maintenance and Reliability **4**, 56-62 (2011)
8. Commission Regulation (EU) No 582/2011 of 25 may 2011 implementing and amending Regulation (EC) No 595/2009 of the European Parliament and of the European Parliament and of the Council. Official Journal of the European Union L167/1
9. P. Bonnel, M. Weiss, A. Provenza, *8th Annual SUN Conference* (Ann Arbor, 2011)
10. www.bluesolution.pl (2012)
11. J. Merkisz, J. Pielecha, S. Radzimirski, *Springer Tracts on Transportation and Traffic* **4**, 67-90 (2014), DOI: 10.1007/978-3-319-02705-0_4
12. E. Król, M. Flekiewicz, Nafta – Gaz **7-8** (1997)
13. J. Merkisz, J. Pielecha, *Springer Tracts on Transportation and Traffic* **8**, 69-92 (2015), DOI: 10.1007/978-3-319-15928-7_7
14. Sensors SEMTECH-DS, On Board Vehicle Emissions Analyzer (2012)
15. The role of natural gas and methane in the fuel mix of the future in Germany (Berlin, 2010)

Flow bench testing of prototype intake manifolds for ultralight aircraft engine

Piotr Swiatek[1], *Pawel* Fuc[1,*], *Artur* Cebula[2], and *Sebastian* Kowalczyk[3]

[1]Poznan University of Technology, Faculty of Machines and Transport, ul. Piotrowo 3,60-965 Poznan, Poland
[2]Cracow University of Technology, Institute of Thermal Power Engineering, al Jana Pawla II 37, 31-864 Krakow, Poland
[3]Ensys SC, ul. Sliwiny 37, 83-110 Tczew, Poland

Abstract. The article describes the research on the intake manifold for ultralight aircraft engine Vaxell 100i. It presents the actual and new redesigned manifold construction and points out the design requirements. The results of previously made numerical simulation of air flow inside the manifold are discussed. Computer analysis confirmed the appropriateness of internal guide vanes usage to improve the uniform air distribution between cylinders. For verification, a flow bench test stand was made for multi-cylinder intake manifold testing. A prototype manifold was built with the possibility of guide vane adjustment. The best variant had almost 5 times better uniformity of air distribution comparing to variant without the guide vane. Flow bench results confirmed the conclusions from numerical simulations.

1 Introduction

The intake system aim is to transport a proper amount of air or air-fuel mixture to the combustion chamber to ensure engine achieves the assumed parameters – power and torque curve, fuel consumption and emission. An intake manifold is located between throttle or carburettor (which is often still used in engines for aircraft propulsion) and head intake ports, and it consists of the main inlet, "plenum" (central air volume) and "runners" (individual tubes leading to cylinder head ports). The construction of intake manifold has to meet specific requirements, of which the most important are [1, 2]:

- low air flow resistance,
- even distribution of air flow between each cylinder intake ports [3],
- runner geometry (length, cross-section shape and dimensions) ensuring the usage of pulsating air wave effect that occur in the intake system,
- tailpipe emission [4, 5].

In the case of manifolds for ultralight aircraft engines, additional assumptions must be met:

- low weight while retaining sufficient strength and stiffness,
- adequate air flow around the head and cylinders for cooling (manifold construction cannot limit it),
- the low overall height of the intake system (maintaining compact size relative to the external dimensions of the engine and its accessories).

Complying with the above requirements makes the ultralight aircraft intake manifold design process complicated and the final design is a result of technical compromises. For these reasons, we performed a numerical flow simulations, followed by a flow bench testing of prototype manifold.

2 Research methodology

2.1 Research object description

The object of the research was an intake manifold used in Vaxell 100i engine, which drives many transportation vehicles including ultralight and light aircrafts (Tab. 1, Fig. 1).

Table 1. Vaxell 100i engine specification.

Number of cylinders	4
Displacement [cm^3]	2276
Cylinder arrangement	boxer
Ignition system	dual
Fuel supply	multi-point injection
Maximum power	90 hp / 3100 rpm
Maximum torque	222 Nm / 2200 rpm

The disadvantages of actual intake system design are large numbers of parts and large dimensions. That plays a significant role in the application for ultralight engines, hence a decision of intake system modification. A 3D model (Fig. 2) of changed manifold was designed with one big throttle on the main inlet, smaller plenum and straight runners – in the further part of the article marked as **manifold 1**. Additionally, variants with guide vanes were designed. This model has been analyzed using Ansys-AIM numerical simulation software [6].

Flow bench testing was performed on newly designed inlet manifold (Fig. 3) with a prototype air tank (Fig. 4) – marked as **manifold 2** in the rest of the paper. The special air tank consists of an aluminum cast (1) with a planned upper surface and welded flange. From

* Corresponding author: pawel.fuc@put.poznan.pl

the top, the plenum was closed with polycarbonate cover (2) mounted using M5 screws (4). In the cover, a 3D printed guide vane (3) was attached using M5 nuts (5). The position of the cover can be adjusted at a 40° angle by 12 mm. Additionally, the guide vane can be attached with 3 different angles: 0 ±12.5°. The polycarbonate cover is transparent, which makes the possibility of carrying out optical studies of the colored air stream inside the tank. Such research is planned in the future works.

Fig. 1. Vaxell 100i engine with actual intake system.

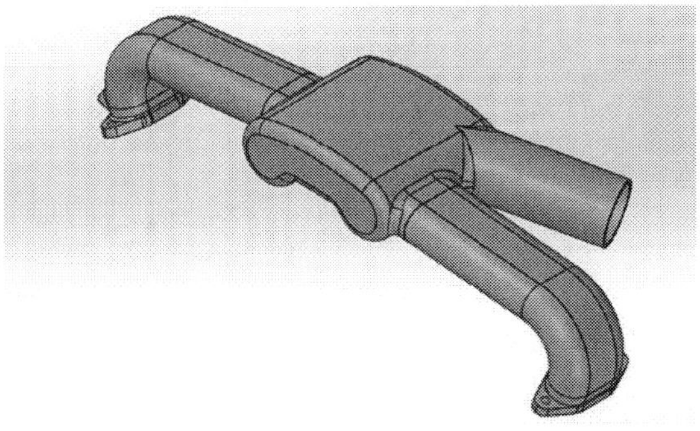

Fig. 2. 3D manifold model for numerical simulations (manifold 1).

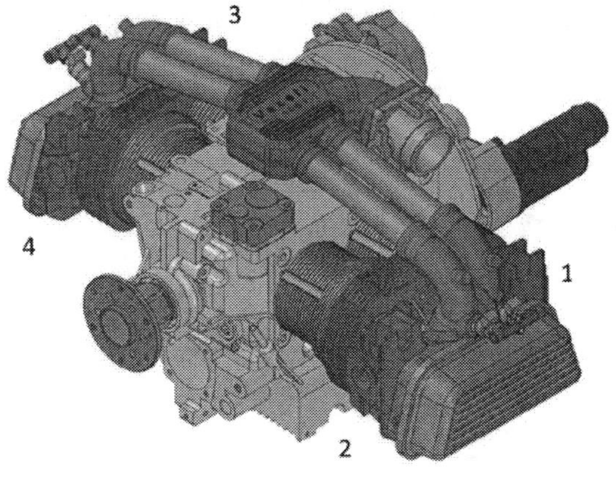

Fig. 2. Vaxell 100i engine with new manifold (manifold 2) with marked numbers of the cylinders.

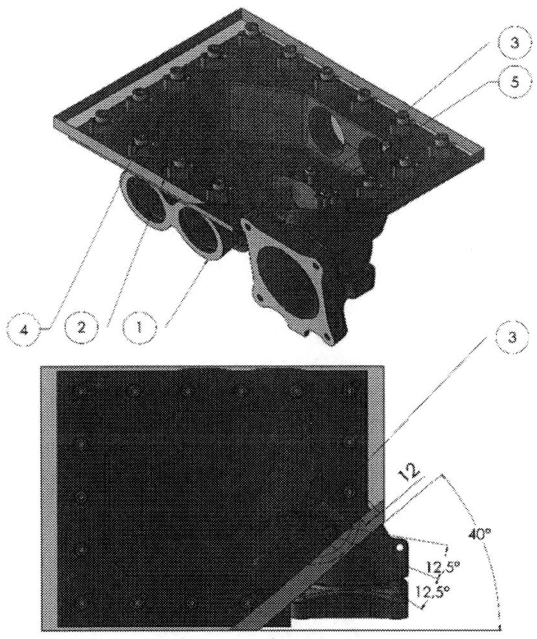

Fig. 4. Prototype air tank for flow bench testing. 1 – bottom part (aluminium cast), 2 – transparent upper cover (polycarbonate), 3 – guide vane (3d printed plastic), 4 – screws mounting the cover, 5 – nuts mounting the vane.

2.2 Flow bench tests methodology

The air flow tests in manifold 2 were performed on a specially built test stand, which is presented in Fig. 5. A fan with regulated speed is mounted in the chamber 1 and provides a partial vacuum. Chamber 2 is used for pressure compensation and measurement, using a piezo-resistive sensor, vacuum gauge and U-tube manometer (6). Four channels with flowmeters (2) are located between chamber 2 and tested manifold (5).

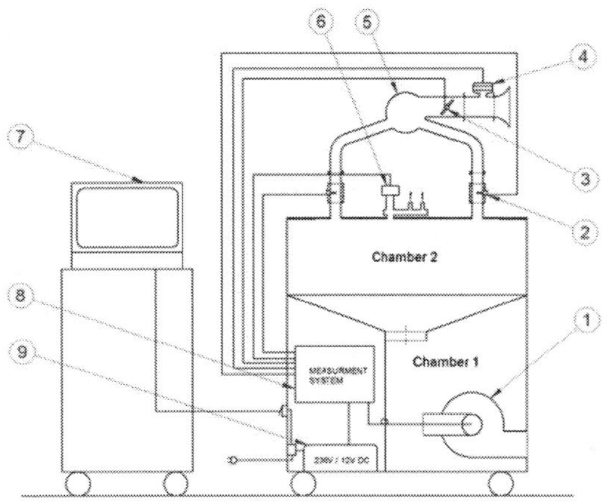

Fig. 5. A schematic of multi-cylinder intake manifold flow bench test stand. 1 – fan, 2 – flowmeters for each cylinder runner, 3 – throttle, 4 – sensors before the throttle, 5 – tested manifold, 6 – pressure sensors, 7 – computer, 8 – measurement system, 9 – power supply.

DOI: 10.1051/matecconf/201711800008

At the manifold inlet, a throttle with position sensor and flowmeter is mounted. Before the throttle, there is located an atmospheric pressure and temperature sensor (4). Flowmeter and sensor data is collected by the measurement system (8) with additional power supply (9). Computer (7) records the data from flow bench during tests.

The test bench fan has a 2000 W motor it is controlled by the inverter. The maximum air flow volume is 1200 m^3/min, which corresponds to flowing air velocity equal 27 m/s. With these setting, the fan creates a 0,1 bar negative pressure in chamber 1.

Five identical HFM5 type flowmeters (Fig. 6) were mounted on the test stand. These are a hot film mass airflow sensors which enable to detect the direction of the airflow. Each one has a measurement venturi with a plastic and wire mesh that ensure a uniform distribution of the air stream. Using a standard anemometer, the flowmeter characteristic was determined – dependency of voltage and mass airflow, presented on Fig. 7.

Fig. 6. HFM5 Flowmeter.

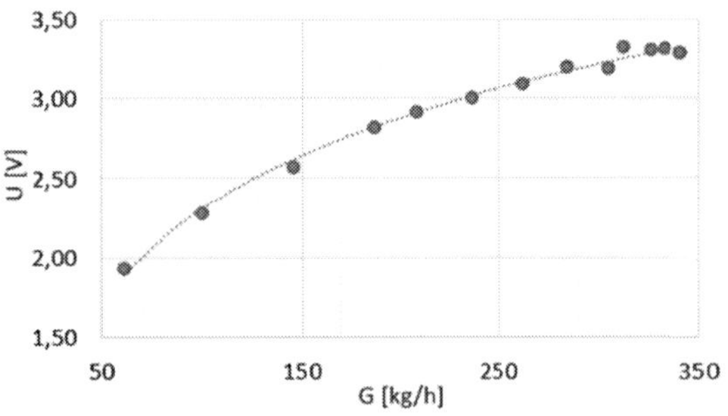

Fig. 7. Flowmeter characteristic.

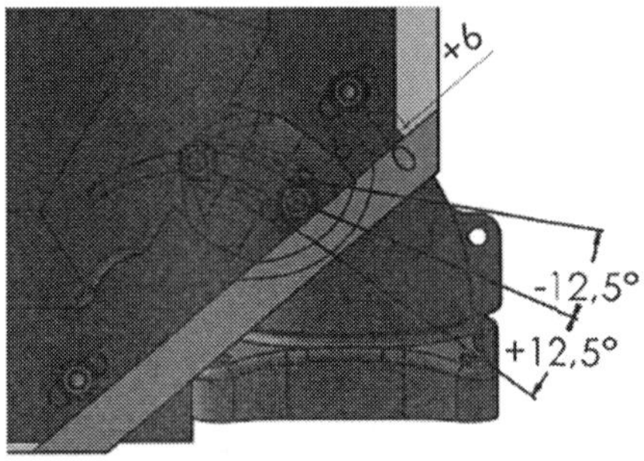

Fig. 8. Range of guide vane position and angle adjustment in prototype manifold.

During prototype manifold testing flow bench fan was set on maximum power and throttle was 100% open. Tests were performed on manifold 2 with prototype air tank (Fig. 4) which enables to change guide vane position and angle (Fig. 8).

For every test, the mass air flow rate for each cylinder was measured and a factor of uneven distribution of airflow was calculated.

3 Results

3.1 Numerical simulations

Details of CFD numerical simulations of manifold 1 are described in the paper [4] and were performed using the commercial Ansys-AIM code. The geometry modifications were carried out with the use of SpaceClaim tool. The boundary conditions are assumed as follows. The air mass flow rate of 60 g/s is set at the inlet. The flowing fluid is air, which is considered to be an ideal gas. The inlet temperature is set equal to 15 °C. On the outlets the opening boundary condition is set where gauge pressure is equal to – 0.1 bar. Since the focus of the work is to optimise the flow path, the thermal effects had not been included in this study. Manifold walls are defined as adiabatic walls which allow setting heat flux across the walls equal to zero ($\dot{q}$ = = 0 W/m^2).

Transport equations of the k–e turbulence model were used in the work, because it is considered as the most suitable to predict conditions inside manifold where the flow is swirl dominated.

Unstructured meshes were generated for each geometry. To reconstruct boundary layer and make it possible to calculate flow gradients near wall regions the inflation layers elements were generated for all manifold and plenum walls. The final mesh size was obtained after the mesh dependency calculations and it has 630 k elements.

Flow analyses were carried out for the baseline case and for proposed solutions representing various guide vanes aimed to provide an even flow split among runners. Fig. 10 shows computational cases that were created by modifying the curvature radius of guide vanes.

The results for manifold without guide vanes showed the uneven distribution of the airflow between cylinders (Fig. 9a). Based on the work [5] the maldistribution coefficient D_g was determined according to the formula:

$$D_g = \sqrt{\frac{1}{n-1}\sum_{i=1}^{n}\left(\frac{\dot{m}_i}{\dot{m}_{ave}} - 1\right)^2} \qquad (1)$$

where: n – manifold outlet number; $\dot{m}_{ave}$ – average mass air flow defined by the formula:

$$\dot{m}_{ave} = \frac{1}{n}\sum_n \dot{m}_i \qquad (2)$$

For manifold 1 without guide vane, the D_g factor was 0.19 (where 0 – best, 1 – worst) and the outlet 3 had the biggest part of the airflow – 30.1 % (Tab. 2). This uneven distribution of airflow causes a smaller volumetric efficiency in cylinder 1 combustion chamber, rough engine work and increase of cylinder temperature. This has a significant impact on the safety of the engine operation. In order to improve the uniformity of airflow distribution between cylinders, use of internal guide vane near the manifold inlet was proposed. Its purpose is to split the airflow and reduction of maldistribution coefficient. Three variants of guide vanes were proposed (Fig. 10) from which variant 3 distributed most evenly air to the outlets (Fig. 9b). For this variant, D_g factor was 0.10 (Tab. 2).

Table 2. The results of computer numerical airflow simulations for manifold 1.

	Variant without guide vane		Variant 3	
	Mass airflow kg/s	Flow split. %	Mass airflow kg/s	Flow split. %
Outlet 1	0.0122	20.3	0.0138	23.0
Outlet 2	0.0133	22.2	0.0146	24.4
Outlet 3	0.0180	30.1	0.0143	23.8
Outlet 4	0.0165	27.5	0.0174	28.9
D_g coeficient	0.19		0.10	

The simulation results confirmed the principle of using a guide vane in the inlet of the intake manifold in order to improve the uniform distribution of the airflow. Therefore a decision to perform a flow bench testing of prototype manifold was made.

a)
b)

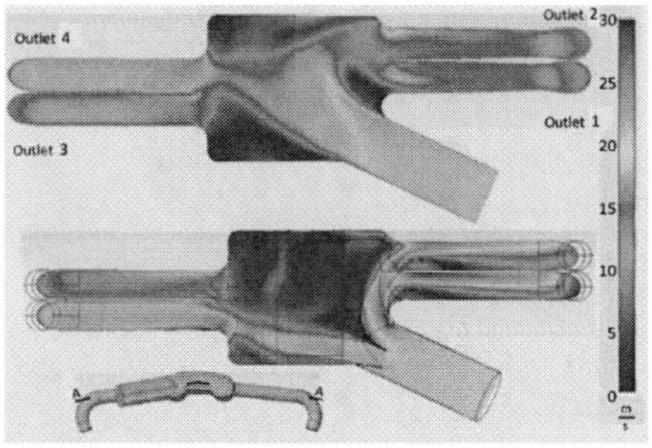

Fig. 9. Comparison of airflow velocity for manifold 1; a) without guide vane, b) with guide vane – variant 3.

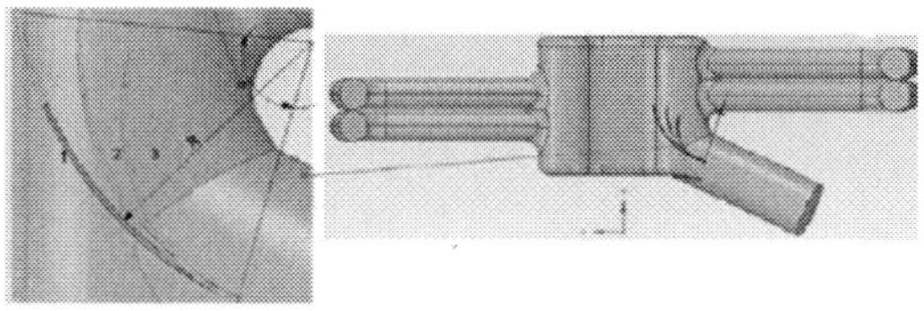

Fig. 10. Three variants of guide vanes in manifold 1.

3.2 Flow bench testing

On the flow bench, air flow measurements were made for the manifold 2 without and with guide vane in 9 different configurations of position and angle (Fig. 8). Without vane, the maldistribution factor D_g = 0.44 was obtained. Results for other configuration are presented in Tab. 3 and the D_g vary from 0.09 to 0.60.

For the configurations: without vane and with guide vane with the best and worst maldistribution factor, additional flow charts of each cylinder are presented in Fig. 11, Fig. 12 and Fig. 13. The numbering of cylinders is presented in accordance with Fig. 3.

Table 3. Results of prototype manifold 2 flow bench testing – maldistribution coefficient.

Angle / Position of guide vane	–6 mm	0	+6 mm
–12.5°	0.50	0.60	0.51
0°	0.20	0.24	0.25
12.5°	0.09	0.17	0.12

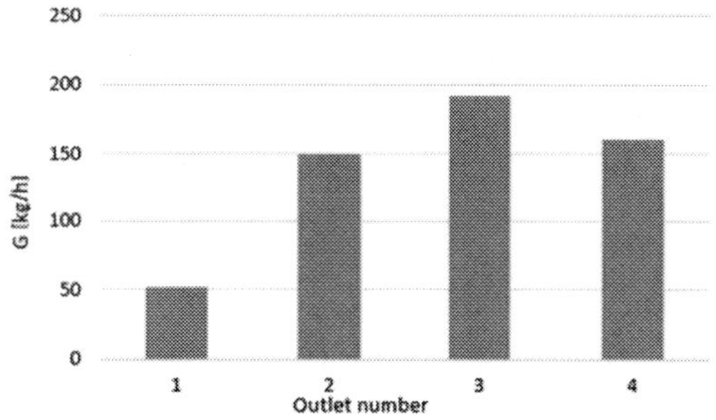

Fig. 11. Mass airflow chart for each cylinder for manifold 2 without guide vane, D_g = 0.44.

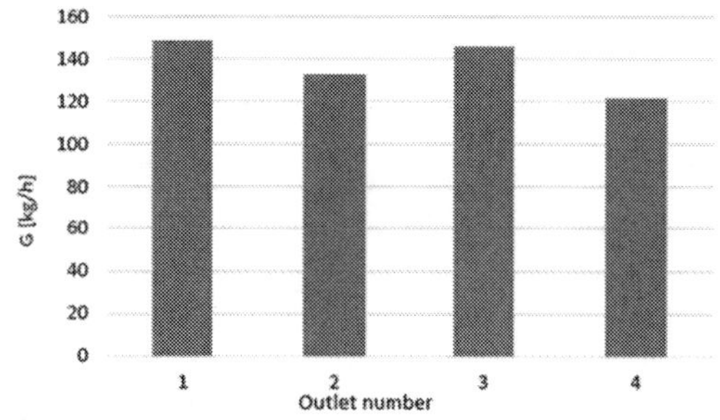

Fig. 12. Mass airflow chart for each cylinder for manifold 2 with guide vane – configuration –6 mm, +12.5°, D_g = 0.09.

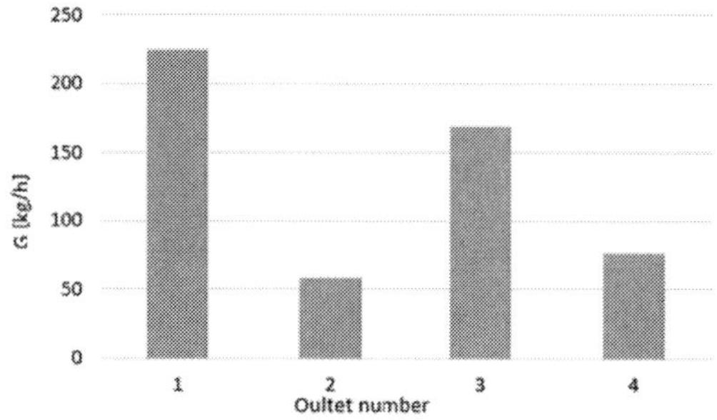

Fig. 13. Mass airflow chart for each cylinder for manifold 2 with guide vane – configuration 0 mm; –12.5°, D_g = 0.60.

4 Conclusions

Research have shown the correctness of design and construction of a special intake manifold flow test bench. It enables to measure the mass airflow in each of the manifold outlets and analyze its distribution for multi-cylinder engines.

The flow bench test results confirmed the numerical simulation conclusion, that application of a guide vane inside the intake manifold improves the uniform distribution of the airflow between the cylinders. For the manifold without vane the maldistribution coefficient D_g = 0.44 was obtained while using guide vane in the best configuration (–6 mm, +12.5°) the factor D_g = 0.09 was obtained. In the worst configuration (0 mm, –12.5°) D_g was 0.60. In this case, the guide vane distributed the majority of airflow to outlet number 1 and 2, so that other outlet received less air flow.

Performed tests do not take into account the wave pulsation effects which occur in the engine intake system. In order to check its influence in the prototype manifold, engine tests on the dynamometer are planned in the future.

Guide vane is an additional element in the intake manifold and it has an impact on airflow resistance. That is why a study on guide vane influence on air flow parameters is planned.

The authors are pleased to acknowledge the financial support from the National Centre of Research and Development in the framework of Innolot contract POIR/01.02.00-00-0015/1510 for the work presented in this paper.

References

1. W.W. Pulkrabek, Engineering fundamentals of the internal combustion engine **249** (Pearson Prentice Hall, 2004)
2. H.P. Lenz, Springer Science & Business Media (2012)
3. L.J. Kapusta, I. Pielecha, K. Wisłocki, A. Teodorczyk, Journal of Thermal Analysis and Calorimetry **123**, 819-828 (2016), DOI: 10.1007/s10973-015-4927-z
4. J. Merkisz, J. Markowski, J. Pielecha, *WIT Transactions on Ecology and the Environment* **174**, 219-230 (2013), DOI: 10.2495/AIR130191
5. J. Merkisz, J. Markowski, J. Pielecha, *IEEE Vehicle Power and Propulsion Conference VPPC* 897-902 (2012)
6. A. Cebula, S. Kowalczyk, P. Swiatek, Proceeding of Modern Power Systems and Units (Krakow, 2016)

Comparison of the combustion engine operating parameters and the ecological indicators of an urban bus in dynamic type approval tests and in actual operating conditions

Lukasz Rymaniak[1,*]

[1]Poznan University of Technology, Faculty of Machines and Transport, ul. Piotrowo 3, 60-965 Poznan, Poland

Abstract. The article presents the considerations regarding a city bus combustion engine performance parameters in dynamic type approval tests and in real operating conditions when servicing an urban bus line. A comparison of the designated engine operating time shares with respect to load and crankshaft rotational speed was made. The analysis included the ETC and WHTC tests, which showed significant discrepancies in the work areas of internal combustion engines in these test when compared to actual driving conditions. The details of the type approval tests used and the method of their denormalization for the drive unit were presented. The vehicle used for this research was an eighteen meter city bus equipped with a CI engine with a displacement of 9.2 dm^3. The latest PEMS mobile equipment technology was used to conduct the road measurements. This allowed the emission indicators for CO, HC, NO_x and PM to be determined, including specific emissions. The obtained values were then compared with the Euro V limits. The analysis of the test results was supplemented with the calculation of fuel consumption using the carbon balance method.

1 Introduction

Currently urban buses most commonly use conventional powertrains, where the compression ignition (CI) internal combustion engine is connected to the gearbox and the transmission. Such systems will most probably continue to be used for several years to come. Due to the shape of the infrastructure, relatively low purchase costs and easy exploitation/repair of the vehicle, when compared to alternative drives and fuel prices. As a consequence, many modern CI engines are designed and engineered to improve their efficiency and emission performance [1-3]. According to the current trends, engine families of similar design are being developed, differing among others with, for example: charging system or control parameters [4-6]. As a result, vehicle and machinery manufacturers have the option of choosing a specific engine that meets their mechanical operating requirements [7].

For heavy-duty vehicles (HDVs), due to their large size and the power generated by the powertrain, the type approval of the exhaust gas emission is made for the internal combustion engine itself at the engine dynamometer. For this reason, the limit values are given in the regulations in g/(kW·h). Within the European Union, Euro standards are marked with numbers from I (introduced in 1992) to VI (currently in force). An additional EEV standard (enhanced environmentally friendly vehicle) for eco-friendly vehicles was introduced in 1999 [8]. It supplemented the regulations from Euro III and was incorporated into Euro V. The specific emission values and the methods of their determination are set out in the directives and regulations of the European Union.

Euro VI is currently in force in the European Union, with the new World Harmonized Stationary Cycle (WHSC) and World Harmonized Transient Cycle (WHTC), replacing the European Transit Cycle (ETC), European Stationary Cycle (ESC) and European Load Response (ELR). The Euro V and Euro VI norms provide mileage and operating time shares corresponding to the normal operation of a heavy duty vehicle. Minimal requirements for durability tests are also provided, depending on the category and maximum mass of the tested vehicle. The records are supplemented by data on the deterioration factors (more precise and clear definitions are given in Euro VI) [9].

Substantial changes in regulations have been made in the field of compliance checks – it is now necessary to perform specific emission measurements of exhaust pollutants in real driving conditions using PEMS (portable emission measurement system). Previously, only the internal combustion engines were tested after being removed from the vehicles. However, it should be noted that the type approval of the vehicle group discussed is carried out on engine dynamometers. It should be noted that, research of exhaust emissions in real driving conditions using PEMS are implemented for all vehicles with combustion engines [10-15]. The article presents the comparison of engine operating conditions

* Corresponding author: lukasz.rymaniak@put.poznan.pl

in real road conditions and dynamic type approval tests. The ETC and WHTC tests are intended to reflect the operation of the combustion engine in the actual moving vehicle.

2 City buses dynamic type approval tests – emission of pollutants

For all Euro standards it is necessary to perform tests at fixed internal combustion engine operating points in static tests. The emission limit values for CO, THC, NO_x and PM (Particulate Matter) for CI engines have been reduced with the introduction of further directives and regulations. Due to the development of internal combustion engines and exhaust aftertreatment systems, it became necessary to introduce limit values for the number of particles emitted designated as PN (particle number) and ammonia NH_3 in Euro VI norms [16]. With the introduction of the Euro III standard in 2000, dynamic braking tests began to apply, in which the internal combustion engine was tested in variable operating parameters, taking into account the presence of an external drive. As in static tests, unit emission limits were gradually reduced for further Euro standards.

In accordance with the Euro V and EEV norms it is necessary to perform ETC, ESC and ELR tests, which are performed in laboratory conditions. The first of these cycles aims at mapping the actual operating conditions of the combustion engine by applying variable operating parameters (dynamic conditions, Fig. 1). Its procedure envisages the possibility of carrying out measurements both on the chassis and engine dynamometers – the simulated operating conditions of the vehicle are converted to the operating parameters of the drive unit. However, it should be noted that only the second variant is tested for type approval. In normal engine operating conditions there can occur negative loads, which is related to the transmission of external torque (simulation of driving down a hill). The test drive cycle is divided into three parts – urban, rural, and motorway. The duration of the test is 1800 s, with each stage contributing 600 s, where the theoretical average vehicle speeds are 6.47 m/s, 20 m/s and 24.44 m/s for each part respectively.

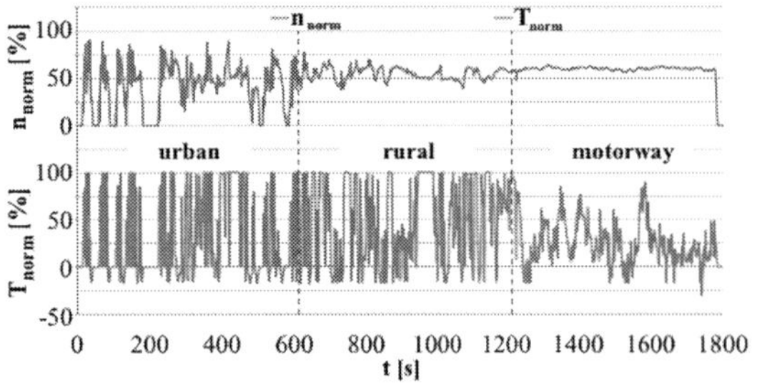

Fig. 1. Torque and engine speed during the ETC test [17].

On the ETC test results, normalized torque and engine speed values are expressed as: n_{norm} [%] and T_{norm} [%]. In order to carry out a real test for a particular internal combustion engine, it is necessary to perform standardization in accordance with the procedure described in Directive 2005/55/EC [17]. Using the full power characteristics the following parameters are determined: n_{idle}, n_{low}, n_{hi} and n_{pref}. The last of these components is the reference speed, at which the torque integral represents 51% of the torque integral in the interval $\langle n_{idle}; n_{95h}\rangle$. It is then possible to calculate the specific values of the combustion engine operating parameters. When determining the actual crankshaft rotation speed, the consecutive test steps use the following formula:

$$n_{actual} = 0.01 n_{norm} (0.05 n_{lo} + 0.95 n_{hi} - n_{idle}) + n_{idle} \quad (1)$$

where:
n_{actual} – actual engine speed [rpm],
n_{norm} – normalized engine speed in the test [%],
n_{lo} – the smallest rotational speed at which the engine reaches 55% of its maximum power output [rpm],
n_{hi} – the highest rotational speed at which the engine reaches 70% of its maximum power output [rpm],
n_{idle} – engine idle speed [rpm],
n_{95h} – the highest rotational speed at which the engine reaches 95% of its maximum power output [rpm].
Detailed guidelines for determining the engine operating parameters are provided in [17].

Calculations of torque value are made using the formula:

$$T_{actual} = (T_{norm} \cdot T_{max})/100 \quad (2)$$

where:
T_{actual} – actual torque [N·m],
T_{norm} – normalized torque in the test [%],
T_{max} – maximum engine torque [N·m].

In the Euro VI standard, in relation to previous regulations, the overall time share of individual WHTC dynamic test phases has changed. The results of the road tests in selected EU countries, Japan and the USA were taken into account when developing WHTC and static WHSC. This was to bring the measured cycles closer to real driving operating conditions around the world. The tests used previously were designed based only on European conditions. Detailed test requirement data is contained in UNECE Global GHS Regulations 4 [18]. Two equivalent methods can be used in both pollutant emissions tests: continuous sampling and constant volume sample (CVS). As with the dynamic cycle of the Euro V standard, WHTC tests can be carried out on chassis and engine dynamometers, with the second variant being the required one for type approval tests (Fig. 2). There are three parts in the test – urban, non-urban and motorway. The test duration is 1800 s, with successive parts lasting 900 s, 481 s and 419 s, where the theoretical average vehicle speeds are 5.92 m/s, 12.11 m/s and 21.31 m/s for each drive part respectively.

In order to carry out tests on a specific engine, it is necessary to denormalize the engine operating conditions. The actual load is determined in accordance with the formula (2) and the actual crankshaft rotation speed is determined by the formula:

$$n_{actual} = 0{,}01 n_{norm} (0{,}45 n_{lo} + 0{,}45 n_{pref} + 0{,}1 n_{hi} - n_{idle}) \cdot 2{,}0327 + n_{idle} \quad (3)$$

where the subsequent variables of the equation are defined just as for the formula (1).

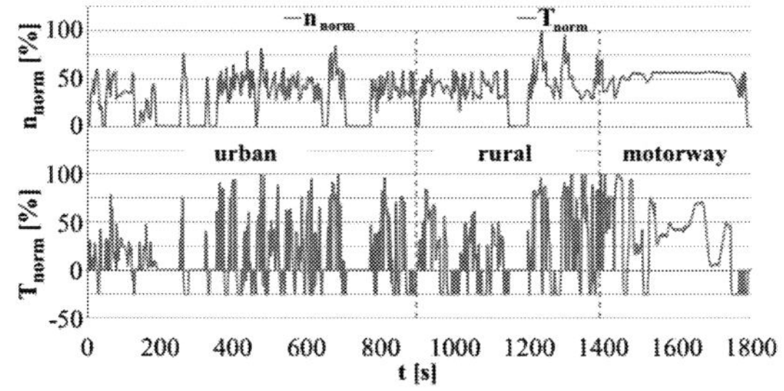

Fig. 2. Torque and engine speed during the WHTC test [18].

3 Research methodology

An eighteen-meter urban bus equipped with a conventional CI engine (Tab. 1) was used. The internal combustion engine had a maximum power of 265 kW at 1900 rpm and a maximum torque of 1450 N·m between 1100 and 1700 rpm. The drive system used a six-speed automatic transmission and a retarder for enhanced braking performance. The vehicle was type approved, complete and technically operational. It had a two-part bodywork and was fitted with four pairs of doors. In order to prepare the test bus, a replacement load was used to reflect the occupants during everyday bus operation. The equipped exhaust gas aftertreatment systems included: diesel oxidation catalyst (DOC), selective catalytic reduction (SCR) and diesel particulate filter (DPF).

Table 1. Tested vehicle technical data.

Parameter	
Engine type/fuel	4-stroke, CI/diesel oil
Displacement [dm^3]	9.2
Cylinder number and setup	6/in-line
Compression ratio	17.5
Maximum power [kW]/[rpm]	265/1900
Maximum torque [N·m]/[rpm]	1450/1100–1700
Exhaust aftertreatment systems	DOC/SCR/DPF
Exhaust emission norm	Euro V and EEV
Vehicle total mass with load [kg]	24,000

SEMTECH DS mobile gas measurement equipment and AVL MSS (Micro Soot Sensor) for solid particle concentration measurement (Fig. 3) were used in the experiments. The portable analyzer SEMTECH DS is coupled with a mass flow meter where a probe is fitted that takes the samples of the exhaust gases. The tested gas volume is transported through a special heated line where the temperature of 191°C is maintained. This solution prevents hydrocarbons condensation before their concentration is measured. Subsequently the sample goes through a filter where particulate matter is eliminated. Now the prepared exhaust gases are accurately measured in individual analyzers (Tab. 2). In the first place the emission of THC is measured in the FID analyzer (flame ionization detector). Another stage is chilling of the sample to 4 °C and sending it to the NDUV analyzer (non-dispersive ultraviolet detector) responsible for the measurement of NO_x. The next stage is the measurement of the CO_x through an NDIR analyzer (non-dispersive infrared detector) [19]. All the analyzers are calibrated and reset prior to the measurement to ensure that the ambient air pollution is taken into account. This takes a few minutes to complete. The sampling frequency range for each analyzer is different. Hence the measurement of all the exhaust components an assumption has to be made for the frequency that all the analyzers can operate at.

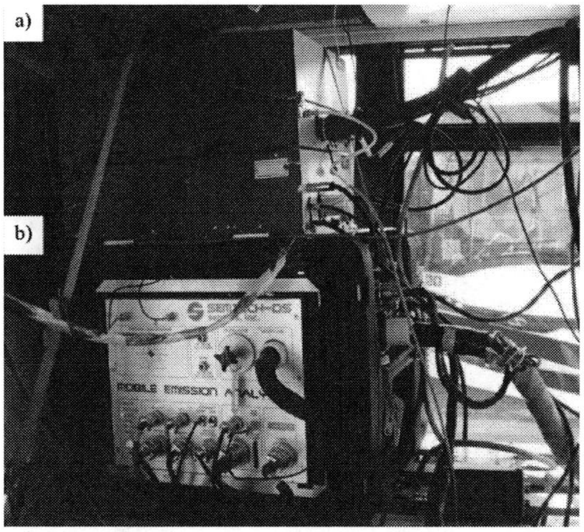

Fig. 3. Picture of the mobile PEMS equipment during bus road tests: a) AVL MSS 483, b) SEMTECH DS.

Table 2. Technical specifications of the portable exhaust emission analyzer SEMTECH DS [19].

Parameter	Measurement method	Accuracy
Component concentration:		
CO	NDIR, 0–10%	±3%
HC	FID, 0–10 000 ppm	±2.5%
NO_x	NDUV, 0–3000 ppm	±3%
CO_2	NDIR, 0–21%	±3%
O_2	electrochemical, 0–21%	±1%
Exhaust gas flow	mass flow rate	±2.5%
	T_{max} to 700°C	±1%
Diagnostic standards	SAE J1850/SAE J1979; SAE J1708/SAE J1587; CAN SAE J1939/J2284	

For the measurement of the PM emission a portable analyzer AVL MSS was used. It allows the continuous concentration measurements of particulate matter in the diluted exhaust gases. The principle of operation is based on the photo-acoustic method and allows measurements of particulates starting from 5 μg/m³. The measurement of the PM (the device measures the main part of the PM) through the photo-acoustic method consists of the absorption of the modulated laser light by the soot particles [20]. This results in a periodical heating up and cooling down of the medium gas (change of the pressure in the measurement chamber) and the generated sound wave is recorded by microphones. Presented PEMS could be used in research of various groups of vehicles [21-23].

For the purposes of the experiment the urban line operated by the local communications operator was used, classed as one of the more busy in terms of both the distance and the number of passengers (Fig. 4). Its total distance was 11.2 km and included twenty-seven stops along with loops. The route started in the northern part of the city (os. Sobieskiego) and travelled through main roads of varying traffic intensity to the Poznan city center (Plac Bernardynski). During the route selection, the EU guidelines on traffic conditions have been incorporated in accordance with EU Regulation 582/2011 [24]. However, the conditions for the minimum work performed by the internal combustion engine were not fulfilled due to the nature of urban buses operation.

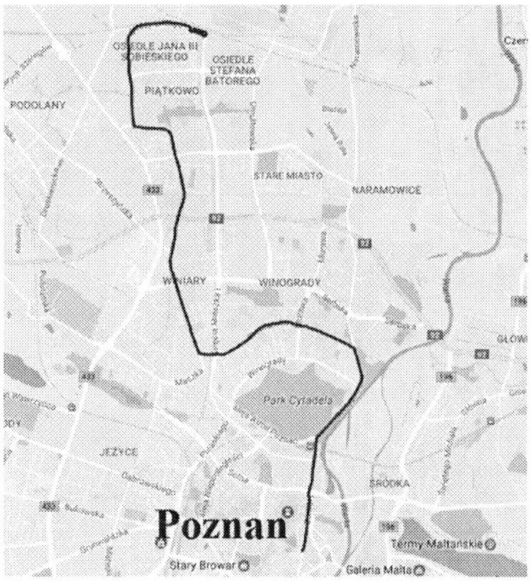

Fig. 4. The route used during the measurements under actual operating conditions (map created using GPS Visualizer).

4 Internal combustion engine operating parameters in type approval test

In order to compare the operating conditions of the internal combustion engine in the type approval tests and test in real driving conditions, the operating time shares of the engine unit were classified into load and crankshaft rotational speed compartments. The analysis quoted the dynamic test cycles ETC and WHTC. These tests have been developed in such a way as to simulate actual engine operating conditions, which is why they have been included in the article's considerations. Based on the performed simulation tests, the total work performed in these tests was determined. It is not only useful to determine the engine performance parameters

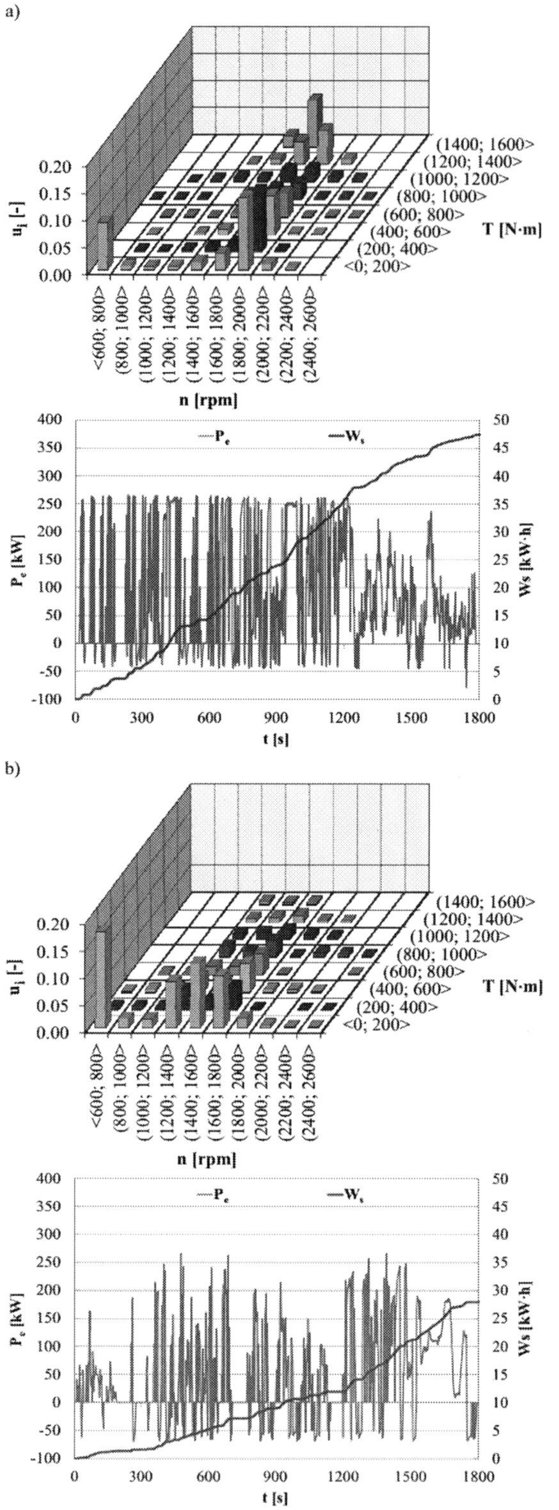

Fig. 5. Shares of rotational speed and torque ranges and power characteristic, together with the work sum of the internal combustion engine of the tested vehicle for tests: a) ETC, b) WHTC.

 DOI: 10.1051/matecconf/201711800009

in the WHTC test for the comparative assessment, but it is also necessary in order to determine the pollutant emission factors for compliance, as reflected in EU Regulation 582/2011 [24]. The procedure it describes assumes using the total energy generated by the internal combustion engine in the WHTC test, which is to be used for determining the measuring windows. However, the research vehicles were Euro V compliant buses, so the ETC test was also included. It should be noted that the obtained values do account for the negative power determined by the nature of the test cycles, whereas the curve representing the sum of the work performed only takes into account the positive values (Fig. 5).

The conventional bus internal combustion engine powered with diesel oil in the ETC test worked most often at crankshaft rotational speed of (1600 rpm; 2000 rpm⟩, in the full load range ⟨0 N·m; 1600 N·m⟩, where the total operating time share reached 73% (Fig. 5a). A significant value of the work area 8.7% also occurred in the range of the smallest rotational speed (the most significant was the range where the torque did not exceed 200 N·m). The instantaneous power generated by the internal combustion engine shows that the test vehicle often reached values close to the rated engine power, which resulted in a significant overall work value of 47.4 kW·h. In the Euro VI dynamic test, the engine operating range also includes a significant area, with the largest share of operating time being in the range of small loads (Fig. 5b). The largest share was 61.6% for rotational speed range of (1200 rpm; 1800 rpm⟩ at a load of up to 1200 N·m, whereas in the area of up to 200 N·m the share of operating time was 29.7%. For idle, the operating time share was 17.6% of the WHTC test duration. Power generation characteristic shows that the vehicle achieved a rated power much less often than the assumed in the ETC test. As a result, total work sum was also smaller at 28 kW·h.

5 Engine operating parameters in real operating conditions

To fully understand the impact of the engine operating conditions and the type of propulsion system used on the performance of the internal combustion engine, a bus test drive was conducted on the urban bus line. The average speed of the tested vehicle was 4.6 m/s. The total sum of operating area shares with the constant velocity of (a = 0 m/s²) reached 3%, while the vehicle stop duration time share was 22.6% of the whole test drive duration (Fig. 6). In the speed range of (0 m/s; 12 m/s⟩ and the acceleration range ⟨–0.8 m/s²; 0 m/s²⟩ and (0 m/s²; 0.8 m/s²⟩ operating time shares were determined to be 23.5% and 36.8% respectively. The characteristic of the entire propulsion system was the most influential on the latter of the mentioned ranges. More dynamic acceleration was difficult to achieve due to the weight of the vehicle and the gear ratios used.

Based on the internal combustion engine operation analysis it can be concluded that the conventional diesel oil powered bus has reached the highest share value of 33% in the idle speed range for torque in the range of ⟨0 N·m; 200 N·m⟩ (Fig. 7). The most significant share in the basic ranges occurred at engine speeds of (800 rpm; 1200 rpm⟩ and a load of up to 200 N·m considering the entire route. The sum of individual time shares were determined as 12.6% and 9%. For the next torque range in the crankshaft rotation range of ⟨600 rpm; 1200 rpm ⟩ the share of engine operating time was equal to 7%, 10.5%, and 17.9%. For the omitted single ranges these shares had values below 4.2%. The maximum crankshaft rotational speed that was recorded in the test cycle was 1687 rpm.

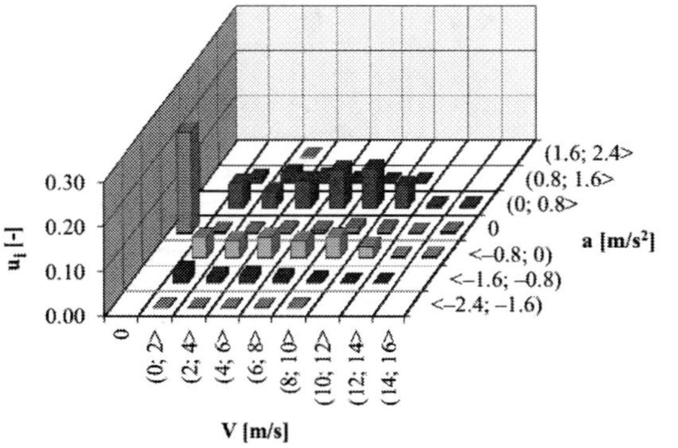

Fig. 6. Share of the urban bus operation time in speed and acceleration intervals during urban road test.

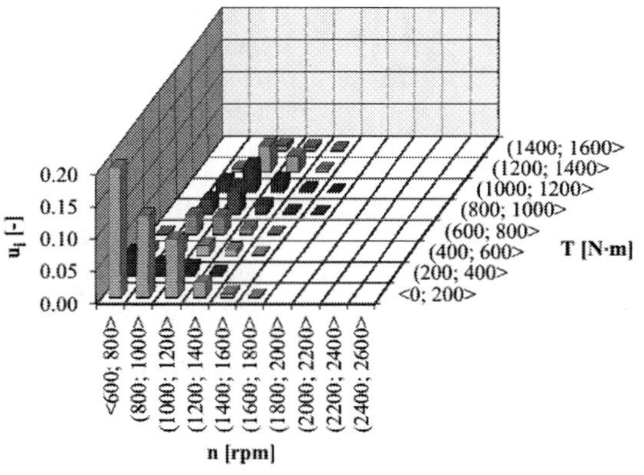

Fig. 7. Percentage share of engine operating time for the tested vehicle in the given speed and torque ranges when measuring driving in urban areas.

Due to the length of the urban bus line and the road conditions in the test drive cycles that occurred during the measurements, the internal combustion engine of the tested vehicle generated a combined work of 27.4 kW·h (Fig. 8). The created acceleration curves indicate the strongest correlation with total work sum value. This is related to the production of a momentary power by the internal combustion engine, characterized by large values. Thus, the acceleration of the vehicle has a strong influence on the total energy increase. For example, a speed variation of 0 to 14.6 m/s occurs in the time period between 1700 s and 1742 s, resulting in an energy requirement of 1 kW·h. When the vehicle is stationary the energy increase is small – it results from the need to supply the functional and auxiliary systems in the bus by the internal combustion engine (such as the pneumatic system, air conditioning, steering, etc.).

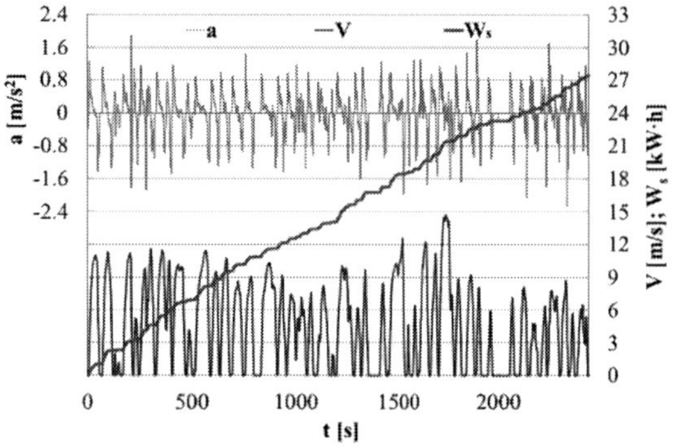

Fig. 8. The instantaneous velocity and acceleration characteristics of the tested vehicle together with the overall work sum of the internal combustion engine in the urban road test.

6 Comparison of combustion engine operating parameters in type approval and road tests

In order to compare the operating conditions of internal combustion engines in the type approval and real driving tests, the difference in the registered operating time shares in the speed and torque ranges was determined (Fig. 9). In both the standardized road tests and the selected test routes the recorded areas of engine operation and their operating time shares are significantly different from those found in the engine dynamometer tests. Dynamic type approval tests were prepared in such a way as to best reflect the actual operation of a heavy-duty vehicle. Therefore, the difference between the two values was taken for the analysis: the operating time share of the examined ranges in the type approval test and the time share of the corresponding compartment in the road tests. A positive value indicates a higher share of operating time in type approval tests compared to road tests. In addition, the operating areas that occurred during the urban test drive are shown on the characteristics, which makes it possible to compare the size of the work area.

Taking into account the ETC test data, the operation area of the internal combustion engine for the bus line test drive was lower within the selected ranges by 43% (Fig. 9a). The average value of absolute differences in engine operating time shares was 2.6%. The largest discrepancies in the presented characteristics were determined in the engine idle range, and also the smallest load range, which was –24.25%. This indicates that in engine dynamometer tests the engine unit operates in these ranges for a too small amount of time, where these conditions are much more often obtained during the actual operation of city buses.

The test vehicle's internal combustion engine gained a larger operating time share in the ETC test than in the road conditions for engine speeds between ⟨1800 rpm; 2000 rpm⟩ and load in the range of ⟨0 N·m; 600 N·m⟩, where for single parameter ranges the difference in operating time share obtained ranged from 7.17% to 13.32%. In addition for ETC tests on engine dynamometer significantly higher rotational speeds were achieved (over 2400 rpm), which did not occur in road tests, due to the characteristics of the propulsion unit and the selection of gear ratios. Based on the determined relationships, it can be stated that in the ETC test conditions, the proportion of operating time in the minimum load range of up to 200 N·m and engine speed of less than 1200 rpm is significantly too small. Extremely high operating time shares are present in the engine speed range of over 50% of the maximum value, with a torque of up to 50% of the nominal value, as well as for the maximum values of both parameters.

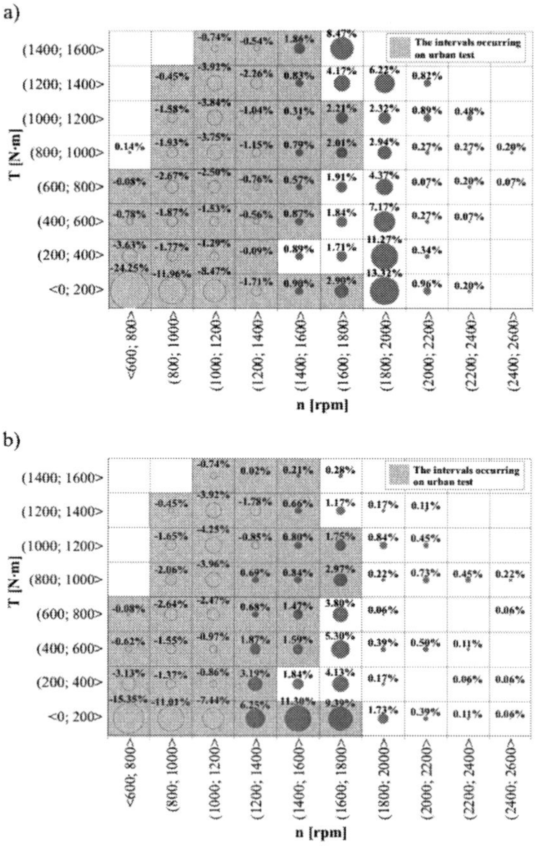

Fig. 9. Difference in engine operating time shares for the tested vehicle in the type approval tests and in the real driving measurements on an urban bus line divided into engine speed and torque compartments: a) ETC test, b) WHTC test.

When comparing the time shares recorded in the WHTC test and in actual operating conditions, minor differences in the individual designated compartments were obtained (Fig. 9b). Based on the size of the work areas obtained, it can be concluded that the Euro VI Dynamic Test conditions include a greater range of engine rotational speeds and loads than is the case in road conditions. Using the characteristics it can be seen that the largest differences in the considered values occurred for a load of up to 200 N·m in the range from idle speed to 1800 rpm. For the first single interval, the operating time share difference was –15.35%; While in other areas, these values varied from –11.01% to 11.3%.

Estimated mean of the absolute difference values was 2.1%. Using this comparison, the WHTC dynamic type approval test better reflects the conditions for the actual operation of a heavy duty urban bus. However, it is characterized by too wide a range of operating parameters – especially the crankshaft rotational speed. In addition, in the lowest load range at a rotational speed of 1200 rpm for CI engines there is an underestimation of operating time shares, and in the ranges above these values are therefore too high.

7 Analysis of ecological indicators and fuel consumption

From the data collected by the mobile exhaust gas analyzer and the GPS system, the emissions of harmful compounds in urban test drive were determined (Fig. 10). CO road emission was 7.11 g/km, and for HC it was 0.46 g/km. The performance characteristics of the CI engine (large λ values) is favorable in terms of emissions of the toxic compounds. The large amount of air contributes to better mixing of fuel with the air in the cylinder, however, it should be noted that under local conditions (e.g. with improperly shaped injection stream), incomplete combustion may occur. The NO_x road emission value of 7.28 g/km was heavily influenced by the exhaust aftertreatment system. In the selective catalytic reduction system, the rate of conversion depends primarily on the temperature and exhaust gas mass flow rate – with a higher flow rate, the conversion is lower. Thus, the variability of the combustion engine operating parameters is not beneficial in terms of exhaust aftertreatment efficiency. The PM emission was 30.74 g/km. A DPF filter was used in the vehicle, which limited the emission of this compound. However, the conditions of vehicle operation (frequent stopping and starting and acceleration dynamics) favored a greater emission of PM.

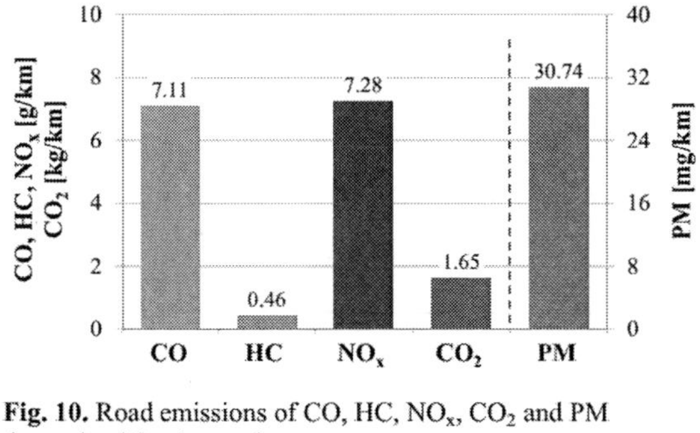

Fig. 10. Road emissions of CO, HC, NO_x, CO_2 and PM determined for the test bus route.

CO_2 road emissions depend significantly on fuel consumption and thus the vehicle's energy needs to travel a specific route. Based on the results of the conducted research, it is possible to determine the fuel consumption of the tested vehicle in road tests. Carbon balance method was used to determine the specific value, taking into account the road emission results of harmful exhaust gases. In this method for diesel vehicles, the formula for the calculation takes the following form:

$$FC = \frac{0,1155}{\rho_{\text{fuel}}} \cdot [(0,866 \cdot \text{HC}) + (0,429 \cdot \text{CO}) + (0,273 \cdot \text{CO}_2)] \tag{4}$$

where:
FC – fuel consumption in [dm^3/100 km],
HC, CO, CO_2 – measured exhaust emissions [g/km],
ρ_{fuel} – test fuel density in the temperature of 15°C [g/cm^3].

The method used allows for accurately calculating the volume of fuel consumed over a given distance (100 km). It is also possible to use data from the on-board diagnostic system or the method associated with the tank volume, but these solutions are burdened with notable uncertainties. Based on the recorded data, the vehicle fuel consumption was 61.7 dm^3/100 km. The result was most affected by the conditions in which the bus was operated. In order to reduce operating expenses by reducing fuel consumption, drivers employed in transport companies often receive a training in eco-driving. Apart from the economic aspect, upgrading and more efficient use of the vehicle fleet contributes to the reduction of environmental impact in urban areas, especially in urban centers of large agglomerations.

The specific emission values for type approval tests are determined using the net engine operating parameters measured at the end of the crankshaft or its equivalent. In actual operating conditions, the internal resistance of the combustion engine is calculated as a percentage – a specified value for the entire engine speed range. It should be noted, however, that this is in a way a simplification that affects the final results of the study. The obtained emission results of a single city bus while traveling on a city bus line are shown in Figure 11. Based on the calculations, the following unit emission results for CO = 2.93 g/(kW·h) and NO_x = 3.04 g/(kW·h). In the PM emission range, a value of 12.68 g/(kW·h) was obtained.

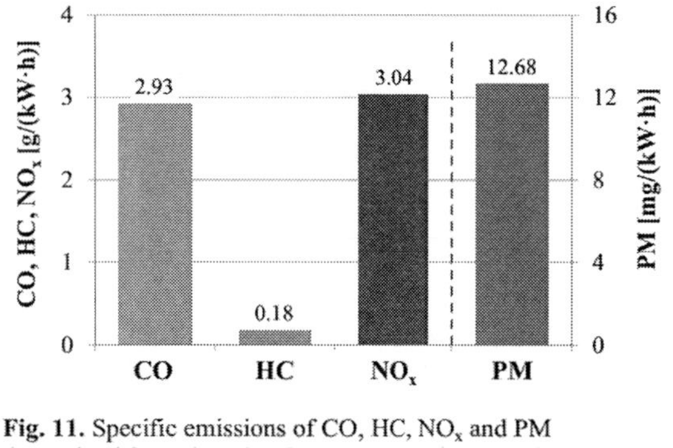

Fig. 11. Specific emissions of CO, HC, NO_x and PM determined from the urban bus route test drive results.

Based on the results of the unit emissions of harmful compounds, their values can be referred to the emission standard supposedly met by the tested vehicle (Euro V and EEV). To this end, the emission factor k_j, should be determined, as was done in [25, 26]. The emission factor is defined as:

$$k_j = \frac{e_{\text{real},\, j}}{e_{\text{limit},\, j}} \quad (5)$$

where:
j – exhaust component, for which the emission factor is determined,
$e_{\text{real},j}$ – specific emission value obtained from road tests [g/(kW·h)],
$e_{\text{limit},j}$ – accepted specific emission limit for Euro V or EEV [g/(kW·h].

The emission limits of the ETC dynamic type approval test were adopted for the analysis, because this test is assumed to reflect actual vehicle operation. The tested vehicle did not exceed the emission limits set for Euro V and EEV for the exhaust emissions of CO, HC and PM (Fig. 12). For these toxic compounds, the emission factor was smaller than one $k < 1$. On the other hand, for NO_x, the emission factor was $k = 1.52$ for both considered limits, which are identical to the standards in question. This means that the emission limits were exceeded by more than 50%. This result was mostly influenced by the exhaust gas aftertreatment system. The SCR system did not reduce NO_x emission continuously, but the conversion rate depended on many factors, including: the exhaust mass flow rate and its temperature. Due to the concerned emission size, the latest vehicle exhaust emission test procedures that already incorporate road emission tests allow for the possibility of exceeding the emission limit. This is also the case for particulate matter emissions.

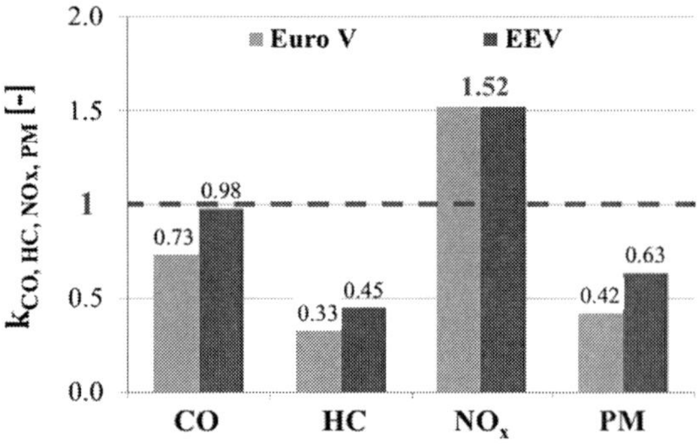

Fig. 12. Emission factor *k* determined for the measurement results on the urban bus line.

8 Conclusions

Due to the size of the urban bus population and the power of their propulsion systems, they have a significant impact on the quality of the environment in urban agglomerations where there are large concentrations of people. The operating conditions of the vehicle group in question are very specific and have a direct bearing on the operating conditions of the drive systems used. Taking into account the operating conditions of the internal combustion engines designated in the type approval tests and under actual operating conditions, it can be concluded that there are significant discrepancies between the obtained results. For the ETC test, valid for the tested vehicle, the biggest differences are in the engine idle speed range. The share of operating time in the compartment greater than 40% of the maximum engine speed in the complete load range is too large. The WHTC test is more similar to the actual operating conditions, but the overall operating area achieved during its implementation is much larger than for real driving, especially for the maximum torque range.

Using the characteristics of the engine operating time share of the tested vehicle and the intensity of pollutant emissions, determined with special assumptions regarding the operation of urban buses, it has been shown that the largest operating time share of these vehicles is for being stationary. The most common acceleration values were in the range of ⟨–0.8 m/s^2; 0 m/s^2) and (0 m/s^2; 0.8 m/s^2⟩ for vehicle driving speed in the range of (0 m/s; 14 m/s⟩. Small shares of operating time were recorded for uniform speed. It was found that the most significant are the compartments describing the crankshaft rotational speed at idle speed up to 800 rpm and the load not exceeding 200 N·m.

The observed operating conditions of the internal combustion engine had a direct impact on the achieved ecological indicators. Emission factors *k* were determined referring the measured results of the harmful gas emissions to the emission limits set by the standards met by the tested vehicle. This allowed to discern the relationship between the results of the road test emissions and the legislative guidelines. The analysis shows that CO, HC and PM emissions in road conditions are lower with respect to Euro V and EEV limits for dynamic tests. It has been demonstrated that NO_x unit emissions significant exceed the emission limit, by about 50%. The fuel consumption of the tested vehicle during urban road tests has reached 61.7 dm^3/100 km, which was determined using the carbon balance method taking into account the road emission of relevant exhaust gases.

The research was funded by the National Centre for Research and Development (Narodowe Centrum Badań i Rozwoju) – research project within the Applied Research Programme (contract No. PBS3/A6/25/2015).

References

1. R. Marquard, G. Topfer, *34th International Vienna Motor Symposium* (Vienna, 2013)
2. I. Pielecha, Journal of Thermal Analysis and Calorimetry **118**, 217-225 (2014), DOI: 10.1007/s10973-014-3956-3
3. W. Schoffmann, H. Sorger, G. Falck, M. Howlett, K. Weihrauch, *34th International Vienna Motor Symposium* (Vienna, 2013)

4. K. Brzozowski, J. Nowakowski, Eksploatacja i Niezawodnosc – Maintenance and Reliability **16**, 407-414 (2014)
5. J. Nowakowski, K. Brzozowski, T. Knefel, *Transport Means – Proceedings of the International Conference* 333-336 (2015)
6. G. Zak, L. Ziemianski, Z. Stepien, M. Wojtasik, Fuel **122**, 12-20 (2014), DOI: 10.1016/j.fuel.2013.12.055
7. J. Szczepaniak, T. Pawlowski, R. Grzechowiak, J. Kromulski, J. Rutkowski, Applied Engineering in Agriculture **31**, 709-716 (2015)
8. J. Merkisz, M. Idzior, J. Pielecha, W. Gis, *WIT Transactions on the Built Environment* **111**, 181-189 (2010), DOI: 10.2495/UT100171
9. J. Merkisz, J. Pielecha, W. Gis, *WIT Transactions on Ecology and the Environment* **120**, 439-449 (2009), DOI: 10.2495/SDP090411
10. P. Fuc, P. Lijewski, A. Ziolkowski, M. Dobrzynski, Journal of Electronic Materials **46**, 3145-3155 (2017)
11. R. Jasinski, J. Pielecha, J. Markowski, *E3S Web of Conferences* **10**, UNSP 00030 (2016), DOI: 10.1051/e3sconf/20161000030
12. J. Merkisz, M. Bajerlein, P. Daszkiewicz, *Lecture Notes in Information Technology* **13**, 106-113 (2012)
13. J. Merkisz, P. Fuc, P. Lijewski, A. Ziolkowski, K.T. Wojciechowski, Journal of Electronic Materials **44**, 1704-1715 (2015), DOI: 10.1007/s11664-014-3522-6
14. J. Merkisz, P. Lijewski, P. Fuc, S. Weymann, Eksploatacja i Niezawodnosc – Maintenance and Reliability **15**, 364-368 (2013)
15. J. Merkisz, J. Pielecha, P. Fuc, P. Lijewski, *IEEE Vehicle Power and Propulsion Conference VPPC* 903-907 (2012)
16. J. Merkisz, J. Pielecha, *Springer Tracts on Transportation and Traffic* **8**, 69-92 (2015), DOI: 10.1007/978-3-319-15928-7_7
17. Directive 2005/55/EC of the European Parliament and of the Council of 28 September 2005 on the approximation of the laws of the Member States relating to the measures to be taken against the emission of gaseous and particulate pollutants from compression-ignition engines for use in vehicles, and the emission of gaseous pollutants from positive-ignition engines fuelled with natural gas or liquefied petroleum gas for use in vehicles.
18. United Nations Economic Commission for Europe: Global technical regulation No. 4: Test procedure for compression-ignition (C.I.) engines and positive ignition (P.I.) engines fuelled with natural gas (NG) or liquefied petroleum gas (LPG) with regard to the emission of pollutants. ECE/TRANS/180/Add.4 (2007)
19. Sensors Inc. Emissions measurement solutions (2010)
20. AVL Micro Soot Sensor. Transient High Sensitive Soot Measurement (2010)
21. J. Merkisz, P. Lijewski, P. Fuc, M. Siedlecki, S. Weymann, Applied Engineering in Agriculture **31**, 875-879 (2015)
22. J. Merkisz, J. Markowski, J. Pielecha, *WIT Transactions on Ecology and the Environment* **174**, 219-230 (2013), DOI: 10.2495/AIR130191
23. J. Merkisz, J. Markowski, J. Pielecha, *IEEE Vehicle Power and Propulsion Conference VPPC* 897-902 (2012)
24. Commission Regulation (EU) No 582/2011 of 25 May 2011 implementing and amending Regulation (EC) No 595/2009 of the European Parliament and of the Council with respect to emissions from heavy duty vehicles (Euro VI) and amending Annexes I and III to Directive 2007/46/EC
25. P. Lijewski, J. Merkisz, P. Fuc, Croatian Journal of Forest Engineering **34**, 113-122 (2013)
26. J. Merkisz, J. Pielecha, P. Lijewski, A. Merkisz-Guranowska, M. Nowak, *WIT Transactions on Ecology and the Environment* **174**, 27-38 (2013), DOI: 10.2495/AIR130031

Turbulent spark-jet ignition in SI gas fuelled engine

Ireneusz Pielecha[1,*], *Wojciech* Bueschke[1], *Wojciech* Cieślik[1], and *Maciej* Skowron[1]

[1]Poznan University of Technology, Faculty of Machines and Transport, ul. Piotrowo 3, 60-965 Poznan, Poland

Abstract. The article contains a thermodynamic analysis of a new combustion system that allows the combustion of stratified gas mixtures with mean air excess coefficient in the range 1.4-1.8. Spark ignition was used in the pre-chamber that has been mounted in the engine cylinder head and contained a rich mixture out of which a turbulent flow of ignited mixture is ejected. It allows spark-jet ignition and the turbulent combustion of the lean mixture in the main combustion chamber. This resulted in a two-stage combustion system for lean mixtures. The experimental study has been conducted using a single-cylinder test engine with a geometric compression ratio $\varepsilon = 15.5$ adapted for natural gas supply. The tests were performed at engine speed $n = 2000$ rpm under stationary engine load when the engine operating parameters and toxic compounds emissions have been recorded. Analysis of the results allowed to conclude that the evaluated combustion system offers large flexibility in the initiation of charge ignition through an appropriate control of the fuel quantities supplied into the pre-chamber and into the main combustion chamber. The research concluded with determining the charge ignition criterion for a suitably divided total fuel dose fed to the cylinder.

1 Introduction

Urbanization and industrialization have led to a sharp increase in transportation demand worldwide. World consumption of energy has reached 12,730 Mtoe, of which 23% is produced using natural gas, which proves its spread and increased popularity as vehicle fuel. There are currently more than 18 million CNG-powered vehicles in the world, of which around 93% are light duty vehicles and trucks. Natural gas can be refueled at over 26,600 stations around the world [1].

Combustion engines powered by natural gas, so called: mono-fuel engines, are equipped with combustion systems that allow for the formation of mainly homogeneous fuel mixtures, often of a composition close to stoichiometric. However, wide flammability range of 4.3-15.2% (v/v), and high self-ignition temperature (above 540 deg. C) allow for the use of CNG in combustion systems using lean fuel mixtures.

The engine thermal efficiency is strongly dependent on the compression ratio [2, 3] and used fuel [4-6]. The high octane number of CNG (120-130) allows for compression ratio of up to 16, without any fear of engine knocking [1]. The gas engines can use either direct or indirect fuel injection systems.

Natural gas combustion takes place differently than the combustion of gasoline or diesel fuel. Natural gas has a higher ignition delay compared to liquid fuels, due to its lower flame propagation speed. As a result, the power of the gas powered engine may be about 5-10% less than that of a liquid powered engine [7, 8]. The comparison of the parameters of main fuels used in combustion engines is presented in Table 1.

Tahir et al. [9] by analyzing a SI engine powered with gas and petrol concluded, that the cylinder pressure in the gas engine was 18.5% lower than in the petrol powered engine, which was attributed to a 23% decrease in the heat output compared to the petrol engine. Such combustion conditions are due to the reduced volumetric efficiency of the engine when powered with natural gas while gas volume displaces air from combustion chamber.

Table 1. Comparison of main parameters of engine fuels [7].

Properties	CNG	Gasoline	Diesel
Octane/cetane number	120–130	85–95	45–55
Molar mass [kg/mol]	17.3	109	204
Stoichiometric (A/F) mass	17.2	14.7	14.6
Stoichiometric mixture density [kg/m^3]	1.25	1.42	1.46
LHV [MJ/kg]	47.5	43.5	42.7
LHV of stoichiometric mixture [MJ/kg]	2.62	2.85–2.75	
Combustion energy [MJ/m^3]	24.6	42.7	36
Flammability limit in air [vol% in air]	4.3–15.2	1.4–7.6	1–6
Flame propagation speed [m/s]	0.41	0.5	–
Adiabatic flame temp. [°C]	1890	2150	2054
Auto-ignition temp. [°C]	540	258	316
Wobbe Index [MJ/m^3]	51–58	–	–

* Corresponding author: ireneusz.pielecha@put.poznan.pl

The use of CNG PFI engine allows for about 30% decreased torque at low engine speeds relative to the SI DI engine [10]. It has been found, however, that the use of the CNG DI system and the optimization of the ignition angle can compensate for torque losses of over 60% at low engine speeds.

Choi et al. [11], by analyzing direct gas injection into the combustion chamber using the so-called homogeneity index, has shown that injection at higher fuel pressure allows for greater mixture homogeneity. Starting fuel injection at 180 deg. TDC at the pressure of 0.6 and 1.4 MPa allows the index value to increase from 0.86 to 0.9 at an angle of 40 degrees before TDC. It has also been found that extending the injection time at lower fuel pressure does not significantly affect the homogenization of the fuel dose. This is confirmed by IMEP studies at various fuel injection pressures. Changes of the observed values oscillate in the range of ±0.02 MPa. The change in engine performance indicators for the direct gas injection engine relative to a petrol fuelled engine with MPI were reported in the research results of Kalam and Masjuki [12]. The CNG DI engine showed about 2 percentage points lower heat efficiency (29.2% versus 32%) compared to the MPI engine. With regards to emissions of exhaust gas components, a 42% reduction in nitrogen oxide emissions was noted, with HC emissions increasing by 25%.

Research performed by Hall et al. [13] on a turbocharged 3-cylinder engine with a displacement of 1.2 dm^3 and a compression ratio of 13.3 in RDE (Real Driving Emission) test showed an increase in CNG engine thermal efficiency of around 2–10% at maximum loads and high engine speed relative to the petrol engine (SI DI, ε = 11.3). The maximum speeding-up of the centre of combustion (CoC) was recorded within the maximum load range and an average crankshaft rotational speed, and ranged from 10 to 18 degrees. Large values result from the engine knock recorded for petrol supply case and the necessary ignition delay. Full engine optimization resulted in the specific fuel consumption being reduced to 200 g/kWh for gas fuelled engine compared to the 240 g/kWh during a conventional gasoline direct injection test.

Natural gas engines emit less carbon monoxide and hydrocarbons (non-methane) compared to conventional gasoline-powered SI engines [14]. It has been observed that in this case over 90% of the unburned hydrocarbons are methane, which is 20 times more potent as a greenhouse gas than carbon dioxide [10]. Early particle number studies (year 2000) [15] generated by gas combustion in a spark-ignition engine indicate the presence of particles below 50 nm regardless of the test conditions (steady-state or transient). Contemporary research (2015) by Alanen et al. [16] shows that the largest number of particles is in the size range of 2-5 nm. The number of particles in the size range over 23 nm is significantly smaller compared to the numbers of the smallest particle size.

Simultaneous use of indirect injection (into the intake manifold) and direct injection allows using prechamber. Combustion of lean fuel mixtures in such systems reduces the maximum combustion temperature as well as reducing the emission of nitrogen oxides. This trend is also supported by the lower flame temperatures when burning gas mixtures (see Tab. 1).

Tests of a prechamber system coupled to the main chamber intended for use in engines of PC or LDV vehicles was presented by Attard et al. [17]. In this injection system, both indirect gas injectors (into the intake manifold) and direct gas injectors (to the prechamber) were used. The use of a prechamber increased the range of the air excess coefficient λ change (from 0.8 to 2.1 with gasoline/propane fuel mixture supply) with respect to combustion in a conventional fuel supply system (gasoline combustion λ = 0.7-1.4).

A comparison of the conventional system and a TJI system (Turbulent Jet Ignition) indicated the possibility of using a higher air excess coefficient (λ = 1.8 for TJI), allowing for a simultaneous ignition delay (up to 5 deg.) and reduced the uneven engine operation CoV(IMEP) = = 1.0% for TJI system compared to 2.5% for conventional gasoline systems). The tests were carried out at a fixed peak cylinder pressure angle (of about 7 deg. on the crankshaft) and almost constant pressure increase after ignition (approx. 1.1 bar/deg).

The commercial version of a combustion system utilizing an additional ignition chamber was also presented by Mahle in 2014 [18]. A portion of the gas was supplied to the prechamber, whose energy accounted for about 2% of the total fuel energy supplied. Combustion of a mixture of about stoichiometric composition took place in the prechamber.

Schumacher and Wensing [19] presented a prechamber system with a gas supply valve. A chamber the size of 1.57% of the entire combustion chamber was equipped with a spark plug M10 and a gas dosing valve. Application of the prechamber increased the engine performance by approximately 5% (λ = 1.6, IMEP = 4.5, n = 1500 rpm) when compared to a petrol engine (λ = 1).

Simulation studies of the liquid fuel injection process conducted with two prechambers were presented by Assanis et al [20]. It was found that the shape of the prechamber and of the piston influenced the velocity of fuel flow from the prechamber. A shape has been obtained, using simulation results, which allowed for complete fuel evaporation in the prechamber and no fuel gathering on the prechamber walls when flowing out of it. Simulation studies for the optimum prechamber shape concept indicated the possibility of burning a mixture with an air excess coefficient of λ = 3 for liquid fuel.

The study of dual-fuel engines with a prechamber was published by Yousefi and Birouk [21]. The specificity of these tests consisted of injecting gas into the intake manifold and diesel fuel into the prechamber. The gas dose injected into the intake manifold was varied. It was found that increasing the share of gas delayed the combustion process, and when its share reaches over 60% it worsens the combustion process leading to misfire.

The use of lean mixtures in combination with prechambers has been researched by Ruter et al. in [22]. The prechamber in such systems represents 1-2% of the entire cylinder volume. The resulting exhaust gases

DOI: 10.1051/matecconf/201711800010

formed in the prechamber leave it in the form of high-energy flame streams covering the main combustion chamber with multiple ignition points. The flame stream has a flash energy of 106 times that of the standard spark plug [22]. Additional simulation studies conducted in [23] demonstrate that the air/fuel ratio of 1/1.09 achieves the lowest combustion time, with a 5.7% variation of the maximum combustion pressure. This confirms the hypothesis that the maximum laminar flame velocity in methane combustion is within the air excess coefficient range of λ = 1/1.05 to 1/1.10 [24].

Optimization of the HDV natural gas fuelled prechamber was discussed by Heuser et al. [25]. Changes in the chamber geometry allowed for combustion of fuel mixtures with λ = 2.1, as well as significantly limiting the combustion duration (by 6 percent with respect to the main chamber). More advantageous distribution and orientation of the outlet channels from the prechamber allowed to reduce the influence of gases flowing from the main chamber, thus increasing the air excess coefficient in the prechamber.

Research conducted by Tanoue et al. [26] using prechamber in a single cycle machine indicates the occurrence of a characteristic pressure increase in the prechamber. The pressure increase rate in the prechamber is all the greater since the spark plug is located at a greater distance from the orifices allowing the flow of prechamber gases into the main combustion chamber. In addition, the authors identified engine knocking with intensity related to the number of injector holes used for the prechamber injection. It has been found that the knocking occurs when the injector has a single hole, and the knocking itself takes place in the prechamber fuel outflow area. Utilizing a fast Fourier transform FFT, knock detection was performed in the 5-35 kHz range. The highest frequencies were found at 6.7 kHz and 11.2 kHz. The maximum value of the combustion pressure change caused by the engine knocking was 0.82 MPa. The temperature value in the vicinity of this phenomenon was estimated to be at the level of 2400-2500 K.

2 Motivation

The search for optimal combustion systems solutions leads to increased interest in natural gas as a fuel that can potentially replace both diesel fuel and gasoline. The authors of this article, using the gasoline burning systems [27, 28], diesel [29-32] and dual fuel systems [33], used a combustion system with a prechamber for the research and analysis of the gas supply system thermodynamic optimization of a CNG spark ignition engine.

The article analyzes the prechamber combustion system in terms of fuel dose ignition. Therefore, an important element of the research work carried out is the fuel dosage rate analysis injected into the prechamber as well as the main chamber. These two parameters describing stratification of the charge and relating to the engine performance indicators are the main research focus of this article.

3 Methodology

3.1 Test stand

Research into determining the flammability limits was carried out on a single-cylinder test engine adapted for natural gas supply (Fig. 1). The measuring stand was equipped with the necessary systems for the conditioning of the engine as well as carrying out the analysis of the combustion process and the emission of toxic exhaust components. The natural gas supply system contained additional damping volumes to ensure the correct indication on the gas flowmeter. For this reason, additional volumes were added before the injectors: a 2 dm^3 volume before the injector in the engine intake manifold and a 3.5 dm^3 volume in the direct injection system.

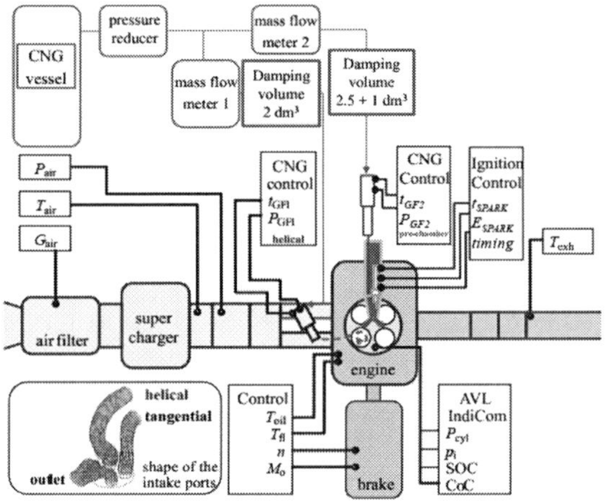

Fig. 1. Engine test stand for direct and indirect injection of natural gas.

The research engine base is an AVL 5804 structure, where the power supply, the piston design (combustion chamber) and the compression ratio have been modified. The standard compression ratio of the compression ignition engine (19.9) was lowered to 15.5 (Tab. 2).

Table 2. Test engine characteristic.

Parameter	Unit	Value
Engine	–	1-cylinder, 4-valve, SI
Cylinder volume	dm^3	0.5107
Bore x stroke	mm	85 × 90
Compression ratio	–	15.5
Fuel system	–	Direct and indirect injection (electromagnetic injectors)
Air system	–	Supercharging

3.2 Research equipment

The engine test stand was equipped with the necessary measuring systems (engine dynamometer) and systems for recording fast processes (Table 3). Due to the use of two gas supply systems, the gas flow measurements were made using two separate flow meters designed for large and small flows. At the same time, apart from the pressure in the cylinder, electrical voltage on the injectors and the voltage in the primary circuit of the ignition coil were measured.

Table 3. Equipment used in research.

Parameter	Producer/name	Description
Engine brake	AVL AMK DW13-170	–50-300 Nm
Air system	Sensycon Sensyfolw	0-720 kg/h
Fuel system	Emerson mCMFS	0.1-2 kg/h
	Bronkhorst 111B	0.1-100 g/h
Oil system	AVL 577	0-150 deg C
Water system	AVL 577	0-150 deg C
Fast-varying processes	AVL IndiSmart	8-channel acquisition (±10 V)
2 parameter	AVL Concerto	Post-processing
Emission	Horiba Mexa 7100D	CO, THC, NO_x

Injection of the fuel dose into the prechamber and the inlet manifold occurred at 160 deg. before the piston TDC. The ignition advance was 15 degrees, but at the maximum fuel dose delivered to the cylinder, the ignition was delayed to 13 degrees to maintain the proper angle of incidence of the peak cylinder pressure. The gas pressure delivered to both injectors was 8 bar overpressure relative to the ambient pressure. The coil charging current was 5.5 A.

3.3 Research plan

Flammability tests of the fuel mixture have been conducted on a wide range of variations in its composition, with differentiated values for the primary dose and the prechamber dose. The test conditions are shown in Table 4. The range of test points was chosen in such a way as to obtain as low a dose of prechamber fuel injected as possible. The injection dose to the prechamber was limited by the minimum operating time (opening) of the injector mounted in the prechamber (a dose of 0.5 mg/inj was found to be the limit in this case).

Table 4. Test conditions.

n [rpm]	*Pin* [bar]	qo_PC [mg/inj]	qo_MC [mg/inj]
2000	1.03	0.5-3.0 variable	ca. 15; ca. 16; ca. 17.5; ca. 19

The study was conducted at 37 measurement points (Fig. 2), maintaining a distribution of measurement points as uniform as possible, which allows for a wide range of load conditions. Fuel injected into the prechamber only serves as the ignition initiator and creates spark-jet ignition of the fuel mixture in the engine main combustion chamber. Therefore, the measurement points were selected so as to minimize its share in the total fuel dose. From the initial analysis of the dose injected into the prechamber, it appears (Fig. 3) that the maximum dose share of the prechamber injection relative to the total dose reaches values greater than 16% (for a small main injection dose) down to about 10% (for a large main dose). The minimum fuel dose share injected into the prechamber obtained in the tests was between 4% and 3.5%. The greater the size of the main dose, the smaller the dose share of the gaseous fuel delivered into the prechamber.

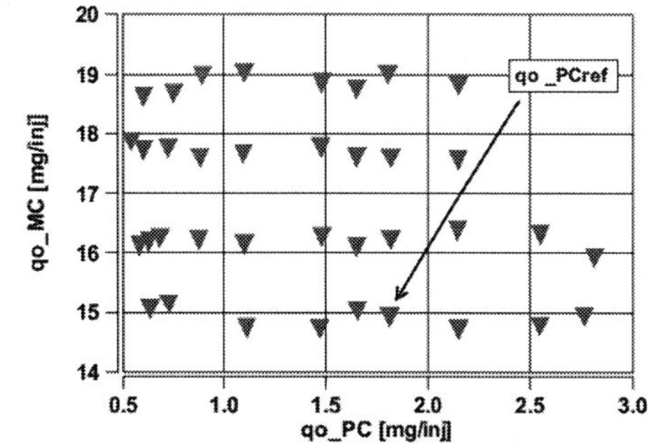

Fig. 2. Measurement points used to determine the fuel mixture flammability.

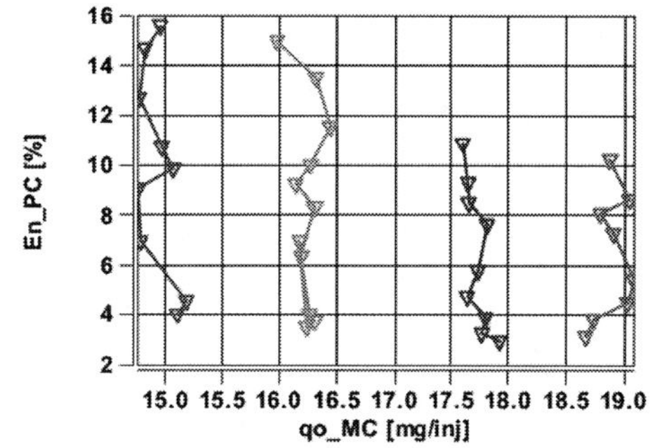

Fig. 3. Energy share of the prechamber dose with respect to the total fuel dose.

The main limitation of prechamber dose reduction was either a very irregular engine operation with a high proportion of misfires or the achievement of the lower fuel dose limit (around 0.5 mg/inj). Research in such conditions has been carried out to determine the limits of the lean gaseous mixtures inflammability.

4 Determining the fuel mixture inflammability limits

The research plan presented in Section 3.3 and a variable signal analysis allowed to determine the Coefficient of Variation (CoV) from IMEP, which is a measure of the engine operation uniformity, which determined the acceptable engine operating limits. It was assumed that the maximum value 5% of this coefficient determines the gas mixture ignition limits.

From the data shown in Fig. 4, it can be seen that the uneven engine operation is related to low fuel doses fed

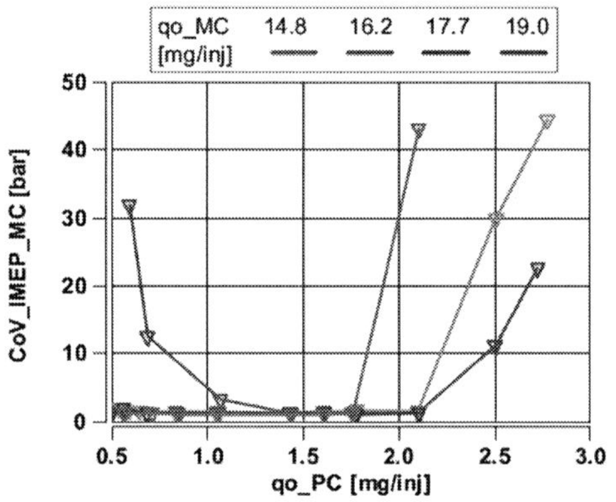

Fig. 4. Determining the relation between the IMEP coefficient of variation and the prechamber dose share with respect to the total fuel dose.

to both the prechamber and the main combustion chamber. Increasing the main chamber dose (with a minimum dose injected into the prechamber) does not result in abnormal engine operation (CoV(IMEP) < 3%). Very big prechamber doses (greater than 2 mg/inj) cause the engine to run in an unstable way and the CoV (IMEP) indicator reaches large values.

Determining the inflammability limits of lean gas-air mixtures required a cylinder pressure analysis, which determined the maximum cylinder pressure values for each engine work cycle. Irregularity of the engine work cycles was determined using the maximum cylinder pressure – CoV(Pm_MC). The subsequent part of this article shows only a part of the study of the minimum (q_PCmin), reference (q_PCref) and maximum (q_PCmax) prechamber fuel dose with the minimum fuel dose being injected into the main combustion chamber (Fig. 5).

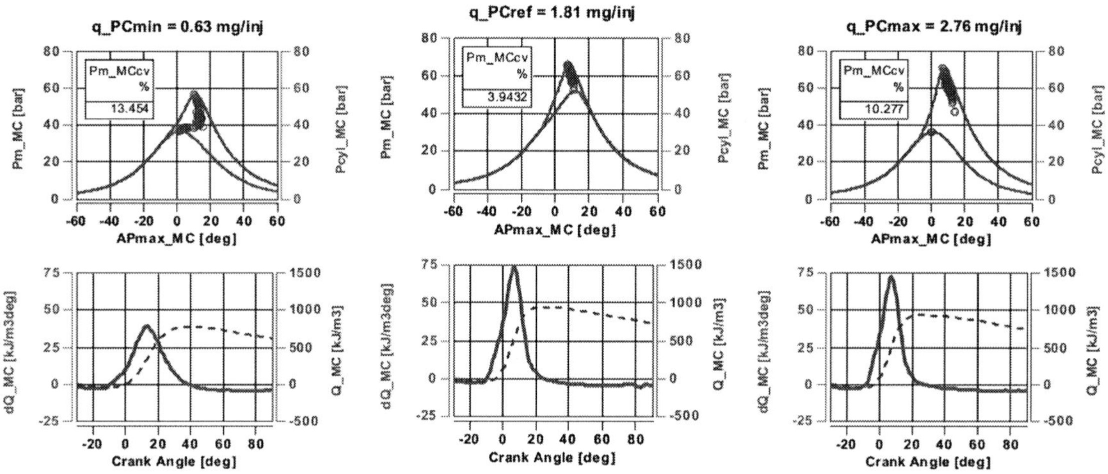

Fig. 5. Analysis of the combustion pressure and the heat release rate for small dose injection into the main combustion chamber (qo_MC = 15 mg).

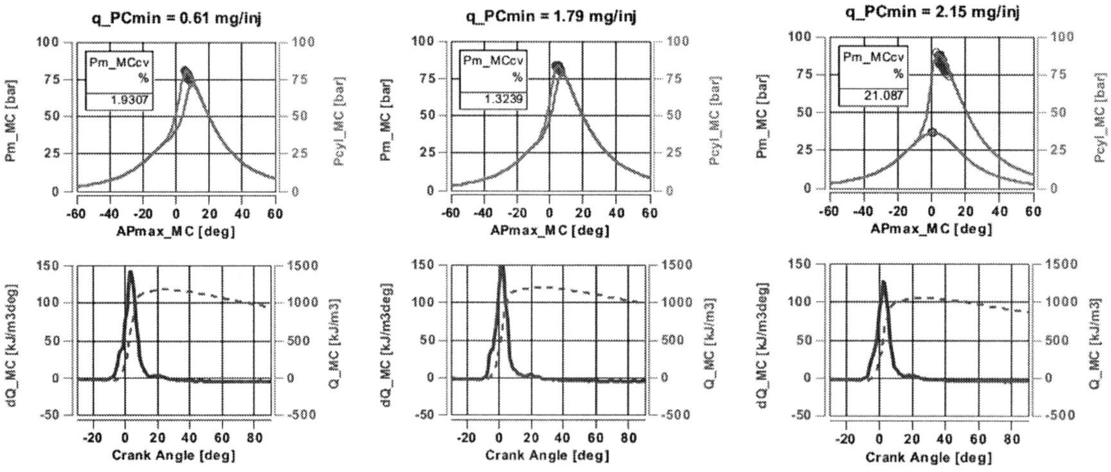

Fig. 6. Analysis of combustion pressure and the heat release rate for large dose injection into the main combustion chamber (qo_MC = 19 mg).

Small prechamber injection dose results in high values of CoV(Pm_MC) – 13%. This is caused by the cylinder misfire, as indicated by the low peak pressure – Pm_MC, without any growth characteristic for a combustion process. Increasing the prechamber dose share improves the main combustion process and its uniformity, as evidenced by the small CoV values (3.9%) and large values of the heat release rate (dQ_MC). Increasing the prechamber dose increases the engine operation irregularity. Among the 100 analyzed cycles, a misfire was detected, indicating an excessive fuel dose. Although there are few misfires in this test run, the variation of the peak cylinder pressure value indicates that this process does not take place as intended.

The pressure traces recorded at the injection of various fuel doses into the prechamber and a large main dose (qo_MC = 19 mg/inj) indicate improved combustion (Fig. 6). The combustion process irregularity, as determined by CoV(Pm_MC), is 1.9% at the minimum fuel dose for the prechamber. Increasing the dose to a reference value decreases this index (which improves the uniformity of the engine operation) and at the same time increases the rate of combustion. Maximum fuel doses (2.15 mg/inj) result in more uneven engine performance. A cylinder misfire, resulting in a high CoV(Pm_MC) value of 21%, was also observed at this measurement point.

Changes in the heat release rate are an important parameter when analyzing fuel combustion in a combustion system equipped with a prechamber. The heat release rate is lower in the first stage, but later it reaches high values (up to the maximum values). Despite the maximum dose administered to the prechamber, the heat release rate decreases, indicating a too rich fuel mixture obtained in the prechamber, as an excessively lean fuel mixture significantly decreases the rate of combustion.

Comprehensive assessment of the lean gas-air mixtures inflammability limits requires studies to characterize changes in engine operating parameters. The assumption about the lean fuel mixtures combustion has been met because the values of the air excess coefficient change are within the range of 1.3-1.8. Increasing the prechamber dose results in a few percent reduction of the global air excess coefficient (Fig. 7a). The mean values of the maximum cylinder pressure (Pm_av_MC) are constant, apart from the minimum doses injected into the prechamber. This trend is independent of the main fuel dose size (however, the larger the main dose, the smaller the change is) – Fig. 7b. The maximum pressure increase rate (dP_MC) in the main chamber indicates the optimum combustion conditions for the small fuel doses fed to the main combustion chamber (Fig. 7c). The optimal doses for the prechamber injection are reduced with the primary dose increase. Also, the upper inflammability limit is determined by the amount of fuel dose injected into the prechamber as the engine load increases. The maximum mean temperatures in the combustion chamber are to a small extent dependent on the size of the fuel dose injected into the prechamber. The dependence of the temperature on the fuel dose is significant at the boundary points where the combustion temperature is significantly reduced. This is partly due to the fact that part of the combustion cycle does not take place, which reduces the mean values of the recorded operating parameters (Fig. 7d).

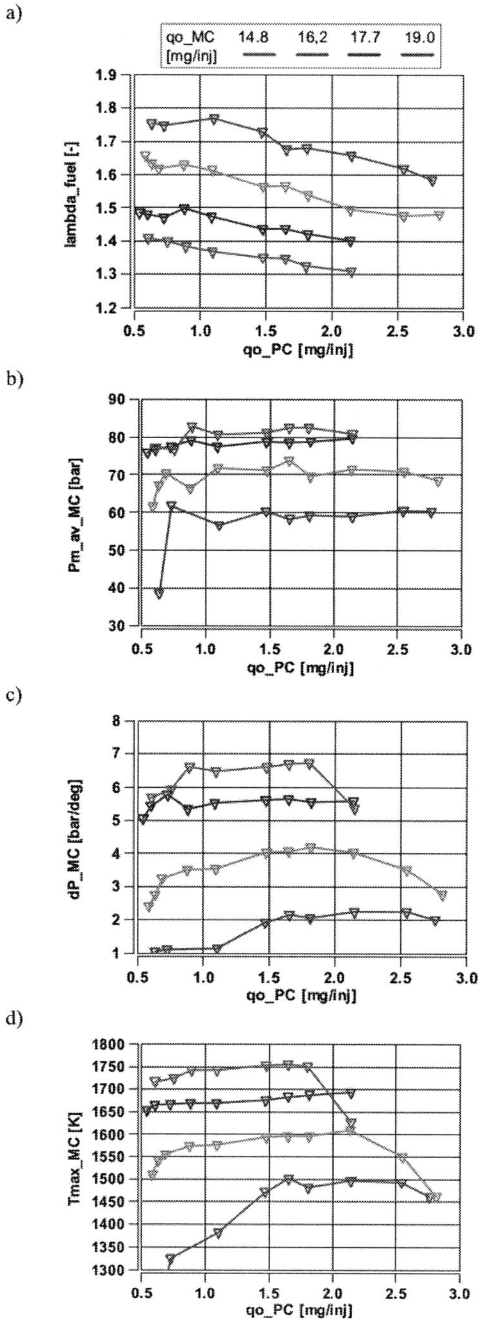

Fig. 7. Analysis of the parameters characterizing the combustion process relative to the prechamber fuel dose: a) air excess coefficient – lambda_fuel, b) maximum cylinder pressure – Pm_av_MC, c) maximum cylinder pressure change – dP_MC, d) peak average cylinder temperature – Tmax_MC.

* Corresponding author: ireneusz.pielecha@put.poznan.pl

An analysis of the combustion process using the indicated mean effective pressure reveals that its value increases with the increase of the prechamber fuel dose (Fig. 8a). IMEP is rapidly falling in the range of high loads and maximum fuel ratios injected into the prechamber. This means that IMEP can be effectively boosted by increasing the prechamber dose.

A slightly different relationship was obtained in the analysis of the peak cylinder pressure. For this parameter (Fig. 8b), a characteristic minimum was obtained indicating an increase in the fuel combustion rate. This means that in the range of qo_PC = 1.5-2 mg/inj, the first part of the combustion process takes place the quickest since the changes in this value range from 1 to 4 degrees with respect to the minimum doses injected into the prechamber.

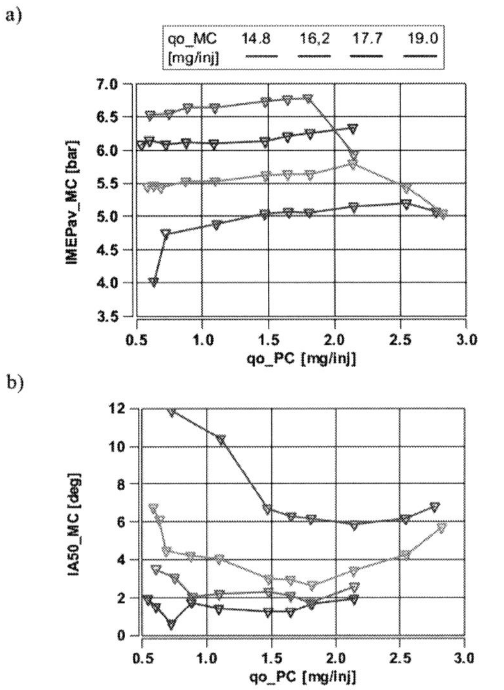

Fig. 8. Combustion process characteristics considering the IMEP values (a) and the IA50_MC (center of combustion) (b), depending on the prechamber dose.

5 Defining the inflammability limits of fuel mixtures using engine maps

5.1 Determination of inflammability limits based on the process thermodynamic indicators

The parameters characterizing the combustion process obtained in the performed tests allowed to create maps defining the fuel dose inflammability limits based on the analyzed engine operating points in the coordinates qo_PC–qo_MC and the combustion indicators (Fig. 9).

The IMEP characteristics indicates the possibility of obtaining higher values with increasing fuel dose in the main combustion chamber and with no change of its value when altering the prechamber dose size (IMEP_MC). A reduction of this value is only observed at small initial doses and a small main dose.

The combustion chamber maximum pressure characteristic (Pm_av_MC) indicates similar relations. However, in this case operating range of high fuel doses injected to both combustion chambers no longer affects the obtained values. The operating area with maximum combustion pressure values is quite large. The pressure changes for low initial fuel doses are minor.

The angle of obtaining 50% of the released heat value (angle of the peak cylinder pressure IA50_MC) appears to be a good measure for accurately determining the dependence of both fuel doses injected into the combustion chambers. The optimal choice for the initial dose size seems to be 1-2 mg/inj irrelevant of the size of the main dose.

The maximum combustion temperatures (Tmax_MC) are well correlated with the IMEP values since the changes in the contour line of the latter correspond to the changes of the first one.

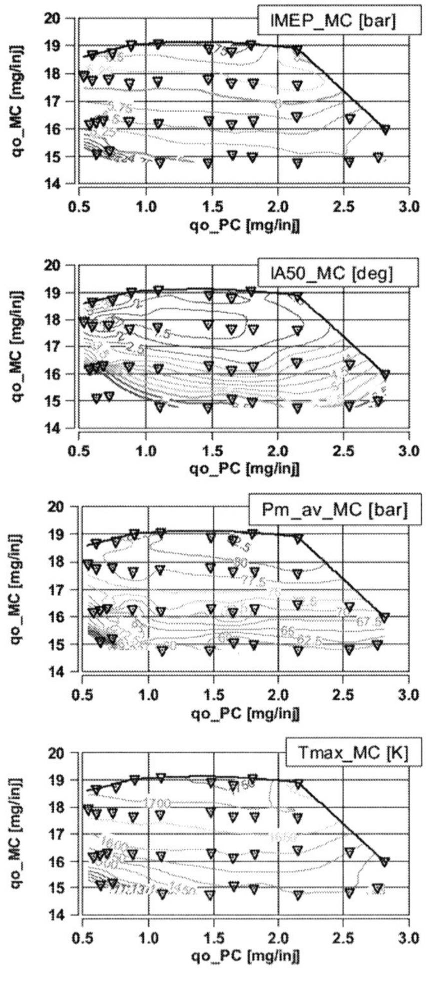

Fig. 9. Combustion indicators maps relative to the dose inflammability characteristics in coordinates qo_PC–qo_MC.

5.2 Determination of inflammability limits considering exhaust emissions

The analysis of the exhaust components emissions allowed determining the dose inflammability maps taking emission of harmful exhaust components into account. Based on the measured CO, THC and NO_x concentrations, the corresponding unit emissions have been determined relative to the power generated by the engine. Using the AVL Concerto software, envelopes have been generated that characterize the varying degrees of emission reduction during the analysis of each exhaust component. The envelopes obtained for the engine performance indicators were identical (Fig. 9).

Exhaust emissions research during engine operation was carried out without the presence of any exhaust aftertreatment systems. It was found that the reduction of the main fuel dose led to a decrease in nitrogen oxide emissions (Fig. 10). The dose injected into the prechamber has a limited effect on the level of nitrogen oxide emissions. Its effect is noticeable only for small fuel doses in the main combustion chamber.

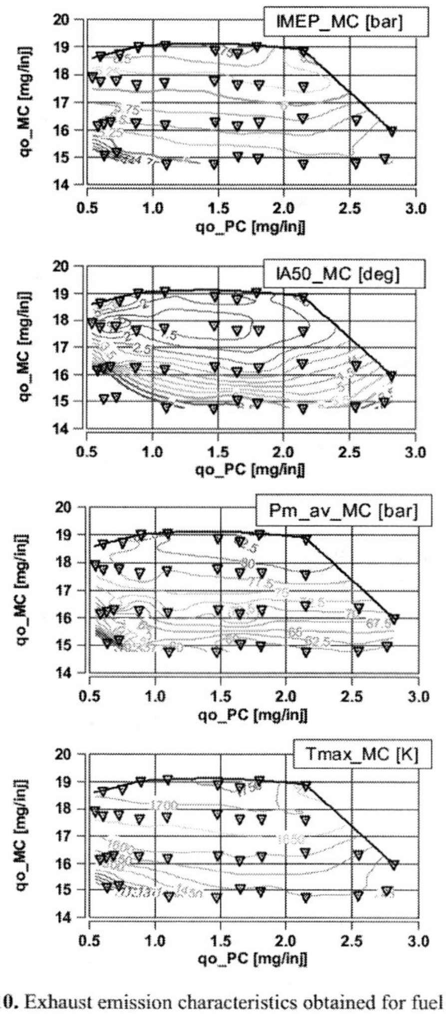

Fig. 10. Exhaust emission characteristics obtained for fuel dose inflammability research.

An emission characteristics of hydrocarbons (including methane) shows a decrease in emissions with the decrease in engine load. This is due to the improvement of the combustion process, but also due to the reduction of the combustion chamber flushing with flue gas that occurs during valve overlap. Highest hydrocarbon emissions occur at the low main dose and at low and high prechamber doses; in that case the emission values are more than double the value obtained for the maximum engine load (19 mg/inj).

Changes in carbon monoxide emissions are inversely proportional the changes in the initial dose size. For a large part of the created emission characteristics its value does not depend on the main dose. This character is evident at very small prechamber doses. Values below 1 mg/inj result in high carbon monoxide emissions. This is partly due to a lack of combustion (small initial doses) and partly from the lean mixtures when the conditions for the combustion process are insufficient.

The main result of the conducted research is determining the indicated overall efficiency value as it was determined based on the specific fuel consumption values. A fairly wide range of fuel dose variability was obtained, corresponding to high overall efficiency values. It has been proven that this efficiency decreases as the doses fed into the prechamber drop to minimum at low main doses. A similar tendency was also observed when injecting maximum doses into the prechamber.

6 Conclusions

Fuel dose inflammability investigation using a new combustion system, such as a single-fuel gas powered engine equipped with a prechamber and a main chamber, has been carried out. Lean mixtures combustion is possible in such an engine when injecting the minimum fuel doses to the prechamber with a minimum gas share of 3%. Below this value, combustion process irregularities are observed, as evidenced by the high CoV values (IMEP) exceeding 10%.

The combustion process research in terms of the fuel dose inflammability leads to the following observations and conclusions:

- the most preferred initial dose inflammability range irrespective of the main dose is observed for doses ranging from 1-2 mg/inj; in this range CoV (IMEP) reaches values up to 3%;
- change in the initial dose share in the total fuel dose has little effect on the change in the global air excess coefficient; but despite the small effect on the global value of this indicator, the effects of these changes have a profound impact both on engine performance and on emission factors;
- the peak cylinder pressure – as one of the best indicators for checking the combustion process accuracy – indicates the existence of an initial dose that can be used to accelerate the combustion process: it is a dose in the range 1-2 mg/inj, with smaller values corresponding to higher loads;
- engine performance maps indicate a small effect of the initial dose on the combustion process; the only

exceptions from this rule are operating points with minimal doses into the prechamber as well as small doses into the main chamber;

- emission characteristics maps confirm the previous conclusions, but the carbon monoxide emissions are the only ones that are clearly dependent on the initial dose size: high engine emission values are observed for engine operation at large initial doses, regardless of the main dose size;
- the overall indicated engine performance contains operating ranges with maximum values that are associated with the initial fuel dose range from 1-2 mg/inj and a base fuel dose of 16-19 mg/inj.

Based on these conclusions, it is not possible to clearly indicate the inflammability limits of the fuel mixture. Incorporating the emission factors and overall engine efficiency allowed to extend the limits of inflammability in a range that was not possible by just using the engine performance indicators (no main dose limits).

The research presented in this paper was performed within the European Research Project Horizon 2020, grant agreement No. 652816, Gas-Only Internal Combustion Engines. This research task was directed by Prof. K. Wislocki, the leader of Research Group. The authors wish to thank the EU for funding the work and express their acknowledgements to all persons, who have had their contribution in this study and the laboratory works.

References

1. M.I. Khan, T. Yasmin, A. Shakoor, Renewable and Sustainable Energy Reviews **51**, 785-797 (2015), DOI: 10.1016/j.rser.2015.06.053
2. M. Noga, B. Sendyka, International Journal of Automotive Technology **15**, 1083-1089 (2014), DOI: 10.1007/s12239-014-0112-9
3. M. Noga, Applied Sciences – Basel **7**, 295 (2017), DOI: 10.3390/app7030295
4. Z. Stelmasiak, D. Pietras, *IOP Conference Series-Materials Science and Engineering* **148**, UNSP 012087 (2016), DOI: 10.1088/1757-899X/148/1/012087
5. W. Smuga, L.J. Kapusta, A. Teodorczyk, Journal of Power Technologies **97**, 1-6 (2017)
6. R. Pyszczek, L.J. Kapusta, A. Teodorczyk, Journal of Power Technologies **97**, 52-60 (2017)
7. M.I. Khan, T. Yasmeen, M.I. Khan, M. Farooq, M. Wakeel, Renewable and Sustainable Energy Reviews **66**, 702-741 (2016), DOI: 10.1016/j.rser.2016.08.041
8. T. Korakianitis, A.M. Namasivayam, R.J. Cro-Okes, Progress in Energy and Combustion Science **37**, 89-112 (2011), DOI: 10.1016/j.pecs.2010.04.002
9. M.M. Tahir, M.S. Ali, M.A. Salim, R.A. Bakar, A.M. Fudhail, M.Z. Hassan, M.S. Abdul Muhaimin, Energy Procedia **68**, 355-362 (2015), DOI: 10.1016/j.egypro.2015.03.266
10. C. Bae, J. Kim, Proceedings of the Combustion Institute **36**, 3389-3413 (2017), DOI: 10.1016/j.proci.2016.09.009
11. M. Choi, J. Song, S. Park, Fuel **179**, 168-178 (2016), DOI: 10.1016/j.fuel.2016.03.099
12. M.A. Kalam, H.H. Masjuki, Energy **36**, 3563-3571 (2011)
13. J. Hall, M. Bassett, B. Hibberd, S. Streng, SAE Int. J. Engines **9** (2016)
14. R. Edwards, H. Hass, J. Larive, L. Lonza, H. Maas, D. Rickeard, European Commission Joint Research Centre Institute for Energy and Transport (2014)
15. Z.D. Ristovski, L. Morawska, J. Hitchins, S. Thomas, C. Greenaway, D. Gilbert, Journal of Aerosol Science **31**, 403-413 (2000), DOI: 10.1016/S0021-8502(99)00530-3
16. J. Alanen, E. Saukko, K. Lehtoranta, T. Murtonen, H. Timonen, R. Hillamo, P. Karjalai-Nen, H. Kuuluvainen, J. Harra, J. Keskinen, T. Rönkkö, Fuel **162**, 155-161 (2015), DOI: 10.1016/j.fuel.2015.09.003
17. W.P. Attard, E. Toulson, A. Huisjen, X. Chen, G. Zhy, H. Schock, SAE Technical Paper 2012-01-0823 (2012)
18. H. Blaxill, *Advanced Fuel and Engine Efficiency Workshop* (2017)
19. M. Schumacher, M. Wensing, SAE Technical Paper 2016-01-2176 (2016)
20. D. Assanis, N. Engineer, P. Neuman, M. Wool-Dridge, SAE Technical Paper 2016-01-2242 (2016)
21. A. Yousefi, M. Birouk, Applied Energy **189**, 492-505 (2017), DOI: 10.1016/j.apenergy.2016.12.046
22. M.D. Ruter, D.B. Olsen, M.V. Scotto, M.A. Perna, Fuel **91**, 298-306 (2012), DOI: 10.1016/j.fuel.2011.06.072
23. D.B. Olsen, J. Adair, B.D. Willson, *ASME Internal Combustion Engine Division 2005 Spring Technical Conference* (Chicago, 2005)
24. D.B. Olsen, A.T. Kirkpatrick, Journal of Engineering for Gas Turbines and Power **130**, 052802-052802-7 (2008)
25. P. Heuser, J. Geiger, S. Lauer, H. Sankhla, G. Ünver, A. Dhongde, *World Automotive Congress FISITA* (2016)
26. K. Tanoue, T. Kimura, T. Jimoto, J. Hashimoto, Y. Moriyoshi, Applied Thermal Engineering **115**, 64-71 (2017), DOI: 10.1016/j.applthermaleng.2016.12.079
27. I. Pielecha, W. Cieslik, Journal of Thermal Analysis and Calorimetry **126**, 815-827 (2016), DOI: 10.1007/s10973-016-5544-1
28. I. Pielecha, International Journal of Automotive Technology **15**, 47-55 (2014), DOI: 10.1007/s12239-014-0005-y
29. I. Pielecha, M. Skowron, W. Bueschke, W. Cieslik, K. Wislocki, *IOP Conference Series-Materials*

Science and Engineering **148**, UNSP 012081 (2016), DOI: 10.1088/1757-899X/148/1/012081

30. I. Pielecha, K. Wislocki, P. Borowski, W. Cieslik, Journal of Thermal Analysis and Calorimetry **122**, 473-485 (2015), DOI: 10.1007/s10973-015-4735-5
31. I. Pielecha, P. Borowski, J. Czajka, K. Wislocki, J. Kazmierowski, Journal of Thermal Analysis and Calorimetry **119**, 695-703 (2015), DOI: 10.1007/s10973-014-4139-y
32. M. Skowron, I. Pielecha, K. Wislocki, *IOP Conference Series-Materials Science and Engineering* **148**, UNSP 012085 (2016), DOI: 10.1088/1757-899X/148/1/012085
33. I. Pielecha, K. Wisłocki, W. Cieślik, P. Borowski, W. Bueschke, M. Skowron, SAE Technical Paper 2016-01-2305 (2016)

MATEC Web of Conferences **118**, 00011 (2017) DOI: 10.1051/matecconf/201711800011
VII International Congress on Combustion Engines

Overview of solutions and analysis of the ability to evaluate the performance parameters of unmanned aerial vehicles propulsion systems

Dominik Karpiński[1,*], and *Marta* Galant[1]

[1]Poznan University of Technology, Faculty of Machines and Transport, ul. Piotrowo 3, 60-965 Poznan, Poland

Abstract. The aim of aircraft engines development is the propulsion which is characterized by high power-to-mass ratio. Therefore, the alternative solutions that provide the required power by the low weight propulsion are sought after. The main advantage of these solutions is improvement of environmental and economic properties. This paper presents the overview of solutions and studies conducted for the unmanned aerial vehicles propulsion. For the purposes of studies a test bench was prepared. Its enables the comparison of the propulsion operating parameters taking into account changes in the values of thrust and propulsion power. The summary includes a proposal to improve the environmental indicators of propulsion systems for unmanned aerial vehicles.

1 Introduction

As a result of the growing popularity of unmanned aircraft, these objects are used increasingly more often. Currently, unmanned aerial vehicles are used both for military and civilian purposes. A significant group of currently used UAVs are used as a part of a hobby. Furthermore, unmanned aerial vehicles are used in the activities of: monitoring disaster, geodetic and meteorological measurements, in search and rescue, forensic finding, aerial photography, inspection of power lines and pipelines, as well as forest fire detection and monitoring [1-4].

The main task of propulsion system for unmanned aerial vehicles is to provide the thrust needed for flight. Small UAVs mostly use lithium-polymer batteries (Li-Po), while larger vehicles rely on conventional airplane engines. As a propulsion system for UAV piston, turbojet, turboshaft, electric and hybrid engines can be used. The type of used engine depends largely on the purpose of the UAV, e.g. as a pilotless target aircraft (high speed required), tactical recognition (high range and altitude), direct field recognition (low sound profile, low speed and altitude). However, the key issue is the reliability of the UAV. Research conducted by the US Department of Defense showed that most of the UAV malfunction was caused by power units and propulsion (38%). Second cause were problems with flight control (Fig. 1). The other are problems with communication, human and ground problems and miscellaneous causes [5].

For commercial and military airplanes alike as well as flying platforms, mainly internal combustion engines are used as propulsion. However, despite such a high popularity of this type of drive, manufacturers are increasingly looking for alternative solutions, which combining an internal combustion engine with an electric motor or, for smaller aircraft (platforms), to completely replace ICE with electric drives. Electric engines compared to ICE are much less complicated to use and operate.

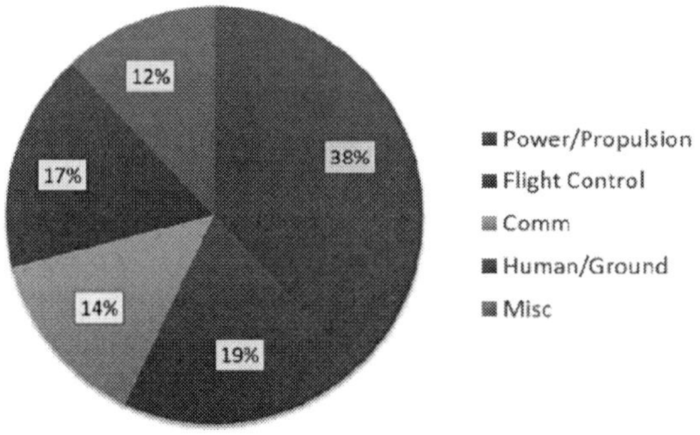

Fig. 1. Average sources of system failures for US Military UA fleet [1, 6].

Fig. 2. Electrical drive application in RQ-11 Raven [6].

* Corresponding author: dominik.p.karpinski@doctorate.put.poznan.pl

2 Prospects for the development of BPL propulsion units

The UAV global military market is dominated by pioneers in the United States and Israel. The US held a 60% military-market share in 2006 [7]. It operated over 9000 UAVs in 2014. From 1985 to 2014, exported drones came predominantly from Israel (60.7%) and the United States (23.9%). Top importers were The United Kingdom (33.9%) and India (13.2%). The leading civil UAV companies are currently (Chinese) DJI with $500m global sales, (French) Parrot with $110m and (US) 3DRobotics with $21.6m in 2014. As of March 2017, more than 770,000 civilian drones were registered with the U.S. FAA, though it is estimated more than 1.1 million have been sold in the United States alone [8, 9].

The introduction of new solutions to the aviation industry is a long term process involving advanced testing and certification procedures. This is due to the existing legislation, taking into account the highest standards of safety and reduce the negative impact of transport on the environment. The requirements imposed on manufacturers and unmanned aerial vehicles are mainly related to the improvement of their propulsion systems. The main goal of engine improvement is to increase economy by reducing service costs, reducing fuel consumption, and achieving the highest power-to-mass ratio. Military and civil aviation also play an important role in protecting the environment, thus reducing the toxicity of exhaust gases and reducing noise level, but in the case of UAV this is less important [10]. Particularly important to the development of propulsion units is to reduce their weight without lowering the received thrust. This is achieved by the use of more and more modern construction materials. In the UAVs weighing more than 50 kg to reduce engine mass boralyn alloys are used. It is a material developed for nuclear rocket heads and used for the production of Formula 1 engine car hulls [1]. This is a relatively inexpensive composite made from aluminum reinforced by boron carbide, making its density comparable to that of pure aluminum. However, in terms of endurance boralyn is second to none. It surpasses titanium, aluminum and steel.

In aviation, security is a very important issue. It is defined as the sum of the reliability of equipment and people. Therefore, in unmanned platforms, redundancy and duplication of certain components and circuits will continue to apply. The UAVs are also equipped with electronics so that when a malfunction is detected, the mission is interrupted and the platform is returned to the starting area.

In the literature [11] one can find the prospect of development of the unmanned actuators towards the use of internal combustion engine with rotary piston or with horizontally-opposed cylinders (boxer type). The engine will be controlled by a digital circuit that will send information about its parameters to the operator in real time. The engine intake system should be equipped with de-icing and dust removal equipment. Care should also be taken to limit noise emission, infrared radiation and harmful exhaust emissions. Seems realistic production of the unmanned aircraft weighing over 50 kg driven by electric motors powered by fuel cells. This will virtually eliminate both the emission of harmful exhaust and the emission of noise [12].

The greatest opportunity for the development of propulsions for the UAV is seen in the use of hybrid drive system. The use of hybrid (ICE-electric) drive in Unmanned Aerial Platforms brings together the advantages of the internal combustion engine (high energy value of aviation fuel) and electric motor (high energy conversion efficiency) [13-16]. Due to the differences between the vehicle's drive cycle and the aircraft flight profile, it is not possible to directly translate the advantages of the hybrid system from road to air transport. The aircraft flight profile is characterized by significantly lower load variations, so that energy recuperation and slow motion phase cannot contribute to improving fuel efficiency. The energy requirement for the aircraft is the highest during the takeoff and climb phases (Fig. 3). This entails the use of conventional aircraft engines to achieve the desired startup power. However, the takeoff and climb phases represent only a few percent of the time in the aircraft flight profile, so the internal combustion engines use only about half of their maximum power. When using a hybrid drive, the role of engines differs depending on the requirements of the propulsion system performance. For example, during an climb phase, the aircraft uses the energy stored in the batteries and the internal combustion engine is an additional emergency drive. It is also possible to use reverse configuration, in which the electric drive acts as an auxiliary unit. In the phases of flight with the highest energy demand, the system draws energy from both engines. These advantages indicate the appropriateness of carrying out work on adaptation of the hybrid propulsion system in the aviation industry.

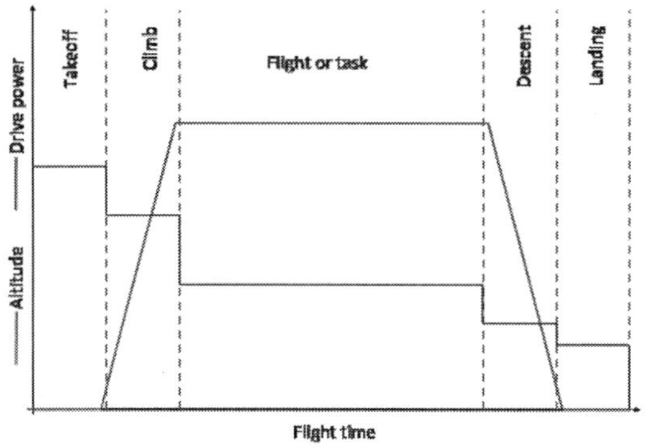

Fig. 3. Traditional aircraft flight profile.

The hybrid system intended for use in a flying object consists of five basic elements, of which there are distinguished: propeller as thrust generator, two power converters (combustion and electric engine), two power sources (hydrocarbon fuel and accumulators). The auxiliary components required for the operation of the drive system are the electric motor controller and the clutch.

DOI: 10.1051/matecconf/201711800011

At the stage of development of the hybrid propulsion system designed to adapt to flying objects, should take into account many factors that affect their opportunities for further exploitation. These include: safety of applied solutions, operating range of the vehicle, criteria and restrictions for such propulsions. Selection of the propulsion system intended for use in aviation is conditioned by criteria such as weight, system efficiency, system complexity, exhaust and noise emissions, manufacturing and operating costs. The mass criterion largely translates into object performance and associated economic (fuel consumption, maximum takeoff mass), logistic (required takeoff and landing procedures) and environmental (exhaust and noise emissions) aspects [17]. Unmanned aerial vehicles require the lowest possible mass of the propulsion system. The technology available today allows the electric motor to be used as a propulsion or generator engine, which reduces the number of required conversion energy sources, and above all reduces the weight of the entire system.

At the time of designing a hybrid propulsion system, a significant factor is the area of application of the flying object and, in particular, its flight profile. The aircraft flight profile is the basis for determining the energy demand in the hybrid propulsion system. Knowledge of the amount of energy and the time required for the particular phase of flight is key information for determining the configuration and operating conditions of the propulsion system. The above-described aircraft flight profile is a contractual profile, which depending on the object application area may be subject to various modifications.

Among the potential benefits of using a hybrid propulsion system are the following: extended flight time, improved energy efficiency, reduced exhaust and noise emissions. In addition, the hybrid propulsion system enables the use of smaller displacement internal combustion engine operating at the speed range corresponding to its highest efficiency. The electric motor used in the propulsion system is a source of additional power, especially during acceleration or climb phase, and serves as a generator in case stabilized flight conditions.

3 Research methodology

Currently, the most commonly used electric motors for unmanned aerial vehicles are brushless AC motors These types of motors are characterized by high efficiency and ease of control and operation. In addition, the advantage of brushless AC motors is the large power to mass ratio as well as the ability to operate in a wide range of engine shaft speed. Due to the requirements for the unmanned aircraft propulsion system, the electric motor should be of small dimensions and low weight and as efficient as possible As a result, three-phase AC electric motor EMP N6354/13 (Fig. 4) was selected for the research. Engine technical specification is shown in Table 1. The external rotor was used to obtain high torque values at small dimensions in the engine.

Fig. 4. Electric motor EMP N6354/1.

Table 1. Technical specification of the EMP engine N6354/13.

Parameter	Value
Power [W]	2500
Engine shaft speed/1 V [rpm]	250
No-load current [A]	1.7
Internal resistance [V]	53
Diameter [mm]	63.4
Length [mm]	57
Output shaft diameter [mm]	10
Engine weight [g]	558

A twin-blade propeller with a diameter of 457 mm and a pitch of 254 mm made of plastic were selected for determining the operating characteristics of the electric motor.

To power the electric motor two lithium-polymer batteries were used, 30 C 6000 mAh 18,5 V, connected in series. These types of batteries are characterized by: low mass, high operating voltage, low capacity loss at negative temperatures. During the tests, the electric motor and shaft speed controller used the same power source. This solution is the most beneficial considering the application of the propulsion assembly on the airplane, as the use of one power source does not increase substantially the entire weight of the power supply. Therefore, an arrangement of the test bench was adopted so as to reflect as much as possible the energy consumption by the device responsible for motor control.

A two-stroke boxer engine was second object (Fig. 5). The engine was manufactured by Zhejiang RCGF Model & Engine and marked RCGF 30. The main advantage of the selected engine is the large power to mass ratio. The technical specifications of the engine are shown in Table 2.

Fig. 5. Combustion engine RCGF 30.

 DOI: 10.1051/matecconf/201711800011

Table 2. Technical specification of the RCGF 30.

Parameter	Value
Engine capacity [cm^3]	30
Output power [W]	2800
Useful engine speed [rpm]	1500–7000
Ignition system	DC-CDI, supply voltage 4,8-6 V
Fuel system	Unleaded fuel + synthetic oil (30:1)
Weight [g]	970

The operating characteristics of the internal combustion engine were determined using a two-blade carbon propeller of the same diameter and pitch of the propeller as in the case of an electric motor propeller.

A specially designed test bench (Fig. 6) was used to determine the characteristics of the motors. During the tests the following parameters were recorded:

- thrust force,
- shearing force,
- engine shaft speed,
- electric current,
- input power of electric motor,
- fuel consumption.

Fig. 6. View of the test bench.

The test bench (Fig. 7) was equipped with two strain gauges capable of measuring the thrust force and the shearing force. Measurement of the electric motor speed was carried out with a sensor using the optical method. The test bench featuring the universal method of mounting the engines makes it possible to conduct tests of other models of electric motors intended for the model aircraft. In order to visualize the results, the sensors used during the measurements were connected to a Wobit programmable indicators MD150T (Fig. 8). These indicators are designed to work with strain gauges for measuring force, enabling the measurements with a resolution of 0.001%.

An optical method was used to measure the engine shaft speed of the electric motor. The engine shaft speed for combustion engine was measured by Hall sensor which supplied by the manufacturer.

The fuel consumption of the combustion engine was measured using the volumetric method.

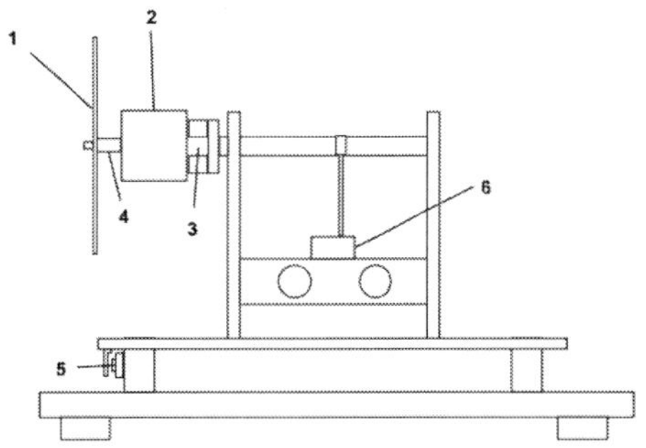

Fig. 7. Schematic of the test bench: 1 – tested airscrew, 2 – electric motor/combustion engine, 3 – shaft speed sensor (electric motor) , 4 – shaft speed sensor (combustion engine), 5 – thrust force sensor, 6 – shearing force sensor.

Fig. 8. View of the data loggers.

The research was divided into two parts. The first part of this study concerned measurements for an electric motor that was conducted for thrust force ranging from 10 N to 50 N. The data obtained out during the measurements was used to determine operating characteristics of the airscrew propulsion. On the basis of the obtained results were compared properties of the airscrews cooperating with a particular type of engine installed on the testing bench. Measurement of transverse forces enabled the determination of torque, which, together with the measured engine speed, was used to determine the output power of the engine.

The second part of the study included the measurements for combustion engine. Measurements were conducted for thrust force ranging from 30 N to 70 N. The assumed range of tested parameters was conditioned by the possibility of ensuring stabilized operation of the internal combustion engine at the test bench. As in the first part, the shearing force was used to determine the torque generated by the airscrew blades.

4 Analysis of the measurement results

The conducted measurements and calculations are presented in the form of dependence of the thrust force, transverse force, electric current, input power and output power on the electric motor shaft speed, and the efficiency as a torque function (Figures 9-12).

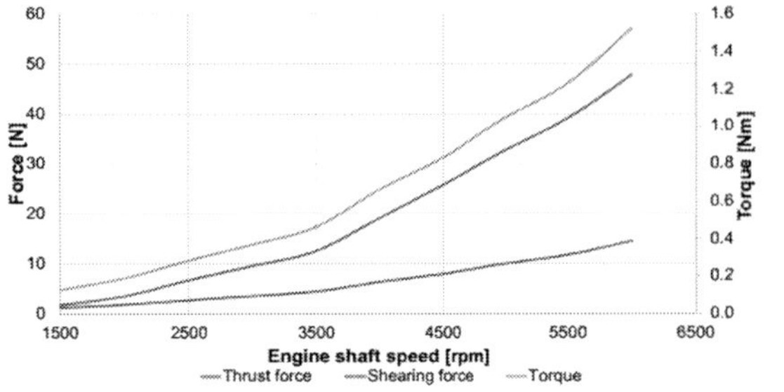

Fig. 9. Thrust, shearing force and torque as the engine shaft speed function.

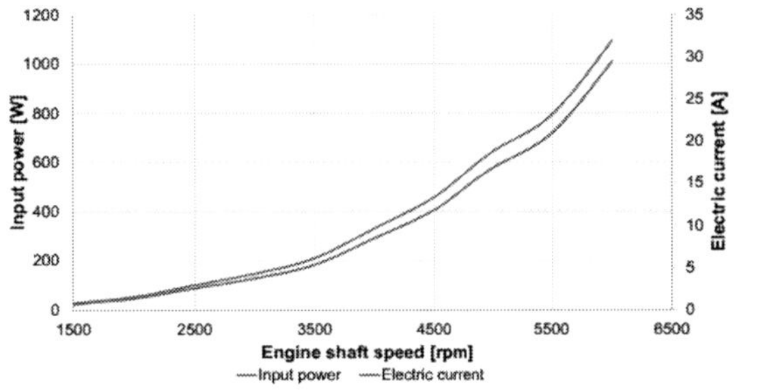

Fig. 10. Input power and electric current as the engine shaft speed function.

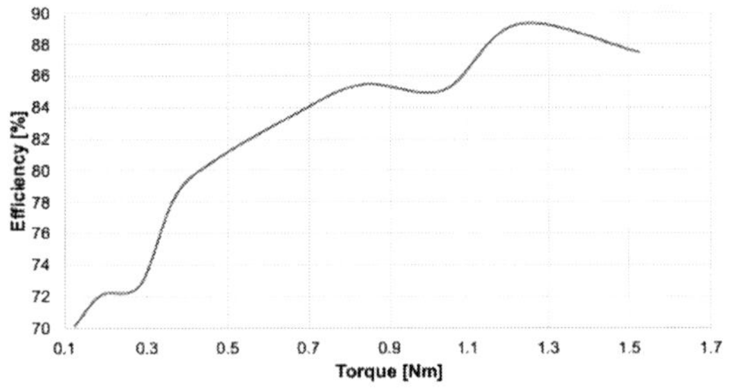

Fig. 11. Electrical motor efficiency as the torque function.

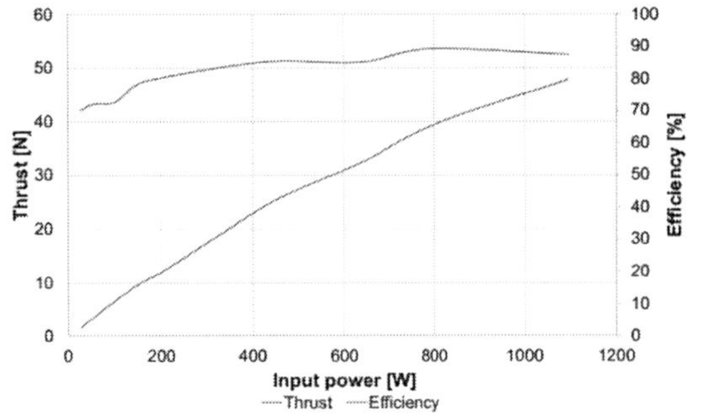

Fig. 12. Thrust as the input power function.

The thrust force is a component of aerodynamic forces generated on the airscrew blades. Obtaining the assumed values of the thrust force required different values of engine shaft speed. For a maximum engine speed of 6000 rpm, a thrust of 47.8 N was achieved using a two-blade propeller. For a maximum engine speed of 6000 rpm, a thrust of 47.8 N was achieved using a twin-blade propeller. For steady-state research, higher thrust and shearing force were obtained. During the measurement, torque was increased with increasing engine shaft speed. The torque curve as the engine shaft speed function allows to determine the impact of airscrew on the torque value. Depending on the parameters of the airscrew, it acts with greater or lesser resistance to the engine. During the tests, the maximum torque reached 1.5 Nm. The obtained thrust, shearing force and torque as the engine shaft speed based on rotational speed are similar to the exponential growth.

By analyzing the obtained input power values depending on the speed of the motor shaft, the highest energy consumption can be observed for electric motor, setting engine shaft speed of 6000 rpm, where the measured power was 1093.4 W and the current was 29.44 A. The energy consumed by motor from battery is due to the geometry of the airscrew, i.e. the coefficient of aerodynamic drag on the airscrew blades. Based on characteristics of the input and output power as a function of the speed differences can be observed between the electrical power supply of the motor, which is higher than the mechanical power obtained at the engine output. The output power of the motor according to the model depends on the torque and the speed of the motor shaft.

As the propulsive efficiency of the airscrew during the engine test in stationary conditions or at a very high flying speed has the value of zero, the propulsive efficiency of the engine was shown in relation to the torque. The obtained values of propulsive efficiency range from 70 to 89%. Lower values of propulsive efficiency were determined for low motor shaft speed settings on the controller. Operation of the engine at such parameters is associated with lower current intensity, below the high efficiency of the engine. The highest efficiency, reaching 89%.

As shown in Figure 12, the dependence of the input power on the produced thrust is an important information at the propeller selection stage. In the case of selecting a airscrew of equal diameter but larger pitch, it is possible to increase the power required to achieve the specified thrust. Also significant is the change in the efficiency curve of the electric motor depending on the type of airscrew mounted.

The conducted measurements and calculations in second part of the research are presented (Fig. 13–16) in the form of dependence of the thrust force, torque, output power, fuel consumption on the combustion engine shaft speed.

During the test, the maximum thrust (70 N) was obtained, for the engine crankshaft rotational speed of 6700 rpm.

Figure 14 shows the torque dependence on the rotational speed of the motor shaft. Analyzing the obtained characteristic curve, an increase in torque can be observed with the increase of the rotational speed of the shaft of the internal combustion engine. The diagram presenting dependence of the transverse force on the speed makes it possible to assess the force with which an airscrew influences the engine housing. The highest torque value of 2.2 Nm was achieved at maximum engine settings. Characteristic of torque to the internal combustion engine as the engine shaft speed function is similar to the exponential growth.

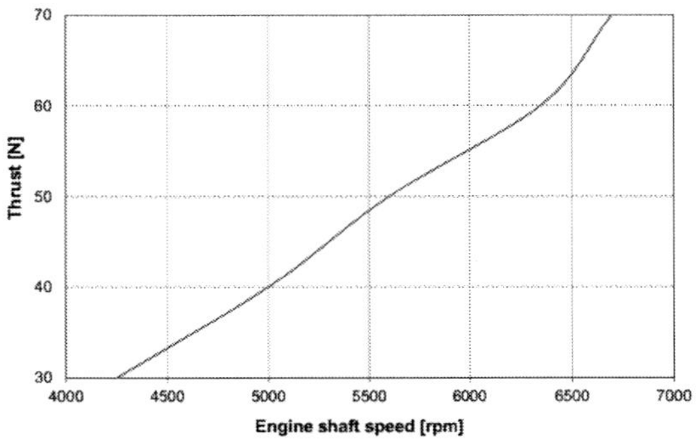

Fig. 13. Thrust as the engine shaft speed function.

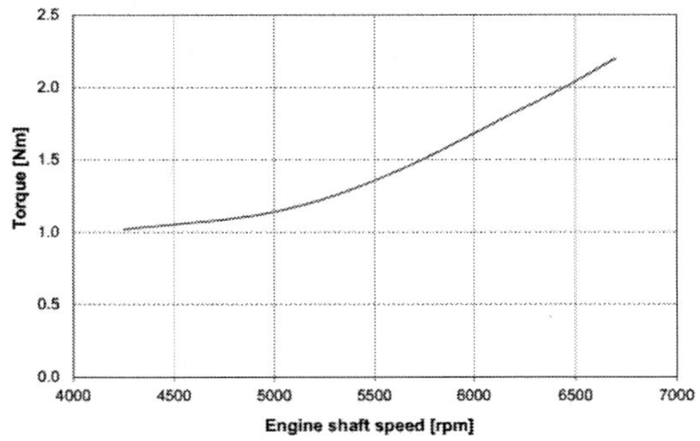

Fig. 14. Torque as the engine shaft speed function.

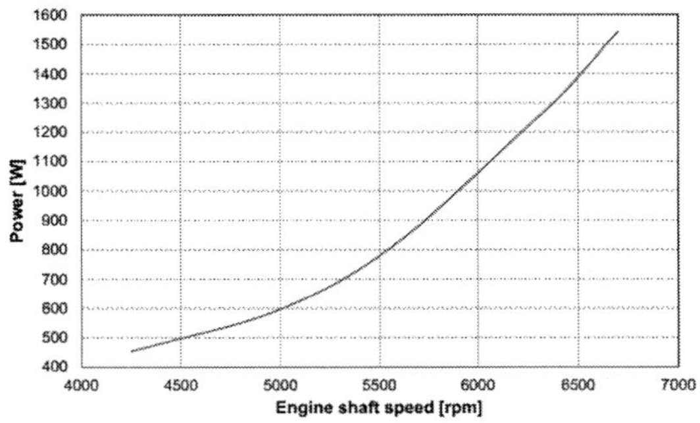

Fig. 15. Engine power as the engine shaft speed function.

The highest output power of the engine was obtained with a maximum shaft speed tested and a torque equal to 2.2 Nm. The value of useful power is closely related to the generated torque and engine shaft speed. The airsrcew propulsion in steady conditions, the power produced by the engine must balance the power absorbed by the airscrew. In this case, the geometry of the airscrew used and the flight conditions of the aircraft are of great importance.

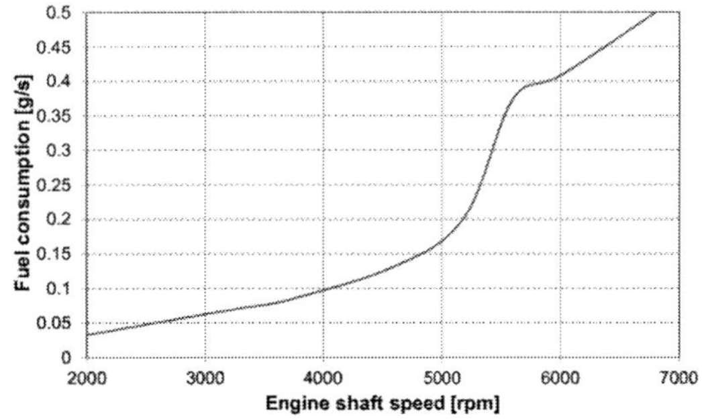

Fig. 16. Fuel consumption as the engine shaft speed function.

By analyzing the fuel consumption per second an increase of this parameter along with the increase of the rotational speed can be observed. The highest fuel consumption (0.5 g/s) was achieved, at the engine shaft speed of 6700 rpm and a torque of 2.2 Nm.

5 Summary

The dynamic development of air transport is a significant factor in the economic development of the world. However, it also generates an increase in nuisance and problems of particular relevance specifically in the local scale, especially near the airports. The source of problems are jet engines of passenger aircraft as well as General Aviation aircraft. However, for the last years a new type of aircraft started being used: UAV (Unmanned Aerial Vehicle). They are usually associated with military applications, but they are also used in civilian environments. In most cases the weight of the units used is in the range of 3 to about 15 kg. Depending on their size and destination, different types of propulsion systems are used. Because of its small mass, most UAVs are powered by electric motors up to 6 kg. In the ecological aspect, unmanned aerial vehicles has a small flying time compared to civil and military aircrafts. Therefore, statistically treated, the unnamed aerial vehicle are not a threat. Exhaust emission of this objects is incomparably lower.

Another aspect of the impact of unmanned aerial vehicle engines on the environment is the noise emission. Due to the low popularity and high independence of this type of unit – there is no platform certification for sports and leisure use – designers are not obliged to adhere to any standards. For military purposes, it is important that the UAV is characterized by the lowest noise emission. In conclusion – the emission of noise from the UAV is not an environment nuisance.

The operation of unmanned aerial vehicles can significantly reduce the global emissions and noise from air transport. In military operation, unmanned aerial vehicles can be particularly useful in tactical identification or direct recognition of a combat field. The use of the UAV possible to limit the flying time of multi-purpose airplanes typically characterized by high emissions and noise levels and increased radar and thermal detection. The costs of buying and using unmanned aerial vehicles are often incomparably lower, and the operation is much simpler and cheaper.

The tests conducted make it possible to specify the scope of required electric power needed to power the airscrew propulsion. Determination of the scope of power consumed by the electric motor powering the airscrew is one of the stages of the research leading to the construction of a hybrid drive system for the aircraft (Fig. 17).

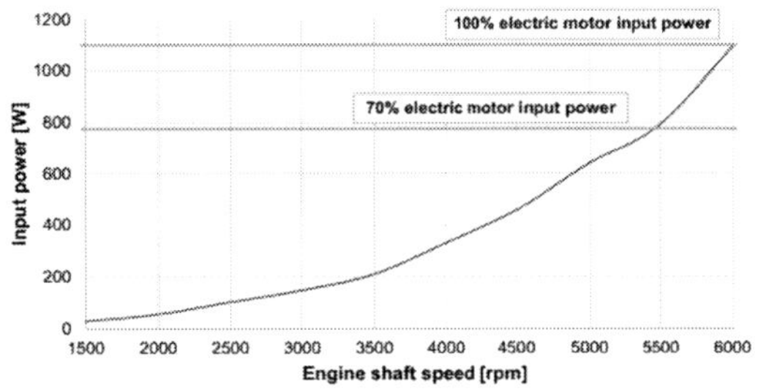

Fig. 17. Input power consumption of electric motor.

On the basis of the conducted measurements, it is possible to formulate the following assumptions:

- supply voltage of electric motor U_{dc} = 32–40 V,
- the energy delivered to the airscrew propulsion should ensure its operation with the maximum power of 1100 W to obtain the thrust force of 50 N,
- the power source should allow the maintenance of a constant electric power of 770 W, assuming that the power needed for flying comprises 70% of the maximum power,
- battery discharge protection below 80% of rated voltage.

On the basis of these requirements, it is possible to propose alternative propulsion system for the unnamed aerial vehicle which is a combination of electric motor and internal combustion engine. Hybrid drive systems shown in Figures 18 and 19 were designed in a CAD software.

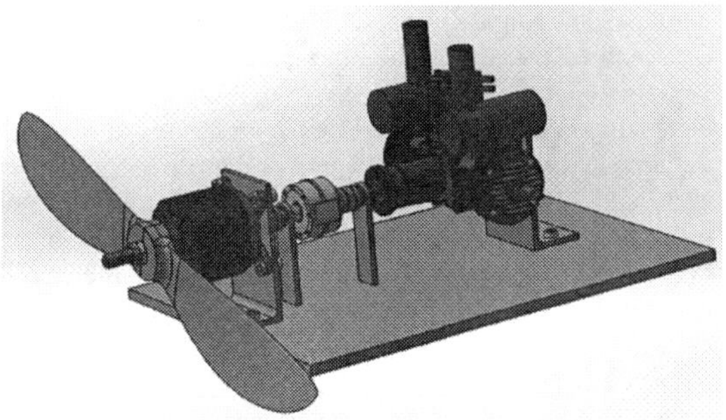

Fig. 18. Hybrid Drive System no. 1.

The configuration used in the hybrid drive system 1 assumes the use of an electric motor as the primary propeller unit. The combustion engine in this connection is an auxiliary power unit driving the airscrew when the electric motor is in generator mode. The use of an electromagnetic clutch allows the airscrew to be driven only by an electric motor powered by the energy stored in the batteries.

The hybrid drive system 2 has a configuration in which the combustion engine is the main drive unit. In this case, the engine should be operation in range of engine shaft speed corresponding to high general engine efficiency. The electric motor can act as an auxiliary unit responsible for replenishing periodic power shortages. The loading process can be accomplished during a stabilized flight profile, where the excess power of the internal combustion engine could be recovered by an electric generator operating in generator mode and converted into electric power to be supplied to the batteries.

Fig. 19. Hybrid Drive System no. 2.

At the stage of constructing the hybrid drive system, special attention should be paid to such system features as:

- maximum energy efficiency,
- minimum weight and size of simplicity of the structure for ease of operation and control,
- limited ecological impact (noise level, exhaust emission).

Also important is the selection of geometric parameters of airscrew to ensure the operation of two alternate energy sources in the widest possible range of engine shaft speed.

The study presented in this article was performed within the statutory research (contract No. 05/52/DSMK/0249).

References

1. R. Charchurski, Journal of KONES Powertrain and Transport **14** (2007)
2. I. Colomina, P. Molina, Journal of Photogrammetry and Remote Sensing **92**, 79-97 (2014)
3. Z. Czyż, M. Gęca, Logistyka **5**, 2015
4. P. Sawicki, Archiwum Fotogrametrii, Kartografii i Teledetekcji **23** (2012)
5. Z. Juda, M. Noga, *IOP Conference Series – Materials Science and Engineering* **148**, UNSP

012042 (2016), DOI: 10.1088/1757-899X/148/1/012042

6. Unmanned Aircraft Systems Roadmap 2005-2030 (Department of Defense USA, 2005)
7. Consumer Drones by the Numbers in 2017 and Beyond (www.newsledge.com, 2017)
8. The Consumer Drone Market: Trend Analysis (emberify.com, 2016)
9. The numbers behind the worldwide trade in drones (www.theguardian.com, 2015)
10. J. Merkisz, J. Markowski, J. Pielecha, *WIT Transactions on Ecology and the Environment* **174**, 219-230 (2013), DOI: 10.2495/AIR130191
11. J.Y. Hung, L.F. Gonzalez, Progress in Aerospace Sciences **51**, 1-17 (2012)
12. D. Taler, A. Cebula, Inzynieria Chemiczna i Procesowa **25**, 2331-2342 (2004)
13. M. Ehsani, Y. Gao, S.E. Gay, A. Emadi, Fundamentals, Theory, and Design (CRC Press, 2005)
14. I. Husain, Electric and Hybrid Vehicles Design Fundamentals (CRC Press, 2003)
15. J. Lieh, E. Spahr, A. Behbahani, J. Hoying, 47th Joint Propulsion Conference & Exhibit
16. J. Xu, APAS R&D Center (Hong Kong, 2007)
17. B. Uragun, I. Tansel, *International Conference On Unmanned Aircraft Systems* (Orlando, 2014)

MATEC Web of Conferences **118**, 00012 (2017) DOI: 10.1051/matecconf/201711800012
VII International Congress on Combustion Engines

Investigations on gas-air mixture formation in the ignition chamber of two-stage combustion chamber using high-speed Schlieren imaging

Wojciech Bueschke[1,*], *Maciej* Skowron[1], and *Krzysztof* Wislocki[1]

[1]Poznan University of Technology, Faculty of Machines and Transport, ul. Piotrowo 3, 60-965 Poznan, Poland

Abstract. Combustion of the lean mixtures in the spark ignition engines provides higher thermal efficiency compared to the combustion of the stoichiometric mixture but is more restrictive to the ignition systems. Due to the limitations of conventional ignition systems, advanced concepts are being used, e. g. spark-jet ignition. Presented research has been carried to determine: 1. The impact of fuel injection pressure on the velocity of mixture formation, 2. Fuel distribution inside ignition chamber in defined phases of chamber filling, 3. Influence of chamber back-pressure on gas jet development. Investigations have been carried using the ignition chamber providing optical access. The visualization has been done with Schlieren-method with "Z"-setup basing on two ϕ = 150 mm parabolic mirrors. Images have been recorded with LaVision HSS5 camera with CMOS transducer. The paper contains a comparison of gas penetration parameters for a different injection pressures and chamber backpressures. The injection into the quasi-static air has been compared to the injection in dynamic conditions. It is stated, that both injection pressure and chamber back-pressure influence gas jet-development in the ignition chamber. The regions of the chamber with increased swirling and therefore providing more efficient micromixing have been identified.

1 Introduction

Big worldwide demand for the fuel energy in combination with limitations in available crude oil resources is a significant fact by the intensification of research into alternative fuels. Between the alternative fuels, important advantages present the group in the gaseous state of matter, for example, relative high calorific value and octane number. It determines the rise of their exploitation as well as motivation to conduct the study on the ignition system for CNG combustion systems [1-3].

Combustions systems can be divided, in terms of air-fuel stoichiometry, into the rich-burn and lean burn. The lean burn combustion process (with high air excess ratio) shows in general higher thermal efficiency than combustion of rich mixtures [4, 5]. However, the concentration of fuel in the mixture impacts the energy demand to initiate the combustion process, as is it presented in Fig. 1.

According to the Fig. 1, the reduction of methane concentration under 8% causes a rapidly rising demand for the energy to ignite the mixture. The maximal portion of the energy transferred to conventional spark ignition system, consist a limitation in the igniting of very lean mixtures, over the $\lambda = 1.7$ [6]. An enhanced ignitability in the combustion system can be obtained using advanced ignition systems, as a laser, corona ignition or turbulent jet ignition. The last mentioned system is especially important due to the implementation of additional chamber resulting in different on-ignition mechanisms.

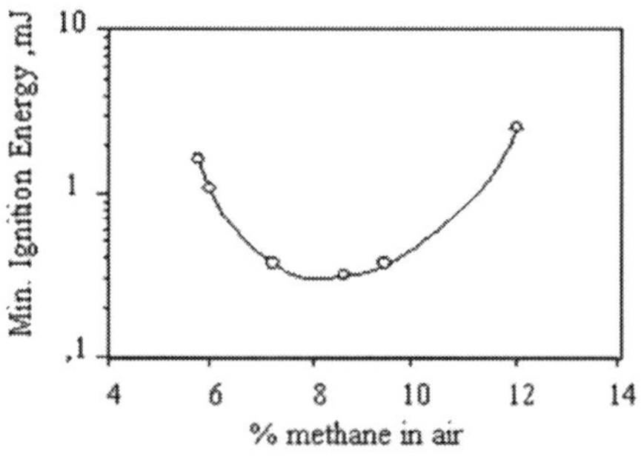

Fig. 1. Minimum ignition energy of the air mixture with methane [7].

The combustion system with spark-jet ignition is equipped with a pre-combustion chamber located in the cylinder head of an engine, presented in the Fig. 2.

As it is presented in Figure 2, ignition chamber (1) is assembled in the cylinder head (2) and located on the top of the main chamber (7). There is a spark plug (4) located in the enclosure. The fuel (5) is supplied to the ignition chamber volume (3), what will be described in next section. The combustion of pre-chamber charge is initiated with a conventional ignition source. The bottom

* Corresponding author: wojciech.bueschke@gmail.com

of ignition chamber consists the nozzles providing the mass transfer between the chambers. During the compression stroke, the charge from the main chamber is being transferred to the ignition chamber. Pre-combustion in this chamber results in pressure rise over the pressure in the main chamber and causes mass transfer into its volume. Hot jets penetrating the main chamber provide inflammation of the main charge.

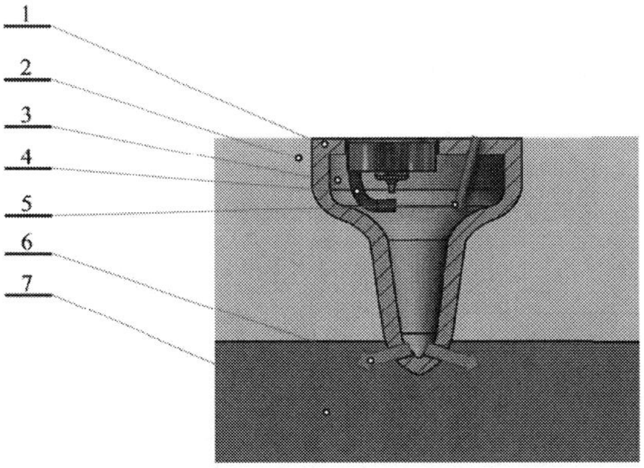

Fig. 2. Combustion system with spark-jet ignition.

Combustion in the main chamber is preceded by several consecutive processes occurring in the ignition chamber, influenced by mixture formation in it. However, the combustion process can be particularly described by indicators calculated from the indicating data, the optical analysis of mixture formation is necessary [8-10]. The registration of methane jet development in the air ambient can't be executed using direct high-speed imaging due to its transparency. Shadowgraph methods are required. The shadowgraph methods are based on the change in refractive index, which is given by:

$$n = \frac{c}{v_l} \tag{1}$$

c – light velocity in the vacuum ambient,
v – velocity of light in the analyzed medium.

The CNG-jet development in terms of variable injection pressure and chamber back-pressure has been analyzed within the study [11]. Jet has been visualized using Schlieren method in "Z" configuration. The stagnation pressure ratios (PR) of 2, 3, 4 and 5 have been examined. The analysis of jet evolution has been conducted, based on its tip – furthest point of the gas jet. Based on the results, in the early (transient) phase of the injection process, during the opening process, penetration is insensitive to nozzle pressure ratio. After the certain time, the velocity significantly rises when PR increased. Significantly shorter time to reach full registration length has been noted for increased PR. It is stated, that injection pressure impacts velocity of jet evolution, but the dependency doesn't show linear character.

As it was mentioned, the implementation of spark-jet ignition leads to the mixture inflammation on the way of another mechanism than conventional spark ignition. On ignition mechanisms of methane/air mixture using a hot turbulent jet have been examined within the study [12]. Schlieren imaging has been coupled with OH* chemiluminescence to distinguish the inflammation mechanisms. Two characteristic processes have been found:

- ignition by a reacting jet (flame ignition),
- ignition by a reacted jet (jet ignition).

In the first mechanism, the flame with active radicals is being transferred into the volume of the main chamber. The second mechanism is being executed, when the hot products of pre-combustion are penetrating the main mixture. The flame is being extinguished while passing through the nozzle, due to the rapid cooling and/or high stretch rate. Both mechanisms show differences in the ignition delay. The inflammation with reacted jets is slower because quenched flame requires more time to be mixed with cold ambient while reacting jet consists intermediate species promoting chain branching reactions. The appearance of mechanism, and therefore the capability to ignition depends on combustion conditions in the ignition chamber and the nozzle parameters. With the increase of nozzle diameter, there is a tendency to flame ignition, while smaller orifice diameter tends to the ignition by reacted jet.

Combustion systems equipped with pre-combustion chamber are known since early 20th century when Harry Ricardo constructed two-stroke SI engine equipped with "bulb" type ignition chamber to provide benefits from the combustion of stratified charges [13]. Currently, combustion systems with ignition chamber are being used to provide enhanced ignitability of ultra-lean mixtures [14]. Within the mentioned study, combustion indicators and flame development in the combustion chamber have been analyzed for the operation of an optically accessible engine equipped with spark ignition, which has been replaced and compared with the turbulent jet ignition. The main chamber has been fed with gasoline in both research campaigns. Ignition chamber of turbulent jet ignition was fueled with gaseous fuel – propane. The stability of combustion has been accessed based on the CoV(IMEP). When the limit in cyclic variations CoV(IMEP) < 5% is assumed, the extension of lean-burn stable operation from $\lambda \approx 1.3$ to $\lambda \approx 2.0$ has been achieved. The rise of thermal efficiency has been confirmed, up to $\lambda \approx 1.8$. From this air excess coefficient value a drop of combustion stability has been noted. In terms of higher mixture dilutions (over $\lambda \approx 1.4$), a rapid drop in NO_x emissions has been noted.

Optical analysis has been conducted to determine the velocity of flame propagation, calculated from registered flame luminosity. With the rise of mixture dilution, the slower rise of the average flame radius has been noted, parallel with smaller flame luminosity. Comparing the images of combustion initiated with conventional and TJI, in terms of $\lambda = 1.4$ and $\lambda = 1.8$ retrospectively, it can be found, that similar luminosity of combustion initiated with TJI in advance, despite higher air excess. Based on the literature review it can be stated, that the intensity of chemiluminescence light correlates with heat release

DOI: 10.1051/matecconf/201711800012

rate [15]. Authors of mentioned study noted and confirmed, therefore, drop of heat release rate with the rise of air excess ratio. The faster burn rate of combustion initiated by hot jets has been also observed.

2 Research methodology

2.1 Test-stand and Schlieren setup

Research has been carried using the experimental optical method. To visualize the development of CNG jet in the transparent air environment, Schlieren method has been used and "Z"-type setup has been formed (as shown in Fig. 3).

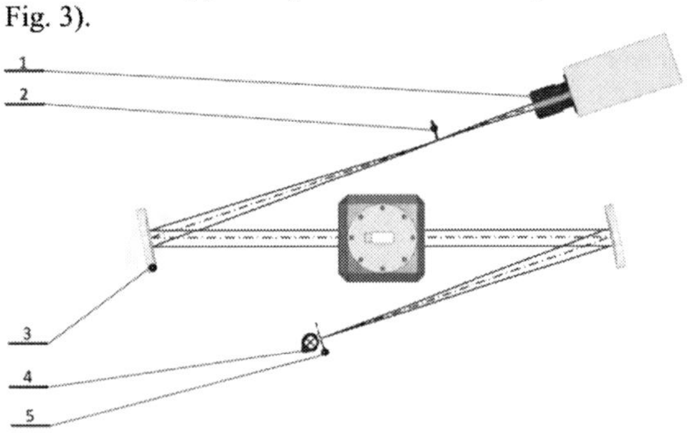

Fig. 3. Optical research setup for Schlieren investigations.

According to the Fig. 3, presented optical setup is based on two parabolic mirrors (3) with diameter $d = 150$ mm and focal length $f = 750$ mm. The registration of the images has been performed using high-speed camera – LaVision HSS5 (1) with a monochromatic CMOS-transducer set on registration frequency of $r = 5$ kHz. The vertically positioned knife edge has been used to create horizontal Schlieren cut-off (2). Light beam has been generated using 40 cd LED (4) with a lens diameter $d = 5$ mm. The beam angle for this diode equals 15 deg. The diode has been coupled with a circular aperture (5). The visualization area is located between parabolic mirrors, where the collimated light stream is provided. In this area, Constant Volume Chamber (CVC) has been mounted to ensure chosen value of back-pressure during the injection. Parameters of the CVC used to the investigations are collected in the Tab. 1.

Table 1. Parameters of CVC.

Parameter	Unit	Value
Volume	dm^3	1.7
Maximal pressure	MPa	10
Material	–	Steel
Optical access	–	$d = 80$ mm, from 5 perpendicular directions
Charge control	–	solenoid valves
Heating system	–	yes, electro-resistive heater

On the top of the CVC, optically accessible ignition-chamber has been mounted. Some constructional parameters of this chamber are described in the further chapter.

2.2 Optically accessible ignition chamber

Ignition chamber is located in the upper part of the combustion system and is connected to the main combustion chamber. Its construction has to provide the possibility to supply fuel to the chamber (scavenged variant) and to deliver external energy necessary to the inflammation of mixture. This auxiliary energy can be transferred to the charge using spark plug or heating plug [16]. The investigation on the variant equipped with spark plug will be discussed within this study. The optically accessible model of such a chamber has been constructed (as shown in Fig. 3).

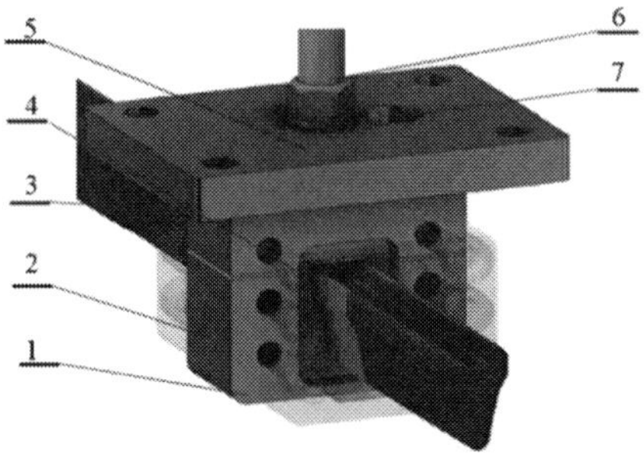

Fig. 4. Isometric CAD-model of optically accessible ignition chamber.

The main shape of the ignition chamber has been machined in the bottom part (no. 1, on the Fig. 3). Due to the implementation of optical access providing use of i. a. Schlieren method, the shape of this model represents an elongated cross-section through the base ignition chamber and provides an optical path (3). The upper part of the chamber (5) contains M12 spark plug (6) from the modern gas engine. Fueling of the chamber has been conducted using gas injector mounted in the slot (7). Mentioned injector slot is connected to the internal volume of the chamber with near-vertical ($\alpha = 3.5$ deg) channel ($d = 1$ mm) and check valve separating internal chamber volume from the fuel supply system. The internal volume of the chamber is restricted with quartz glass windows with thickness $g = 20$ mm. Components are sealed with PTFE-gaskets. The bottom part consists nozzles ($d = 1.5$ mm) providing mass transfer between interiors of main chamber and ignition chamber. The construction of chamber ensures the possibility to conduct the research on injection and acyclic combustion.

2.3 Scope of research and data analysis

Research has been carried to investigate the impact of the injection pressure (p_{inj}) and the pre-chamber back pressure (p_{ch}) on the mixture formation. The experiment to detect the impact of the first mentioned parameter on mixture formation has been conducted for three values of pressure p_{inj} measured upstream the injector:

- 5 bar abs,
- 7 bar abs,
- 9 bar abs.

P_{inj} was adjusted using pressure reducer mounted on the vessel with methane N35. The influence of the p_{ch} has been investigated in terms of injection to the chamber with atmospheric pressure and 2 bars overpressure. Raw images consist a monochromatic flat representation of the volumetric processes occurring in the pre-chamber. Two-dimensional images have been post-processed in the Davis software to amplify optical signal and reduce background noise. Images exported to a format compatible with Fiji ImageJ software have been converted to volumetric representation to provide better identification of a front of the gas jet and share of fuel-rich regions in the whole interior of the chamber (compare with Fig. 5a). The analysis of four operating points has been conducted, based on the traces of parameters calculated for jets tip, presented in the Fig. 5b.

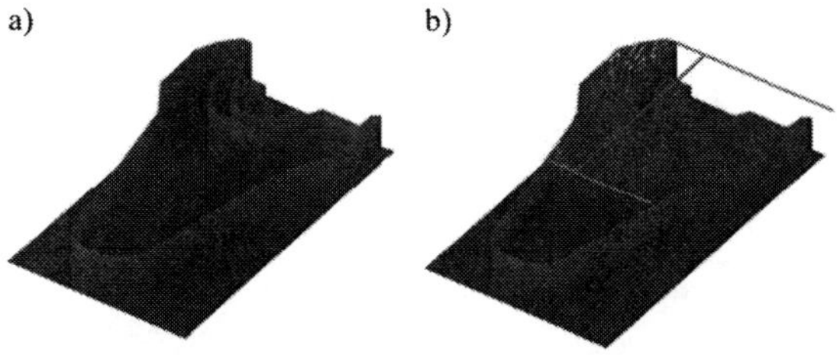

Fig. 5. Extruded outline of cross-section through chambers interior (a), representation of jets vertical penetration Ly (b).

Regarding Fig. 5b, jets length L_y has been calculated from the jets furthest point y_{tip} and the horizontal top surface y_0 of the chamber's internal volume:

$$L_y = y_{tip} - y_0 \qquad (2)$$

and resolved over time.

The point of fuel delivery is located on the horizontal top surface, on the side of spark plug's mass electrode (compare with Fig. 4). The velocity of jets tip has been calculated from the formula

$$V_y = \frac{dL_y}{dt} \qquad (3)$$

where dt – time period between registration of two consecutive pictures, and resolved over L_y. It leads to the characterization of jets dynamics in relation to the position in the chamber.

Three main sources of uncertainty have been noted in this research:

- repeatability of injectors operation,
- the turbulent behavior of jet,
- post-processing uncertainty.

Fuel injector was operated using open, full-variable control system with steady temperature conditions and acyclic operation, it's high repeatability can be assumed. The impact of gas-dynamic phenomena has been reduced to repeating the operating points and calculation of mean value. However, the optical data has been presented from the single operating point. The uncertainty of post-processing depends on the definition of furthest point of registered density inhomogeneity. This uncertainty has been rated as 5 pixels, what represents 1.7% of jets length in the middle of chamber vertical fulfillment.

3 Analysis of optical data

Based on the Schlieren-theory, the optical signal (shadows and enlightened areas) occurs when the local inhomogeneity lead to the refractive deflection of incident light and change in index n [17]. This change is observed in terms of temperature differences, pressure gradients or variations in chemical compositions. According to this study, Schlieren signal is positively influenced by the differences in temperature caused by the drop of CNG pressure. The further reason of occurrence of Schlieren-signal is placed in the gradients of pressure caused by the expansion of gas supplied with higher pressure compared to the pressure of charge air in the volume of the chamber. Jets development has been presented in the Fig. 6.

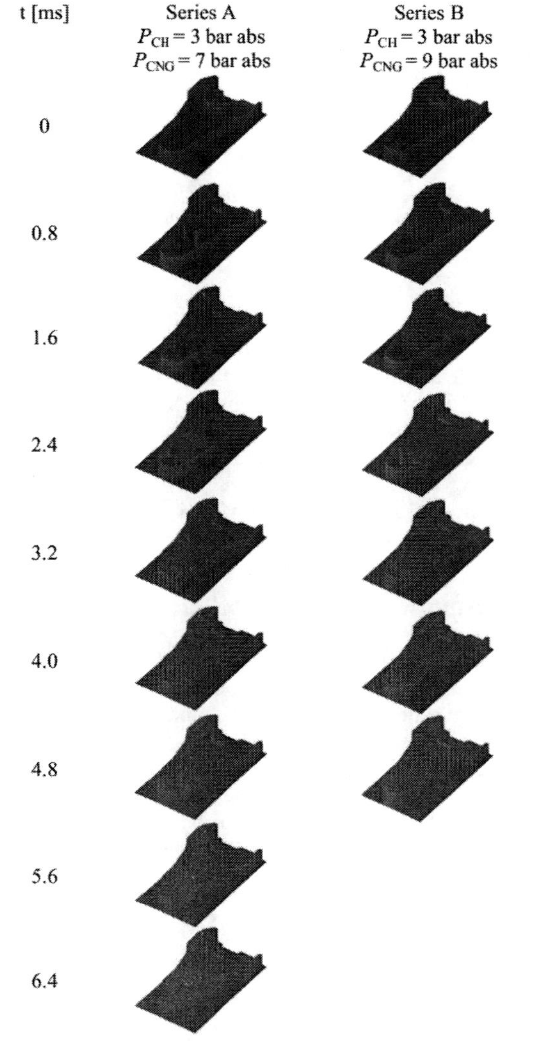

Fig. 6. The development of gas jet in the ignition chamber.

Series located on the left side (series A) has been registered in terms of chamber overpressure of 2 bars and 7 bars absolute pressure in the CNG rail. Series presented on the right side (series B) shows jets evolution under CNG-rail pressure increased to 9 bar abs. The image on the beginning of injectors control signal has been shown in the previous chapter and represents the outline of ignition chamber with spark-plug elements, which consists an obstacle on the way of expanding fuel. The first occurrence of the signal with luminosity differences has been marked as a timestamp 0. In this moment, a wider area of the chamber has been covered with fuel for the series B, where the CNG rail pressure was higher. From the analysis of pictures registered 0.8 ms later it can be stated, that in the case of series B fuel expands deeper in the horizontal direction, while comparable vertical penetration. It means intensified expansion of fuel in the area of the spark plug and the faster beginning of mixture formation in the spark gap volume under increased pressure in fuel supply. This character of horizontal penetration causes fuel-enriched regions occurring close to the wall of the chamber, which can be also observed in further steps. Comparing the total series of registered images, faster expansion in the volume of the chamber can be observed for the operating point executed under increased rail pressure. The bottom has been achieved in shorter time period. As it was mentioned, the intensity of Schlieren-signal is influenced by the pressure gradients. The regions of highly intensified luminosity have been identified, mainly in the top of chambers volume. The occurrence of these spots is combined with the constructional feature of the ignition chamber – organization of fuel supply. Additionally, this effect can be caused by the turbulent movement around the mass electrode of the spark plug. Next regions with identified higher luminosity are located in the middle and lower part of ignition chamber, where the area of horizontal chambers cross section becomes smaller. It leads to the rise of pressure combined with rise of flow velocity. The dynamics of gas jet expansion will be assessed in details in the next chapter.

4 Dynamic behaviour of gas jet in the chamber

The dynamics of gas jets motion has been assessed based on the position of the furthest point, which has been calculated from registered Schlieren-signal and named jets-tip. Vertical positions of this point, resolved over the time, have been presented in the Fig. 7a. Analyzing the diagram of injection into the chamber with atmospheric pressure under p_{inj} = 5 bar abs (blue line), the calculated time to reach the bottom of the chamber equals t_b = 2 ms. In the case of operating point with increased p_{inj} (green line), the time period to reach the bottom of the chamber was significantly shorter, while the early stage of injection shows comparable character. This situation finds the justification for intensified horizontal expansion in terms of increased CNG-rail pressure.

The rise of p_{ch} represents a real engine conditions in retarded moment of injection. Analyzing the data, significant changes in jets evolution (case marked with a red line on Fig. 7a) are observed – approximately 4.5 – fold longer time period is required to cover the chamber with expanding fuel. Similar to the atmospheric case, increased injection pressure (orange line) leads to shorter expansion time combined with a comparable vertical course jets development in the early phase. Location-resolved values of velocity have been presented in the Fig. 7b. The rapid rise of the V_y has been found in the initial phase of fuel expansion in the chamber. In the lower part of the chamber, lower V_y has been noted. The source of this drop can be the rise of chamber pressure in terms rapid expansion of additional mass into the volume of the chamber.

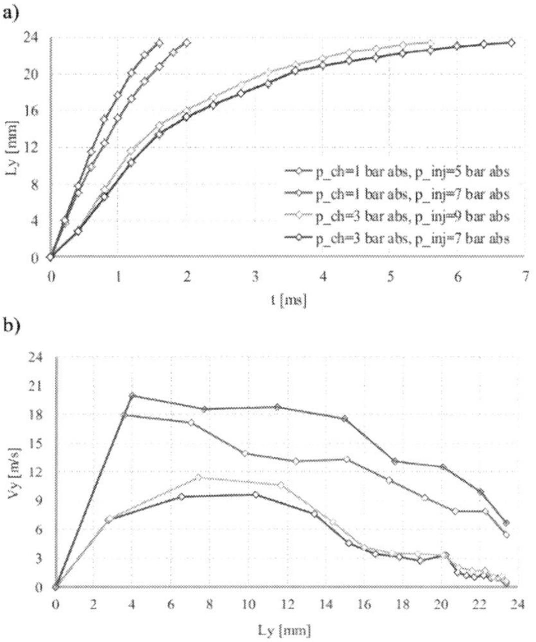

Fig. 7. Time-resolved vertical length of the gas jet (a) and distribution of vertical component of velocity over the position in the chamber.

Alternatively, the drop of supply pressure combined with wave-effect, but to confirm this phenomena research including fast analysis of pressure in fuel supply is required. Higher velocities of the jets tip have been calculated for the injection into the lower p_{ch} due to the smaller density of ambient charge and therefore smaller static mass breaking the expanding gas. Independently from chamber back-pressure, increased p_{inj} resulted in shifted (for later) point of maximal velocity. In the lower part of the chamber (17 mm of distance from the fuel supply point), the slower drop of the velocity has been noted. This has a connection with the issue identified on optical data and finds its justification in the convergent construction of lower part of the chamber causing the increase in pressure and expansion velocity.

5 Summary and conclusions

Within the study, the optical research method has been applied to the mixture formation analysis in the ignition chamber of the spark-jet ignition system. The applicability of the Schlieren method in "Z"-configuration to the research on injection of gaseous fuel into the ambient air has been confirmed. Based on the optical data, the parameters of jets tip have been calculated and dynamics of gas expansion in the chambers internal volume has been assessed. The impact of both chamber back-pressure and pressure in fuel supply system on the mixture formation has been evaluated. Shorter time to cover the chamber volume with fuel, promoting better homogenization of mixture has been found in terms of increased pressure in fuel supply. The regions of intensified motion of the charge have been identified.

The research can be continued and complemented by the fast measurement of pressure in the fuel supply system, and its analysis to understand better the impact of gas-dynamic phenomena on mixture formation.

The study presented in this article was performed within the statutory research (contract No. 05/52/DSPB/0246).

References

1. BP Statistical review of World energy (2014)
2. J. Merkisz, J. Pielecha, W. Gis, *Proceedings of the Ninth Asia-Pacific International Symposium on Combustion and Energy Utilization*, 477-482 (2008)
3. J. Merkisz, J. Pielecha, W. Gis, *WIT Transactions on Ecology and the Environment* **120**, 439-449 (2009), DOI: 10.2495/SDP090411
4. S.K. Chen, SAE Technical Paper 2003-01-0638 (2003)
5. M. Noga, B. Sendyka, International Journal of Automotive Technology **15**, 1083-1089 (2014), DOI: 10.1007/s12239-014-0112-9
6. Y.T. Kato, K. Saeki, H. Nishide, JSAE **222**, 365-368 (2001)
7. D. Bjerketvedt, J.R. Bakke, K. Van Wingerden, Gas explosion handbook **52**, 24 (1997)
8. M. Noga, Applied Sciences – Basel **7**, 295 (2017), DOI: 10.3390/app7030295
9. I. Pielecha, International Journal of Automotive Technology **15**, 47-55 (2014), DOI: 10.1007/s12239-014-0005-y
10. I. Pielecha, Journal of Thermal Analysis and Calorimetry **118**, 217-225 (2014), DOI: 10.1007/s10973-014-3956-3
11. A. Haji, D.M. Edgington-Mitchell, D. Honnery, S. Mosafa Mirsalim, Fuel **184**, 100-109 (2016)
12. S. Biswas, S. Tanvir, H. Wang, L. Qiao, Applied Thermal Engineering **106**, 925-937 (2016)
13. D. Morrison, The Piston Engine Revolution 153-176 (2011)
14. W.P. Attard, E. Toulson, A. Huisjen, X. Chen, G. Zhu, H. Schock, SAE Technical Paper 2012-01-0823 (2012)
15. M. De Leo, A. Saveliev, L.A. Kennedy, S.A. Zelepouga, Combust Flame **149**, 435-447 (2007)
16. A. Hanenkamp, S. Terbeck, S. Köbler, MTZ **67** (2006)
17. W. Merzkirch, *Flow visualization* (Academic Press, 1974)

MATEC Web of Conferences **118**, 00013 (2017) DOI: 10.1051/matecconf/201711800013
VII International Congress on Combustion Engines

Analysis of design parameters of pistons and piston rings of a combustion engine

Grzegorz Koszalka[1,*], and *Andrzej* Suchecki[2]

[1]Lublin University of Technology, Faculty of Mechanical Engineering, ul. Nadbystrzycka 36, 20-618 Lublin, Poland
[2]Automotive Research and Development Institute BOSMAL, ul Sarni Stok 93, 43-300, Bielsko-Biała, Poland

Abstract. This paper analyzes the effect of gaps in the lock of the first and second sealing rings and selected dimensions of the ring belt of the piston on the piston-ring-cylinder seal. The research was performed using a mathematical model of the seal system. Simulation results have shown that a significant reduction in the exhaust gas flow to the crankcase – without compromising other parameters characterizing the performance of the seal – would require increased production rigor. Reducing the blow-by without the need to narrow the tolerance intervals of the engine components is possible, but this would be accompanied by increased reverse gas flow from the inter-ring space towards the combustion chamber. The research has also shown that the ring seal of the automotive compression ignition engine, which was the starting point for testing, was optimized for minimizing gas flow towards the combustion chamber at the expense of increased flow of exhaust gases to the crankcase.

1 Introduction

The choice of dimensions for the ring part of the piston and of the piston rings that determine the size of the clearance between the rings, the piston and the cylinder is crucial to the proper functioning of the ring pack of an internal combustion engine. Due to the various functions the sealing system must fulfil and the variable working conditions, optimization of this system is difficult. Dimensions that cause a lower exhaust gas flow rate are not the most advantageous in terms of oil consumption, or resistance to motion and engine durability [1]. Therefore, it is necessary to seek compromise solutions. Improving the piston-ring-cylinder (PRC) structure only via empirical research is very expensive and does not always produce satisfactory results. This is due to the high cost of producing samples and the large number of variants to examine, as well as difficulties with regard to the assessment of changes made to the engine. Assessed parameters such as blow-by, quantitative engine oil consumption, exhaust gas pollutant concentrations, mechanical losses, etc. are characterized by low measurement repeatability, and taking into account the relatively small expected changes resulting from structural changes made to the system, obtaining reliable results requires extended testing, ideally carried out on multiple units of the test engine. Therefore, in design work mathematical modeling is employed. Currently, advanced models allow testing of other structural parameters on the PRC under different engine operating conditions [2-12].

Several mechanisms for the transport of engine oil towards the combustion chamber are mentioned in the literature: entrainment in the form of oil mist together with gases flowing through the seal of the PRC towards the combustion chamber; oil scrapping through the piston rings towards the combustion chamber; oil throwing due to inertial effects in the vicinity of TDC. The oil entering the combustion chamber leads to increased concentrations of hydrocarbons and particulate in the exhaust gas [1, 12-15].

Some researches point out that the oil entrainment in the form of oil mist with the gases flowing in the direction of the combustion chamber may be the main cause of oil consumption [16-17]. To limit return flows, efforts are made (inter alia) to reduce the pressure in the gap between the first and second compression rings. These actions can, however, cause increased blow-by.

Many modern engines are characterisied by relatively large flows of exhaust gas into the crankcase (blow-by). The engine that was the starting point for the results presented in this work had the following blow-by: at maximum torque more than 30 dm^3/min per dm^3 of the engine's swept volume (in terms of power – 1 dm^3/min/kW) and at maximum power: 20 dm^3/min/dm^3 and 0.5 dm^3/min/kW, respectively [18].

The existence of this relatively large blow-by was the motivation for undertaking research with the aim of evaluating the engine design solutions adopted. This paper is limited to the assessment of the impact of selected dimensions of the ring part of the piston and the clearance in the compression ring locks on the blow-by and the intensity of the reverse gas flowing from the ring seal towards the combustion chamber (blow-up).

* Corresponding author: g.koszalka@pollub.pl

DOI: 10.1051/matecconf/201711800013

2 Research object and methods

2.1 Research object

The research engine was a 4-cylinder, turbocharged diesel engine with a displacement of 1.3 dm^3 with intake air cooling and exhaust gas recirculation. According to its technical specifications, the maximum power of the engine is 55 kW, obtained at 4000 rpm and the maximum torque is 190 Nm at 1750 rpm. The engine has 4 valves per cylinder and a Common Rail injection system with direct fuel injection. The engine has a cast iron block with cylinders of nominal diameter 69.6 mm; the stroke of the piston is 82 mm. The aluminum pistons have a cast iron insert under the first compression ring. The ring seal consists of three rings: firstly a rectangular compression ring with a chrome-plated barrel face surface; secondly a tapered compression ring; and a double-lipped oil ring with a chromium-plated face surface and a helical spring. Nominal gaps in the ring locks are: first compression ring – 0.25 mm, second compression ring – 1.25 mm and oil ring – 0.375 mm.

2.2 Mathematical model

The mathematical model of PRC sealing presented in [19] was used for the simulation studies. This model is an integrated model of gas flow through the apertures of the PRC system, displacement and twisting of the rings in the piston grooves and the oil film between the rings and the cylinder liner. The calculations presented in this study do not take ring twisting into account and the oil film sub-model was not employed; frictional forces between the rings and the cylinder were calculated from empirical correlations (see [19]).

The input data for the calculations were determined on the basis of the engine's documentation, MES calculation (thermal deformation of the piston and the rings) and actual measurements (combustion chamber pressure). Calculation of the cylinder liner's thermal deformation was taken into account, but it was simplified to assume that the diameter of the deformed liner was constant over its entire height.

2.3 Simulation testing

The dimensions of the clearance in the locks of the compression rings and the dimensions of the piston shown in Fig. 1 were varied in the execution of the simulations. The influence of these dimensions on the blow-by and blow-up, as well as the axial displacement of the rings in the grooves and the pressure in the inter-ring spaces (output values) were assessed.

In the first step, the impact of each dimension on the output values was evaluated, with the other dimensions remaining unaltered. In the second phase, taking into account the results obtained in the first stage, the effect of simultaneous changes of several dimensions on the performance of the sealing system was assessed. In the simulation individual dimensions varied over a wide range – but remained within the limits of realistic implementation.

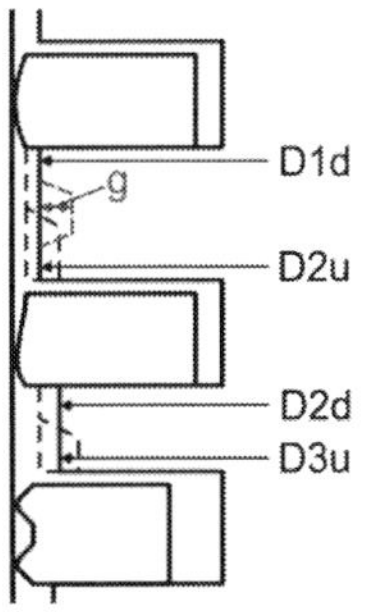

Fig. 1. Dimensions of the piston used as variables in the simulation tests.

3 Results and discussion

3.1 Impact of selected dimensions on PRC sealing

The effects of the clearance in the lock of the first and second compression rings on blow-by and by-up under rated conditions (full load at 4000 rpm) are shown in Figures 2 and 3. In the figures the larger symbols show the values corresponding to the starting dimensions, based on the documentation provided by the engine manufacturer. As expected, reducing the clearance in the locks causes the blow-by to decrease, with the effect of the clearance in the lock of the first ring being much larger than the clearance in the second ring. Reducing the clearance in the lock of the first ring also reduces blow-up, while reducing the clearance in the lock of the second ring results in very high increase in the rate of blow-up. The reason of this increase in the blow-up is the increase in pressure in the space between the first and second rings caused by the reduction of the clearance in the second ring lock (Fig. 4). The pressure increase in the first inter-ring space above the pressure in the combustion chamber forces the gas flow from this space towards the combustion chamber.

Pressure changes in the inter-ring spaces also affect the displacement of the rings within the grooves. With low clearance values in the lock of the second ring, the first compression ring changes its axial position in the groove during the compression stroke and at the end of the power stroke (Fig. 5). It is considered that displacement in these strokes of the cycle is unfavorable and may lead to so-called collapse of the ring, which is accompanied by a near-complete loss of its gas-tightness [4, 20]. In turn, higher values of the pressure in the first inter-ring space may positively affect the displacement of the second ring. At certain values of these pressures (as for second ring clearance equal to 0.25 mm, hence the same as the clearance of the first ring in the engine design), the second ring does not move at all and during the entire engine work cycle it is adjacent to the lower edge of the groove (Fig. 5).

DOI: 10.1051/matecconf/201711800013

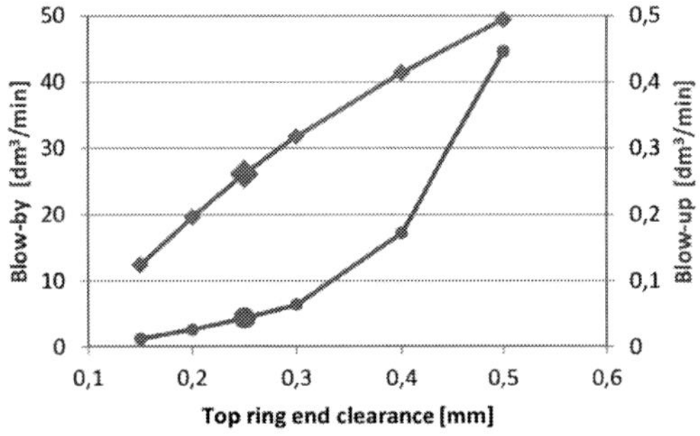

Fig. 2. The influence of the clearance in the lock of the first compression ring on the blow-by and blow-up.

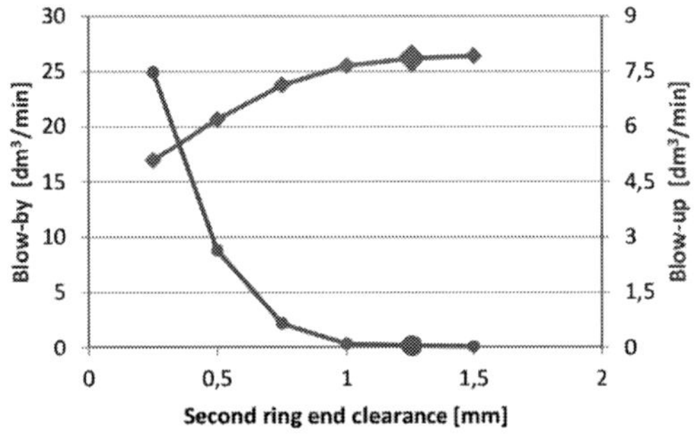

Fig. 3. The effect of the clearance in the lock of the second compression ring on the blow-by and blow-up.

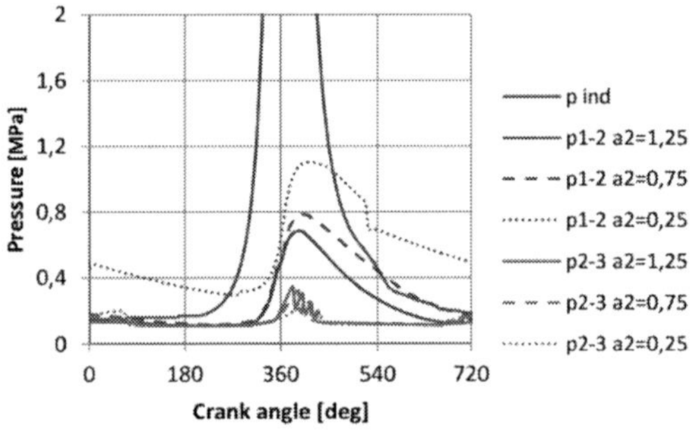

Fig. 4. Impact of clearance in the lock of the second compression ring on pressure in the inter-ring spaces.

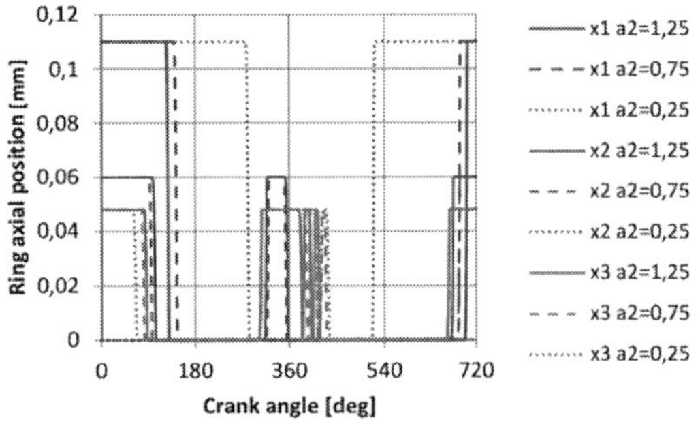

Fig. 5. Impact of clearance in the lock of the second compression ring on the axial displacements of the rings in the grooves.

The diameters of the piston in the vicinity of the ring grooves determine the cross-sectional areas through which the gas can flow through the ring locks. Their impact on the gas flow is therefore somewhat similar to the impact of the clearances in the ring locks. The greater the diameter of the piston under the first groove *D1d*, the lower the blow-by and blow-up (Fig. 6). Increasing the diameter of the piston under the second groove *D2d* also reduces the blow-by, but much less than in the case of *D1d*. Unfortunately, increasing this diameter is highly detrimental to blow-up (Fig. 7).

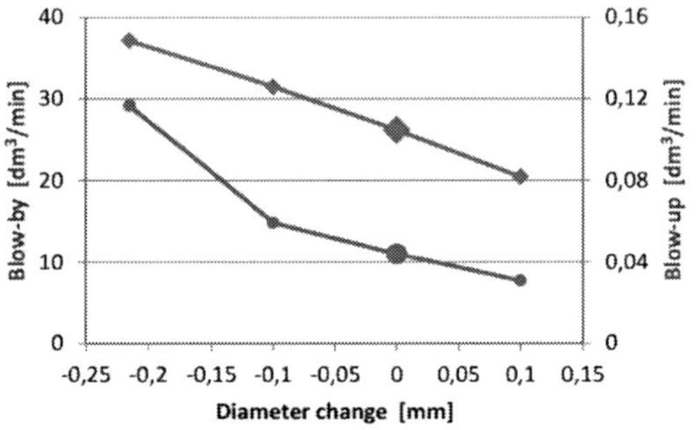

Fig. 6. Influence of piston diameter under the first compression ring *D1d* on the blow-by and blow-up.

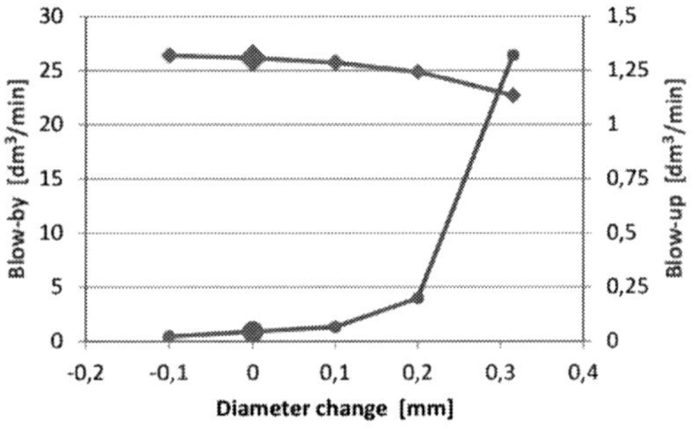

Fig. 7. Influence of piston diameter under the second compression ring *D2d* on the blow-by and blow-up.

The diameter of the piston over the second groove *D2u* has a significant effect on the volume of the first inter-ring space, and hence on the pressure at that location. On the other hand, it affects the active cross section of the gas flow through the lock of the second ring when it is adjacent to the upper flank of the groove. The combined effect of these factors – despite the significant changes in the pressure in the inter-ring space and in the displacement of the second ring (Fig. 9) – makes the impact of this diameter on blow-by and blow-up relatively small and non-monotonic (Fig. 8).

The effect of using an additional groove made in the inter-ring space to increase the accumulative capacity of this space was also simulated (see blue line in Fig. 1; such an approach is sometimes used in engines, especially larger ones). Increasing this volume results in a reduction of the maximum pressure in the inter-ring space, which has a positive effect on the blow-by. At the same time, however, the values of this pressure decrease much more slowly during the exhaust stroke, which adversely affects the blow-up (Fig. 10).

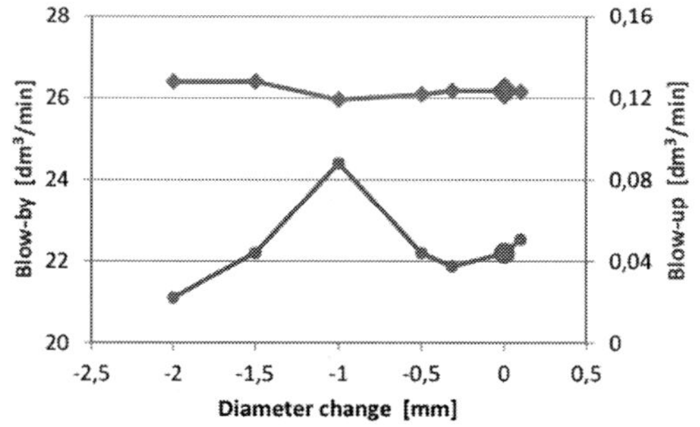

Fig. 8. Influence of piston diameter over the second compression ring $D2u$ on the blow-by and blow-up.

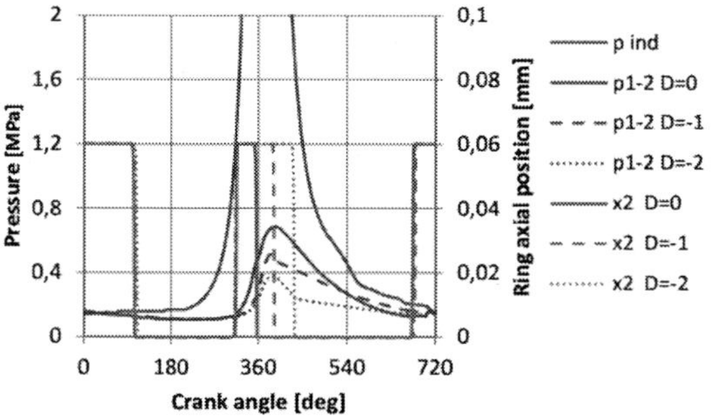

Fig. 9. Influence of piston diameter over the second compression ring $D2u$ on pressure in the inter-ring spaces and the axial displacement of the rings in the grooves.

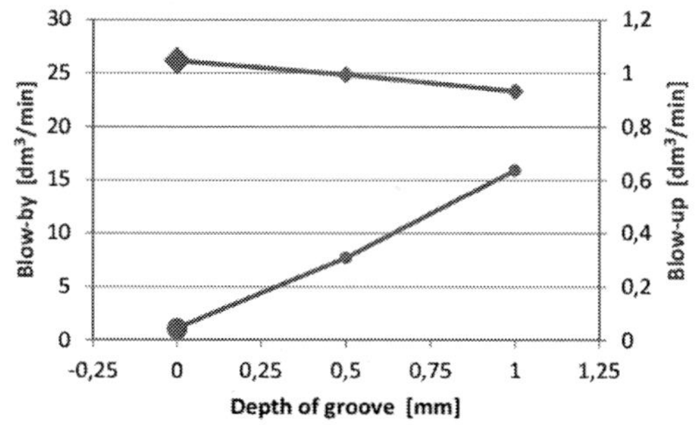

Fig. 10. The effect of the groove depth in the space between the first and second ring on the blow-by and blow-up flows.

3.2 Combined impact of different dimensions on the PRC seal

The influence of different combinations of piston and ring dimensions on blow-by and blow-up were analyzed. The analysis has shown that the greatest benefit, both from the point of view of blow-by and blow-up, can be achieved by reducing the clearances in the region of the first compression ring. Reducing the clearance in the lock of the second ring and increasing the diameter of the piston below it does not have such a large beneficial effect on blow-by, but at the same time adversely affects blow-up. Similarly, the use of an additional groove to increase the volume of the inter-ring space reduces blow-by, but with a simultaneous increase in blow-up, with the negative impact of the groove on the blow-up being greater when the clearance in the lock of the second ring is smaller. The impact of different dimensional combinations on both blow-by and blow-down is shown in Figure 11. Unless otherwise stated in the figure or its caption, this means that for a given variant the given dimension is equal to the initial dimension (similarly in Fig. 12).

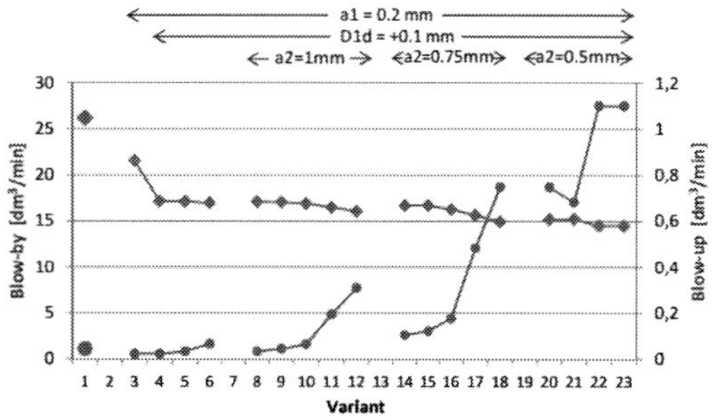

Fig. 11. Effect of different variants on the blow-by and blow-up variant 1 – initial dimensions accepted by the manufacturer; $D2u$ = +0.1 mm for variants: 5-6, 9-12, 15-18 and 21-22; $D2d$ = +0.1 mm for variants: 10, 12, 16, 18, 22 and 23; g = 0.5 mm for variants: 6, 11-12 and 17-18.

Reducing the clearance in the lock of the first ring by 0.05 mm and increasing the diameter of the piston below it by 0.1 mm would reduce blow-by and blow-up by more than a third. At the same time, reducing the clearance in the lock of the second ring and changing the dimensions of the piston in its surroundings allow a further reduction of blow-by, but at the expense of increased blow-up (Fig. 11). Unfortunately, such a reduction in the clearance in the lock of the first ring and an increase in the diameter of the piston under it could lead – at an unfavorable combination of elements and under very unfavorable operating conditions (heavy load, cold engine) – to elimination of the clearance altogether, causing engine damage. The introduction of such a change would probably require an increase in production rigor (tightening of the piston, ring and liner tolerances).

As a result of the above, in the next step an analysis was conducted to improve the performance of the seal without reducing the clearance of the first ring lock and increasing the diameter of the piston between the first and second rings. The analysis showed that there was a possibility of significant reduction in the blow-by, but at the expense of increased blow-up (Fig. 12). Reducing the clearance in the lock of the second ring to make it equal to the clearance of the first ring and increasing the diameter of the piston below it to dimensions that should not hinder the draining of oil, and making the groove in the inter-ring space would reduce the intensity of blow-by by more than a third. Unfortunately, such changes would result in a massive increase in blow-up, to a numerical value close to that of the blow-by. In addition, such changes would adversely affect the axial displacement of the first ring in its groove. It should be emphasized that in this study the influence of only some structural parameters on the effect of the seal were analyzed. Other dimensions, in particular dimensions of the rings other than clearance in the locks and the dimensions of the piston ring grooves may also significantly affect the performance of the PRC seal.

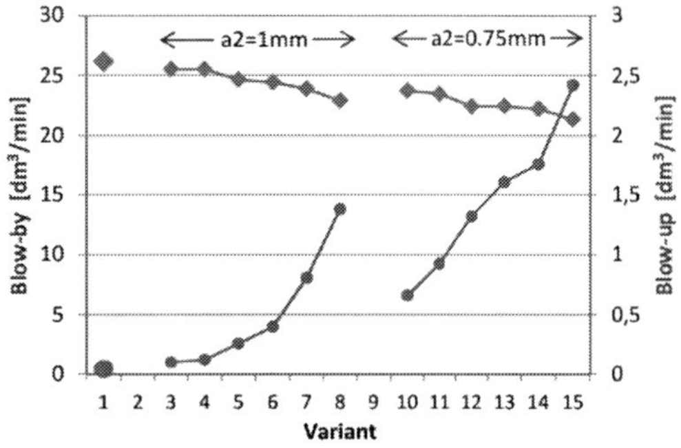

Fig. 12. Effect of different variants on the blow-by and blow-up variant 1 – initial dimensions; $D2u$ = –0.2 mm for variants: 4, 6, 11 and 13; $D2d$ = +0.1 mm for variants: 5, 6, 8, 12, 13 and 15; g = 0.5 mm for variants: 7-8 and 14-15.

4 Summary

This paper presents the results of simulations of the impact of clearance in the compression ring locks and of selected dimensions of the ring belt of the piston on the functioning of the piston-ring-cylinder system. The study was made for a small automotive Diesel engine. The simulation results have shown that a significant reduction in the rate of gas flow from combustion chamber into the crankcase (without deteriorating other parameters characterizing the functioning of the ring pack) would require a reduction of the clearance in the first ring lock and the piston diameter under this ring. Unfortunately, such a change would require a tightening of engine component production tolerances.

Reducing the blow-by – without needing to increase the rigor of the production process – is possible by reducing the clearance in the lock of the second ring and the dimensions of the piston in its vicinity. However, this would be accompanied by an increased reverse flow of gas from the inter-ring space towards the combustion chamber. This could therefore lead to increased oil consumption and thereby higher concentrations of hydrocarbons and particulate matter in the exhaust gas.

The results of the research also showed that the dimensions of the rings and piston analyzed for the test engine had been optimized to minimize blow-up, at the expense of increased blow-by.

References

1. A. Ito, K. Tsuchihashi, M. Nakamura, SAE Technical Paper 2011-01-1402 (2011)
2. G. Koszalka, Eksploatacja i Niezawodnosc – Maintenance and Reliability **48**, 72-81 (2010)
3. G. Koszalka, Eksploatacja i niezawodnosc – Maintenance and Reliability **51**, 40-44 (2011)
4. P. Lyubarskyy, D. Bartel, Tribology International **104**, 352-368 (2016)
5. J. Merkisz, J. Mizera, M. Bajerlein, L. Rymaniak, P. Maj, Experimental and Applied Mechanics **518**, 102-107 (2014), DOI: 10.4028/www.scientific.net/AMM.518.102
6. J. Merkisz, J. Pielecha, W. Gis, *WIT Transactions on Ecology and the Environment* **120**, 439-449 (2009), DOI: 10.2495/SDP090411
7. W. Serdecki, P. Krzymień, Combustion Engines **161**, 53-63 (2015)
8. A. Usman, C.W. Park, Energy Conversion and Management **117**, 115-131 (2016)
9. M.P. Veettil, F. Shi, SAE Technical Paper 2011-01-1406 (2011)
10. A. Wolff, Tribology Transactions **57**, 653-667 (2014)
11. P. Wróblewski, A. Iskra, Combustion Engines **167**, 38-52 (2016)
12. E. Yilmaz, T. Tian, V. Wong, J. Heywood, SAE Technical Paper 2004-01-2909 (2004)
13. A. Oliva, S. Held, Tribology International **101**, 98-109 (2016)
14. I. Pielecha, W. Cieslik, Journal of Thermal Analysis and Calorimetry **126**, 815-827 (2016), DOI: 10.1007/s10973-016-5544-1
15. B. Thirouard, T. Tian, SAE Technical Paper 2003-01-1952 (2003)
16. P.J. Burnett, SAE Technical Paper 920089 (1992)
17. C. De Petris, V. Giglio, G. Police, SAE Technical Paper 961216 (1996)
18. G. Koszalka, A. Suchecki, Combustion Engines **154**, 34-39 (2013)
19. G. Koszalka, M. Guzik, Polish Maritime Research **21**, 66-78 (2014), DOI: 10.2478/pomr-2014-0043
20. S. Przesmitzki, T. Tian, SAE Technical Paper 2008-01-0794 (2008)

DOI: 10.1051/matecconf/201711800014

Preliminary analyzes in terms of the possibility of reducing energy consumption by the SM42 locomotive used in track works

Pawel Daszkiewicz[1,*], and *Maciej* Andrzejewski[1]

[1]Rail Vehicles Institute TABOR, ul. Warszawska 181, 61-055 Poznan, Poland

Abstract. Track operations require specialized machinery and equipment. Among them, the SM42 locomotive is an indispensable vehicle. Locomotives of this type have a significantly different nature of work compared to locomotives used in passenger and freight traffic. A large percentage share of the work is engine running under idle and under low load. The nature of the load is, among other things, the cause of high fuel consumption. Such a situation calls for a number of necessary technical and organizational activities, leading, inter alia, to reducing energy consumption. The article provides preliminary operational analyzes aimed at identifying the potential for reduce energy consumption by the SM42 locomotive, supporting track works. As a result of reducing the locomotive's downtime with running engine by half, fuel consumption was reduced by approximately 16%.

1 Introduction

For decades, the primary power source for vehicles used in shunting operations, in most cases for locomotives, were compression ignition (CI) engines of obsolete designs, characterized by high values of effective power and torque, but which often do not meet any emission standards and generate high maintenance costs. Taking into account the vehicles designed for the toughest applications, their engines are further characterized by a large displacement volume, which results in higher fuel consumption and more exhaust gases [1]. It follows, therefore, that the operation of shunting locomotives can cause high energy consumption and substantial emissions, hence the need to pay special attention to issues of reducing their negative environmental impact [2-4].

Performing works within the modernization of railway lines requires the use of specialized machinery and equipment, often equipped with large-volume compression-ignition internal combustion engines with high power outputs. Inevitably, this involves significant fuel consumption and significant emission of harmful pollutants in the exhaust gases [5, 6]. Negative environmental impacts of activities performed during the construction or modernization of railway infrastructure, further enhances the often very outdated machinery with combustion engines that meet the old and fairly liberal emission standards [7, 8].

Therefore, any action that improves the efficiency of track machines in terms of fuel consumption and emission of harmful exhaust gases is valuable [9, 10]. The authors of this paper have made an assessment of the possibilities of improving the operation of these machines. The paper presents an analysis of working conditions and fuel consumption by the SM42 diesel shunting locomotive, which is exactly used for rail works (Fig. 1).

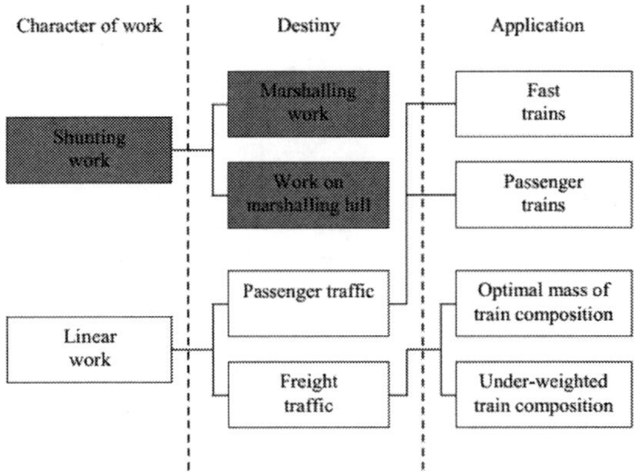

Fig. 1. The basic use of the SM42 diesel shunting locomotive.

It should be noted in the discussed aspects that during operation of diesel locomotives in real conditions, the energy utilization rates of the locomotive are significantly different from those achieved at full power, especially when there is a very high proportion of engine idling. It is also important that in the control system of the locomotive driving system, the control is discrete, determined by the position of the driving adjuster.

* Corresponding author: p.daszkiewicz@tabor.com.pl

2 Shunting work

2.1 Regulations

According to railway regulations, based on the Instruction in the techniques of shunting operations (R-34) shunting includes all intended movements of railway vehicles and related operations on railway tracks with the exception of the train entry, exit and transit. Adjusting the train position at the station from one track to another, pushing it into the desired position (reversing, pulling) after previously stopping on the track, should be considered as shunting maneuvers. A shunting locomotive set are locomotives that are coupled with the rail vehicle performing the shunting maneuvers or multiple unit performing shunting operations. The shunting vehicle may be behind, before or between other units. If the station carries out planned shunting maneuvers in coupling, decoupling and remodeling of train sets, as well as the use of station and track sidings associated with it, such stations are usually assigned their own shunting locomotives to carry out these operations [11].

Shunting locomotives generally have one cabin, which can be used for driving in both directions. These vehicles could be with electric drives or combustion engines.

The goals of shunting operations include [12]:

1. moving carts into separate tracks,
2. coupling of cart sets,
3. changing a group of carts in a set,
4. moving carts into position for loading and unloading,
5. bringing and taking back of empty passenger carts onto tracks next to platforms and parking them,
6. moving carts between tracks,
7. picking out selected empty carts with distinct characteristics,
8. moving carts into position for special operations such as: cleaning, washing, decontamination, weighing, repairs, etc. and bringing these cars back onto platforms after they are done,
9. taking carts out of trains sets when they are faulty and would impact the train travel safety.

2.2 Shunting methods

The following shunting methods and maneuvers are used [12]:

1) positioning, where the operation (shunting) involves the moving the cart set onto the desired track (place) and the decoupling of any selected carts from the set after it comes to a stop,
2) loose shunting (singular or set), where singular loose shunting means that the shunting locomotive, when pushing a set of carts, stops after reaching the intended speed, which causes the unconnected carts to drift apart and roll onto the desired track; set loose shunting on the other hand involves speeding up of the shunted set after the decoupling of the first connection up to a speed that ensures the shunted set reaches its destination. Then the locomotive stops to allow the decoupled cart set to move away, after which it speeds up again to allow for the detachment of another shunted cart or set,
3) hump shunting, it involves pushing carts off an inclination or a marshalling hill.

Pushing the carts can take place after completion of shunting on the track, or to connect to the cart set in the process of coupling the train. Placement (shunting) can be performed by a: road vehicle, shunter, shifter, or other technical device designed for the movement of rail vehicles [13].

2.3 Shunting speed

According to the instructions Ir-9/R34/ shunting maneuvers are performed most frequently on the tracks occupied by the rolling stock, and therefore to be accomplished with the speed needed to maintain the safety of people, vehicles and cargo.

Shunting vehicle speed must not exceed 25 km/h, with the exception of the following cases [12, 14]:

1) 40 km/h – when driving a rolling railway vehicle or a train set when the vehicles are pulled on an empty track, of which the driver has been notified; does not apply to driving through crossovers,
2) 20 km/h – when moving through unmonitored crossings,
3) 15 km/h – when moving a train set forward on a main track at a downward slope of over 2.5 thousandth, and when the shunting vehicle could not be placed on the lower end of the whole set, and when pushing a set towards a marshalling hill when signaled to do so,
4) 10 km/h – when performing maneuvers with carts occupied by passengers, with dangerous goods, with the exception of marked labels No. 8 or 15 according to the RID, and tanks marked with an orange strip, for which the speeds are smaller, as well as for carts that exceed gauge or are extremely heavy weighing more than 60 tons, in one set after first notifying the driver by the shunting operations head. Speed does not apply to carts occupied by passengers when the shunting is performed on a secured track [15, 16],
5) 5.4 km/h – for uncoupled carts coming down from a marshalling hill towards a stationary set at automated stations,
6) 5 km/h – if the technical instructions do not specify further speed reduction:
   - when pushing carts off of a marshalling hill after the signaling permitted operation "at moderate speed",
   - when uncoupling carts with the uncoupling rod,
   - when shunting carts loaded with dangerous goods, marked with a warning label No. 8 or 15 according to the RID, and tanks marked with an orange strip,
   - when driving the shunting train set forwards when an employee could not take the place in the first pushed cart or vehicle,
   - when shunting carts using a road-based vehicle with combustion engine,
7) 3.6 km/h – when carts sets pushed off a marshalling hill approach a stationary train set,
8) 3 km/h:

DOI: 10.1051/matecconf/201711800014

- when the control signal shows "push slowly",
- when approaching a stationary train set with a locomotive or a shunted cart set,
- when shunting carts using mechanical shunting devices.

3 Research object and conditions of the analysis

The research object was a vehicle of the NRMM (Non-Road Mobile Machinery) group – mentioned SM42 diesel locomotive (Fig. 2). It is essentially designed for shunting and marshaling works. It can also be used for light freight trains. The main operational parameter for this locomotive is the maximum engine power that is 588 kW (800 hp).

Fig. 2. SM42 locomotive – view from the driver's cab.

For the purpose of the article, the analysis was carried out, among other things, to assess the operating conditions of the SM42 locomotive and the fuel consumption of its engine during track works supporting the modernization of the railway line leading from Poznan to Pila. On this working day, the SM42 locomotive was used to operate five self-unloading wagons, including three "dumpcar" type (with a tipping box) wagons to deliver a new crushed stone to the construction site (Fig. 3 and 4). This crushed stone was loaded onto wagons using a wheel loader.

Detailed works schedule which were made using the SM42 locomotive is presented in Figure 5. Indirectly from the presented information result various ways and conditions of operation of the SM42 locomotive. On the basis of the recorded measurement data, the percentage share of the individual activities performed by the locomotive in the whole cycle of its work (about 4 h) was determined (Fig. 6).

An attention is drawn to the large share of activities during which the engine of the SM42 locomotive was idling. Overall it was 155 minutes, which accounted for about 64% of the total operating time of the locomotive engine (Fig. 7, Tab. 1) [17-19]. In this case, the greatest potential for improving the efficiency of the propulsion system of the locomotive is achieved (achieving greater combustion efficiency). As a result of the various organizational and technical activities, the minimum downtime of the machine with the running engine should be ensured.

Fig. 3. The part of the train pulled by SM42 locomotive (self-unloading wagons).

Fig. 4. The part of the train pulled by SM42 locomotive ("dumpcar" type wagons).

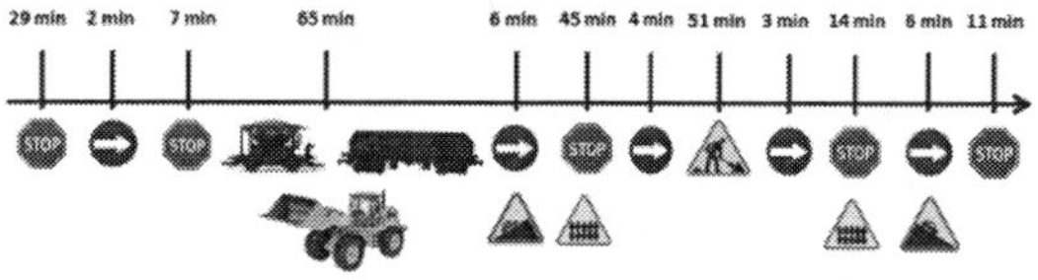

Fig. 5. The way of performing works using the SM42 locomotive.

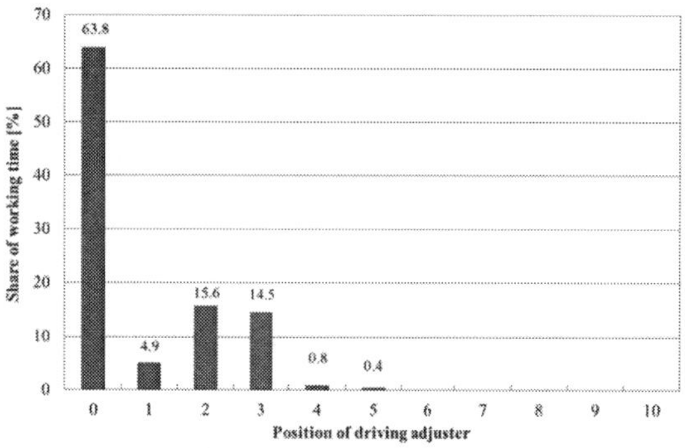

Fig. 6. The histogram of the engine load of the analyzed SM42 locomotive (4 h of work).

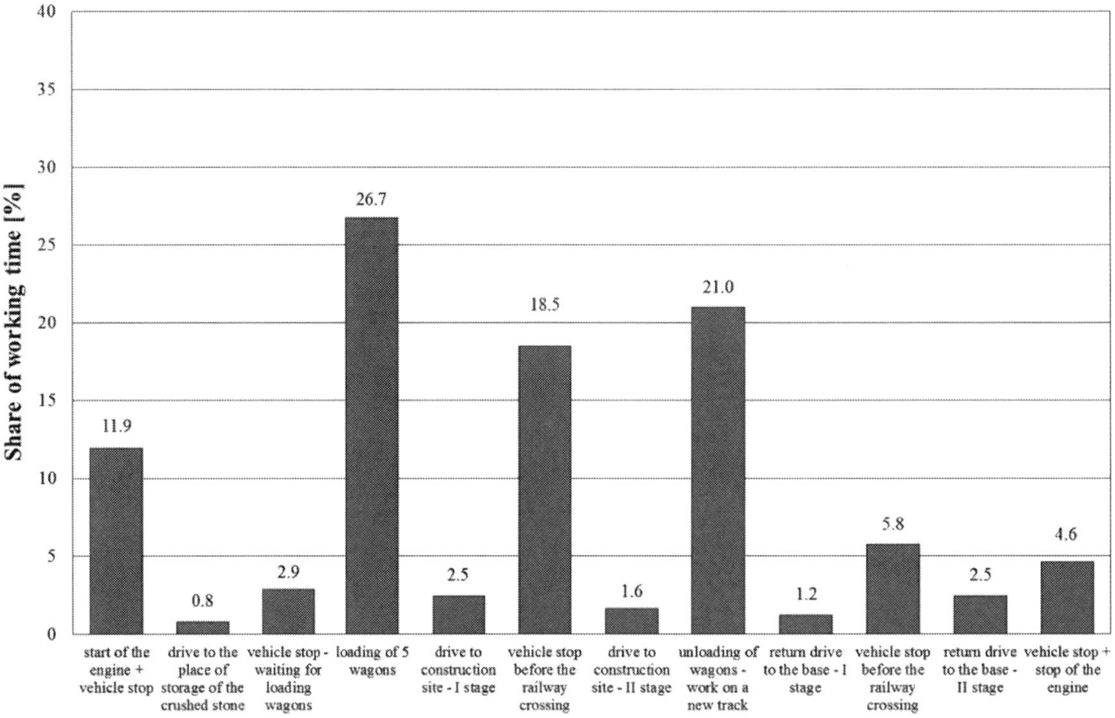

Fig. 7. The share of the individual activities in the whole cycle of work of SM42 locomotive (4 h).

On the basis of the analyzes, it can be observed that the engine of the locomotive was operating in a narrow range of loads – mainly on the 2 and 3 position of the driving adjuster. Higher positions were used only because of the need to overcome the greater movement resistance while traveling with full wagons under a small hill (occurs on a short segment of the route).

Table 1. The average power of the SM42 locomotive on the individual positions of the driving adjuster.

Position of the driving adjuster [–]	Average usable power [kW]
1	54
2	108
3	162
4	216
5	270
6	324
7	378
8	432
9	486
10	540

4 Analysis of the possibilities of improving the work efficiency of the SM42 locomotive

Improving the work efficiency of the SM42 locomotive in terms of reducing energy consumption, the authors looked at the possibility of shortening the duration of some activities carried out within track works (Fig. 5 and 6). In order to estimate fuel savings (as part of improving work organization), it was proposed to reduce engine idling time by 50%. This reduced the work time by 45 minutes, resulting in a reduction in fuel consumption of around 16% (Fig. 8). It should be noted that data on the amount of fuel used by the locomotive were derived from the electronic fleet management system of the owner of this locomotive. The confirmation of the above effects of the proposed organizational activities is a significant reduction of engine idling time from about 64% to 55% – the suggested work cycle (Fig. 9). As could be expected, most contributes to this significant reduction of the first stoppage time of the locomotive before the railway crossing (by more than 20 minutes).

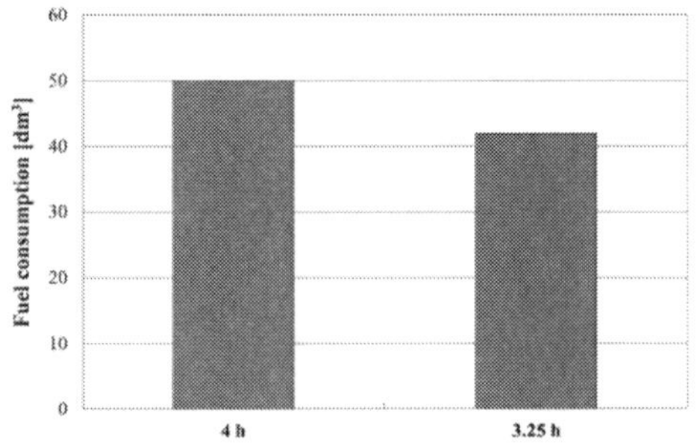

Fig. 8. The amount of fuel consumed by the locomotive engine at the given time of operation.

* Corresponding author: p.daszkiewicz@tabor.com.pl

This resulted in a decrease in the share of stopping in total activity performed by the locomotive from 18.5 to 11.5% (Fig. 10).

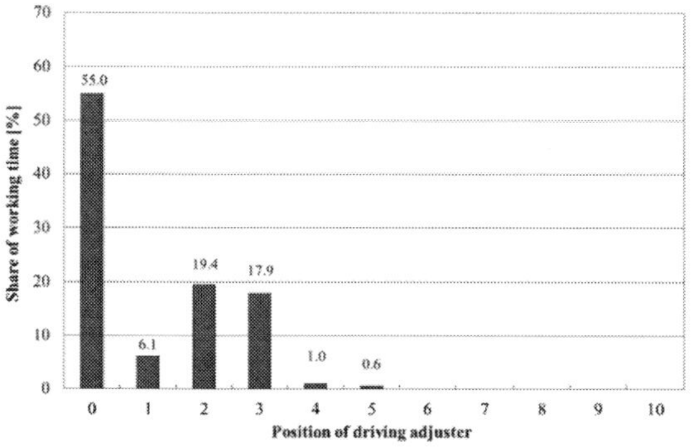

Fig. 9. The histogram of the engine load of the analyzed SM42 locomotive including improved work organization (3.25 h of work).

5 Summary

Among the initiatives aimed at increasing the efficiency of rail transport, the activities related to reducing the costs of shunting vehicles operation play an important role. The way to achieve this is to modernize diesel locomotives or replace shunting locomotives with more economical rail-road vehicles, additionally taking into account the emission and noise considerations. For decades, the main source of propulsion for vehicles used in shunting works, in most cases locomotives, are diesel engines with obsolete structures, characterized by high efficiency and torque values, which often do not meet emission standards and generate high maintenance costs. If you consider vehicles for the heaviest applications, their propulsion units are also characterized by a high cylinder volume value, which also results in high fuel consumption and high toxic exhaust emission. From the above it follows that the operation of shunting locomotives can cause high energy consumption and significant emission of pollutants, so special attention should be paid to the problem of reducing their negative impact on the environment.

The analyzes were conducted to determine the operating conditions and energy consumption of the SM42 locomotive used during the modernization works on the railway line. There is some energy wastage in track works caused by long and unproductive stoppages of machines with running engines. As a result of the proposed organizational actions, the share of inefficient engine idle time was reduced by almost 9% (Fig. 7 i 9), resulting in a reduction in fuel consumption of 8 dm^3 (Fig. 8). In order to further improve the use of chemical energy provided in the fuel, after completing the supplementary tests, an energy balance will be prepared to develop appropriate operating guidelines for operators and users of the SM42 locomotive.

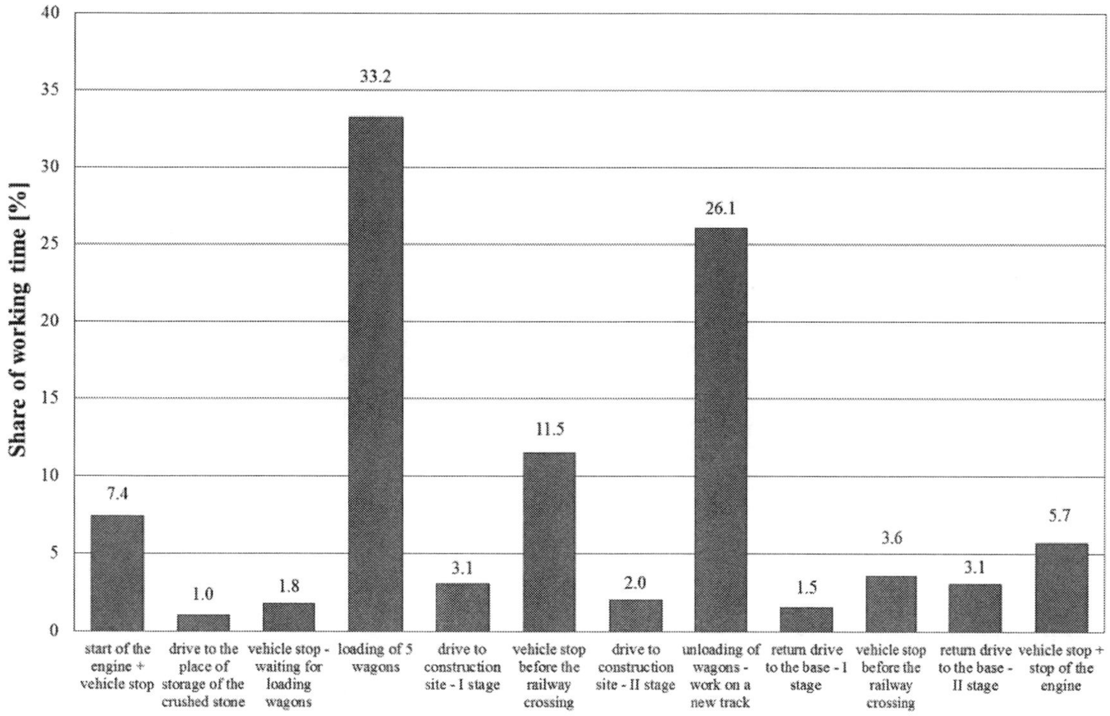

Fig. 10. The share of the individual activities in the whole cycle of work of SM42 locomotive including improved work organization (3.25 h).

References

1. P. Lijewski, J. Merkisz, P. Fuc, A. Ziolkowski, L. Rymaniak, W. Kusiak, European Journal of Forest Research **136**, 153-160 (2017), DOI: 10.1007/s10342-016-1015-2
2. P. Fuc, L. Rymaniak, A. Ziolkowski, *WIT Transactions on Ecology and the Environment* **174**, 207-218 (2013), DOI: 10.2495/AIR130181
3. J. Merkisz, M. Bajerlein, W. Kozak, *Urban Transport and the Environment in the 21st Century* **14**, 325 (2008)
4. J. Merkisz, P. Fuc, P. Lijewski, A. Ziolkowski, M. Galant, M. Siedlecki, Journal of Electronic Materials **45**, 4028-4037 (2016), DOI: 10.1007/s11664-016-4543-0
5. S. Krezo, O. Mirza, Y. He, S. Kaewunruen, J.M. Sussman, *Civil-Comp Press* **288**, 1-14 (2016)
6. J. Szkopiński, Archives of Transport **29**, 65-75 (2014)
7. P. Daszkiewicz, M. Andrzejewski, A. Merkisz-Guranowska, H. Stawecka, D. Gallas, Combustion Engines **162**, 666-671 (2015)
8. M. Feo-Valero, L. Garci´A-Mene´Ndez, S. Del Saz-Salazar, Transportation **43**, 101-122 (2016)
9. P. Daszkiewicz, M. Andrzejewski, A. Merkisz-Guranowska, H. Stawecka, D. Gallas, Combustion Engines **162**, 666-671 (2015)
10. M. Feo-Valero, L. Garci´A-Mene´Ndez, S. Del Saz-Salazar, Transportation **43**, 101-122 (2016)
11. J. Merkisz, J. Pielecha, P. Molik, M. Nowak, *2014 IEEE Vehicle Power and Propulsion Conference,* Coimbra (2014)
12. Materials of Polish State Railways (PKP)
13. J. Merkisz, P. Lijewski, P. Fuc, S. Weymann, Eksploatacja i Niezawodnosc – Maintenance and Reliability **15**, 364-368 (2013)
14. A. Suchecki, J. Nowakowski, T. Knefel, *Transport Means – Proceedings of the International Conference* 194-197 (2014)
15. J. Merkisz, P. Lijewski, P. Fuc, M. Siedlecki, A. Ziolkowski, *IOP Conference Series-Materials Science and Engineering* **148**, UNSP 012077 (2016), DOI: 10.1088/1757-899X/148/1/012077
16. J. Merkisz, P. Lijewski, P. Fuc, M. Siedlecki, S. Weymann, Applied Engineering in Agriculture **31**, 875-879 (2015)
17. J. Markowski, J. Pielecha, R. Jasinski, T. Kniaziewicz, P. Wirkowski, *E3S Web of Conferences* **10**, UNSP 00140 (2016), DOI: 10.1051/e3sconf/20161000140
18. J. Merkisz, S. Tarkowski, Eksploatacja i Niezawodnosc – Maintenance and Reliability **2**, 50-58 (2011)
19. W. Gis, J. Pielecha, J. Waskiewicz, M. Gis, M. Menes, *IOP Conference Series-Materials Science and Engineering* **148**, UNSP 012040 (2016), DOI: 10.1088/1757-899X/148/1/012040

The latest technical solutions in rail vehicles drives

Maciej Andrzejewski[1,*], *Dawid* Gallas[1], *Pawel* Daszkiewicz[1], *Agnieszka* Merkisz-Guranowska[2], and *Hanna* Stawecka[1]

[1]Rail Vehicles Institute TABOR, ul. Warszawska 181, 61-055 Poznan, Poland
[2]Poznan University of Technology, Faculty of Machines and Transport, ul. Piotrowo 3, 60-965 Poznan, Poland

Abstract. The article discusses the latest trends and solutions used in the offers of rail vehicles manufacturers. The study is mainly concerned with the technical solutions used in hybrid rail vehicles, whose development has become one of the priorities for the development of rail vehicles in the European Union in recent years. Stricter emissions standards for harmful compounds in the European Union have forced manufacturers to use increasingly sophisticated technology, including hybrid drives and alternative fuels. The products and solutions offered by the major manufacturers on the market, along with their capabilities and future applications are described. The predicted trends in the development of propulsion technology for rail and road-rail vehicles are also indicated in view of the current legislative aspirations among the EU Member States.

1 Introduction

From the invention of rail transport and rail vehicles their drive systems have evolved the most of all modern forms of transport. From steam engines through diesel engines and electric motors all the way to alternative drive technologies, hybrid drives and drives based on fuel cells. There are certain advantages to rail vehicles that makes them more compatible with a wider variety of drive systems. This serves as an explanation why rail vehicles were more likely to see innovations in vehicle propulsion technology tested on them so often. This is all mostly due to the right restrictions on size, weight, power, fuel types, and operating parameters for the engines of rail vehicles. Cars were not able to use the early steam engine technology effectively, although steam-powered cars have been tried. The overall size and weight requirements for an efficient steam engine have made the solution simply not viable. Steam engines have also been tried on planes, but again, due to their relatively high weight to power ratio this was done mostly as a proof of concept than a viable solution. As a result in transport steam engines have only seen widespread use in rail vehicles and ships [1]. Drive systems and engine units that are either too big for standard road vehicles, too heavy for airplanes, or too low power for ships [2, 3] can all find middle ground use in rail vehicle propulsion in some forms. Today rail transport is still boasts a huge variety of drive system types among all means of transport. Thus testing new drive systems and new fuel types happens commonly in rail vehicles, enabling manufacturers to test the economic viability and technological reliability of their novel ideas and new advanced propulsion technologies. While the majority of rail vehicles are equipped with either electric drives or conventional high-power CI combustion engines attempts have been made to test the less obvious solutions in hopes of finding more modern alternatives. This effort is mostly driven by a combination of economy and ecology, and is forced by the legislations passed in many countries around the world.

The rail vehicle drives currently in use rely primarily on fossil fuels [4]. This is a limitation that directly contributes to the emission of harmful compounds to the atmosphere. In addition to carbon dioxide, these include carbon monoxide, hydrocarbons, nitrogen oxides and particulate matter. The European Union's new and more stringent Stage standards are characterized by a reduction in emissions limits in each successive version (Fig. 2). Meeting these limits requires high-tech solutions inside the engine as well as aftertreatment solutions. Unfortunately, these solutions have limited ability to reduce emissions of harmful compounds, additionally their introduction is economically viable only to some extent. Moreover, continuous changes to the limits would require regular modifications and upgrades to the systems in use. To meet these challenges, as well as to improve air quality and reduce emissions, a more complete solution would be to use electric drives, or in cases where the internal combustion engine is necessary, hybrid drives. The most modern types of hybrid drives can significantly reduce the harmful emissions from internal combustion engines, thus allowing them to operate in accordance with the described standards [5]. An example of such a solution is the hybrid rail vehicle called "Evolution Hybrid" (Fig. 1), offered for purchase by General Electric.

There are many barriers to the introduction of state-of-the-art technology for rolling stock vehicles in the EU. Apart from costs or emissions reduction, it is necessary to take into account the laws and technical

* Corresponding author: m.andrzejewski@tabor.com.pl

DOI: 10.1051/matecconf/201711800015

specifications required by individual member states. In addition, vehicles and solutions in manufacturers' offers must meet not only the current emissions or safety requirements but also take into account possible future changes. As the average life span and operating life of rail vehicles is relatively high, the investment and costs incurred by the train operator must be justified, whereas the purchase of vehicles whose exploitation after a few years may cost the operator more than the manufacturer intended is undesirable. Hence many manufacturers of rail vehicles and vehicle drives are planning their commercial offers from the perspective of technologies that are just now entering the market. Examples include the latest hydrogen-fueled fuel cell technology. These are solutions that in their present form are too expensive to use individually or sporadically. It is certain, however, that as the infrastructure develops, these solutions will become increasingly competitive with today's conventional drives. And while a complete overhaul of all rail vehicle drive systems is neither economically viable nor financially feasible at this time, mostly due to the lack of sufficient support and supply for the new drive systems, as well as their proper maintenance and the necessary infrastructure, these new technologies are still an important indicator of the current trends in the development of rail vehicles.

Fig. 1. Hybrid locomotive on offer from GE [6].

2 The available hybrid drives

2.1 Modular systems

Among the available solutions there are modular constructions, which allow for the choice of the drive system elements depending on the client's needs. With the battery system for powering the electric drive, it is also possible to use the pantograph to power the traction where available, provided the battery technology has a sufficient capacity and lifespan [7].

One of the main disadvantages of modular systems is the problem of properly utilizing all the available space. Ergonomics in modular solutions must be adapted to include the ability to change and select any module so that it can be integrated into a functional unit. Manufacturers more often offer modular systems for rail vehicles, rather than conventional freight or passenger locomotives. One of these vehicles is the Modalrac rail vehicle made by Schalke. It offers a choice from a range of power supply and drive solutions, including CI engine, battery pack, pantograph system and third rail power supply (Fig. 3). This allows a significant extension of the operational range for modular rail vehicles. The key feature offered by the manufacturer is the ability to exchange the selected module for another within the span of 60 minutes. This allows the rail vehicle to be used with a set of modules chosen based on the needs and limitations relating to the assigned task. It means that when working in urban or populated areas for example, the vehicle can be powered using the energy stored in the batteries or through pantographs, while work performed further in the field or off the grid would be performed when powered using a conventional CI engine. The option of making such modifications translates directly into a reduction of both harmful components and noise emissions in the populated urban areas, reducing the inconvenience as well as the health hazard posed to the local population.

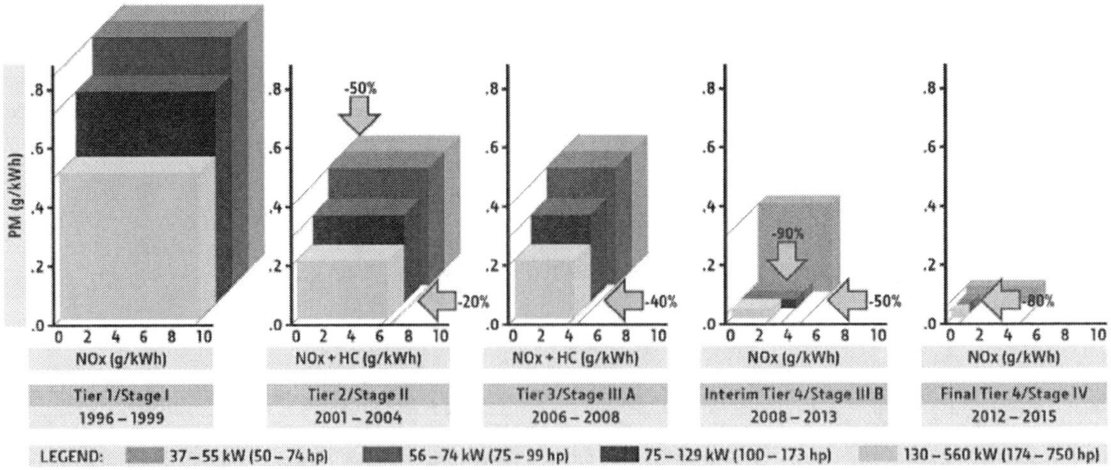

Fig. 2. Emission standards for toxic exhaust gases of non-road vehicles according to John Deere in UE and EPA [8].

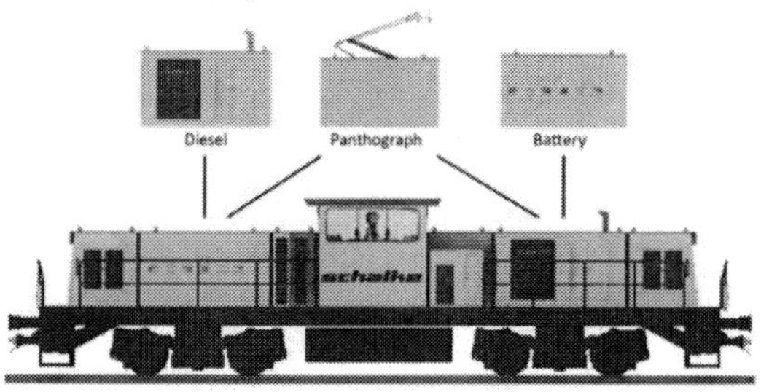

Fig. 3. An example of a specialist rail vehicle equipped with a modular drive system [9].

MTU has developed a modular hybrid drive unit for rail buses (Fig. 4 and 5), consisting essentially of an internal combustion engine (1) and a set of high-efficiency energy storage units (6, 7). These trays can be located both on the roof of the vehicle as well as in its frame. In addition, the assembly includes an electric motor (2) with a power transmission (3). In order to meet the current European Stage IIIB emission standard for Non-Road Mobile Machinery (NRMM) most vehicles need to be tested extensively and require aftertreatment systems and engine modifications [10-13]. In comparison this vehicle is equipped with a technologically advanced selective catalytic reduction system (4), which is used to reduce the nitrogen oxide content of the exhaust gas using a urea solution, under the commercial name AdBlue (5 – tank with this liquid). The extended control unit (8) oversees the correct operation of the entire propulsion system.

The MTU hybrid drive unit can operate in 4 modes. The first mode is to drive using only an internal combustion engine (Fig. 6a). When the demand for energy rises (greater movement resistance), an electric motor powered by high-efficiency energy storage is additionally activated by the control unit. In situations where low emission of pollutants in exhaust gases (passage through highly urbanized areas, tunnels, operation in railway stations) is required, the energy needed to drive the vehicle is derived only from the above mentioned batteries (Fig. 6b). The fourth mode of operation of the MTU drive unit is the recuperative braking mode, during which the energy storage units are charged (Fig. 6c).

Fig. 4. Hybrid drive railbus from MTU company [14].

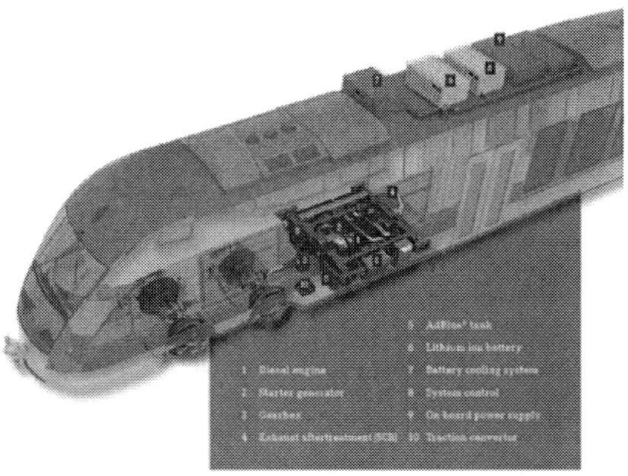

Fig. 5. Hybrid drive system used in the VT 642 rail bus [14].

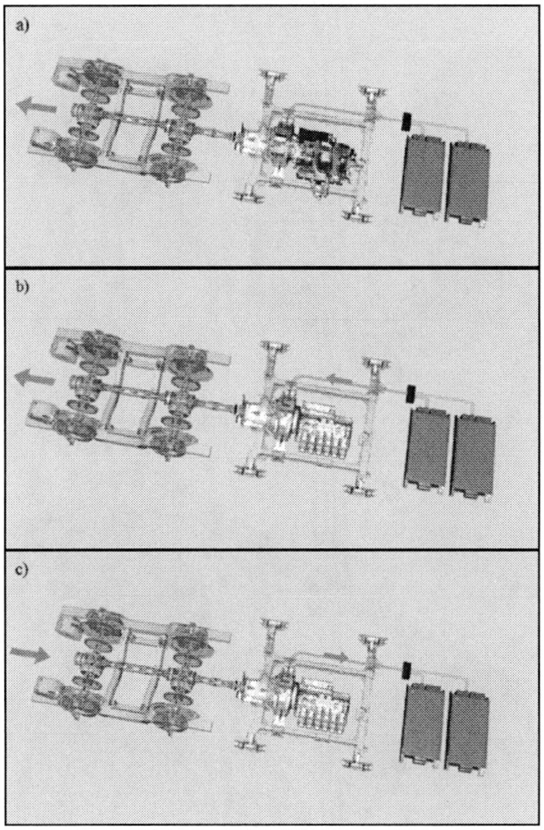

Fig. 6. Selected modes of MTU hybrid drive system operation [14]: a) driving mode – power supply from internal combustion engine, b) driving mode – power supply from batteries, c) recuperative braking mode – charging batteries.

2.2 Types of hybrid drives

In rail applications, the same type of hybrid drive division is used as for road vehicles. Series hybrid and parallel hybrid drives are offered as standard. Examples include hybrid drives provided by Toshiba that offer complete solutions for series and parallel-series drives for locomotives.

DOI: 10.1051/matecconf/201711800015

The series hybrid drive is used in some rail vehicles, where it allows the engine to be stopped while the vehicle is stationary instead of leaving it idle. This solution, similarly to the start-stop system known in road vehicles, significantly reduces or even eliminates the share of operating time of the CI internal combustion engine spent in the operating areas responsible for the most significant part of harmful compounds emissions. In this case, the use of an electric drive avoids the need for the engine to run continuously. Series is the most rational solution, because the internal combustion engine can operate at a narrower speed range, thus speed and load changes can be avoided so that the engine runs with its optimal parameters [5]. The solution offered is a full drive system, including batteries (Fig. 7). This arrangement also allows for the recovery of some of the energy during the braking process [15], increasing the overall vehicle efficiency (Fig. 8).

Fig. 7. Series hybrid drive system offered by Toshiba [16].

Fig. 8. Energy recovery using a hybrid drive offered by Toshiba [16].

Tests of this solution performed by the manufacturer have shown that the use of a series hybrid drive reduces emissions of nitrogen oxides by 61%, noise emissions by 22%, fuel consumption by 36%, as well as reducing the operating and maintenance costs of the vehicle. This is a significant improvement with respect to conventional drives, especially given the difficulty in achieving greater emissions reduction improvement through standard exhaust gas aftertreatment solutions. Including those performed using retrofitting.

Similar solutions are used in special-purpose rail vehicles. Platform Basket uses a hybrid serial drive to power its road-rail vehicle for work on electrical traction or other difficult to reach places (Fig. 9). This solution uses two inver connected to two electric motors. One of the electric motors drives the vehicle, while the other powers the arm (Fig. 10).

Fig. 9. Road-rail specialist platform with a hybrid drive system by Platform Basket [17].

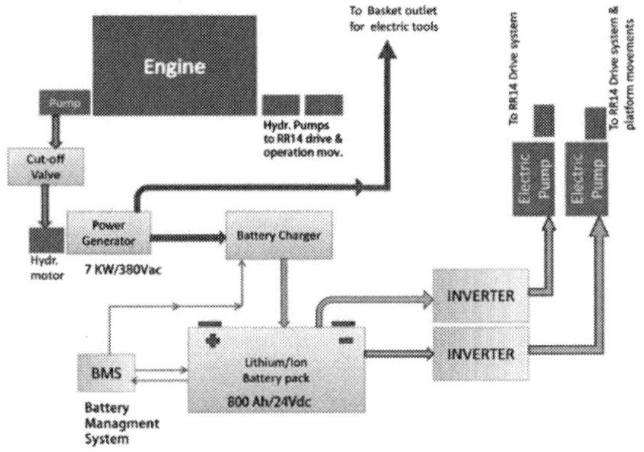

Fig. 10. The series hybrid drive setup of a Platform Basket cherry picker vehicle for on-rail operations [17].

Thanks to the use of the engine in series hybrid setup in this drive system, the manufacturer has significantly reduced the environmental damage caused by the vehicle's operation. This is the solution that has resulted in the popularity and competitiveness of the vehicle. It should be noted that in addition to the general directives and laws imposed by the European Union referred to as the Technical Specifications for Interoperability (TSI), all Member States continue to have diversified national laws, including noise emission limits, safety, operability, compatibility, efficiency or operating parameters. These directives are the obstacles to achieving full interoperability of rail systems in all the EU Member States. This is particularly important in the case of Poland because, as noted in the European Union Agency for Railways Interoperability Report for year 2015 [18], Poland is ranked 2nd among the countries with the greatest number of national rules and specifications in the EU. In Poland in the year 2015 there were 1113 country-specific rules for rail and road-rail vehicles (Fig. 11). For comparison, the only country with more national rules was the Netherlands (1270). It is more than double the average of the EU countries, especially the western countries like Germany (541) and France (574), many times more compared to Scandinavian countries like Sweden (319) or Norway (259). As a result of these differences in regulations and specifications between EU countries, it is difficult to provide rail vehicle operators in different countries with systems and vehicles that meet all these requirements uniformly. Hence, the technical solutions visible in the Platform Basket vehicles are tailored to meet these specifications and requirements for as many Member States as possible. For RR 14 EVO vehicles, the technical solutions have been selected to meet both the European and the country-specific requirements, in this case for Germany, Sweden, France and the United Kingdom.

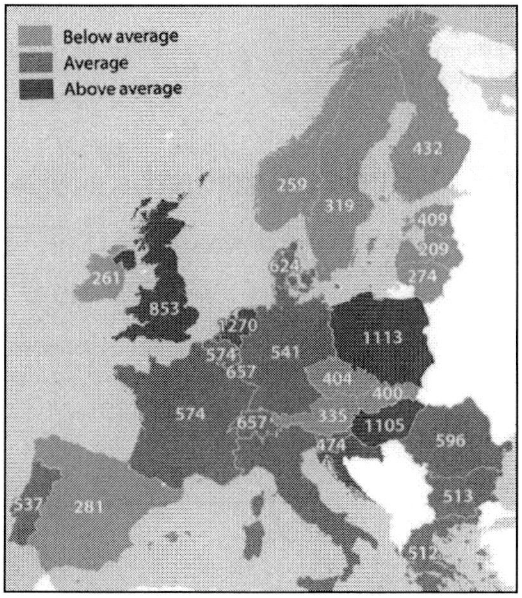

Fig. 11. National rules per EU member state in 2015 [18].

Another option available is the hybrid drive in serial-parallel setup. This solution allows to power the wheels using either the internal combustion engine or the electric motor. This way it becomes possible to avoid the constraints of powering the wheels with an electric motor and, depending on the needs, transfer power to the wheels directly from the combustion engine (Fig. 12). This is of course at the expense of an increased fuel consumption and increased emissions, as it is not always possible to provide optimal operating conditions for the internal combustion engine when it is used to drive the vehicle directly. It is, however, possible to use its full power, which is a major advantage of this solution. Due to the operating characteristics of the series-parallel hybrid solution, it is used primarily in rail shunting vehicles. The CI internal combustion engine can still be regulated in such a way that it can be pushed towards its optimum operating conditions for as long as possible. As a result, the series-parallel hybrid drive system still offers significantly better emission rates than the conventional versions, including fuel consumption, which directly impacts the operating costs of hybrid vehicles. Hence, their purchase becomes an increasingly competitive offer for rail transport companies and operators.

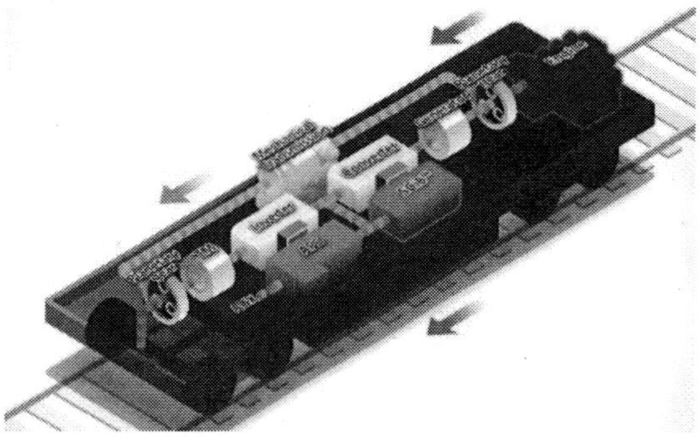

Fig. 12. Series-parallel hybrid drive design for a rail vehicle [16].

For passenger services, the most common solution is the Diesel-Electric Multiple Unit (DEMU). This solution makes it easier for the manufacturer to design a vehicle, allowing a design where several units can be combined and share certain elements, so that propulsion and axillary systems and elements can be deployed more ergonomically to maximize the space available to passengers. As an example, one of Woojin's latest designs utilizes the TeC1-M-T-TeC2 four-unit drive system (Fig. 13).

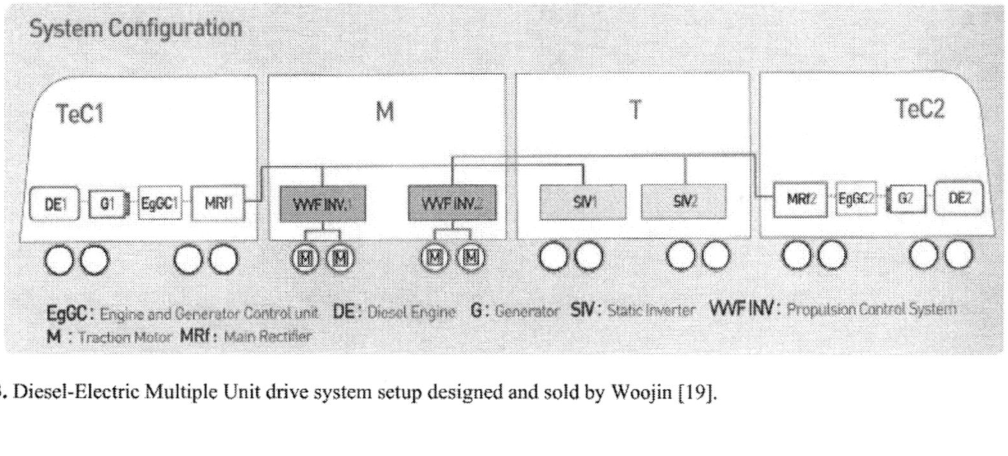

Fig. 13. Diesel-Electric Multiple Unit drive system setup designed and sold by Woojin [19].

3 Rail vehicle fleet modernization possibilities

Electric and hybrid drives are used in shunting vehicles increasingly more often. They are, however, less powerful, often requiring more operations to perform the same shunting work. Nevertheless, the costs and total harmful exhaust component emissions of each of these operations are significantly lower. This makes the solution even better than conventional shunting locomotives in terms of their environmental impact. The main factor in this case is the engine idling time, which is very high for maneuvering vehicles and constitutes a large share of the engine operating time [20]. Thus eliminating a key aspect responsible for a significant portion of the total exhaust emissions from an internal combustion engine in the engine working time.

The existing sales offers of new hybrid drive locomotives have only been on the market since relatively recently. The first hybrid locomotives were built on the principle of modernization of old conventional locomotives by replacing the drive unit. Over time, conventional diesel locomotives are expected to be replaced with electric and hybrid locomotives, at least to as high an extent as it is possible. Due to the costs associated with the modernization this process is slow in Poland, but it is noticeable nevertheless. Investments in this field are particularly important because of the very high average age of diesel locomotives used on Polish railroads. These locomotives need to be modernized or replaced with new ones, hence the widening range of new hybrid drive rail vehicles is a good and welcome sign. Still, the number of locomotives waiting for replacement or modernization is high. In Poland many older locomotives that would be discarded in other European countries are still being used, although retrofitting and modernization options are usually available (Fig. 14). An example of the process of vehicle modernization is the list of operated rolling stock offered by the PKP Cargo carrier and presented in Table 1. It is important to note that the total number of rail vehicles in operation decreased, which means that not all of the 424 rail vehicles taken out of service have been replaced with newer vehicles, such as rail vehicles with hybrid drives.

Fig. 14. SD59MX locomotive after retrofitting [21].

Numerous studies have researched the effects of diesel locomotive modernization on the vehicle's emissions, including the replacement of the engine unit with newer ones, the addition of new exhaust aftertreatment systems, and hybridization of the vehicle drive system [20, 22, 23]. Rail vehicle operators usually make the choice between these options based on the available funds, age and technical condition of the vehicles, the vehicle operating requirements and the necessity caused by the legal emission limits. All of these solutions can reduce harmful exhaust emissions and increase vehicle efficiency and profitability, but operators often limit themselves to the cheapest and easiest solutions out of all available options. It is therefore important for manufacturers to further develop hybrid drive rail technologies in order to allow the elimination of the conventional combustion engine drive vehicles from operation over time. This is also done through legislative means, urging operators to act in the name of environmental protection and reduce emissions of harmful exhaust components from the high power CI engines used to drive freight and passenger locomotives. Obtaining a CO_2-free transport technology is one of the main and most ambitious objectives of the European Union, so further pursuit of ever cleaner and more efficient vehicle drives will remain an important aspect in the decision and law making process.

4 Hydrogen-based drives

In recent years vehicles with modern alternative drives are becoming more and more common in the offers of rail vehicle manufacturers. One of the main examples are hydrogen-fueled locomotives. These types of solutions have been around for some time now, but the costs associated with their purchase and operation effectively reduce the demand for these types of vehicles. Vehicles using fuel cell have very similar performance characteristics to electric rail vehicles and series hybrids. The difference is mainly the source of electric power. As with hybrid electric vehicles, the source of energy generated for the electric motor is an internal combustion engine operating in uniform speed and load ranges, and for vehicles powered with hydrogen fuels, the source of electric energy are the fuel cells. Combining hydrogen and oxygen in the fuel cells generates electricity based on instantaneous needs, controlled by the fuel cell hydrogen feed systems. This is by far one of the cleanest solutions available on the market, where the final product is the production of small amounts of water. Of course, the approach to pollutants generation should be holistic, hence the costs of producing, delivering and storing hydrogen as fuel for this type of a vehicle should be weighed into the calculation when assessing these new drive systems. In this case, hydrogen is produced by other companies, using mostly electricity from the power grid.

Keep in mind that energy losses occur at every step of production, processing, transportation, and ultimately operation. Despite the non-emissive nature of this type of fuel itself, its production can have a negative effect on the environment if its production uses electricity generated using fossil fuel based power plants. This problem is the same between hydrogen vehicles and electric vehicles. In addition, there are barriers to the full and expansive application of such technologies in rail transport. The main problem of these modern solutions are the initial costs of the vehicle as well as the infrastructure. Transporting and storing hydrogen as a fuel is much more difficult than other fuels, it is more expensive and more dangerous than handling diesel, petrol or even natural gas [24]. Hence, the fuel supply of hydrogen-powered rail vehicles also requires proper adaptation of the infrastructure to allow safe and efficient storage and refueling of hydrogen at the stations where it will be supplied to power the vehicles. It is very important in this case to protect the hydrogen fuel from leakage. The storage methods for hydrogen fuel are more demanding than for LPG fuel. The main problem is to provide air-tightness and stability from ambient temperature fluctuations for hydrogen that is normally stored in tanks at a pressure of 350-700 bar. The leakage for this kind of fuel tank involves the risk of explosion, while ensuring sufficient strength of the tank under this pressure requires the use of state-of-the-art high-strength and durability materials. The solutions currently on the market mostly use specialized roof tanks for hydrogen storage. This is the main place in the rail vehicles considered by manufacturers to be the safest and easiest to exploit in the existing rail vehicle designs (Fig. 15). It should be noted, however, that with the development of these technologies and with the increase in popularity of hydrogen as vehicle fuel new technologies and solutions will inevitably follow. Thus the current solutions are not truly optimized to take advantage of the new drive solutions, but the later iterations will build on the previous experiences, giving the manufacturers of vehicles the time they need to adjust their previous designs and concepts to the changes in vehicle drive technologies.

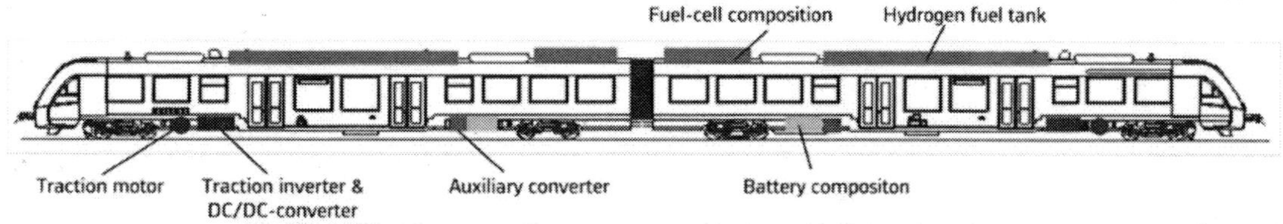

Fig. 15. An example of solving the problem of placing hydrogen tanks and fuel cells in Alstom's fuel cell locomotive design [25].

Table 1. Comparison of the change in vehicles in operation by the PKP Cargo [26].

Number on 31.12	SP32	SP42	ST43	ST44	ST45	SU42	SU45	SU46	Sum
2001	102	119	174	103	–	40	160	36	734
2002	102	119	174	103	–	40	157	36	731
2003	102	119	174	103	–	40	157	36	731
2004	15	116	165	103	–	40	143	36	618
2005	15	116	165	103	–	40	156	36	631
2006	15	76	139	82	–	40	149	36	537
2007	15	79	148	82	–	40	150	36	550
2008	2	79	146	93	–	–	80	36	436
2009	2	51	141	92	1	–	76	36	399
2010	2	2	133	90	4	–	44	35	310

5 Conclusions

Purely electric drives are most commonly used in medium-power locomotives on connections where electric traction or catenary is available. This is the simplest solution, but at the same time it significantly limits the use of this type of drive depending on the quality and availability of the appropriate infrastructure. As a result, a large number of conventional diesel locomotives remain in service. Engine and drive manufacturers are aware of this, so there are also numerous options offered of modernizing existing diesel locomotives. It should be noted that modernization was one of the easiest methods of introducing the first locomotives with hybrid drives to operation. Despite some initial costs, modernized rail vehicles are proving to be a good investment for train operators. While visible economic effects of such activities are expected over many years, the ecological effects are measurable fairly quickly [27].

Hybrid drives are also not the ultimate solution to the problem of high exhaust gas emissions. But they are now regarded as one of the main steps towards eliminating harmful substances from the exhaust gases emitted by means of transport. There are also a number of other new technologies that are gradually gaining popularity and competitive edge on the market. These include biofuel powered vehicles, or hydrogen powered fuel cells. Unfortunately, the latest of these technologies are not yet economically and practically usable on an industrial scale. Nevertheless, the future of transport, both rail and road-rail, should be considered. Hence, the main challenge facing the drive systems in the future is to create the right infrastructure for their full and proper exploitation. Due to the initial cost of such a project, cheaper and easier to use hybrid drives can fill the transition gap between the conventional and today's most technologically advanced rail vehicle drives.

References

1. J. Markowski, J. Pielecha, R. Jasinski, T. Kniaziewicz, P. Wirkowski, *E3S Web of Conferences* **10**, UNSP 00140 (2016), DOI: 10.1051/e3sconf/20161000140
2. Z. Korczewski, J. Rudnicki, *Computational Methods in Marine Engineering VI* 490-501 (2015)
3. C. Specht, J. Rudnicki, Polish Maritime Research **23**, 20-27 (2016), DOI: 10.1515/pomr-2016-0028
4. J. Merkisz, W. Stawecki, Pojazdy Szynowe **3** (2015)
5. M. Michnej, M. Szkoda, Technika Transportu Szynowego **10**, 38-40 (2007)
6. www.getransportation.com (2017)
7. Z. Juda, M. Noga, *IOP Conference Series – Materials Science and Engineering* **148**, UNSP 012042 (2016), DOI: 10.1088/1757-899X/148/1/012042
8. www.deere.pl (2017)
9. www.eickhoff-bochum.de (2017)
10. P. Fuc, P. Kurczewski, A. Lewandowska, E. Nowak, J. Selech, A. Ziolkowski, International Journal of Life Cycle Assessment **21**, 1438-1451 (2016), DOI: 10.1007/s11367-016-1104-y
11. J. Merkisz, P. Lijewski, P. Fuc, M. Siedlecki, S. Weymann, Applied Engineering in Agriculture **31**, 875-879 (2015)
12. J. Merkisz, P. Lijewski, P. Fuc, M. Siedlecki, A. Ziolkowski, *IOP Conference Series-Materials Science and Engineering* **148**, UNSP 012077 (2016), DOI: 10.1088/1757-899X/148/1/012077
13. J. Merkisz, P. Lijewski, P. Fuc, S. Weymann, Eksploatacja i Niezawodnosc – Maintenance and Reliability **15**, 364-368 (2013)
14. MTU company materials (2017)
15. P. Kortas, J. Kropiwnicki, Combustion Engines **3**, 631-638 (2015)
16. Toshiba company website (2016)
17. Platform Basket company website (2017)
18. Progress with railway interoperability in the European Union, European Union Agency for Railways Interoperability (2015)
19. wjis.co.kr (2017)
20. R. Konsek, A. Mężyk, Maszyny Elektryczne **1**, 19-23 (2014)
21. www.up.com (2017)
22. M. Ehsani, Y. Gao, A. Emadi, Technology & Engineering (2009)
23. K. Lewandowski, *XX Międzynarodowa Konferencja Naukowa Pojazdy Szynowe* 2012 (Poznan, 2012)
24. W. Gis, J. Pielecha, J. Waskiewicz, M. Gis, M. Menes, *IOP Conference Series-Materials Science and Engineering* **148**, UNSP 012040 (2016), DOI: 10.1088/1757-899X/148/1/012040
25. Alstom company website (2017)
26. PKP Cargo rail vehicles in operation (2017)
27. W. Morten, S.C. Jorgensen, K. Sorenson, International Journal of Vehicle Design **20**, 1-4, (1998)

MATEC Web of Conferences **118**, 00016 (2017) DOI: 10.1051/matecconf/201711800016
VII International Congress on Combustion Engines

Exhaust emissions from small engines in handheld devices

Piotr Lijewski[1,*], *Pawel* Fuc[1], *Michal* Dobrzynski[1], and *Filip* Markiewicz[1]

[1]Poznan University of Technology, Faculty of Machines and Transport, 60-965 Poznan, ul. Piotrowo 3, Poland

Abstract. The paper presents the results of investigations on the exhaust emissions carried out under real operating conditions of gasoline engines operating in a power generator and a chainsaw. During the operation of these devices the authors measured the following exhaust emissions: CO, HC, NO_x and CO_2. For the measurements the authors used a portable exhaust emission analyzer SEMTECH DS by SENSORS. This analyzer measures the concentrations of the exhaust gas components in an on-line mode while the engine is running under real operating conditions (road, field etc.). The exhaust emissions tests of non-road engine applications are performed on engine test beds in the NRSC (ISO 8178) and NRTC tests. The presented method is a new solution in determining of the exhaust emissions from such engines. The obtained results were compared with the applicable emission requirements. Besides, based on the performed investigations, the authors attempted an evaluation of the possibilities of the use of the measurement method for development works related to the reduction of the emission from small gasoline engines.

1 Introduction

Combustion engines in their various applications still remain the main source of propulsion. Aside from typical applications in on-road vehicles, combustion engines are also fitted in non-road vehicles and machines. A numerous group are small engines used in handheld devices (handheld chainsaws, lawnmowers etc.), small power generators or scooters. The number of these small engines grows rapidly, the example of which are scooters. The global market for motorcycles, scooters, and mopeds is projected to reach 61.2 million by 2020, driven by population growth, urbanization and expanding middle class population countries [1]. Unfortunately, this is followed by an increased global fuel consumption (CO_2 emission) and exhaust emissions. It is noteworthy that small engines are a serious threat to the natural environment. The exhaust emission limits for these vehicles are more liberal compared to other on-road engine applications.

Homologation tests related to exhaust emissions from small spark ignition engines, used to power handheld devices are carried out in stationary tests i.e. for constant engine speeds and loads (NRSC – Non Road Stationary Cycle, Fig. 1) [2]. In recent years, tests performed under engine actual operating conditions are gaining in importance (RDE – Real Driving Emissions). These tests are performed with the PEMS equipment (Portable Emission Measurement System). The fundamental advantage of such tests is the coverage of all actual conditions of operation and aspects of a combustion engine, which is hardly reproducible under laboratory conditions during tests performed on engine test beds. When analyzing the literature, we need to stress the RDE tests are performed for a variety of vehicle categories (on-road vehicles [3-8], non-road vehicles [9-14] or aircraft [15-17]. In the said works, particular attention was drawn to several aspects such as the research methodology, exhaust emissions level and engine operating parameters. The advancement of RDE tests made them commonplace and resulted in their introduction in the homologation procedures of on-road as well as non-road vehicles (NRMM – Non Road Mobile Machinery). It is to be stressed that the exhaust emissions tests on small engines operating under actual conditions of operation are rarely performed and the results are rarely published. In few relevant publications the authors indicate the potential problems related to the testing of small engines [18, 19]. Expect exhaust emission important issue for non-road and small engine is vibration in occupational lives [20, 21]

The largest problem is the lack of appropriate testing equipment of compact, light design adapted to the testing of small engines in terms of the measurement range. The adaptation is also the case for such parameters as concentration of the exhaust components or the measurement of the exhaust mass flow [19, 22].

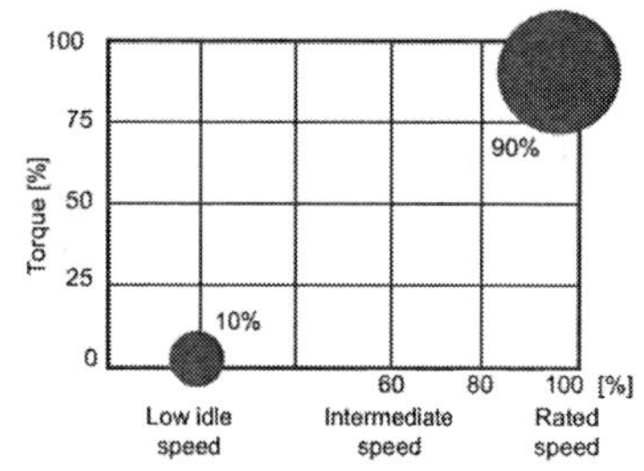

Fig. 1. NRSC (ISO 8178-G3) cycle for handheld rated speed applications.

* Corresponding author: piotr.lijewski@put.poznan.pl

Another problem is that the recording of the engine operating parameters is impossible (engine speed and load), which, in the case of on-road vehicles (thanks to the OBD system fitted) is very simple [18]. Hence, it appears fully justified to initiate research in terms of exhaust emissions from small engines. Eventually, the RDE tests performed on small engines should be included in the homologation procedures, as is in the case of engines of regular on-road vehicles. This, however, requires resolving of a variety of issues, developing methodology, and adapting engines and the measurement equipment. This paper is one of the first works devoted to the said problems.

2 Testing methodology and measurement equipment

The exhaust emission tests were performed under actual engine operating conditions. The investigated equipment (handheld chainsaw, power generator) performed typical work not different from that carried out on a daily basis. The technical specifications of the tested engines have been presented in Table 1 and the view of the investigated equipment in Figure 2.

Table 1. Technical specifications of the tested engines.

	Handheld chainsaw	Power generator
Engine type	2-stroke, air cooled, spark ignition	4-stroke, air cooled, spark ignition
Displacement	64 cm^3	170 cm^3
Power output	1.8 kW/12000 rpm	4.25 kW/3000 rpm

Fig. 2. View of the equipment during the tests (handheld chainsaw and power generator).

The engines in the investigated equipment operate mainly at two engine speeds: idle and maximum (or close to maximum). The tests were, thus, performed according to a scenario of a test presented schematically in Figure 3. The entire test for each device lasted approx. 30 min. The measurements of the concentrations of the exhaust components and the exhaust mass flow were performed on a continuous basis and were initiated immediately after the start of the engines. The schematic of this procedure was typical of regular operation of these machines. The measurements carried out in this way allow for the engine cold start and non-stationary operating conditions – changes in the engine speed (from idle to the operating engine speed) and load.

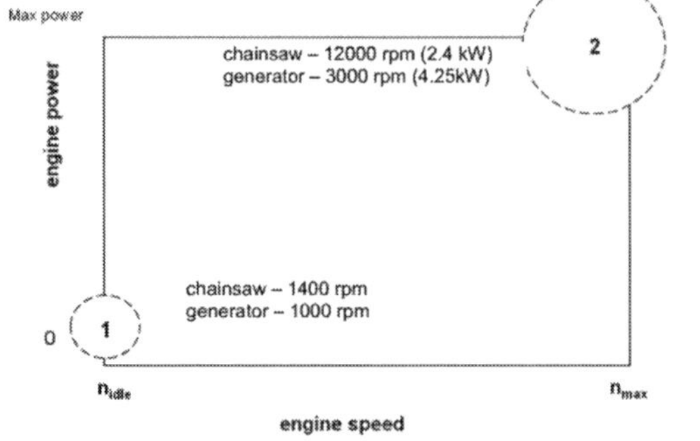

Fig. 3. Course of the cycle according to which the tests were performed.

One of the main test objectives was the application of the Semtech DS measurement equipment. However, due to the high concentrations of the exhaust components emitted by the power generator that exceeded the measurement range of the Semtech DS analyzer, it was necessary to use the Testo 360 device. Even then, the emission was still out of the measurement range, which is why the results of the tests related to this component were not included in the paper. The typical PEMS analyzer (Semtech DS.) was used only for the testing of the handheld chainsaw. The technical specifications of both analyzers have been presented in Tables 2 and 3.

Table 2. Characteristics of Semtech DS, a portable exhaust emissions analyser.

Parameter name	Measurement method	Accuracy
1. Emissions		
CO	NDIR, range 0–8%	±3%
HC	FID, range 0–10.000 ppm	±2%
NO_x ($NO + NO_2$)	NDUV, range 0–2500 ppm	±3%
CO_2	NDIR, range 0–20%	±3%
O_2	Electrochemical, range 0-25%	±1%
2. Data storage capacity	Over 10 hours at 1 Hz data acquisition rate	
3. Interface	SAE J1850 (PWM), SAE J1979 (VPW)	

Table 3. Characteristics of the Testo 360 exhaust analyser.

Parameter name	Measurement range
CO_2	0–25 Vol %
NO	0–3000 ppm (0–6160 mg/m^3)
NO_2	0–500 ppm (0–1030 mg/m^3)
NO_x ($NO + NO_2$)	0–3500 ppm (0–7190 mg/m^3)
CO	0–10000 ppm (0–12560 mg/m^3)
SO_2	0–5000 ppm (0–14650 mg/m^3)
O_2	0–21 Vol. %
With integrated absolute pressure measurement	40–1200 hPa
Temperature FT	–40–1200 °C

For the measurement of the exhaust mass flow, Semtech EFM-HS (High Speed Exhaust Flow Meter) was applied. Its technical specifications have been presented in Table 4.

Since the investigated engines were not equipped with on-board diagnostic systems enabling the recording of the engine speed, an AVL DiTEST Speed 2000 device was applied.

Table 4. Characteristics of a High Speed Exhaust Mass Flow Meter – Semtech EFM-HS.

Parameter name	Value
Exhaust temperature range	–5 to 700 °C standard, higher temp capability available upon request
Exhaust temperature accuracy	±1% of reading or ±2°C, whichever is greater
Flow measurement linearity	< 1.0% of full scale (exceeds 1065 Subpart D and ISO 16183 standards)
Flow measurement accuracy	±2% of reading or ±0.5% of full scale, whichever is greater
Warm-up time	< 5 minutes at 20 °C ambient
Data rate	5 Hz standard
High speed option data rate	500 Hz (100 Hz display rate)
Flow tube outer diameter	25 mm
Flow tube length (length including extension)	508 mm
Flow rate at 100 °C	6.9 – 85 kg/h
Flow rate at 400 °C	10.4 – 64 kg/h

3 Results and analysis

Figures 4, 5 and 6 present the hourly exhaust emission of HC, CO and NO_x. These are selected fragments of the entire test cycle. Thanks to these curves we may analyze the influence of a variety of factors (e.g. engine speed and load) on the emission changes. For the engine of the chainsaw, the curves showing the emission of HC, CO and NO_x we can clearly see the influence of particularly the engine speed and load on the exhaust emissions. The changes of the engine load result from the specificity of the chainsaw operation. This load is variable, which results from the chain resistance in contact with timber. In this case, the load changes very dynamically in a wide range.

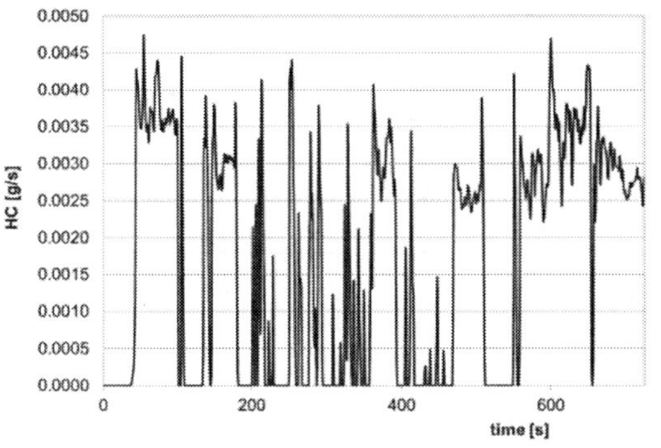

Fig. 4. HC emission from the chainsaw in the test cycle.

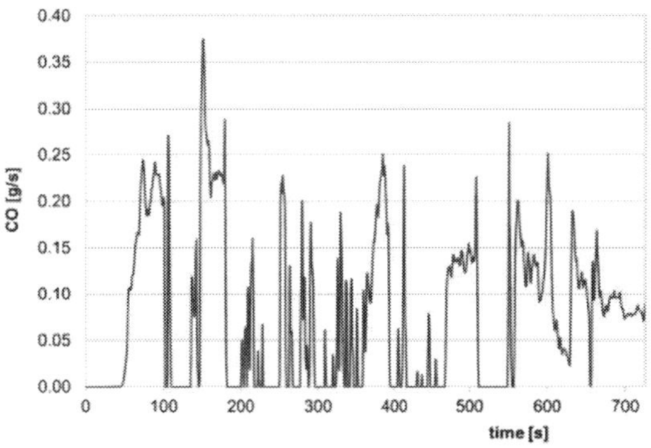

Fig. 5. CO emission from the chainsaw in the test cycle.

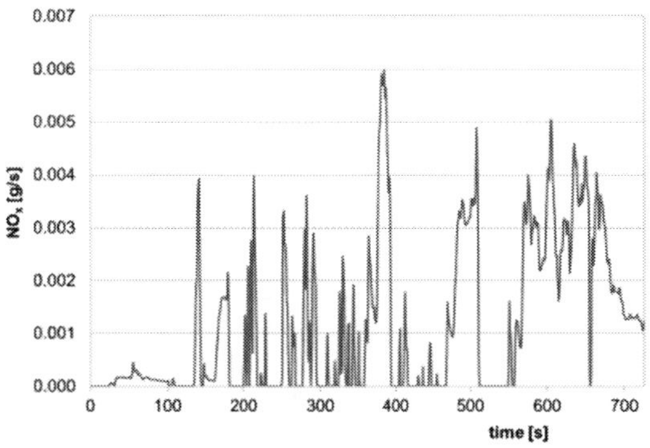

Fig. 6. HC emission from the chainsaw in the test cycle.

These changes are most visible in Fig. 7 showing the curve of the mass flow and temperature of the exhaust. The emission of all three exhaust components was close to zero when the engine idled (the fragment from approx. 500 to 550s). Besides, it is noteworthy that in the initial phase of operation (up to approx. 100 s) a relatively high emission of CO and HC occurred while the emission of NO_x was rather low. This was the start of the engine operation and it most likely resulted from the engine thermal state (engine was cold).

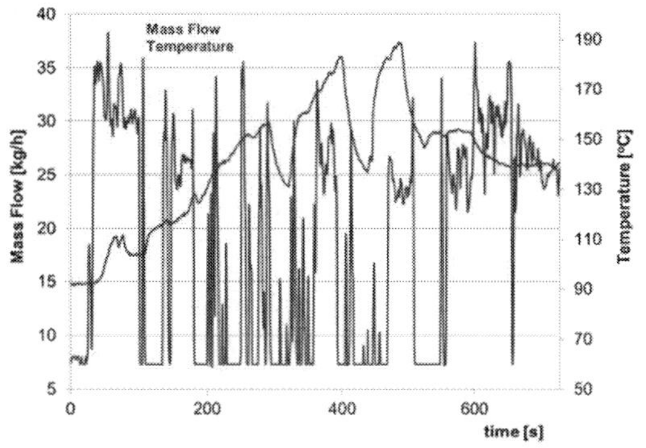

Fig. 7. The curves of the exhaust mass flow and temperature of the chainsaw engine during the tests.

In the case of the engine fitted in the power generator, the hourly emission of HC and CO changes proportionally to the engine load (Fig. 8 and 9). The engine load was determined based on the energy consumption by the electrical receivers. The power generator operated at a steady engine speed of 3000 rpm. Similarly to the chainsaw engine, in the initial phase of operation we may observe an increased emission of HC and CO.

The nature of the emission changes along the changes in the engine load is most visible for HC, for which the change of load immediately results in an increased emission of this exhaust components and the temperature (Fig. 8). The changes in the emission of CO are not as great as it is in the case of HC.

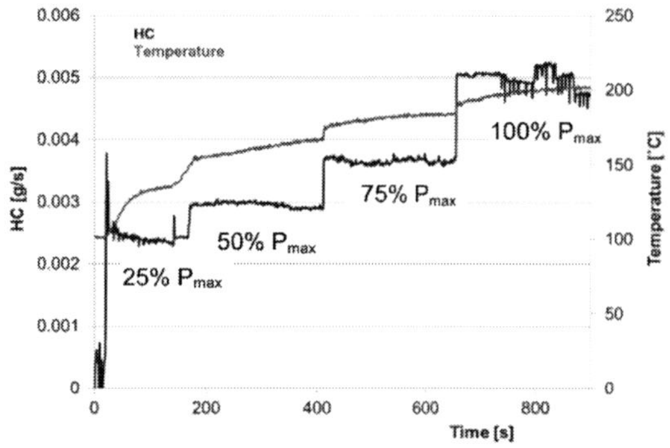

Fig. 8. HC emission from the power generator and temperature changes in the test cycle.

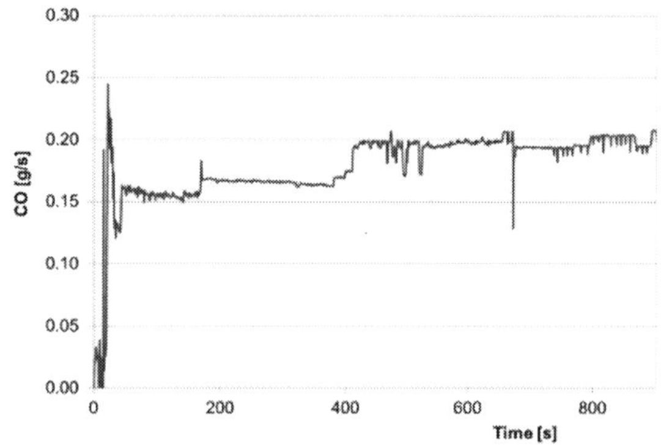

Fig. 9. CO emission from the power generator in the test cycle.

Figure 10 present the results of the measurements of the exhaust emissions and the applicable Stage III emission limits (standard applicable to the investigated equipment). It should be noted that the tests according to the European standards are performed under different conditions – these are laboratory tests carried out on measurement test stands, which is why the comparison is for information purposes only. Additionally, since the engine load of the handheld chainsaw was not measured, it was arbitrarily assumed at 80% of the nominal power output throughout the entire test. In this case, the results are for information purposes as well. From the presented comparisons it results that the emission of CO and $NO_x + HC$ under actual operating conditions of the chainsaw is greater than the admissible values similarly to the emission of HC and CO from the power generator (NO_x not included, because the concentration was out the measurement range).

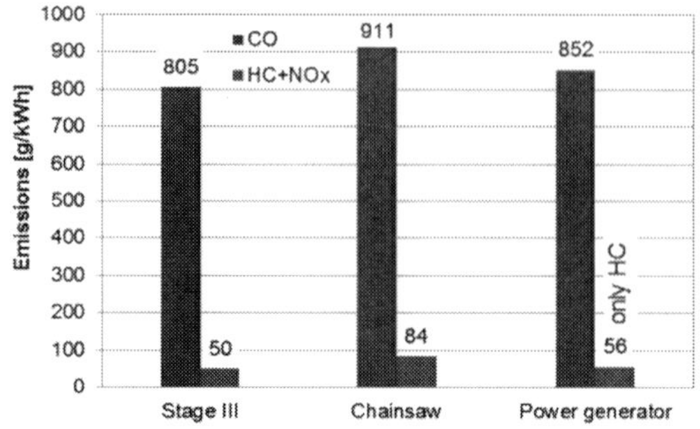

Fig. 10. Comparison of the exhaust emissions from the handheld chainsaw, the power generator and Stage III limits.

4 Conclusions

The results of the investigations presented in the paper and their analysis confirm that the selected method of testing under actual conditions of operation may be successfully applied in works on the advancement of small engines fitted in handheld devices. It is however, necessary to continue works on the improvement of not only the measurement equipment dedicated for the engine of these devices but also the engines themselves. The measurements carried out on small engines are different from the measurements performed on the engines of on-road vehicles, which results from their specificity, conditions of operation as well as their design. One of the more important issues the needs to be resolved is the measurement of the engine speed and engine load that, contrary to on-road engines, are currently impossible to measure in majority of engines from the small engine category. Another issue is the adaptation of the PEMS measurement equipment. Majority of these devices have the exhaust components and exhaust mass flow measurement ranges out of the range that would correspond to the category of small engines. An additional problem is the need of a compact design of the measurement equipment. The development of the testing methodology for small engines operating under actual conditions, allowing for the specificity of operation of these devices also needs to be addressed.

The conducted research indicates that the engine load and speed have a significant impact on the exhaust emission of the tested engines. The operation of the chainsaw engine is characterized by changes in engine speed and engine load, which is expected to be the cause of the relatively high exhaust emissions. The impact of the engine load on the exhaust emission is clearly visible for the power generator engine, especially for the HC emission. Measured in real operating conditions, both HC and CO emissions are higher than the permissible values, which is likely due to different engine operating conditions. Therefore, changes in research test procedures (type approval) should be sought that aim to better reflect actual operating conditions of the tested units.

While carefully following the current trends in engine exhaust emissions, it appears fully justified to undertake emission research related to small engines working under actual operating conditions. In the future such research should be included in the homologation procedure.

The study presented in this article was performed within the statutory research (contract No. 05/52/DSPB/0244).

References

1. Global Industry Analysts Inc., Report MCP-2708 (2015)
2. Worldwide Emissions Standards, Delphi Innovation for the Real World 2012/2013 (2012)
3. M. Bajerlein, L. Rymaniak, P. Swiatek, A. Ziolkowski, P. Daszkiewicz, M. Dobrzynski, Applied Mechanics and Materials **518**, 108-113 (2014), DOI: 10.4028/www.scientific.net/AMM.518.108
4. M. Bajerlein, L. Rymaniak, Applied Mechanics and Materials **518**, 96-101 (2014), DOI: 10.4028/www.scientific.net/AMM.518.108
5. J. Merkisz, J. Pielecha, *Springer Tracts on Transportation and Traffic* **8**, 131-139 (2015), DOI: 10.1007/978-3-319-15928-7_10
6. J. Merkisz, J. Pielecha, W. Gis, *WIT Transactions on Ecology and the Environment* **120**, 439-449 (2009), DOI: 10.2495/SDP090411
7. J. Merkisz, J. Pielecha, S. Radzimirski, *Springer Tracts on Transportation and Traffic* **4**, 1-170 (2014), DOI: 10.1007/978-3-319-02705-0
8. J. Merkisz, J. Pielecha, W. Gis, *WIT Transactions on Ecology and the Environment* **120**, 439-449 (2009), DOI: 10.2495/SDP090411
9. S. Abolhasani, H.C. Frey, K. Kim, et al., Journal of the Air & Waste Management Association **58**, 1033-1046 (2008).
10. M. Benjamin, 2013 PEMS Conference & Workshop University of California, Riverside (2013)
11. P. Bonnel, J. Kubelt, A. Provenza, JRC Scientific and Technical Reports, Publications Office of the European Union (2011)
12. P. Fuc, P. Kurczewski, A. Lewandowska, et al., The International Journal of Life Cycle Assessment **21**, 1438-1451 (2016), DOI: 10.1007/s11367-016-1104-y
13. C. Fulper, *2011 PEMS Conference & Workshop* (Riverside, 2011)
14. A. Mamakos, P. Bonnel, M. Carriero, A. Perujo, *2013 PEMS Conference & Workshop* (Riverside, 2013)
15. J. Merkisz, J. Markowski, J. Pielecha, *WIT Transactions on Ecology and the Environment* **174**, 219-230 (2013), DOI: 10.2495/AIR130191
16. R. Jasinski, J. Pielecha, J. Markowski, *E3S Web of Conferences* **10**, UNSP 00030 (2016), DOI: 10.1051/e3sconf/20161000030
17. J. Pielecha, J. Merkisz, J. Markowski, R. Jasinski, *E3S Web of Conferences* **10**, UNSP 00073 (2016), DOI: 10.1051/e3sconf/20161000073
18. H. Edtmayer, A. Trattner, A. Schmidt, et al., SAE Technical Paper 2013-32-9124 (2013)
19. J. Merkisz, P. Lijewski, P. Fuc, SAE Technical Paper 2011-32-0511 (2011)
20. J. Szczepaniak, W. Tanas, T. Pawlowski, J. Kromulski, Annals of Agricultural and Environmental Medicine **21**, 403-406 (2014), DOI: 10.5604/1232-1966.1108613
21. J. Kromulski, T. Pawlowski, J. Szczepaniak, W. Tanas, A. Wojtyla, M. Szymanek, J. Tanas, W. Izdebski, Annals of Agricultural and Environmental Medicine **23**, 373-376 (2016), DOI: 10.5604/12321966.1203908
22. Lai Ji-Ying, Lin Chih-Wei, Hsu Chia-Wei, et al., *2015 PEMS Conference & Workshop* (Riverside, 2015)

MATEC Web of Conferences **118**, 00017 (2017) DOI: 10.1051/matecconf/201711800017
VII International Congress on Combustion Engines

Various aspects of research of the SI engine with an additional expansion process

Marcin Noga[1,*]

[1]Cracow University of Technology, Institute of Automobiles and Internal Combustion Engines, al. Jana Pawla II 37, 31-864 Krakow, Poland

Abstract. The paper presents an analysis of the results of the both experimental results and theoretical works on the SI engine with additional expansion of exhaust gases, also known as five-stroke engine. The engine like this was constructed at Cracow University of Technology as a retrofitted in-line four cylinder engine in which outer cylinders (1st and 4th) work as fired cylinders and inner cylinders (2nd and 3rd) work as volume for the additional expansion. The aim of development of such an engine is to gain higher energy recovery ratio of the combusted fuel through the second expansion of exhaust in a separate cylinder. The operating parameters of the engine in various versions were analyzed: as naturally aspirated, supercharged using a turbocharger with a waste-gate valve and a turbocharger with variable nozzle turbine. Selected results of the indicating measurements of the engine with special emphasis on the indicated thermal efficiency were presented. The results pointed out the directions of further optimization of the engine. These results are all the more important, because according to the author's knowledge, the research on the real object of this type are carried out in only one science center in the world besides Cracow University of Technology.

1 Introduction

The use of internal combustion engines to drive vehicles is inevitably associated with emissions of toxic exhaust gas and carbon dioxide into the atmosphere. The number of motor vehicles on the roads of the world is growing year by year. In order to reduce pollution of the natural environment, more stringent emission limits for toxic gases and CO_2 emissions are introduced. The vehicle approval testing procedures are changed to match the actual road conditions [1, 2]. However, despite the introduction of advanced technical solutions for the engines themselves as well as the exhaust aftertreatment systems, it is not possible to completely eliminate the problem of emissions from the powertrains with combustion engines. Consequently, interest in electric vehicles has increased over the past 20 years [3]. The pure electric drive for vehicles has the advantage that the vehicle does not emit toxic components or CO_2 in the area where it is used. Some negative connotations are connected with the way of generating electricity in a given area [4]. In any case, it is true that the use of electric cars in crowded city centers will significantly improve the quality of the air there.

However, electric drives for road vehicles encounter many problems in their development. The most important problem is the extremely limited capacity of electric energy storage system using contemporary batteries, which in turn causes very significant problems in obtaining a range acceptable to the user while maintaining a reasonable price for such a vehicle [5]. The next constraint in development is the long battery charging time and the troubles of developing a common standard of charging device for all electric car manufacturers [6]. Such a state of affairs makes that a reasonable solution for the coming years is to improve and further disseminate the various hybrid drive systems for road vehicles [7]. First of all the use of a hybrid system gives the ability to reduce fuel consumption through energy recovery during braking and enables the engine to operate in a limited area of the operational map in which it is highly efficient. On the one hand, electrical systems are developed. Innovative high-performance electric motors are emerging, innovative electronic solutions such as SiC-based transistors are used in inverters and voltage converters. On the other hand, internal combustion engines for hybrid drive systems are also subject to continuous improvement [8].

2 Engines with an extended expansion process – the background

As with all automotive combustion engines, also for engines for hybrid vehicles, the development is now primarily driven by emission reduction and fuel consumption considerations. The engines of hybrid drive systems work in slightly different regime than in conventional propulsion systems. This causes that they can be optimized for a narrower area of the work map, to

* Corresponding author: noga@pk.edu.pl

work mainly in this optimized area [9]. One of the ways to make better use of the part of energy of the burned fuel that is lost in the exhaust system in a normal engine is to use an increased expansion ratio of the exhaust gases in relation to the compression ratio [10-12]. A comparison of the thermodynamic cycle of the classical SI engine and the cycle with expansion of exhaust gases to the ambient pressure is presented in a Figure 1.

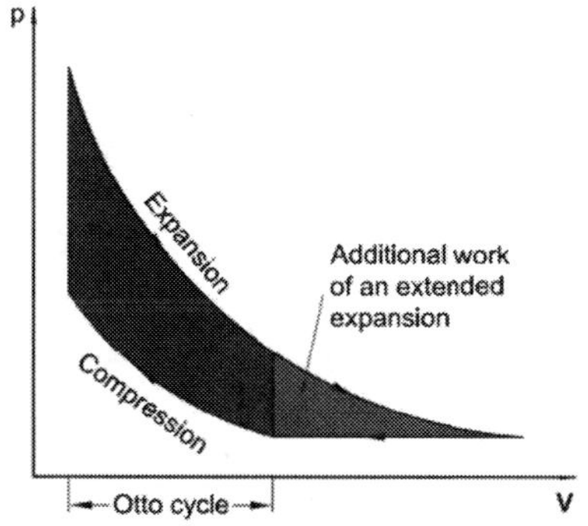

Fig. 1. Thermodynamic cycle with an additional expansion process.

As it is shown in the figure above, the cycle with extended expansion of the charge with the same amount of heat supplied to the working medium offers an additional portion of useful work, which contributes to a significant increase in the efficiency of the engine working according to this principle [13, 14].

The first attempts to implement this idea into a heat engine include the James Atkinson engine [15], in which the intake and compression stroke were shorter than the power (expansion) stroke and exhaust stroke. Miller presented a different approach for a 4-stroke supercharged engine [16]. He proposed closing the intake valve before the end of the intake stroke, which, in addition to cooling the expanding cylinder charge, reduced the effective compression ratio. A similar effect is achieved by closing the intake valve far after the BDC, what was used in Japanese car engines as early as the 90s of the 20th century [17]. In this type of engine, the expansion of the charge occurs as in the classical engine, which results, with the shortened effective part of the compression stroke, in a significant increase in the expansion ratio over the compression ratio. In recent years, the popularity of engines realizing Atkinson/ Miller cycle at least in part of their map has increased [18]. Research and development works on engines with increased expansion ratio are also carried out in several scientific centers in Poland [19-21].

However, it should be noted that increasing the efficiency of internal combustion engines in this way is not free of disadvantages. The first one is the reduced maximum performance [22] resulting from reduced volumetric efficiency. Another is pumping back the fuel-air mixture to the intake manifold at the start of the compression stroke with the intake valve open, which is not a loss-free process [23].

A slightly different approach to the concept of extended charge expansion was presented by Gerhard Schmitz [24] in his engine, which proposed that additional (second) expansion of exhaust gas occurs in a separate low pressure cylinder with a volume significantly higher than the volume of the fired cylinder in which a four-stroke cycle is performed. The engine with additional expansion of the exhaust in a separate cylinder has the advantage that the design parameters of the two cylinders can be optimized for their functions independently of each other. This mainly concerns the compression ratio of the working cylinder and the volume of additional expansion of the working medium. The first working model of this engine was built by Ilmor Engineering and named as a five-stroke engine due to the fifth stroke of an additional expansion added to the classic four-stroke cycle. Gerhard Schmitz's five-stroke engine has two working cylinders with 360 deg CA phase shift of the operation. Exhaust gases from these cylinders feeds alternately the additional expansion cylinder placed in the middle of engine, which in this situation operates in a two-stroke mode. The five-stroke engine design is shown in Figure 2.

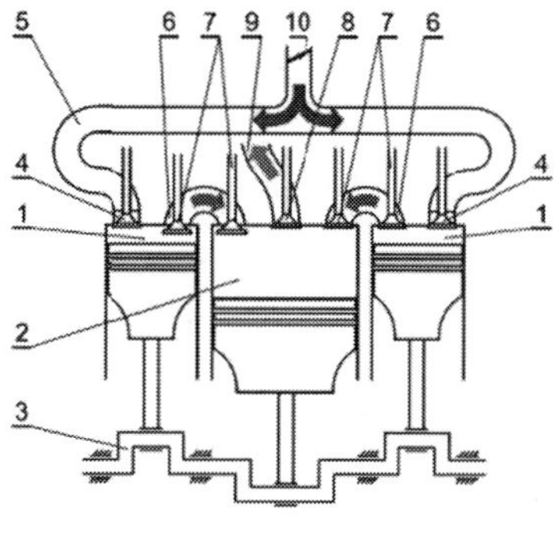

Fig. 2. Scheme of the three cylinder five-stroke engine; 1 – fired cylinders, 2 – additional expansion cylinder, 3 – crankshaft, 4 – intake valves, 5 – intake manifold, 6 – transfer port, 7 – transfer port valves, 8 – exhaust valve, 9 – exhaust port, 10 – throttle.

Currently, the inventor of the five-stroke engine develops his concept in a team working on the use of this engine for a hybrid powertrain [25, 26]. Generally, the issue of the five-stroke engine raises a lot of interest among researchers from all over the world. This is evidenced by publications on various modelling and simulations of the 5-stroke engine [27-31], however, the results of the experimental work of such an engine were presented by the team whose member is the originator of the engine [25, 26] as well as by the author of this paper [10, 13, 32, 33].

3 The five-stroke engine developed at Cracow University of Technology

In 2011-2013, at the Cracow University of Technology (CUT) a research project N N509 559040 funded by the National Science Center of Poland was carried out.

Professor Bronisław Sendyka was head of the project and the author of this article was a principal investigator. Within this project, an expanded-expansion engine was developed based on an in-line four cylinder engine. The developed engine has two fired cylinders (No. 1 and No. 4), while the central cylinders (No. 2 and No. 4) are connected by an internal port in the cylinder head and form a volume of additional expansion of working medium. The project required significant modifications to the cylinder head ports and designing of new camshafts and intake and exhaust manifolds. As shown in Figure 3 the exhaust gases from the cylinders No. 1 and No. 4. are now returned to the cylinders respectively No. 2 and No. 3, where they undergo an additional expansion process. Only after this process exhaust gases leave the engine. This causes that the camshaft, which originally controlled only the intake valves of the engine, now controls the intake valves of the fired cylinders 1 and 4 and the exhaust valves of the cylinders of an additional expansion process No. 2 and No. 3. The camshaft which was an exhaust camshaft for the 4-stroke engine now controls the valves of the transfer ports between cylinders 1 and 2 as well as 4 and 3. It should also be noted that all engine exhaust valves open at the same time once for each rotation of the crankshaft (2-stroke mode of work of the additional expansion cylinders).

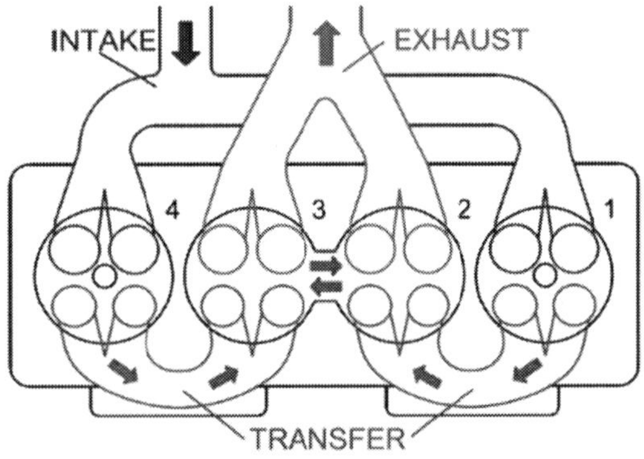

Fig. 3. Scheme of a cylinder head flow of the five-stroke engine developed at Cracow University of Technology.

After the preliminary tests, where the engine worked as a naturally aspirated engine, it was equipped with a turbocharger. Experimental research of the engine equipped with a turbocharger with a boost pressure controlled by a waste-gate valve (WG) upstream the turbine, followed with a turbocharger with a variable-nozzle turbine (VNT) were carried out. Selected results of this work will be presented later in the article. The specifications of the five-stroke engine developed at CUT are presented in a Table 1.

Control of the engine work is based on a programmable engine management system that allows real-time settings to be changed. In addition, the engine has a direct injection system. Fuel injection pressure during tests, the results of which are presented here, was set at 7 MPa. The engine was supplied with a mixture of stoichiometric composition (rel. AFR = 1.0) with a wideband exhaust oxygen sensor in a feedback loop. This was due to the desire to match the engine to the requirements of the classic three-way catalytic converter.

Table 1. Technical data of the five-stroke engine developed at CUT.

Parameter	Value
Number of cylinders and arrangement	4 in-line, 2 fired and 2 for an additional expansion
Bore	82.5 mm
Stroke	92.8 mm
Displacement of a fired cylinder	496 cm^3 each
Displacement of additional expansion cylinders	992 cm^3 both
Compression ratio	10.5
No. of valves	4 per cylinder
Turbocharger type	KP39 (with WG), BV35 (with VNT)
Control system model	AEM EMS 30-1010
Injection method	Direct injection
Injector driver	Denso 131000-1041
Control method of high-pressure of fuel	pressure relief valve

A valve timing of the test engine are listed in Table 2. The intake valve timing of the fired cylinders is fixed and correspond to the center position of the camshaft in the base engine, which originally had a system of variable valve timing. The exhaust valve timing of the fired cylinders and the additional expansion cylinders are the same as in the base engine. The intake valves of the additional expansion cylinders work synchronously with the exhaust valves of the adjacent fired cylinders, i.e. 1 together with 2 and 4 together with 3.

Table 2. Valve timing of the engine with an additional expansion process developed at CUT.

Valve timing	Fired cylinders	Additional expansion cylinders
EVO	28° BBDC	28° BBDC
EVC	8° BTDC	8° BTDC
IVO	7° ATDC	28° BTDC
IVC	17° ABDC	8° BBDC

The presented valve timing is given in accordance with the manufacturer's base engine guidelines. This means that the opening and closing angles of the valves are given for the valve lift of 1.0 mm. The engine tests was carried out in the Laboratory of Combustion Engines at Cracow University of Technology. The test stand is equipped with an electronically controlled Eddy-current dynamometer, with a system for measuring and recording engine operating parameters, a gravimetric fuel flowmeter and an exhaust gas analyzer. The general view of the 5-stroke engine test bench is shown in Fig. 4.

Fig. 4. General view of a test-stand of the engine with additional expansion process developed at CUT.

4 Selected research aspects of the engine with an additional expansion process

4.1 Thermodynamic cycle of the engine with an additional expansion process

For initial assessment of the potential of improving the efficiency the theoretical cycle for the engine with an additional expansion of exhaust gas and the expansion ratio higher twice than the compression ratio has been developed [10, 13]. This is the cycle based on the Otto cycle, but with an extended expansion process to a volume equal to the total volume of the two engine cylinders. For the analysis air as an ideal gas was used, which is the working medium of the cycle. Due to the simplification of the analysis, the cycle is carried out in one working volume, and the pressure drop at the opening of the valves of transfer ports connecting the fired cylinder to the additional expansion cylinder is not taken into account. In the aforementioned works, the influence of selected parameters of the cycle on the obtained thermal efficiency of the cycle was analyzed. It is important to point out that, unlike the Otto cycle, the thermal efficiency of the cycle with an additional expansion depends on the amount of heat supplied to the cycle. Figure 5 shows the theoretical cycle of the engine with an additional expansion of exhaust gases developed at CUT.

If the amount of heat supplied to the cycle is high enough (process 2-3), the expansion process ends up above the ambient pressure – point 4_{5s}. Then the process of heat rejection at a constant volume occurs - process 4_{5s}-5_{5s} and the process of heat rejection at constant pressure of 5_{5s}-1. In the particular case, when the amount of heat supplied to the cycle is accordingly lower (process 2-3'), the expansion of the charge ends exactly at 5_{5s}, i.e. at atmospheric pressure. When the amount of heat supplied to the cycle is even smaller (process 2-3") the expansion of the working medium to the volume of $2\times(V_{cyl} + V_{ch})$ results in a pressure drop below the ambient pressure p_{amb} – process (3"-4"$_{5s}$). Then there is the process of additional heat supply to the cycle at constant volume (process 4"$_{5s}$-5_{5s}), followed by the heat rejection at a constant pressure.

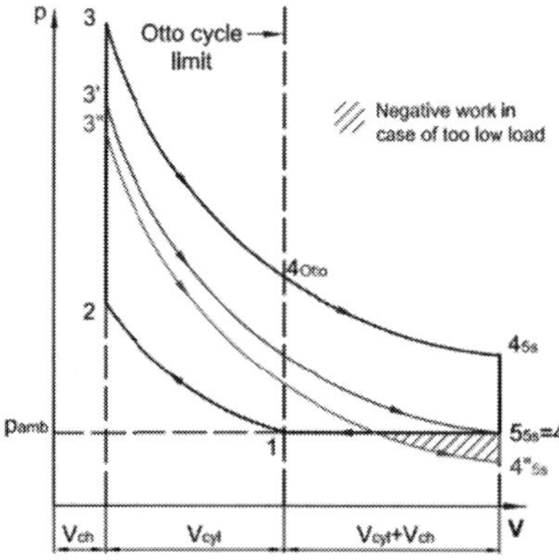

Fig. 5. Thermodynamic cycle of the engine with an additional expansion process to the volume of $2\times(V_{cyl}+V_{ch})$.

The graph shows the area of a negative work appearing at that time. The paper [13] presents the formula for the thermal efficiency of the proposed cycle:

$$\eta_{5s} = \frac{\alpha - 1 + \varepsilon_{cp}^{1-\kappa} \cdot (2 - \kappa - \alpha \cdot 2^{1-\kappa})}{\alpha - 1} \tag{1}$$

where:
η_{5s} – thermal efficiency of the proposed cycle,
α – heat addition pressure ratio, $\alpha = {p_3}/{p_2}$,
κ – isentropic exponent,
ε_{cp} – compression ratio, $\varepsilon_{cp} = \frac{V_1}{V_2} = \frac{V_{ch}+V_{cyl}}{V_{ch}}$.

The formula (1) for calculating thermal efficiency of the cycle is valid for the amount of heat supplied to the cycle allowing to reach the pressure of the point 3' and higher. Table 3 shows the dataset used for the analysis of the proposed cycle.

Table 3. Parameters of the proposed cycle used for presented analysis.

Parameter	Symbol	Unit	Value
Specific heat at const. pressure	c_p	[kJ/(kg·K)]	1.005
Specific heat at constant volume	c_V	[kJ/(kg·K)]	0.718
Isentropic exponent	κ	[–]	1.4
Mass of the working medium	m	[kg]	0.636×10^{-3}
Initial pressure (ambient)	p_1	[MPa]	0.1
Initial temperature (ambient)	T_1	[K]	300
Cylinder displacement	V_{cyl}	[m^3]	496×10^{-6}
Volume of a comb. chamber	V_{ch}	[m^3]	52.2×10^{-6}
Overall expansion ratio	ε_{dcp}	[–]	21

The expansion ratio is exactly twice of the compression ratio ε_{cp}, because the final volume of working medium is equal to the displacement volume of the two cylinders and their two combustion chambers, thus the formula (2) is true:

$$\varepsilon_{dcp} = \frac{2\cdot(V_{ch}+V_{cyl})}{V_{ch}} = 2\varepsilon_{cp} \quad (2)$$

If one takes into account the data set from Table 3, the minimum amount of heat supplied to the cycle Q'_{H_min} allowing to reach the pressure of point 3' can be determined from the set of following equations (3)-(7):

$$Q'_{H\_min} = m \cdot c_V \cdot (T_{3'} - T_2) \quad (3)$$

$$T_{3'} = T_2 \cdot \frac{p_{3'}}{p_2} \quad (4)$$

$$p_{3'} = p_{4'5s} \cdot \left(\frac{V_{4'5s}}{V_{3'}}\right)^{\kappa} = p_{4'5s} \cdot \left(\varepsilon_{dcp}\right)^{\kappa} \quad (5)$$

$$T_2 = T_1 \cdot \frac{p_2 \cdot V_2}{p_1 \cdot V_1} = T_1 \cdot \frac{p_2}{p_1} \cdot \left(\varepsilon_{cp}\right)^{-1} \quad (6)$$

$$p_2 = p_1 \cdot \left(\frac{V_1}{V_2}\right)^{\kappa} = p_1 \cdot \left(\varepsilon_{cp}\right)^{\kappa} \quad (7)$$

Finally, after the insertion of data and execution of the calculations, the minimum amount of heat supplied to the cycle Q'_{H_min} was equal to 575.1 J. For such a limit of the supplied heat, the heat-addition pressure ratio (or explosion ratio) α' equals (8):

$$\alpha' = \frac{p_{3'}}{p_2} = \frac{7.097526}{2.689460} = 2.639 \quad (8)$$

Figure 6 shows a curve of the thermal efficiency η5s of the theoretical cycle of the engine with an additional expansion of the working medium to the total volume of the two cylinders as a function of the amount of heat supplied to the Q_H cycle. The figure also shows the thermal efficiency of the Otto cycle, which has a constant value, independent of the amount of heat supplied to the cycle.

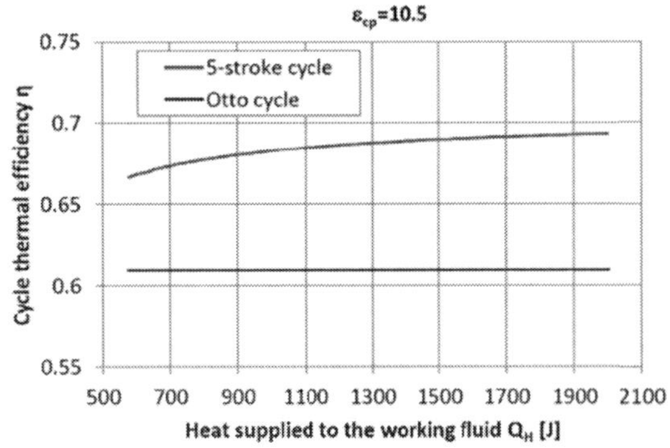

Fig. 6. Comparison of the thermal efficiency of Otto and 5-stroke engine cycle as a function of amount of heat supplied to the working fluid.

For the data being analyzed, Otto cycle efficiency is about 0.61 and is constant, while the efficiency of the cycle of the engine with an additional expansion process increases with the increase in the amount of heat supplied to the working medium Q_H, which for a real engine means increased load.

4.2 Operational parameters of the naturally aspirated engine

In the first stage of development of the engine with additional expansion of the exhaust gases, tests of the engine in the naturally aspirated (NA) version were conducted. Figure 7 shows the charts of selected engine operating parameters as a function of rotational speed. The engine was fueled by a stoichiometric mixture (rel. AFR = 1.0) and worked at wide open throttle (WOT). The ignition timing angle was adjusted to obtain maximum brake torque (MBT).

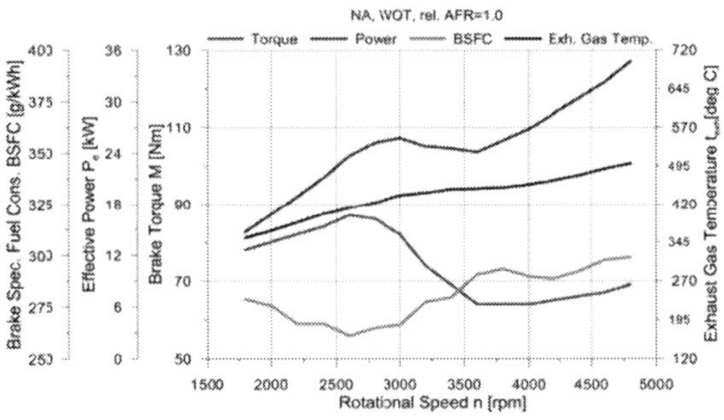

Fig. 7. Operational parameters of the naturally aspirated (NA) engine with an additional expansion process as a function of rotational speed.

During the tests the engine achieved a maximum effective power of 35 kW at 4800 rpm. The tests was not continued for a higher rotational speed because the brake specific fuel consumption (BSFC) has already reached a high level of 300 g/kWh. The exhaust gas temperature after the extra expansion process is low. During the tests it did not exceed 500 °C. It can be seen in the figure above that, although the maximum value of the torque is relatively high, the torque curve is irregular and different from that encountered for conventional naturally aspirated engines. In order to verify the initial assumptions regarding the explanation of this state, the volumetric efficiency was calculated. Figure 8 shows the curve of the volumetric efficiency as a function of engine speed. The torque curve of the engine has been also showed in this figure to facilitate the analysis.

As it is shown in the figure above, the volumetric efficiency reaches a relatively high maximum value. This is caused by the use of air filter, intake pipe and throttle from the base engine (2.0 dm^3 of displacement, with turbocharging) but above the rotational speed of 2800 rpm η_V begins to decrease significantly. This phenomenon may be associated with a disturbed process of exhaust in the engine, in which the exhaust gas after the power stroke does not discharge into the atmosphere, but flows into a next cylinder for further expansion.

 DOI: 10.1051/matecconf/201711800017

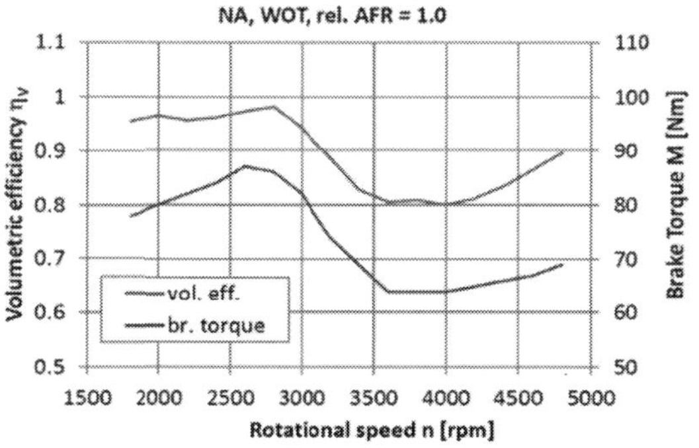

Fig. 8. Volumetric efficiency of the naturally aspirated engine with an additional expansion process.

From a rotational speed of 4200 rpm, η_V starts to rise slightly again, which in turn can be explained by the occurrence of wave phenomena in the intake manifold pipes with a relatively low length to diameter ratio. The described changes in the course of η_V as a function of rotational speed are reflected in the brake torque curve of the engine. Minor differences between waveforms of the volumetric efficiency and the brake torque are most likely due to changes in the thermal efficiency and mechanical efficiency of the engine, which was not analyzed at this initial stage of project.

4.3 Operational parameters of the turbocharged engine with a turbocharger with WG Valve and with a turbocharger with VNT

The results of the initial tests of the engine in the naturally aspirated version indicated that the minimum BSFC was increased with respect to the original 4-stroke engine, where the best achieved value was slightly below 240 g/kWh. On the one hand, this is contrary to the aim of the undertaken research work, on the other hand, at the initial stage of the development of the engine in the naturally aspirated version it was an expected phenomenon. The process of additional expansion of the exhaust gases allows to increase the thermal efficiency of the cycle, but one should remember that the mechanical efficiency of the engine, in which in two of four cylinders only additional expansion process takes place, will be undoubtedly lower than for the four-stroke four-cylinder engine.

The five-stroke engine's theoretical cycle exhibits an increase in thermal efficiency along with an increase in the amount of heat supplied, so that, for satisfactory performance, the test engine was equipped with a turbocharger with a boost pressure control by means of waste-gate valve (Fig. 9).

Figure 10 presents selected test results for a turbocharged engine with the turbocharger with the WG valve. The boost pressure (p_{bst}) was limited to about 0.35 bar by directly feeding the boost pressure to a pneumatic actuator of the WG valve.

Fig. 9. Turbocharger with a Waste Gate valve mounted on the engine with an additional expansion process.

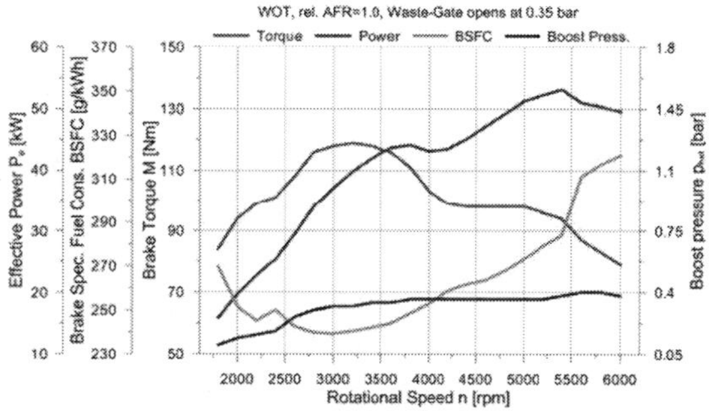

Fig. 10. Operational parameters of the turbocharged engine with an additional expansion process as a function of rotational speed.

Under these conditions, the engine achieved maximum torque of 119 Nm at 3200 rpm and a maximum effective power of 53 kW at 5400 rpm. The minimum BSFC was 239 g/kWh at 3000 rpm, which is comparable to the best result of the 4-stroke base engine. Above the engine speed of 3600 rpm, engine brake torque drastically decreased despite maintaining a constant boost pressure. This is due to the limitation of the engine volumetric efficiency due to the increased flow resistance of exhaust of the engine (Fig. 11).

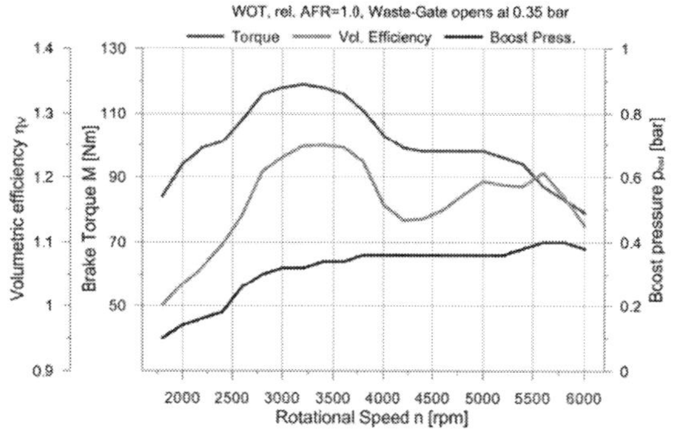

Fig. 11. Volumetric efficiency of the turbocharged engine with an additional expansion process.

From the rotational speed of 4200 rpm, the volumetric efficiency increases again but no increase in the torque is observed. The reason for this is that, similarly to a naturally aspirated engine, the mechanical and thermal efficiency of the engine running at increasing rotational speed decreases.

On the other hand, the turbocharger with WG did not provide the expected boost pressure at low crankshaft rotational speed. The boost pressure of 0.3 bar was reached above the rotational speed of 2800 rpm. This meant that the turbocharger did not match the engine characteristics. For this reason, it was decided to investigate how the use of the VNT-type turbocharger would affect performance of the engine with an additional expansion of exhaust gases. After estimating the expected exhaust gas flow it was decided to use the BV35 turbocharger used in CI engines with a displacement of 1.3 and an effective power of about 66 kW. Figure 12 shows a view of a BV 35 turbocharger mounted to the test engine.

Fig. 12. View of the VNT-type turbocharger mounted to the tested engine.

Figure 13 shows the torque and the BSFC curves as a function of the boost pressure of the test engine running at 2000 rpm and supercharged with a VNT turbocharger. The boost pressure was increased by rotating the turbine nozzle vanes by means of a vacuum actuator.

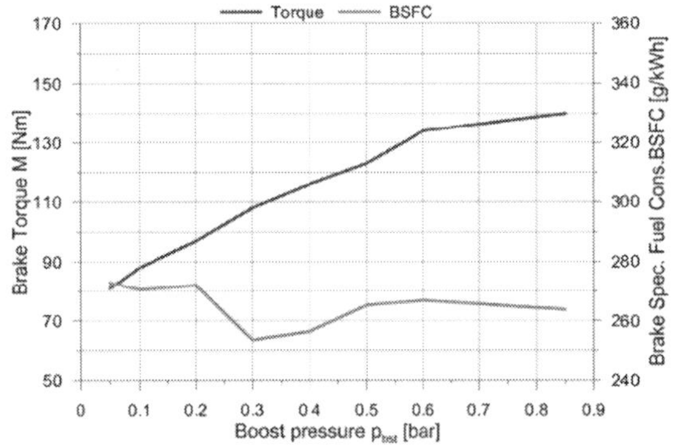

Fig. 13. Brake torque and BSFC as a function of boost pressure of the engine with an additional expansion process working at 2000 rpm with a VNT-type turbocharger.

The maximum torque of 2000 rpm was 140 Nm at 0.85 bar, which is a significant increase in comparison to the performance of the engine with a classic turbocharger (94 Nm at 0.14 bar). On the other hand, the BSFC is in the entire range higher than that recorded for the engine with a turbocharger with a conventional turbine (251.4 g /kWh at 0.14 bar). The minimum value of BSFC with the VNT-type turbocharger was achieved for a boost pressure of 0.3 bar – 253.5 g/kWh.

Figure 14 shows the torque and the specific fuel consumption curves as a function of the boost pressure p_{bst} for the engine with additional expansion and with the VNT-type turbocharger. The engine worked this time with a rotational speed of 2800 rpm.

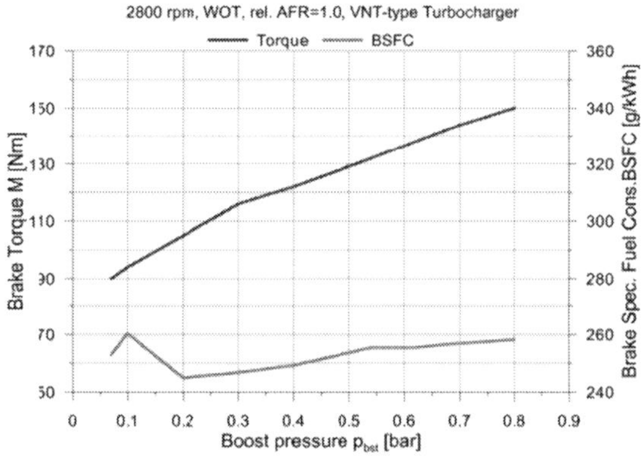

Fig. 14. Brake torque and BSFC as a function of boost pressure of the engine with an additional expansion process working at 2800 rpm with a VNT-type turbocharger.

For a rotational speed of 2800 rpm, the situation is similar to that of 2000 rpm. The use of the VNT turbocharger allows for a significant increase in boost pressure and, consequently, gives higher engine torque, however, the engine efficiency is reduced again. For the engine equipped with a turbocharger with WG the BSFC of value 239.7 g/kWh (at p_{bst} = 0.3 bar and M = 116 Nm) was obtained. With the VNT-type turbocharger, the minimum BSFC was 244.8 g/kWh at 0.2 bar of a boost pressure and 105 Nm of torque. For the same torque 116 Nm as for the turbocharger with WG, the engine had BSFC 246.7 g/kWh.

The use of the VNT turbocharger allowed to obtain the expected boost pressure even at lower rotational speeds, but the minimum BSFC was higher in both cases than with supercharging by the turbocharger with WG valve. The use of a turbocharger with adjustable nozzle vanes system increases the backpressure of exhaust gases, which in turn has an effect on reducing exhaust gas energy recovery in additional expansion cylinders. In addition, the variable nozzle of a turbine suppresses the effect of the pulsating gas flow into the turbine, which process is not disturbed in a turbocharger with a classic turbine.

DOI: 10.1051/matecconf/201711800017

4.4 Indicated thermal efficiency of the turbocharged engine

One of the research stages on the developed engine included in-cylinder pressure measurements. The pressure sensors were located in the fired cylinder No. 4 and in the adjacent additional expansion cylinder No. 3. The research was carried out on a turbocharged engine with WG valve. Measurements were made using Optrand optoelectronic pressure sensors [34].

In the paper [33] an analysis of indicating test results of the engine with additional expansion of exhaust gases is presented. The results of calculating the mechanical efficiency of the engine are presented. The obtained results do not deviate from the only known results available for a five-stroke engine [26]. The maximum value of mechanical efficiency was 85.5% [33], which is rather lower than that of today's four-stroke engines. The work also outlines the relationship between the power indicated in the fired cylinders and in the additional expansion cylinder assembly and the relationship between the brake mean effective pressure of the engine and the indicated mean effective pressure of the additional expansion cylinders. This allowed to determine the energy recovery rate in the additional expansion cylinders and to determine the load level of the engine from which the additional expansion cylinders deliver power to the engine crankshaft. The results of the analysis indicated that the additional expansion cylinders deliver power to the engine crankshaft at a load of about 60 Nm and higher. Until this value they receive power from the crankshaft, which is due to the fact that, with a low amount of cylinder charge, it is expanded in additional expansion cylinders below ambient pressure (see theoretical cycle). The results of calculating the indicated thermal efficiency of the cycle of engine with additional expansion process will be presented in this paper.

Indicating measurements were carried out at selected points in the engine working area. These points are marked in Figure 15. During testing, the WG valve of the turbocharger was closed in order to obtain maximum performance.

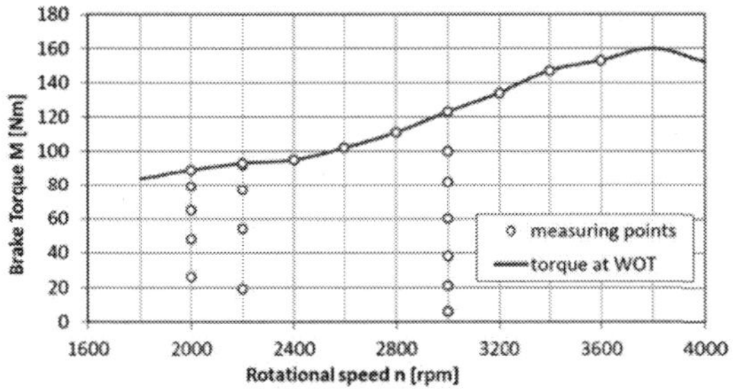

Fig. 15. Points of indicating measurements.

Figure 16 shows an example of the in-cylinder pressure waveforms as a function of crank angle registered for both cylinders at a rotational speed of 2800 rpm and a load of 111 Nm.

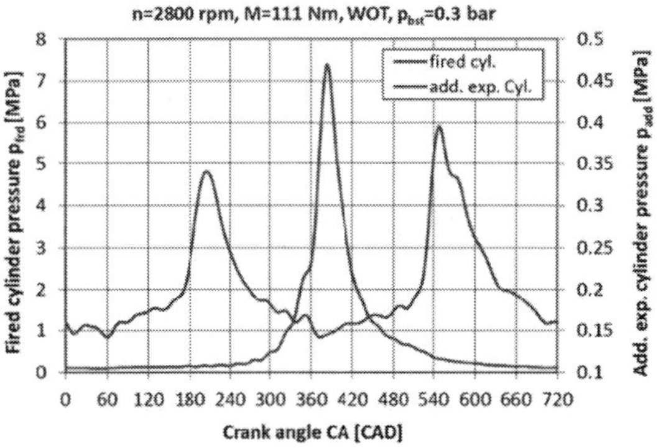

Fig. 16. Indicator diagram of the engine with additional expansion process.

It can be seen in figure above that the pressure course in the additional expansion cylinder No. 3 is not identical for the two presented cycles of this cylinder - covering 360 degrees of the crank angle (CAD). The right part of the chart (360-720 CAD) corresponds to the situation when this cylinder is fed directly from the cylinder No. 4 (red line), while the left part refers to the situation where the exhaust gases flow from the cylinder No. 1 to cylinder No. 2 and then, through the port in the cylinder head feed the cylinder No. 3. The phenomenon of throttling of flow in the port connecting the cylinders No. 2 and 3 explains the difference in the curves of pressure and its maximum value for the additional expansion cylinders. Based on the recorded pressure waveforms obtained as averaged over dozens of cycles, the indicated mean effective pressure (IMEP) was calculated for the fired and additional expansion cylinders. The indicated power was then calculated for both groups of cylinders. It has been taken into consideration that additional expansion occurs in the volume of two cylinders and in the two-stroke mode. Obtaining the indicated power values has allowed to calculate the indicated thermal efficiency of the engine at each measurement point.

For a classical combustion engine, the indicated thermal efficiency can be determined from the formula (9) [35, 36]:

$$\eta_i = \frac{P_i}{G_e \cdot LHV} \tag{9}$$

where:
P_i – indicated power, kW
G_e – fuel consumption rate, g/s
LHV – lower heating value of the fuel, MJ/kg

For the purposes of this analysis, the value of LHV equal to 43.0 MJ/kg was assumed. In the case of the engine with an additional exhaust expansion in a separate cylinder, the situation is somewhat different from that of the classic engine. If the engine load is sufficiently high, the exhaust gas recovery takes place in the additional expansion cylinders. In this situation, the indicated mean effective pressure of these cylinders $IMEP_{add}$ has a value higher than zero. The indicated power of the engine P_{i_5s} is the sum of the indicated power of the fired cylinders P_{i_frd} and in the additional expansion cylinders P_{i_add}. It may also happen that the load on the engine is so low that, in the process of additional expansion, the pressure in these cylinders falls significantly below the exhaust pressure. This is related to the occurrence of the indicated mean effective pressure in the cylinders of additional expansion $IMEP_{add}$ of a value lower than zero. This means that in the additional expansion cylinders the part of the power resulting from the combustion of the fuel in the fired cylinders is lost. If this is the case, the power indicated in the additional expansion cylinders P_{i_add} of a negative value should be omitted for calculating the indicated thermal efficiency of the cycle η_{i_5s}, as these cylinders act then as an inherent load e.g. a water pump or an oil pump in a conventional combustion engine. Naturally, this will lower mechanical efficiency of the five-stroke engine. Equations (10) and (11) show the method of calculating the indicated thermal efficiency of the engine with an additional expansion process considering the above description.

for $P_{i\_add} > 0$: $$\eta_{i\_5s} = \frac{P_{i\_frd}+P_{i\_add}}{G_e \cdot LHV} \qquad (10)$$

for $P_{i\_add} \leq 0$: $$\eta_{i\_5s} = \frac{P_{i\_frd}}{G_e \cdot LHV} \qquad (11)$$

Figure 17 shows the curves of the indicated thermal efficiency and the effective efficiency (fuel conversion efficiency) of the engine with an additional exhaust gas expansion as a function of the rotational speed at wide open throttle and closed WG valve. Figure 17 also shows the variation of boost pressure and BMEP during the test.

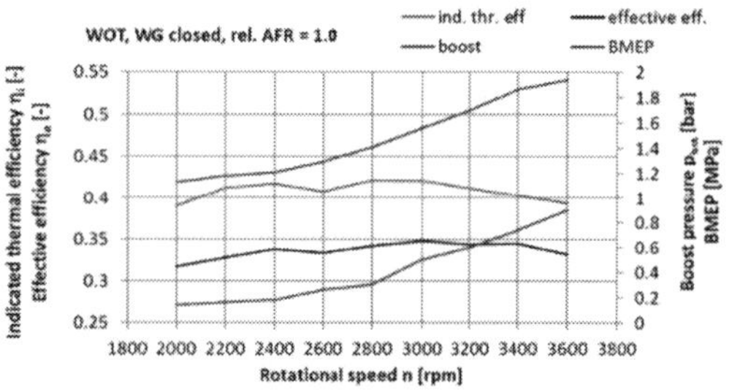

Fig. 17. Indicated thermal efficiency of the engine with an additional expansion process developed at CUT as a function of rotational speed.

The maximum recorded value of the indicated thermal efficiency was approximately 0.42 at 2800-3000 rpm. This is a relatively high value and is the result of an efficient energy recovery process in the additional expansion cylinders. For a rotational speed of 3000 rpm, the maximum effective efficiency of the engine operating with the closed WG valve is also recorded.

Figure 18 shows the charts of the indicated thermal efficiency of the engine with the additional expansion of exhaust gases for three constant values of the rotational speed and a variable load.

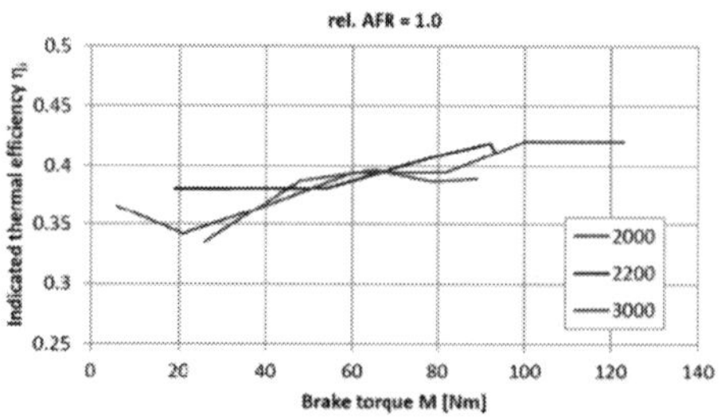

Fig. 18. Indicated thermal efficiency of the engine with the additional expansion process as a function of engine load.

It can be seen that the indicated thermal efficiency increases with the engine load to a maximum value of 0.42, and also acquires similar values for the various rotational speeds in the analyzed part of the engine operating map.

5 Conclusions and summary

On the basis of the results described in the paper, the author tried to formulate the following detailed conclusions, but also guidelines for further work on the subject:

- For the theoretical cycle of the engine with an additional expansion of exhaust gases, a significantly higher thermal efficiency is obtained than for the Otto cycle. In addition, the efficiency increases with increasing the amount of heat supplied to the cycle. For a real engine, a high indicated thermal efficiency has been achieved, but one should remember that the mechanical efficiency of a five-stroke engine must be lower than for a four-stroke engine. Improving the mechanical efficiency of an existing engine will be difficult. It is possible, for example, to change the method of high fuel pressure control from a relief valve to a supply "on demand" what means that only necessary amount of fuel is delivered at high pressure. On the other hand the indicated thermal efficiency may be increased in this engine, what will be described in a detailed form in the next few bullets.
- The engine with additional expansion of the exhaust gases in the naturally aspirated version achieves lower efficiency than the engine in the original four-stroke version. Such a result was generally expected. A participation of friction losses in the naturally aspirated five-stroke engine is so increased (transition from four fired cylinders to two fired and two additional expansion cylinders) that it is not compensated by the exhaust gas energy recovery during extra expansion. In order for the engine with an additional expansion process to operate effectively, it needs to be equipped with a turbocharger. It also follows from the thermodynamic cycle, where the efficiency gain is the higher the higher the amount of heat supplied to the working medium.

- The use of a turbocharger with the WG valve enabled a significant increase in engine performance and a reduction in the BSFC to a level of the original 4-stroke engine, however, the turbocharger did not provide sufficient boost pressure at low engine speeds.
- For the above reason, and because of the low exhaust gas temperature downstream the additional expansion cylinders, it was decided to use the VNT-type turbocharger. Initial tests of this solution have shown that the torque has increased significantly, but the BSFC was increased relative to the situation where the engine worked with the classic turbocharger with a WG valve. This is due to the increased back pressure of the exhaust gas in the additional expansion cylinders caused by the variable nozzle turbine. In the near future, a new turbocharger with a WG valve will be selected with a characteristic for the engine with a displacement of about 1.0 dm^3, but with exhaust gases of lower energy than that of a conventional engine, due to the additional expansion.
- The indicated thermal efficiency η_i of the engine with additional exhaust energy recovery is relatively high, is slightly higher than that obtained for the classic four-stroke engine. On the other hand, η_m is lower, so the effective efficiency η_e is now similar to that of the classical engine.
- Thermal efficiency would certainly increase thanks to the increasing of the cross section of the port connecting the cylinders No. 2 and No. 3, the reducing the volume of former combustion chamber in cylinders No. 2 and 3 and the shortening of transfer ports between fired and additional expansion cylinders. The simplest method is to increase the port cross-section between the additional expansion cylinders. It would be slightly more difficult to reduce the chambers in additional expansion cylinders by using pistons with higher crown. However, the modification of the transfer ports between the fired and additional expansion cylinders is practically impossible for the cylinder head coming from an existing four-stroke engine.
- It is also planned to optimize the valve timing using the 0D/1D modeling and simulation. Currently the valve timing is adopted as for the original 4stroke engine, so an improvement of the efficiency of the engine with an additional expansion process is still possible.
- Once the amendments to the existing design have been made, an exhaust gas aftertreatment system will be selected taking into account the specifics of the operation of the engine with an additional expansion (i.e. mainly low temperature downstream the turbine).

The study results show that the five-stroke engine developed at Cracow University of Technology achieves good performance, especially taking into account that it has been developed on the basis of an existing four-stroke engine, what creates some constraints (e.g. two separate cylinders for the additional expansion processes and the cylinder head design). The best results are obtained when working at high load and at about 3000 rpm. The result is that the engine would not be a good source of propulsion for the vehicle with classic drive system, but it will be suitable for applications to vehicles with hybrid drive of series type or extended range electric vehicle, or for stationary applications. A further development of the design is planned in the nearest future to increase the fuel conversion efficiency of the engine while maintaining the stoichiometric air-fuel mixture to allow the use of a simple system of a classic three-way catalytic converter.

References

1. J. Merkisz, J. Pielecha, S. Radzimirski, *Springer Tracts on Transportation and Traffic* **4**, 1-170 (2014), DOI: 10.1007/978-3-319-02705-0
2. W. Gis, J. Pielecha, J. Waskiewicz, M. Gis, M. Menes, *IOP Conference Series-Materials Science and Engineering* **148**, UNSP 012040 (2016), DOI: 10.1088/1757-899X/148/1/012040
3. L.J. Kapusta, I. Pielecha, K. Wisłocki, A. Teodorczyk, Journal of Thermal Analysis and Calorimetry **123**, 819-828 (2016), DOI: 10.1007/s10973-015-4927-z
4. P. Staniak, W. Moćko, A. Wojciechowski, Journal of KONES Powertrain and Transport **19**, 139-148 (2012)
5. H. Golas, A. Mazur, J. Gruszka, *ACSR-Advances in Computer Science Research* **57**, 257-264 (2016)
6. Z. Juda, Materials Science and Engineering **148**, 012043 (2016), DOI: 10.1088/1757-899X/148/1/012043
7. European Commission. Horizon 2020 Work Programme 2016-2017, Smart, green and integrated transport, 91
8. Y. Ota, Y. Ito, A. Kawamura, H. Nishiura, S. Matsuo, Toyota Technical Review **62**, 54-59 (2016)
9. G. Koszalka, Eksploatacja i Niezawodnosc – Maintenance and Reliability **16**, 133-139 (2014)
10. M. Noga, B. Sendyka, International Journal of Automotive Technology **15**, 1083-1089 (2014), DOI: 10.1007/s12239-014-0112-9
11. I. Pielecha, W. Cieslik, Journal of Thermal Analysis and Calorimetry **126**, 815-827 (2016), DOI: 10.1007/s10973-016-5544-1
12. I. Pielecha, M. Skowron, W. Bueschke, W. Cieslik, K. Wislocki, *IOP Conference Series-Materials Science and Engineering* **148**, UNSP 012081 (2016), DOI: 10.1088/1757-899X/148/1/012081
13. M. Noga, B. Sendyka, Bulletin of the Polish Academy of Sciences: Technical Sciences **62**, 349-355 (2014), DOI: 10.2478/bpasts-2014-0034
14. I. Pielecha, Journal of Thermal Analysis and Calorimetry **118**, 217-225 (2014), DOI: 10.1007/s10973-014-3956-3
15. C.L. Cummins, *Internal Fire* (Carnot Press: Wilsonville, 2000)

16. R.H. Miller, ASME Transactions **69**, 453-464 (1947)
17. T. Goto, K. Hatamura, S. Takizawa, N. Hayama, H. Abe, H. Kanesaka, SAE Technical Paper 940198 (1994), DOI: 10.4271/940198
18. S. Harada, T. Shinagawa, T. Kondo, K. Togawa, M. Kudo, W. Matsubara, Toyota Technical Review **62**, 54-59 (2016)
19. K. Grab-Rogaliński, S. Szwaja, Combustion Engines **162** (2015)
20. I. Pielecha, J. Czajka, P. Borowski, K. Wisłocki, Combustion Engines **154**, 517-524 (2013)
21. I. Pielecha, International Journal of Automotive Technology **15**, 47-55 (2014), DOI: 10.1007/s12239-014-0005-y
22. J.B. Heywood, *Internal Combustion Engine Fundamentals* (Mc Graw-Hill, 1988)
23. V. Gheorghiu, *10th Biennial Conference on Engineering Systems Design and Analysis* (2010), DOI: 10.1115/ESDA2010-24031
24. G. Schmitz, US Patent No. 6553977 (2003), http://www.google.com/patents/US6553977 (accessed on 17th of March 2016)
25. C. Ailloud, B. Delaporte, G. Schmitz, A. Keromnes, L. Le Moyne, SAE Technical Paper 2013-24-0095 (2013), DOI: 10.4271/2013-24-0095
26. Kéromnès, A., Delaporte, B., Schmitz, G., Le Moyne, L. Energy Conversion and Management **82**, 259-267 (2014), DOI: 10.1016/j.enconman.2014.03.025
27. T. Li, B. Zheng, T. Yin, Energy Conversion and Management **103**, 448-458 (2015), DOI: 10.1016/j.enconman.2015.06.078
28. T. Li, B. Wang, B. Zheng, Energy Conversion and Management **123**, 140-152 (2016), DOI: 10.1016/j.enconman.2016.06.038
29. A. Melin, D. Kittelson, W. Northrop, SAE Technical Paper 2015-01-1752 (2015), DOI: 10.4271/2015-01-1752
30. M. Palanivendhan, H. Modi, G. Bansal, International Journal of Control Theory and Applications **9**, 5855-5862 (2016)
31. K. Stuart, T. Yan, J. Mathias, SAE Technical Paper 2017-01-0633 (2017), DOI: 10.4271/2017-01-0633
32. M. Noga, B. Sendyka, Journal of KONES Powertrain and Transport **20**, 239-246 (2013)
33. M. Noga, Applied Sciences **7**, 295 (2017), DOI: 10.3390/app7030295
34. M.T. Wlodarczyk, Combustion Engines **151**, 3-8 (2012)
35. R. Van Basshuysen, F. Schäfer (eds), Internal Combustion Engine Handbook: Basics, Components, Systems, and Perspectives (SAE International 22, Warrendale, 2004)
36. I. Pielecha, K. Wislocki, P. Borowski, W. Cieslik, Journal of Thermal Analysis and Calorimetry **122**, 473-485 (2015), DOI: 10.1007/s10973-015-4735-5

MATEC Web of Conferences **118**, 00018 (2017) DOI: 10.1051/matecconf/201711800018
VII International Congress on Combustion Engines

Exhaust gas emissions evaluation in the flight of a multirole fighter equipped with a F100-PW-229 turbine engine

Jarosław Markowski[1,*], *Maciej* Kazimierczak[1], *Paweł* Benedict[1], and *Damian* Olejniczak[1]

[1]Poznan University of Technology, Faculty of Machines and Transport, ul. Piotrowo 3, 60-965 Poznan, Poland

Abstract. The issue of exhaust gas emission generated by turbine engines described in ICAO Annex 16 of the International Civil Aviation Convention includes a number of procedures and requirements. Their implementation is aimed at determining the value of the engine's environmental parameters and comparing them to the values specified in the norms. The turbine engine exhaust gas emission test procedures are defined as stationary and the operating parameters values are set according to the LTO test. The engine load setting values refer to engine operating parameters that occur when the plane is in the vicinity of airports. Such a procedure is dedicated to civilian passenger and transport aircraft. The operating conditions of a multirole fighter aircraft vary considerably from passenger aircraft and the variability of their flight characteristics requires a special approach in assessing its environmental impact. This article attempts to evaluate the exhaust gas emissions generated by the turbine engine in a multirole fighter flight using the parameters recorded by the onboard flight recorder.

1 Introduction

Aircraft and passenger aircraft with turbine engines are normally operated in flight conditions at a selected altitude and at a selected speed, which translates into established operating parameters of the engines. There are several papers tackling the problems with addressing the exhaust emissions assessment issue for turbine engines in established flight conditions.

One of the characteristics of a multirole aircraft is the ability to perform various aviation tasks related to pursuit, air combat at various altitudes, interception and aerial reconnaissance or patrol [1, 2]. All of these tasks differ from one another in terms of operating parameters: the altitude, speed or type of maneuvers performed (Fig. 1). These parameters are reflected in the operating characteristics of the drive unit, and thus of harmful exhaust components emissions from the engine [2-4].

All flight parameters listed above are reflected in propulsion system performance characteristics, and thus in emissions of toxic compounds contained in exhaust gas. State-of-the-art emission measurement technologies do not allow measuring toxic emissions in actual flight conditions. Collection of exhaust gas samples for analysis and safe installation of measurement equipment are the key challenges in this regard. Therefore, emissions of toxic compounds in exhaust gas are measured in a stationary LTO test. However, tests of this kind do not allow one to determine emission rates of each toxic compound in exhaust gas in actual operation conditions [2, 5-7]. Therefore, efforts are being made to develop new methods enabling a more accurate assessment of toxic emissions from multi-purpose aircraft. One of the concepts proposes to assess toxic emissions on the basis of functional connections between operation parameters and emission parameters. Therefore, it is of key importance to analyze operation parameters recorded by on-board flight recorders.

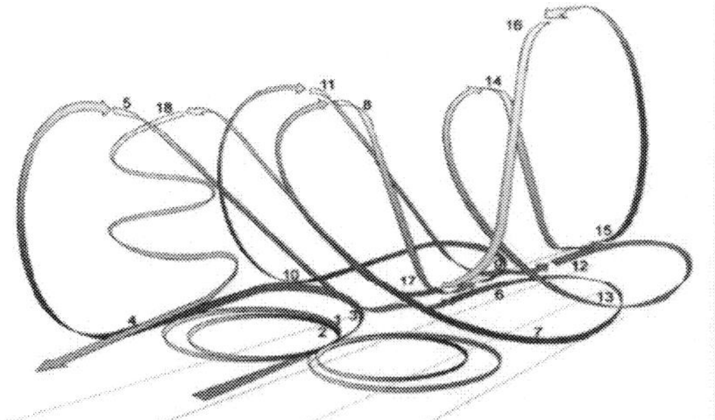

Fig. 1. Maneuvers of a multi-purpose aircraft while attacking ground targets [1]: 1 – turn by 45°, 2 – turn by 60°, 3 – half turn left, 4 – roll-off-the-top left, 5 – dive, 6 – half turn right, 7 – roll-off-the-top right, 8 – dive, 9 – half turn left, 10 – roll-off-the-top left, 11 – dive, 12 – half turn right, 13 – roll-off-the-top right, 14 – dive, 15 – Immelmann, 16 – dive, 17 – pull up, 18 – spiral.

Some of the many parameters recorded in this way may be helpful in analyzing operation conditions of the aircraft in general and its propulsion system in particular. For instance, to assess aircraft operation conditions one can use recorded changes in velocity (v), flight altitude (h_{fly}), thrust lever positions (P_p), exhaust gas temperature as well as afterburner on/off information, relative 1st and

* Corresponding author: jaroslaw.markowski@put.poznan.pl

2nd stage turbine speeds (n_1 and n_2), relative opening of the exhaust gas exhaust nozzle (WO), and instantaneous fuel consumption (G_e) (Fig. 2).

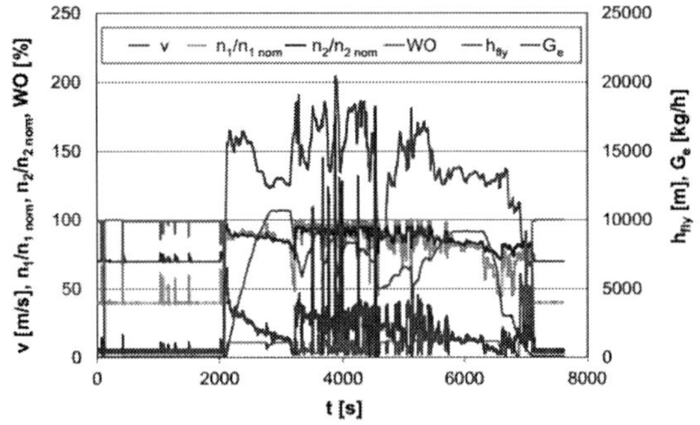

Fig. 2. Selected operation parameters of a multi-purpose aircraft and its engine recorded during the flight.

Once the operation conditions of an aircraft and its propulsion system have been measured, they can be linked to exhaust gas emission parameters [2, 8-10]. Subsequently, an analysis aimed at assessing emissions from the aircraft's engine can be carried out. To do so, it is necessary to develop a working methodology.

The variable operating conditions of a multirole fighter aircraft and the variable operating conditions of its turbine engine cause a problem in assessing exhaust emissions in such aircraft operation. Therefore, this article makes an attempt to develop a method of estimating the exhaust emissions from multirole fighter aircrafts in their actual operating conditions. An important issue in assessing the exhaust gas emission of a turbine engine is the assessment of the engine exhaust flue gas generation in real flight conditions.

2 Methodology

The concept of the internal combustion engine operation is aimed at obtaining operating parameters that translate into propulsion. For turbine jet engines the flow of the exhaust stream is essential to generating thrust. According to the basic equation defining thrust, the mass and velocity of the exhaust gas flow is important. These indicators relate to operating parameters such as air intake and fuel consumption, which translate into the total mass of the exhaust gas generated by the engine (Fig. 3). Knowing the instantaneous value of the fuel consumption and measuring the chemical composition of the exhaust gases, to enable determination of the instantaneous value of the air-fuel ratio λ, the air mass fed to the engine in time can be estimated.

The sum of fuel consumption and air intake corresponds to the flow of exhaust gas generated:

$$\dot{m}_{air} + \dot{m}_f = \dot{m}_{exh} \qquad (1)$$

where:
$\dot{m}_{air}$ – air mass flow,
$\dot{m}_f$ – fuel mass flow,
$\dot{m}_{exh}$ – exhaust gas mass flow.

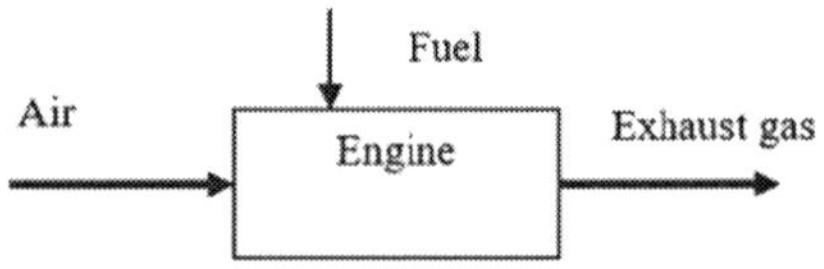

Fig. 3. Diagram of the input and output of a turbine jet engine.

Assuming that it is possible to measure two of these three values, the third value can be calculated. Therefore, a study was conducted with the aim of measuring the fuel consumption and the size of the stream of exhaust emitted and simultaneously measuring the concentrations of harmful exhaust compounds and air-fuel ratio λ.

The assessment of exhaust gas emission in real operating conditions was performed on the basis of the F100-PW-229 engine performance record (Fig. 4) recorded during the F-16 multirole fighter flight (Fig. 5).

Fig. 4. F100-PW-229 engine cutaway drawing [1].

Fig. 5. F-16 multirole fighter aircraft.

The Pratt & Whitney F100-PW-229 engine is a two-shaft unit with a hydraulically adjustable nozzle. It is equipped with a three-stage low-pressure compressor and a 10-stage high-pressure compressor. The combustion chamber is annular. Its maximum thrust values are 79.13 kN – without the use of afterburner and 128.91 kN with afterburners enabled.

The assessment of the exhaust gas during aircraft operation required an analysis of many parameters in terms of relating the engine performance with the characteristics of the generated exhaust gases. The procedure is illustrated in an example analysis.

3 Result analysis

Analyzed data were recorded by on-board flight recorders during over ten patrol and training flights. Figure 6 shows an example of changes in engine operation parameters recorded during one of such flights. Considering the relative thrust force values (K/K_n) depending on a number of engine operation parameters and present in each operation phase, it is possible to present those values as a function of the relative value of engine speed (n_1). By linking each operation phase to values recorded during the flight, the authors were able to determine operation areas defined by thrust force as a function of engine speed (n_1).

Flight profiles can be described on the basis of the distribution of points corresponding to given phases. For the most part, points corresponding to the engine start-up and warming up phase are located in the area within K/K_{max} = 40% and corresponding to the low-pressure stage speed $n_1/n_{1\ nom}$ = 70%. A number of points from this particular phase are located outside the said area – they denote the verification of correct operation of exhaust nozzle opening controller. The identified operation area also contains a number of points assigned to the taxiing phase and partially to the landing phase. The latter can be divided into two stages.

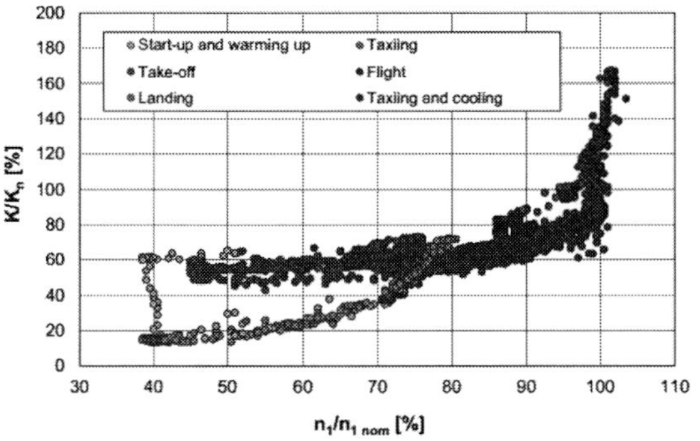

Fig. 6. Analyzed parameters in the engine operation range, recorded during the flight.

The first stage corresponds to the approach to land, whereby engine operation parameters stay in the range typical of actual flight (K/K_{max} = 40–80%, $n_1/n_{1\ nom}$ = = 70–80%). The other stage is landing, whereby engine parameters are lower (K/K_{max} = 40%, $n_1/n_{1\ nom}$ = 70%). A similar division can be proposed for the take-off phase. In the first stage, i.e. acceleration and take-off proper, one can identify an area corresponding to the range of K/K_{max} = 100% and $n_1/n_{1\ nom}$ = 96%. The other stage corresponds to the ascent after take-off (K/K_{max} = = 80–90% and $n_1/n_{1\ nom}$ = 85–90%). The example of relative thrust force and fan speed values during the take-off relates to a plane taking off with the afterburner off and with a small load. If a plane takes off fully loaded, the afterburner is usually off. The afterburner increases the engine's propulsion force by approximately 70%. The extra force comes mostly from the mass of fuel additionally combusted in the afterburner chamber, located downstream of the engine's turbine assembly. The energy generated from fuel delivered to the afterburner chamber increases the energy of exhaust gas ejected from the exhaust nozzle, and thus the overall thrust force. Simultaneously it needs to be emphasized that afterburner status does not affect engine speed parameters (n_1 and n_2). The flow of air through the engine corresponds to values occurring at maximum thrust obtained with the afterburner off. The range of engine operation with the afterburner on covers relative thrust force points in excess of K/K_{max} = 105%. The largest operation area covers points from the flight phase (K/K_{max} = 40–170% and $n_1/n_{1\ nom}$ = 45–105% of the low pressure stage speed). It is evident that this phase contains diverse values of the relative thrust force as a function of engine speed (n_1). The above analysis shows that operation of a low dual flow jet engine fitted in a multi-purpose aircraft stays in the following ranges: $n_1/n_{1\ nom}$ = 38–105% and K/K_{max} = 40–170%.

The above division of relative thrust force values (K/K_{max}) as a function of relative engine speed ($n_1/n_{1\ nom}$) made it possible to identify the ranges of operation phases. Relative thrust force values below K/K_{max} = 40% and relative engine speed up to $n_1/n_{1\ nom}$ = 70% usually correspond to engine start-up, warming up, taxiing and cooling. Most of the range of K/K_{max} = 40–100% and $n_1/n_{1\ nom}$ = 45–105% corresponds to take-off and flight, as well as to the approach to land stage of the landing phase. Points in the operation range of K/K_{max} = 100–170% and $n_1/n_{1\ nom}$ = 95–105% correspond to afterburner on situations: these points represent the take-off phase and the maximum engine operation parameters during special maneuvers.

The example distribution of the engine operating parameters, expressed as relative power lever setting values, as a function of the relative shaft speed values n_1 indicate the potential for different engine load values for the individual shaft speeds. Consequently, additional parameters must be used to determine the operating status of the engine.

Among the recorded flight parameters and engine performance parameters, the parameters closely related to the engine load are the speed of the exhaust gas generator shaft n_2, the exhaust gas temperature before the turbine T_3, and the fuel consumption G_e. By comparing these values with each other and the value of exhaust gas temperature T_3, you the load factor of engine W_p can be established (2).

$$W_p = \frac{n_2}{T_3} \quad \left[\frac{\text{rpm}}{\text{K}}\right] \tag{2}$$

where:

$n_2 - n_1$ shaft rotational speed,

T_3 – exhaust gas temperature before the turbine.

By substituting the example values for the minimum load n_2 = 10,064 rpm, T_3 = 666 K and maximum load n_2 = 13,590 rpm, T_3 = 1276 K the following equations were obtained:

$$W_{p\,min} = \frac{10064}{666} = 15.11 \quad \left[\frac{\text{rpm}}{\text{K}}\right] \tag{3}$$

$$W_{p\,max} = \frac{13590}{1276} = 10.65 \quad \left[\frac{\text{rpm}}{\text{K}}\right] \tag{4}$$

In this way a certain kind of inverse value is obtained, namely the minimum load factor is characterized by a large value and the maximum load factor by small. Additional information about the engine load can be obtained taking into account the excess air ratio [4]. The value of this parameter was determined on the basis of stationary assessment of the exhaust gas composition carried out for the analyzed engine performed at an engine dynamometer. These studies have shown the dependence of the excess air ratio value on the load [4]. Its values are respectively $\lambda_{p\ min} = 8.5$ at the smallest engine load and $\lambda_{p\ max} = 6.5$ at the highest load. Values of the following parameters: engine load W_p and air excess ratio λ, are characterized by similar relations relative to the engine load (5, 6).

$$\frac{W_{p\,min}}{W_{p\,max}} = \frac{15.11}{10.65} = 1.41 \quad [-] \tag{5}$$

$$\frac{\lambda_{p\,min}}{\lambda_{p\,max}} = \frac{8.5}{6.5} = 1.31 \quad [-] \tag{6}$$

Consequently, the excess air excess ratio λ must be linked to the determined load factor of the engine W_p. Hence the relation (7, 8) was proposed as:

$$Z_{w-\lambda} = \frac{W_{p\,min}}{\lambda_{min}} = \frac{15.11}{8.5} = 1.70 \quad \left[\frac{K}{rpm}\right] \tag{7}$$

$$Z_{w-\lambda} = \frac{W_{p\,max}}{\lambda_{max}} = \frac{10.65}{6.5} = 1.63 \quad \left[\frac{K}{rpm}\right] \tag{8}$$

Taking into account the similarity in the obtained relationships, the mean value can be used for further calculations $Z_{w-\lambda} = 1.67 \left[\frac{K}{rpm}\right]$.

As a result, the following equation (9) can be established:

$$\lambda = \frac{W_p}{Z_{w-\lambda}} \quad \left[\frac{\frac{rpm}{K}}{\frac{K}{rpm}}\right] = [-] \tag{9}$$

This equation is complementary to the equation for the determination of the air mass flow using fuel consumption (10):

$$\dot{m}_p = G_e \cdot \lambda \cdot 14.5 \quad \left[\frac{kg}{h}\right] \tag{10}$$

Where the above equation takes the form (11):

$$\dot{m}_p = G_e \cdot \frac{W_p}{Z_{w-\lambda}} \cdot 14.5 \quad \left[\frac{kg}{h}\right] \tag{11}$$

The obtained relation in equation (1) allows to determine the value of the emitted flue gas based on the operating parameters recorded by the onboard flight recorder using the equation (12).

$$G_e \cdot \frac{W_p}{Z_{w-\lambda}} \cdot 14.5 + G_e = \dot{m}_{exh} \quad \left[\frac{kg}{h}\right] \tag{12}$$

Using the listed equations, an analysis of the exhaust gas flow estimation of a multirole fighter aircraft turbine engine during a default flight was performed.

According to the presented set of equations, the values of the following operating parameters were used in the calculation: temperature before turbine T_3, shaft speed of flue gas generator n_2, and fuel consumption G_e. The results are graphically presented in Figure 7. Parameter changes described as time functions show a notable similarity between the T_3 temperature characteristic and the shaft rotational speed of the flue gas generator n_2. Changes in these parameters result from changes in the fuel consumption G_e, as they are a consequence of fuel combustion in the engine combustion chamber. Depending on the combustion process resulting from the load of the engine, the airflow change characteristics G_p is obtained. From the described relationship, it can be identified as the main factor determining the value of the exhaust gas stream flowing from the engine G_{exh}.

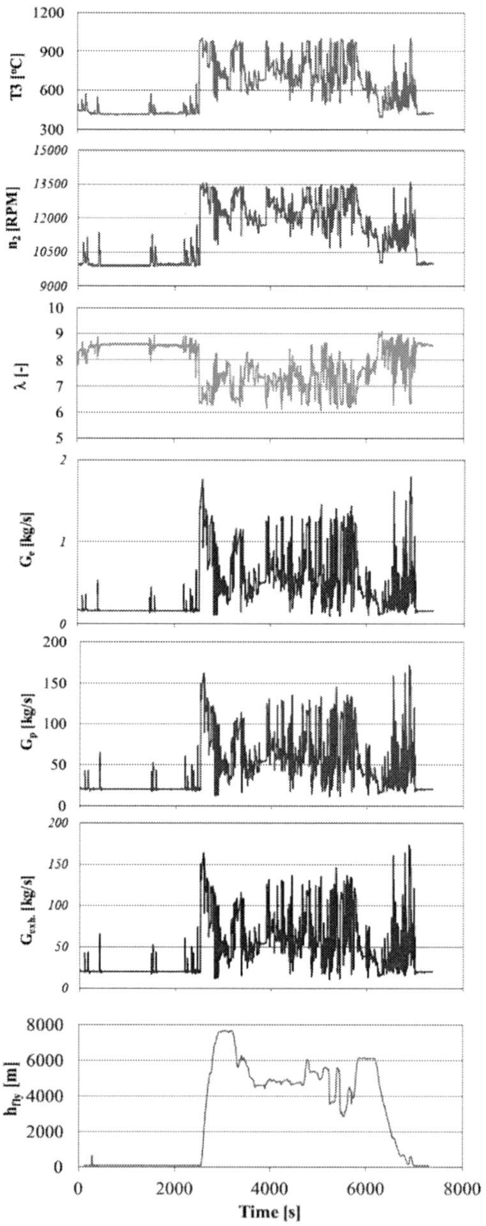

Fig. 7. Variable values of operating parameters as a function of time; Recorded parameters: T_3, n_2, G_e, h_{fly}; calculated parameters: λ, G_p, G_{exh}.

The changes of these parameters along with the h_{fly} flight characteristics presented in the bottom graph, highlight the flight characteristics that are specific to multirole fighter aircraft.

Taking into account the 1 second time resolution of the recorded parameters, the individual values can be summed up during the entire recorded flight performed by the aircraft. As a result, it was reported that during a flight lasting 7368 seconds, that is 2 hours, 2 minutes and 48 seconds, the F100-PW-229 engine powered aircraft consumed 3297 kg of fuel, used 348,003 kg of air and generated 351,301 kg of exhaust.

4 Conclusions

The analysis of the interdependence of the recorded operating parameters was carried out to evaluate the possibility of determining the instantaneous value of the flue gas generated by the turbine engine when operating in high load parameters. The analysis showed that there are relationships between the parameters recorded by the flight recorder during the aircraft flight. The analysis has also shown that information derived from the engine emission characteristics is required in the form of an excess air ratio λ value, which is necessary to establish the relationship of the combustion process in the engine to its load. The emission characteristics of the engine in the form of harmful exhaust gas emissions concentration values will be used in the next steps to determine the instantaneous emission values of the pollutants during the aircraft flight.

The study presented in this article was performed within the statutory research (contract No. 05/52/DSPB/0244).

References

1. P. Wygonik, Journal of KONES (2012)
2. J. Merkisz, J. Markowski, J. Pielecha, *Selected Issues in Exhaust Emissions from Aviation Engines* (Nova Science Publishers, 2014)
3. V. Ramanathan, Y. Feng, Atmospheric Environment **43** (2009)
4. S. Capoccitti, A. Khare, U. Mildenberger, Journal of Technology Management & Innovation **5**, 2 (2010)
5. A.K. Chaturvedi, Journal of Analytical Toxicology **34** (2010)
6. M. Masiol, R.M. Harrison, Atmospheric Environment **95** (2014)
7. ICAO Annex 16, Vol. 2 – Environmental Protection – Aircraft Engine Emissions. ICAO (2008)
8. J. Merkisz, J. Pielecha, S. Radzimirski, *Springer Tracts on Transportation and Traffic* **4**, 1-170 (2014), DOI: 10.1007/978-3-319-02705-0
9. J. Merkisz, J. Pielecha, S. Radzimirski, *Springer Tracts on Transportation and Traffic* **4**, 67-90 (2014), DOI: 10.1007/978-3-319-02705-0_4
10. J. Merkisz, J. Markowski, J. Pielecha, D. Karpinski, M. Galant, *18 ETH-Conference on Combustion Generated Nanoparticles* (Zurich, 2014)

Motor transport in Poland – hydrogenization

Wojciech Gis[1,*], *Stanislaw* Kruczynski[1], *Maciej* Menes[1], and *Jerzy* Waskiewicz[1]

[1]Motor Transport Institute, ul. Jagiellonska 80, 03-301 Warsaw, Poland

Abstract. In paper is presented programs of hydrogen technology development in the road transport, in the world. Also is presented: status and trends of the road network changes in Poland, with particular emphasis on the Polish sections of the TEN-T network, production technologies and hydrogen production in Poland, preliminary indications of the future hydrogen refuelling stations locations in Poland, assumptions for the development of hydrogen-powered vehicle fleet in Poland by the 2050, expert forecast of the demand for hydrogen by the road transport in Poland by the 2050 and the ecological effects of the use of hydrogen fuel in the road transport in Poland by the 2050.

1 Introduction

The advantages of hydrogen as an automotive fuel is the lack of pollutants emission from motor vehicles' engines, which is especially important in crowded city centers and with the possibilities of its local production. The use of hydrogen fuel in the road transport to a large degree brings about independence from the import of crude oil and crude oil derived fuels. In the case of producing hydrogen by water electrolysis using electricity from renewable energy sources, the result is the use of "clean" energy [1].

In practice, so far, on the relatively mass scale, Toyota launched the production of fuel cell cars in the autumn of 2014. In 2014 of the Mirai model produced in 700 units, and for 2015 envisaged production of 3500 cars. Cars with fuel cells are also manufactured by Hyundai. Around the half of the first decade of the twenty-first century the countries technologically advanced in automotive production, began to develop draft projects of multi-annual work on the use of hydrogen in transport and new fuel cell technologies [2-5].

The development of hydrogen technology in the road transport in the EU countries is recommended, among the others, in the Directive of the European Parliament and of the Council 2014/94/EU of 22 October 2014 [6]. Under the provisions of the said Directive, it is recommended to EU countries to progressively ensure accessibility to hydrogen cars on their territories, and above all to ensure the possibility of driving hydrogen vehicles between the member States.

One of the major international projects concerning hydrogenization of motor transport is the "Hydrogen mobility Europe – Creating the European Visio for hydrogen transport" project – H2M [7].

The project has a budget of EUR 170 million, of which EU funds amount to EUR 67 million, and brings together 40 entities of 8 countries. Within the framework of the project 41 hydrogen refuelling stations (HRS) are to be launched within years 2015-2022: 20 in Germany, 12 in France, 10 in Scandinavia, 6 in the UK and 1 in the Nether-lands and approx. 1400 hydrogen passenger cars and trucks are to be tested.

According to preliminary results of studies the level of costs per vehicle/km of a conventional vehicle (ICE) equal to 23 cents, a hydrogen vehicle may achieve at the price of 1 kg of hydrogen of EUR 5 and if exempted from taxes [8].

With a price of hydrogen of 10 EUR/kg the cost per vehicle/km shall increase to 32 cents. The hydrogenization of motor transport is the subject of many international re-search projects, e.g. „Clean Hydrogen in European Cities" – CHIC where 23 partners of 8 countries invested EUR 81 million in HRS and 28 hydrogen buses (5 in Bologna, 8 in London, 5 in Oslo, 6 in Hamburg and 4 in Cologne) or „Hydrogen Acceptance In the Transition Phase – Public and Stakeholder Acceptance of Fuel Cell Electric Vehicle In Europe" – HYACINTH. This project brings together 11 partners from 5 countries and a budget of EUR 1 million, of which EU funds total EUR 661 thousand [9-11].

The project comprised questionnaires among over 7 thousand residents of EU and interviews with 455 representatives of entities of potential hydrogen transport market players.

Of special interest is the lecture regarding the plans of hydrogenisation of the British road transport prepared with E4tech [12].

The plan assumes that to 2020 the UK will have 30-65 hydrogen refuelling stations (HRS) and 50-100 hydrogen buses will be used and approx. 500 passenger cars. In 2025 the number of stations is expected to expand to approx. 150 and the number of used hydrogen vehicles should be several thousand. After year 2025 from 10 to 100 HRS, 100 buses and from 10 thousand to several thousand other vehicles with hydrogen drive should appear every year [13].

* Corresponding author: wojciech.gis@its.waw.pl

The issue of hydrogen in road transport was also addressed by the lecture prepared by the Japanese Institute of Advanced Technologies [14]. Interesting is the fact that in Japanese condition it is more profitable to import hydrogen from Australia (10,000 km) or even Norway (20,000 km) than to produce it from gas reforming or electrolysis.

The territorial accessibility for hydrogen vehicles is determined by the availability of hydrogen refueling infrastructure, in the first place along the TEN-T network [15].

As a result of verification, under Polish conditions, of the original method developed for determining the initial location of the hydrogen refueling station in Poland, in the pre-commercial phase (2020-2030), the said location has been indicated along with the order of investment, taking into account above all the freedom to move around Poland of cars powered by hydrogen visiting Poland and transiting our country between other EU countries.

Despite the strategic importance of developing hydrogen filling stations infrastructure, in the available materials, including various national implementation programs for hydrogen propulsion technology developments, the explicitly formulated programming methodology for the development of these stations, has not been encountered.

2 National implementations plans

National implementation plans are in place in a number of countries, e.g. Belgium, France, the Netherlands, Sweden, etc. In Poland the methodology developed is of multi-stage character. Individual steps leading to the designation of the location of hydrogen refuelling stations in Poland (as the methodology alone seems to be of universal character) are as follows [16]:

Stage I: Method allowing to identify regions in which the hydrogen refuelling stations should be located in the first place.

Stage II: Method allowing to identify urban centres, in which should be located the said stations.

Stage III: Method for determining the area of the station location.

The Table 1 shows method allowing to identify regions (Stage I), urban centres (Stage II), area of the station location (Stage III).

Additionally, while pre-indicating subsequent hydrogen station locations taken into consideration were:

- average passenger car traffic intensity and average traffic volume projected for 2020,
- development of hydrogen filling stations network in the country,
- development of hydrogen refuelling stations in areas with potentially high demand for hydrogen fuel also by fleets of buses and taxis.

The order of investment: while taking into account the initial hydrogen refuelling stations locations (Stages I-III), in the first place included were:

- existing hydrogen refuelling capabilities in the neighbouring countries,
- the assumed future hydrogen refuelling stations locations in the Baltic Sea countries,
- new stations at the distances up to approx. 300 kilometres away from the existing stations or sequentially from the newly-opened stations.

With the above criteria, the order the construction of a hydrogen refuelling stations in Poland: 1 – Poznań, 2 – Warsaw, 3 – Białystok, 4 – Szczecin, 5 – the Lódź region, 6 – the Tri-City region, 7 – Wrocław, 8 – the Katowice region, 9 – Kraków (Fig. 1).

3 Results

As the first city to build a hydrogen car tanks filling station was selected Poznan because it is located at a distance of less than 300 km from the nearest hydrogen refuelling station in Berlin. This would be the first hydrogen refuelling station on the Polish territory located in the North Sea – Baltic Sea TEN-T corridor (E30/A2) leading then via E67/S8 road through Kaunas to Riga and further to Helsinki.

Table 1. Method allowing to identify regions (stage I), urban centres (stage II), area of the station location (stage III) [16].

Stage I	Stage II	Stage III
Average GDP per capita	Average distance of the cities, over 250 thousand inhabitants, from the place of hydrogen manufacture (acquisition)	Results of locating the stations in question obtained in stage I and II, and relying on the results of the measurement of average traffic volume of passenger cars on the roads leading to these cities or on selected road junctions located near these cities
Average population density (inhab./km^2)	Average distance of the cities, over 250 thousand inhabitants from the nearest hydrogen refuelling station located outside Poland	
The number of people living in the largest cities in the region, out of the cities with > 250 thousand inhabitants	Number of taxies in the city	Traffic flow intensity of passenger cars was included on the roads leading to the following cities: Warsaw, Poznań, Kraków, Wrocław, Katowice (Upper Silesia conurbation), Tri-City, Szczecin, Lódź, Białystok
Average traffic volume of passenger cars on the national roads of international significance running through the region (passenger cars/24 hours)	Number of municipal transport buses	

Fig. 1. Map of Poland with marked sites of the proposed public hydrogen refuelling station locations [16].

A second location of the proposed hydrogen refuelling station is in Warsaw. This would be second station in Poland in the TEN-T North Sea - Baltic Sea (E30/A2) corridor and further along the E67/S8 in the same corridor leading to Helsinki. This location is justified by the distance between Warsaw and Poznan, of little more than 300 km.

The third location of hydrogen refuelling stations in the TEN-T North Sea – Baltic Sea corridor should be in Białystok. Białystok is located in the North Sea – Baltic Sea (E67/S8) corridor. The road distance of Bialystok from Warsaw is about 200 km and from Kaunas is about 250 km.

The fourth location of hydrogen filling station is proposed in the Szczecin area in the Baltic – Adriatic TEN-T (the E65/S3) corridor. This hydrogen station will be able to be used, among the others, by the users of hydrogen cars moving through ferry crossings (Świnoujście) and Scandinavia (e.g. Trelleborg, Ystad). Besides Szczecin is away from Berlin at distance of about 150 km and from Poznań (suggested location of the first hydrogen refuelling station in Poland) at the distance of about 240 km (stretch of the TEN-T Baltic – Adriatic corridor: road E65/S3 and the section of the TEN-T North Sea – Baltic Sea corridor: E30/A2).

Fifth hydrogen refuelling station should be built at the junction of the TEN-T Baltic – Adriatic corridor (E75/A1) and TEN-T North Sea – Baltic Sea corridor (E30/A2), near Łódź. Location of the station in this region would shorten the distance between refuelling stations from Poznań to Warsaw (the distance from Poznań to Łódź is 200 km and Łódź–Warsaw is 130 km). Additionally, this location would be a condition for the integration of Tri-City in the emerging hydrogen refuelling network in Poland.

The sixth hydrogen refuelling station on the Polish territory should be located in the Tri-City area. The station would allow passage of vehicles between Tri-City and Stryków (TEN-T Baltic – Adriatic corridor: E75/A1) and then in the TEN-T North Sea – Baltic Sea corridor (E30/A2) and drive to Poznań (stretch of E75/A1 and a part of the E 261 road). The road distance to Łódź region (e.g. northern part of Łódź) via TEN-T Baltic – Adriatic corridor (E75/A1) slightly exceeds 300 km. The road distance between Tri-City and Poznań (stretch of E75/A1 and a part of the E261 road) is approximately 300 km.

The choice of the next, seventh hydrogen station location in Wrocław would provide connection with the city of Poznań (the E 261 road – about 200 km), Prague, Dresden (about 300 km).

Location of the eighth, in turn, hydrogen station near Katowice is due to the heavy traffic and the anticipated increase in traffic intensity of passenger cars between Wrocław and towns of Upper Silesia in the Baltic - Adriatic corridor (road E40/A4). The distance between these two cities is the order of 200 km. A high volume of traffic is currently taking place and will occur in the future between Katowice and Warsaw (20,000 and more vehicles per day) along the stretch of the Baltic – Adriatic corridor: E75/A1 and the section of North Sea – Baltic Sea corridor: E67/S8). Within less as 300 kilometres from Katowice is located: Łódź (in the Baltic – Adriatic corridor E75/A1), Warsaw (in the Baltic – Adriatic corridor: stretch of E75/A1 and section of E67/S8) and Kraków (in the Baltic – Adriatic corridor E 40/A4).

Locating of the ninth hydrogen refuelling stations on the Polish territory could possibly be in the area of Kraków. It is a city located directly in the Baltic – Adriatic corridor: E40/A4 and E77/S7).

On the Fig. 2 shown the movement area of cars using fuel cells based on 5 hydrogen refueling stations situated on the national TEN-T road network by the 2025.

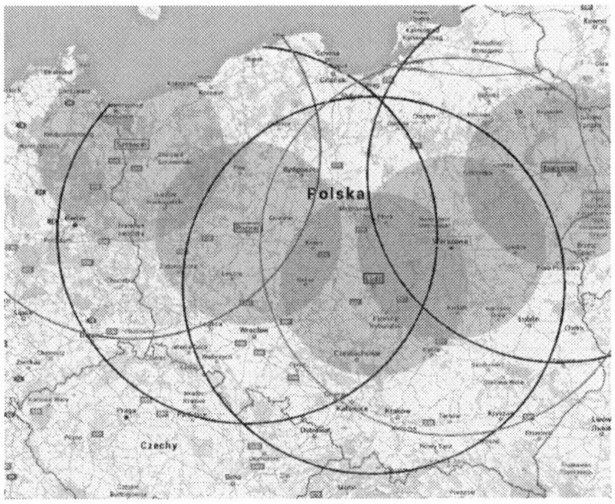

Fig. 2. Penetration area of cars using fuel cells based on 5 hydrogen refuelling stations situated on the national TEN-T road network by the 2025 [16]; when driving in one direction (diameter of large circles – to approx. 600 km), when driving there and back (diameter of small, shaded circles – to approx. 300 km).

On the Fig. 3 shown the movement area of cars using fuel cells based on 9 hydrogen refueling stations situated on the national TEN-T road network by the 2030.

* Corresponding author: wojciech.gis@its.waw.pl

DOI: 10.1051/matecconf/201711800019

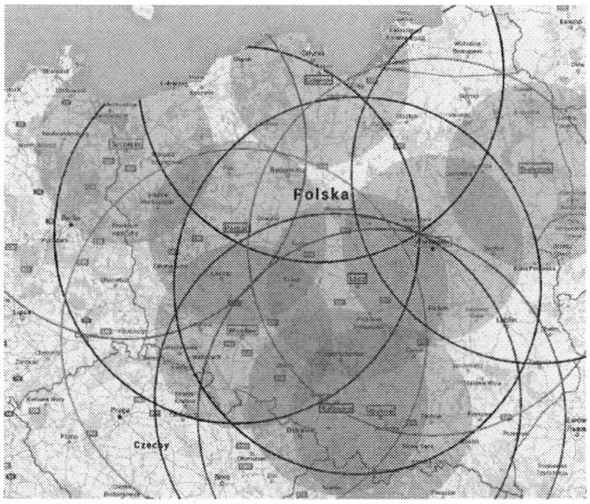

Fig. 3. Penetration area of cars using fuel cells based on 9 hydrogen refuelling stations situated on the national TEN-T road network by the 2030 [16]; when driving in one direction (diameter of large circles – to approx. 600 km), b) when driving there and back (diameter of small, shaded circles – to approx. 300 km).

Full commercialization of hydrogen technology in Poland (the years 2040-2050):

- Respectively: 200-600 hydrogen refueling stations,
- Financial expenditures for the construction of hydrogen refueling infrastructure in the order of 155-190 million €,
- The assumptions: 600 thousand respectively – 2 million passenger cars, 500-1000 buses, 100-300 thousand cars transiting Poland annually.

On the Fig. 4 shown forecast of demand for hydrogen by the vehicles equipped with fuel cells in Poland by the 2050.

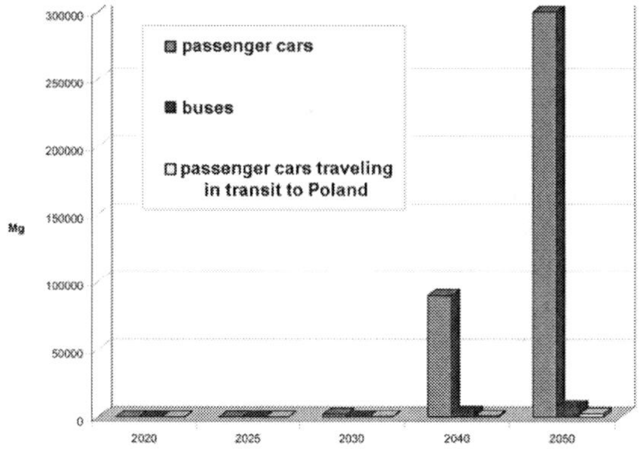

Fig. 4. Forecast of demand for hydrogen by the vehicles equipped with fuel cells in Poland by the 2050 Mg [4].

The implementation of hydrogen technology in the road transport allows to reduce the consumption of petroleum fuels, presented in the Fig. 5.

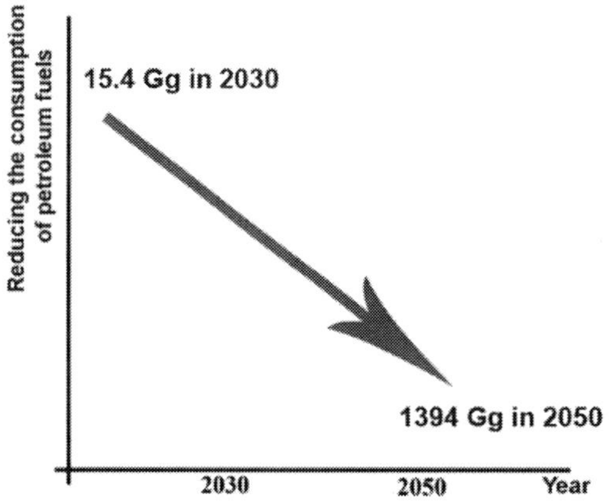

Fig. 5. The implementation of hydrogen technology in the road transport allows to reduce the consumption of fuels.

Reduction in consumption of petroleum-based fuels by the road transport in Poland as a result of the accepted scenario of the hydrogen technology development will affect the reduction in emissions (Tab. 2).

Table 2. Reduction in consumption of petroleum-based fuels by the road transport in Poland as a result of the accepted scenario of the hydrogen technology.

	in 2030	in 2050
Carbon dioxide	47.8 Gg	4314 Gg
Methane	4.4 Mg	426 Mg
Nitrous oxide	1.9 Mg	182 Mg
Carbon monoxide	1055 Mg	109.7 Gg
Nitrogen oxides	186 Mg	15.2 Gg
Non-methane volatile organic compounds	65 Mg	6.1 Gg

The avoided costs related to air pollution as a result of the hydrogen technology development will amount to:

- 2030 – 2772 thousand PLN (649 thousand €),
- 2050 – 287 million PLN (67.3 million €).

4 Summary

In summary it can be stated that:

- The advancement of hydrogen supplied fuel cells technology.
- Hydrogen as an automotive fuel is the lack of pollutants emission from motor vehicles' engines, which is especially important in crowded city centres and with the producing electrical energy used by automobile engines offers a real opportunity for the global automotive industry.
- The advantages of possibilities of its local production.
- The use of hydrogen fuel in the road transport to a large degree brings about independence from the import of crude oil and crude oil derived fuels.

- The source of obtaining hydrogen can be technological processes in the chemical industry, where this gas is, among the others, a by-product or autonomous hydrogen production processes e.g. electrolysis of water, hydrogenation of methane or bioprocesses.
- In the case of producing hydrogen by water electrolysis using electricity from renewable energy sources, the result is the use of "clean" energy.
- Efficient use of electricity produced during periods of excess production of energy or outside of peak electricity demand, can rely on its retention in the form of hydrogen, which is then used subsequently for different purposes, e.g. to power electric vehicles equipped with fuel cells.
- In Poland, at the moment, despite producing more than 1 million tons of hydrogen a year, there is no hydrogen of purity class required by the automotive industry. The production of hydrogen in industrial plants located on the Polish territory, for vehicles with fuel cells would require additional investments, which, however, would be conditional on cost-effectiveness of the operation of new installations.
- The development of hydrogen technology in the road transport in the EU countries is recommended, among the others, in the Directive of the European Parliament and of the Council 2014/94/EU of 22 October 2014 [6]. Under the provisions of the said Directive, it is recommended to EU countries to progressively ensure accessibility to hydrogen cars on their territories, and above all to ensure the possibility of driving hydrogen vehicles between the member States.
- The territorial accessibility for hydrogen vehicles is determined by the availability of hydrogen refuelling infrastructure, in the first place along the TEN-T network.
- The ten-year time sequence of the development of hydrogen technology in relation to the classic electric drive technology, is deepened in the case of Poland, by circa ten years quality and technological backwardness of the motorism.
- As a consequence, the study assumed that pre-commercial phase of the development of hydrogen technology in Poland will take place in the years 2020-2030, and its full commercialization will be possible only in the years 2040-2050.
- As a result of verification, under Polish conditions, of the original method developed for determining the initial location of the hydrogen refuelling station in Poland, in the pre-commercial phase, the said location has been indicated along with the order of investment, taking into account above all the freedom to move around Poland of cars powered by hydrogen visiting Poland and transiting our country between other EU countries.
- The places where it is proposed to built hydrogen refuelling stations should be (in order of their creation, along the TEN-T corridors): 1 – Poznan, 2 – Warsaw, 3 – Bialystok, 4 – Szczecin, 5 – Lodz region, 6 – Tri-City area, 7 – Wroclaw, 8 – Katowice region, 9 – Krakow.
- Under the adopted assumptions, taking into account both the progressive development of the hydrogen car fleet in Poland (15,000 cars, vans and 1,000 buses) and transit traffic (60,000 cars), the construction by the 2030 of thirty refuelling hydrogen stations (including nine bases HRS along the TEN-T corridors, should require a financial outlay of 12-15 million Euro.
- In the full commercialization phase (years 2040-2050) in the country there should operate approximately 200-600 hydrogen fuelling stations serving 600 thousand and 2 million passenger cars respectively, 500-1000 buses and 100-300 thousand cars transiting Poland.
- Due to the innovativeness of the introduction of hydrogen technology in transport, it is expected that the economic efficiency of the actions undertaken will appear only at full commercialization of this technology.
- The pre-commercial phase of development of hydrogen technology will require the use of various instruments to implement the assumed political strategy. They may be varied instruments of economic and administrative nature, addressed both to the users of electric vehicles with fuel cells, and those addressed to the users of vehicles with conventional engines.
- The implementation and spread of hydrogen technology in the Polish transport requires proper lobbying, including the development of a multi-stage information – educational program.

References

1. Z. Juda, M. Noga, *IOP Conference Series – Materials Science and Engineering* **148**, UNSP 012042 (2016), DOI: 10.1088/1757-899X/148/1/012042
2. Assessments of the implementation of European alternatives fuels strategy and possible supportive proposals. Project TREN/R1/350-2008-Lot-3 COWI (2012)
3. A.Ki. Byung, IPHE Meeting (2011)
4. K. Hirose, Toyota approach to sustainable mobility and fuel cell vehicle development, Toyota
5. H2 Mobilite France, Study for a fuel cell electric vehicle national deployment plan
6. Directive 2014/94/EU of the European Parliament and of the Council of 22 October 2014 on the deployment of alternative fuels infrastructure (2014)
7. A. Stewart, *Hybrid and Fuel Cell Electric Vehicle Congress*, EEVC-207 (2017)
8. J. Merkisz, M. Bajerlein, P. Daszkiewicz, *Lecture Notes in Information Technology* **13**, 106-113 (2012)
9. A.H. Gema, *Hybrid and Fuel Cell Electric Vehicle Congress*, EEVC-2017 (2017)

10. M. Bajerlein, L. Rymaniak, Experimental and Applied Mechanics **518**, 96-101 (2014), DOI: 10.4028/www.scientific.net/AMM.518.96
11. J. Merkisz, J. Pielecha, *Springer Tracts on Transportation and Traffic* **8**, 131-139 (2015), DOI: 10.1007/978-3-319-15928-7_10
12. F. Lehner, *Hybrid and Fuel Cell Electric Vehicle Congress*, EEVC-2017 (2017)
13. P. Fuc, P. Kurczewski, A. Lewandowska, E. Nowak, J. Selech, A. Ziolkowski, International Journal of Life Cycle Assessment **21**, 1438-1451 (2016), DOI: 10.1007/s11367-016-1104-y
14. Y. Kudoh, *Hybrid and Fuel Cell Electric Vehicle Congress*, EEVC-2017 (2017)
15. P. Lijewski, J. Merkisz, P. Fuc, A. Ziolkowski, L. Rymaniak, W. Kusiak, European Journal of Forest Research **136**, 153-160 (2017), DOI: 10.1007/s10342-016-1015-2
16. W. Gis, E. Menes, J. Waśkiewicz et al. Circumstances of the national plan for hydrogenization of road transport in Poland. Report prepared as part of the HIT-2-Corridors project

MATEC Web of Conferences **118**, 00020 (2017) DOI: 10.1051/matecconf/201711800020
VII International Congress on Combustion Engines

Analysis of hypocycloid drive application in a high-pressure fuel pump

Mateusz Bor[1], *Tomasz* Borowczyk[1], *Marek* Idzior[1], *Wojciech* Karpiuk[1,*], and *Rafał* Smolec[1]

[1]Poznan University of Technology, Faculty of Machines and Transport, ul. Piotrowo 3, 60-965 Poznan, Poland

Abstract. This paper undertakes the problem of applying a hypocycloid mechanism in the drive of pumping sections of high-pressure pumps for Diesel combustion engines. The authors have reviewed existing pump drive mechanisms, analyzed the weak links in existing solutions and discussed the latest trends in common rail systems. Further sections describe the design of a pump utilizing a hypocycloid mechanism, demonstrate beneficial functional features and present selected results of the assembly's dynamic simulation. As a result of work, a pump model was developed, with operating parameters that do not deviate from those of the latest solutions of leading manufacturers, simultaneously eliminating the greatest flaws of existing pumps that lead to their premature wear.

1 Introduction

Common rail (CR) injection systems have been universally applied in passenger car engines for twenty years. Moreover, they are applied on a wide scale in trucks, machinery and stationary engines. The term common rail refers to the fuel rail of the injection system. Its architecture is based on three primary elements: a high-pressure pump, fuel rail and fuel injector (Fig. 1). The system owes its most important and beneficial feature to its design, which is the capability of injecting fuel under high pressure regardless of the engine's crankshaft position.

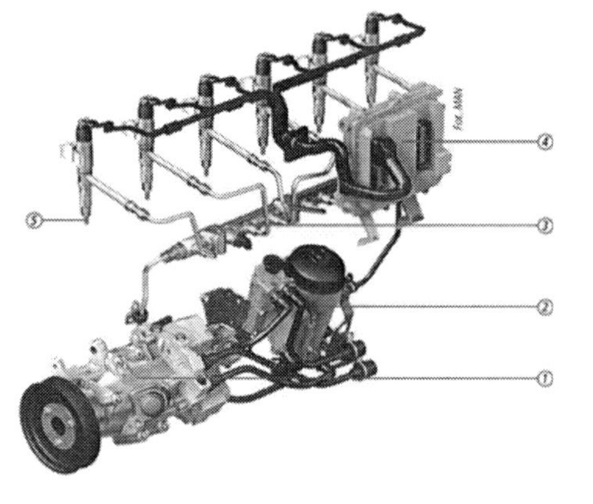

Fig. 1. Common rail system in a truck with its most important subassemblies: 1 – high-pressure pump, 2 – fuel filter, 3 – fuel rail, 4 – control system, 5 – injector [1].

On the European market, common rail (CR) systems displaced volumetric pumps (VP) and UIS/UPS injector systems. In contrast to other generations, CR enables multiple division of a batch, regardless of the crankshaft's position and speed. Batches can be divided into as many as 10 parts with the help of the latest solutions (CRS3-27 from Bosch). This precise injection control system, effected by electromagnetically or piezoelectrically controlled injectors, makes it possible to significantly reduce emissions into the atmosphere and interoperate with external emissions purification systems. In engines with large capacities, designed for heavy-duty machinery and trucks, injection pressures reach up to 300 MPa. High pressure makes it possible to obtain very small fuel droplets, which combust more easily due to increased droplet surface area. High pressure also reduces the duration of individual injection phases, leading to better control of the combustion process. The fuel pump is responsible for generating high pressure in the system. In common rail systems, these pumps are designed as radial sectional pumps with one, two or three pumping sections. In the case of tri-sectional pumps, the angle of aperture between cylinders is 120°, bi-sectional pumps usually have a Continental system or cylinders are positioned on perpendicular axes. Pumps with a greater number of pumping sections, positioned in rows, are also encountered in trucks. Depending on the application, CR pumps can be integrated with a low-pressure feed pump. Two possible types of radial sectional pumps can be distinguished according to design:

- with an external cam (Delphi DFP1, Denso HP2),
- with an internal cam (numerous solutions, including the latest designs, e.g. Bosch CP4i).

The use of fuel as a lubricant is a common feature of high-pressure pumps applied in passenger vehicles from all manufacturers. Over the course of the pump's work, fuel is supplied to pumping sections and the drive system (camshaft). Thus, the proper lubricating parameters must

* Corresponding author: wojciech.karpiuk@put.poznan.pl

be ensured in order to generate a film and separate contacting surfaces. Numerous analyses and studies of damage and wear on contacting parts, conducted by the authors, have led them to conclude that insufficient fuel quality is the primary cause of pump defects. However, it should be noted that improper design is also a factor fostering the initiation and progression of pump damage, with respect to both the feed pump and high-pressure pump. The most important causes of pump wear are [2]:

- improper quality (lubricity) of fuel at fuel stations,
- presence of abrasive grains originating from parts of injection systems,
- application of fuel filters of insufficient quality, or insufficiently frequent filter replacement.

The lubricity of fuels is determined using the HFRR method. It is based on determining the maximum diagonal of an impression formed by a ball wear test, using the tested fuel as a lubricant. This chart (Fig. 2) presents the dependency between runtime and wear of pump elements (expressed as loss of material layer in μm). As can be seen, there is a strong dependency between the HFRR parameter and the lifetime of pump parts. At HFRR > 600 μm there is very sudden wear growth, leading to the pump's destruction. For this reason, it is very important for the fuel sold at fuel stations to meet the requirements of standard PN-EN 590:2013-12, according to which the upper limit of the HFRR parameter is 460 μm. The introduction of FAME additives to Diesel effectively improves its lubricating properties due to the high fatty acid methyl ester content in rapeseed oil, which is below 200 μm [3]. The considerations above show that increasing pump lifetime can be achieved by:

- providing independent lubrication of the sectional drive system,
- changing or modifying the cam system used until now in order the reduce the influence of the applied lubricant.

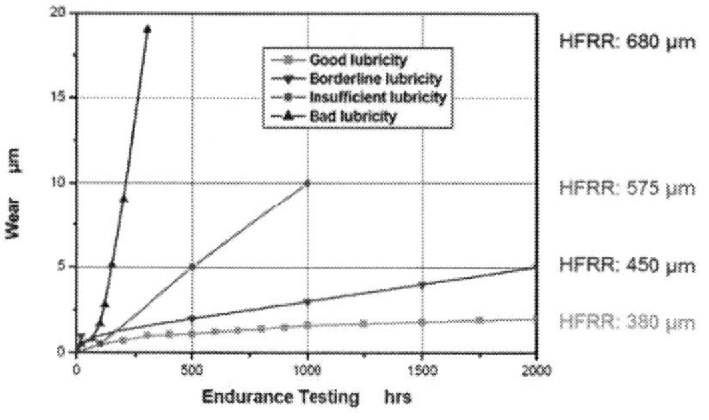

Fig. 2. Wear of pump parts over the test's duration [1].

2 Analysis of selected design solutions

2.1 Bosch pumps

Bosch is among the leading manufacturers of CR systems. The first generation of Bosch common-rail systems made its debut in 1997, in motor vehicles sold by Fiat. More high-pressure pump solutions appeared as the field grew, and they include:

- CP1 – also in the CP1H variety, with 160 MPa maximum operating pressure,
- CP2 – used in trucks,
- CP3 – modernized design of CP1, 200 MPa maximum pressure,
- CP4 – the latest mono- or bi-sectional pump, 270 MPa maximum pressure.

The functional problems of CP1 have been discussed at length in the literature of that field. The defective design of the piston drive system, which are additionally loaded by a lateral force, often leads to destruction of the entire pump. Workshop practice shows that the tri-sectional CP3 from Bosch is the most durable pump, where the flaws of CP1 were eliminated by introducing an additional sectional guide.

CP4 is currently applied in vehicles equipped with the latest Bosch injection systems. In contrast to earlier generations, it has one or two pumping sections with a 90° angle of aperture, performing two work cycles per revolution of the crankshaft (Fig. 3). The compact dimensions achieved by limiting the number of pumping sections and changing the drive method are characteristic.

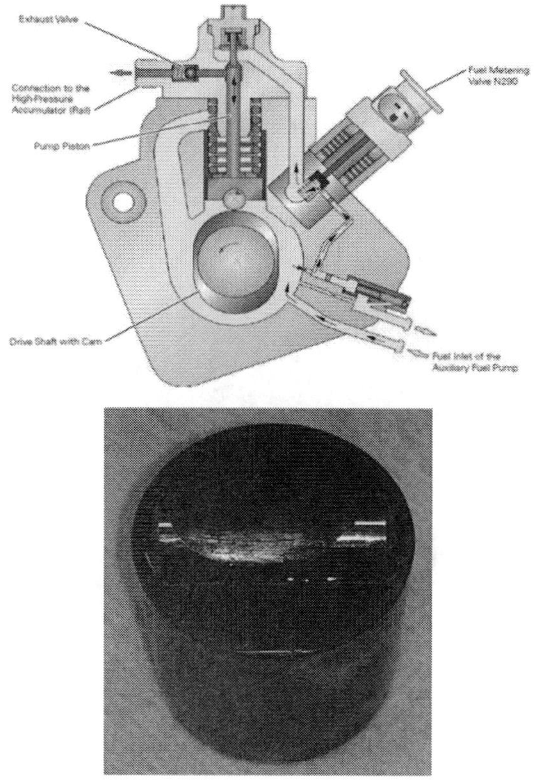

Fig. 3. Cross-section of CP4 and damage of push roller [4].

In the case of vehicles with low power, limiting the number of sections to one is very beneficial on account of the friction losses in the system. Furthermore, the reinforced structure of the body, head and elements of the drive system (including larger bearings) makes it possible to generate pressures reaching up to 270 MPa (e.g. CP4-27/2). CP4 is selected to match the engine so that there is one fuel compression cycle per engine work cycle. This is made possible by ensuring stable high pressure in the fuel rail and reducing pressure pulsations significantly. The pump's discharge is adjusted analogously as in previous generations, using the fuel batching valve.

An improper cam-roller-pusher solution is a fundamental flaw of this generation of pumps. The applied roller significantly contributed to reducing forces in the mechanism by utilizing rolling friction, however the pusher with a circular cross-section had a tendency to rotate, particularly when contaminants were present, friction was elevated by inferior fuel quality or insufficient fuel quantity. When the roller's position changes to perpendicular relative to the shaft's axis, rolling friction changes to sliding friction, which exponentially accelerates the mechanism's wear. Metal filings from the damaged roller destroy inter-operating elements of the pumping section, and cause seizing when they penetrate into injectors [4, 5]. When replacing the fuel filter in vehicles equipped with CP4, it is particularly important to deaerate the fuel system properly using a diagnostic tester. Improper deaeration leads to fuel deficiencies in the drive mechanism and can cause the roller's position to change. The scale of the problem is large enough that an independent company (2microntech) has developed a set dedicated for the CP4, including a filter purifying fuel flowing from the drive system to the pumping section (Fig. 4). The set also contains a filter indicator and makes it possible to replace the filter insert after filtering holes become clogged. Thanks to the application of an additional filter, it is possible to avoid failures of other elements of the injection system.

Fig. 4. Additional filter with filtering efficiency up to 2 μm for application in CP4 [4].

2.2 Delphi pumps

Among the many manufacturers of CR systems, only Delphi and Denso decided to apply rotary pump systems in the first generation. Individual pump varieties are abbreviated as follows:

- DFP1 – rotary injection pump for first-generation CR,
- DFP3 – bi- or tri-sectional radial pump,
- DFP6 – the latest generation of monosectional pumps, used in HDi DW10F engines .

Due to the low number of applications and short production period, no typical operational problems of DFP6 have been defined at this time. However it should be noted that DFP3, boasting a longer presence on the market, did not exhibit serious defects and can be considered a durable design. The functional problems of these pumps are limited to replacement of the discharge adjustment valve, in which the fuel mesh filter is usually damaged.

In the rotary DFP1, similarly as in Bosch's CP4 solution, a system with a roller was used to drive the piston. In this case, the head and pumping sections are stationary, while the pistons' reciprocating motion is forced by an external cam generating four piston work cycles per cam revolution. As a result of abrupt cam ascension and under conditions of insufficient lubrication, the surface of the material is damaged (flaking) (Fig. 5). This phenomenon is exacerbated by the high number of work cycles, which also increases the material's fatigue wear. Filings only 2μm in size are transported into the fuel rail and injectors along with fuel, leading to damage. Fuel also enters the fuel vessel via overflow. For this reason, if a DFP1 is damaged, all injection apparatus must be replaced, including fuel conduits and the fuel vessel. Failure to do so carries the risk of leaving filings in the system, which will damage new elements [5].

Fig. 5. Worn rollers in DFP1 [1].

2.3 Continental/VDO pumps

Pump generations are not distinguished by any symbolic designation in the Continental group. Nevertheless, these pumps can be separated into generations on the basis of design changes:

- old pumps, tri-sectional,
- new pumps, bi-sectional.

In the case of Continental pumps, feed pump failure is the primary source of damage. This is a vane pump, essentially consisting of a body, inner wheel and vanes that are slid into it. The pump is mounted on the main shaft and vanes are pressed toward the body by the centrifugal force of rotation. Their light weight, in combination with contaminants in the fuel, is the cause

DOI: 10.1051/matecconf/201711800020

of jamming, when the centrifugal force is lower than the forces of cohesion/friction between a vane and the wheel. Loss of discharge occurs as a result of leaks [6]. Reduced discharge leads to deterioration of lubrication in the drive and reduces the pump's volume output. This is slightly corrected by the discharge adjustment valve, which places additional load on the pump. The effect is seizing of contacting surfaces of the cam and pivot, blocking of the pump's pumping sections and damage to injectors.

Fig. 6. Feed pump with vanes and blocked pumping section [1].

2.4 Directions of development

Increasingly high requirements are being posed towards modern Diesel engines due to emissions standards imposed on manufacturers. The current Euro 6 standard requires the application of numerous emissions purification systems external to the engine, such as solid particle filters and SCR reactors. The injection system plays an equally important role in reducing exhaust emissions, and at the same time, it should ensure stable interoperation with emissions purification systems. In reference to injection pumps, this means creating the possibility of feeding fuel under high pressure regardless of revolutions. At the same time, one must account for the resistance arising from interoperation of parts, which grow as pressure increases. Modern pump designs should also be less susceptible to damage caused by fuel.

A tendency to reduce the number of pumping sections to one while simultaneously reducing exterior dimensions can be observed in the general design of pumps from all leading manufacturers. Figure 7 presents selected pump designs, which, besides CP4i, are already applied in engines meeting the Euro 6 emissions standard. Two essential goals are achieved by reducing the number of sections. First of all, the contacting surface of interoperating elements is significantly reduced, thanks to which internal friction is greatly reduced. Thanks to this friction reduction, the power consumed by the pump is strongly limited, up to as much as 5 kW, which is a significant portion of the power generated by the vehicle's engine.

The second goal of reducing the number of pumping sections is to facilitate synchronization of the pumping process with the injection process. In solutions utilizing tri-sectional pumps applied until now, non-uniform injection pressure was caused by a lack of equality between the number of pumping cycles and injection cycles, and over the course of exploitation – by varying efficiency of each section. In effect, varying spray jet atomization parameters were obtained as early as the fuel feeding stage. Monosectional pumps can be adapted to a 3, 4, 6 or 8-cylinder engine by changing only the transmission ratio between the crankshaft and the pump's shaft. For 8-cylinder engines, bisectional pumps with a fork configuration are also encountered (Bosch CP4.2), in which each section feeds one row of cylinders.

Fig. 7. Latest solutions of pumps for motor vehicles (from the left) : Denso HP5S, Delphi DFP6, Continental DHP1, Bosch CP4i [3].

The problem of pump lubrication, extremely important from an operational perspective, was solved by using motor oil in the CP4i design. The solution of lubricating the pump with motor oil is commonly applied in trucks, where the kilometerage between repairs may reach up to several hundred thousand kilometers. Motor oils have much better lubrication and viscosity parameters, thus enabling significant reduction of cases where pump elements are worn pathologically. Ensuring that pumping sections are tight (leakproof) is a significant problem, and CP2, used in trucks, is a good example of this, as motor oil dilution was a frequent phenomenon in this design. At this time, the details of the CP4i design are unknown, nevertheless the manufacture declares motor oil leaks at a level not exceeding 1 ml per million cycles, with 95% hydraulic efficiency [7].

To improve pump efficiency, manufacturers are planning to change the method of discharge control. In traditional CR solutions, a discharge adjustment valve is installed in the body (also visible in the Delphi DFP6 solution in Fig. 7). This valve is controlled by PWM, or pulse-width modulation. Depending on the desired discharge, pulse-width is increased or decreased, causing the valve to reduce or increase the cross-section of the fuel channel feeding the pumping section. The new concept calls for direct control of section filling in every work cycle. During the suction phase, the valve is open and makes it possible to fill the pumping section completely. After the piston passes through its extreme lower position, the valve is closed only when the requirement amount of fuel is found in the section. As the piston ascends in its stroke, excess fuel return to the feed conduit via the inlet channel. The advantages of this solution should be sought, above all, in the capability of real-time pump control. It is possible to freely increase or decrease discharge in consecutive compression cycles, which, in combination with synchronization of the pump shaft's position with the crankshaft, can be applied to reduce harmful volumes harmful to the fuel rail (reduction of hydraulic phenomena). Furthermore, pumps performing two work cycles per section per shaft revolution are under high loads, both mechanical and thermal. Complete filling of a section contributes to the internal cooling of the pump's cylinder.

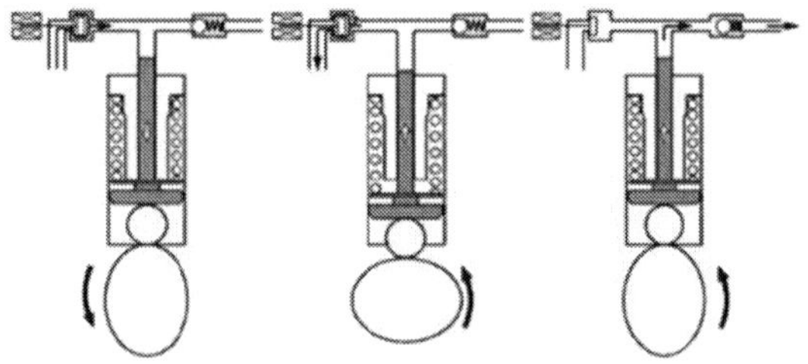

Fig. 8. Pumping diagram of a Denso GP5S with a direct-action valve [5].

The changes proposed in prototype CR systems fall outside of the subject of pump optimization and also call for the application of pumps with completely different hybrid designs. Many manufacturers have presented a solution based on the application of components typical for UIS/UPS systems and CR (Fig. 9). Fuel vessels of significantly diminished volume, interoperating with injectors typical for a CR system, are a characteristic feature of this approach. Pumps are situated in the head and powered in a manner typical for a unit injector system. Another solution is a hybrid system, in which some injectors are classical CR injectors and the rest (usually in a 1:1 ratio) are electronically controlled unit injectors connected to the CR injectors by the fuel vessel. In such a system, the unit injector performs the role of both individual injection unit for a given cylinder and pump for the rest of the system. An undoubted advantage of solutions like this are their limited overall dimensions, which make it possible to situate all subassemblies under the valve cover. In their new CR pump designs, Delphi and Bosch provided the possibility of installation directly in the head. The pumping section with the roller and control valve constitute a module that can be placed in the body of a traditional pump or directly in the head, with a camshaft drive (Fig. 9). This method of fitting the pump enables easy lubrication of the drive system through the crankshaft's oil system while simultaneously maintaining small exterior dimensions of the drive unit. It should be emphasized that the presented solutions are prototypes, and the actual benefits of applying such systems are still unknown. Considering the presented problems and directions of development, pressure will continue to rise in future systems, and pump designs will be simplified, with an oil lubrication system and perhaps systems being a hybrid of existing solutions will be developed.

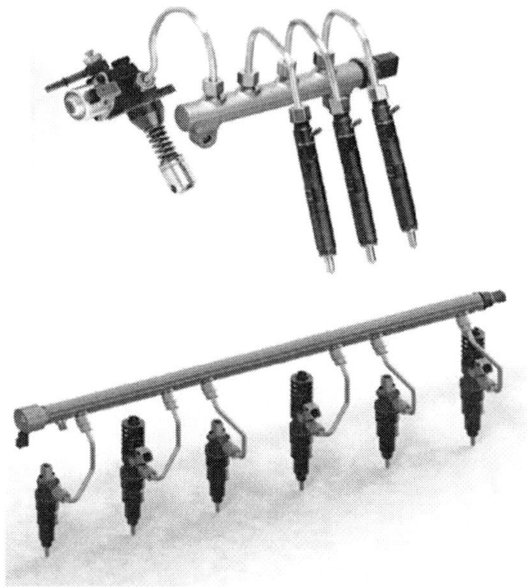

Fig. 9. Prototype Delphi systems : UP-CR on the bottom, common rail on the top with modular pump interoperating with the crankshaft [1].

3 Utilizing a hypocycloid drive in a high-pressure fuel pump

3.1 Origins

The hypocycloidal transmission with interior toothing was already patented in the 19th century. Depending on the selected wheel diameters, it made it possible to obtain various curves resulting from its resultant motion. Its popularity was limited by design and economic aspects. At the time, these drives and engines did not transmit high powers or torques, and the lifetime of subassemblies was very limited. So, the problem of optimizing the design of a system converting reciprocating motion to rotation was not as significant as it is today. Moreover, in reference to the traditional slider mechanism, manufacturing a hypocycloid transmission would be more expensive and more difficult due to the high requirements relating to its teeth.

For many years, designers have sought possibilities of applying a hypocycloid gear in combustion engines. This system would replace the slider-crank mechanism used until now. The absence of lateral forces acting on the guided element is a great advantage of hypocycloids on account of the crank system. It is accepted that lateral forces make up approx. 10% of the axial force generated by the pressure in the cylinder. Significant lateral forces generate very high powers, and pressures, from one cylinder. Eliminating the effects of their action is a very large field of study for engineers concerned with combustion engine design [8, 9].

One example of a hypocycloid transmission is the Wiseman engine patented in 2001 (Fig. 10). As part of work performed at the company, an engine with 10, 20 and 30 HP was developed and an engine with approx. 1 HP was made and tested. The design was based on the classical slider-crank engine. The experiment's results showed power and efficiency parameters that were very similar to the base engine, however at the same time, high potential for optimization and improvement of indicators was observed.

Fig. 10. Wiseman engine with hypocycloid transmission in cross-section [6].

In addition, the engine's overall dimensions were limited significantly by using one of the hypocycloid's beneficial features, which is high pitch compared to the system's size. Unfavorable efficiency and power parameters were obtained for the two-stroke engine, due to the large volume of the crankcase, resulting in low pressure. Numerous patent applications relating to designs utilizing a hypocycloid drive demonstrate the system's high potential and benefits of its application. Many attempts to implement the hypocycloid are being undertaken, in order to replace the aforementioned slider system or the cam drive. The hypocycloid can transmit power in two directions, which means it can be the system driving the working element and the power take-off (in the case of an engine – the piston), which is a significant advantage in comparison to the cam – pusher system.

3.2 Description of the drive

The hypocycloid transmission consists of two wheels – the larger wheel (R) has interior toothing and the smaller wheel (r) has exterior toothing. Torque is applied to the smaller wheel, making it turn, however the larger wheel cannot rotate around its axis. The smaller wheel moves over the circumference of the larger wheel, and any point on the smaller wheel's radius moves in a curve called the hypocycloid. To create the working drive of the pump, a hypocycloid transmission with a gear radius ratio of $R/r = 2$ was applied. This selection of wheels makes it possible to achieve resultant linear motion, as per Copernicus's Theorem: "Consider two circles of radii R and $R/2$ with the smaller one rolling inside the bigger circle without slipping. Any fixed point on the circumference of the small circle traces a straight line segment – a diameter of the big circle." Figure 11 presents a diagram of the transmission, and the line on which the mechanism's sliding element moves is marked.

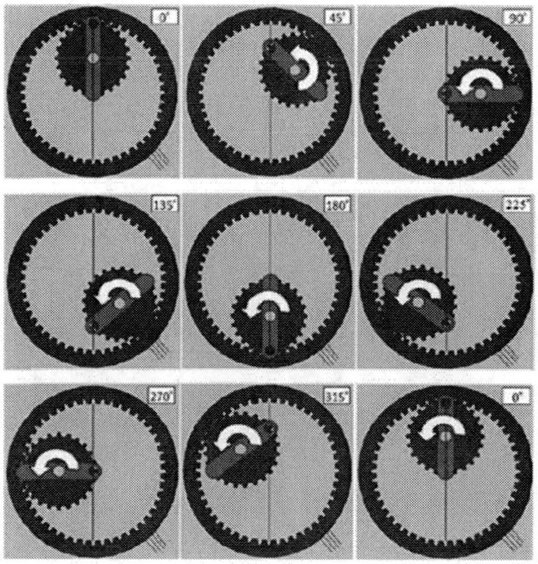

Fig. 11. Principle of operation of hypocycloid transmission [6].

Analysis of the motion of the working element indicates that it performs movement in which the speed can be described using the sine function, similarly to a slider mechanism. The parametric equation determining the point's position as a function of angle of rotation can be defines as follows:

$$x(\alpha) = (R-r)\cos(\alpha) + r\cos\left(\frac{R-r}{r}\propto\right) \quad (1)$$

$$y(\alpha) = (R-r)\sin(\propto) - r\sin\left(\frac{R-r}{r}\propto\right) \quad (2)$$

3.3 Benefits of applying a hypocycloid drive in a fuel pump

The operational problems of pumps in common rail systems, described in the first two sections, indicated their high sensitivity to fuel quality as well as tendency for accelerated wear in the area of the drive due to lateral forces. The proposed solution of a pump with hypocycloid drive is free from these flaws and simultaneously makes it possible to utilize other beneficial features of the hypocycloid. Figure 12 presents a concept cross-section of the designed pump, which was modeled using Autodesk Inventor software. The pump is driven by the main shaft (1), in which the countershaft (2) is mounted in a manner enabling their reciprocal rotation. A toothed wheel with exterior toothing is found on the countershaft and meshes with toothed wheel (3), which is immobile and bolted to the pump's body. The drive shaft has bearings on the exterior surface of wheel 3, bearing 4. A mandrel (9) is mounted eccentrically in the countershaft, and the plunger (8) along with the guide element is connected to this mandrel via bearing and positioned so that it can slide in the cylinder (7). Furthermore, the countershaft has a support (5) with bearings in the pump's body – bearing (6).

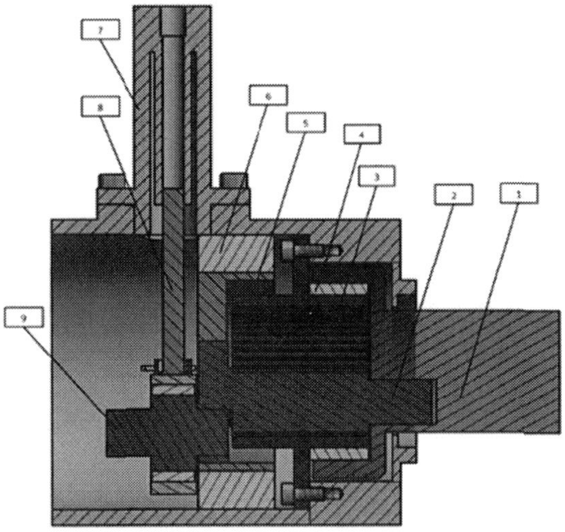

Fig. 12. Scheme of high-pressure pump with hypocycloid drive: 1 – main shaft, 2 – countershaft, 3 – toothed wheel with interior toothing, 4 – bearing of main shaft, 5 – support of countershaft, 6 – support bearing, 7 – cylinder of pumping section, 8 – piston (plunger) with guiding element, 9 – mandrel.

One of the most important benefits of the pump's design is the high pitch of the working element, equal to the diameter of the large wheel's pitch circle diameter. In traditional CR pumps, the pitch of the working element is no greater than 7-8 mm. For the module under consideration, with a transmission ratio of 1 and 40 teeth, the pitch of the working element will be 40 mm. Multiplying the piston's pitch while maintaining its diameter makes it possible to obtain much greater discharge per work cycle. In practice, this solution leads to two benefits:

- the possibility of obtaining very small pump dimensions relative to discharge in comparison with conventional cam pumps,
- the possibility of significantly reducing forces present in the pump mechanism by using high discharge to generate high pressures by means of a simple, mechanical hydraulic amplifier. Besides reducing the pump's overall dimensions, reducing forces will make it possible to apply cheaper (poorer) constructional materials.

Another effect of high piston pitch is the possibility of easy utilization and control of the implosion effect. This effect is based on introducing a liquid or gas into a space and dissolving it in the volume of that space by increasing pressure (which reduces volume). During the injection process, when there is a pressure drop, gas is released from the liquid and has a beneficial effect on the fuel jet atomization process, improving fuel spraying. Thanks to increased pitch, and thus, greater sectional volume, the parameters of the air-fuel mixture can be controlled easily.

The problem of lateral forces in the drive system has been eliminated in the proposed solution. In the case of a cam drive, the lateral force on the piston reaches values up to 300 N. Besides elevated wear in the cam – piston (plunger) area, unfavorable chamfering of the piston takes place, leading to increased wear of the cylinder's side surface and thus reducing the section's tightness. An additional guiding element loaded with lateral force is applied in many pumps, however this is an unfavorable solution due to the increased number of friction pairs and the inertia of moving parts. Moreover, a defective guiding system can be the cause of deteriorating pump parts, as shown by the example of CP4. In relation to the absence of lateral forces, it seems beneficial to apply ceramic engineering materials for contacting surfaces of the piston and cylinder. In the case where fuels of reduced quality or heavy fuels are pumped, hard ceramic surfaces contribute to eliminating the wear on the piston-cylinder pair. Besides the possibility of applying many other modern fuels, attention should be paid to the suitability of designs for fueling an engine with rapeseed oil and its esters. In the case of traditional pumps, during standstill, esters "glue" interoperating elements together, which is often the cause of many failures damaging the entire system, as shown in the example of vane blocking in the Continental pump, or sliding friction between the roller and cam, as in Delphi DFP1.

The applied drive solution should also increase the pump's resistance to heavy fuels. In the presented pump design, the pumping section was separated from the drive system, under the assumption that it would be lubricated using motor oil or transmission oil. Due to the piston's high pitch and the pumping section's length, it is possible to apply double labyrinth sealing along with a channel draining leaking fuel into injectors' overflow conduit. This will make it possible to achieve the minimum, acceptable leakage of drive oil into the lubricating fluid. Substituting fuel with lubricating oil will have a beneficial impact on the lifetime of interoperating elements, with regard to both sliding bearings and applied toothed wheels [10].

The absence of a return spring should also be counted as one of the pump's benefits. Pump solutions utilizing a cam system for each section have a return spring that ensures that the pusher is constantly pressed into the cam. Besides obviously increasing the weight of the system, the spring is an element that receives energy and gives it off in the form of heat to a small extent. In addition, considering a high number of cycles, return springs are susceptible to damage caused by material fatigue. The presented pump design is characterized by a

DOI: 10.1051/matecconf/201711800020

journal-bearing joint for the mandrel and working element, so two-way movement is determined by the resultant of these two components.

The very high discharge from the pumping section that was achieved fits well into the latest trends in fuel injection systems. The concept injection systems mentioned in the second section are characterized by the pursuit of the minimum number of elements while maximizing the engine's economic and environmental indicators. The combination of high piston pitch and a direct control valve makes it possible to eliminate the fuel vessel entirely. In CR systems, the existence of a common fuel rail is a necessity, however it is problematic from the perspective of precise spraying control, above all, control of injected fuel, due to complex hydraulic phenomena. By using a direct-action valve, the start and end point of injection can be determined easily, and a conventional valve with the same design as a pressure regulator in a common rail can be used to adjust pressure. Such a system largely resembles a UPS assembly, nevertheless, the type of pump and control system are significant differences.

3.4 Selected operating parameters of pump with hypocycloid drive

The presented pump design has been filed under the patent application P.418961, which also contains reservations concerning the further development of the design, such as the application of engineering ceramics or the introduction of gas into the chamber for the purpose of obtaining the implosion effect. Preliminary strength calculations have been conducted and confirm the correctness of selected materials. Additional strength calculations using the finite element method were conducted using Femap software, running on the basis of the NX Nastran module. The countershaft is the element most under load in the system due to the complex state of flexural and torsional stresses. The authors created a model of the design in Autodesk Inventor software and conducted dynamic simulation of the assembly's operation on the basis of this model. Figure 13 presents the progression of loading force acting on the pump piston in the dynamic simulation. Pressure was set to 1500 bar, and the 75% of the section's volume was filled with fuel. In effect, a force of 5300 N acting on the piston was obtained. The rapid growth of force to the maximum value during the compression phase is due to the fluid's low compressibility. When pressure reaches 1500 bar, the valve is opened and a batch of fuel is pumped outside of the system. The force progression is similar to the curves obtained for cam pumps with the same compression pressure and piston diameter, in terms of the value and profile of the curve. Decidedly greater values were achieved, however, for the piston's linear speed, which reaches a maximum of 3.14 m/s. This speed is directly proportional to angular speed and to the distance of the center of rotation from the center of the working element (in a cam system – eccentric, in a hypocycloid – distance of the mandrel's axis from the axis of the wheel with interior toothing). The relationship between angular speed and speed of reciprocating motion is defined by the formula :

$$V(\propto) = \omega r \sin(\propto) \qquad (3)$$

The increase of piston speed is linked to deterioration of the cylinder filling efficiency parameter due to elevated flow resistance, which is related to both elevated pumping resistance on conduits feeding the section and to the valve's design. For this reason, it seems advantageous to apply new-generation direct-action valves, which must be characterized by low flow resistance due to the two-way flow of compressed fluid. Due to the proportional dependency between speed and eccentricity, the hypocycloid system clearly benefits interoperating parts, particularly at low speeds. Increasing linear speed, according to lubrication theory, makes it possible to generate a layer of lubricating film in the piston-cylinder pair, which effectively separates the two interoperating elements. In relation to the absence of lateral forces which cause chamfering of the piston, very beneficial parameters of interoperation between the side surfaces of the cylinder and piston (plunger) are obtained.

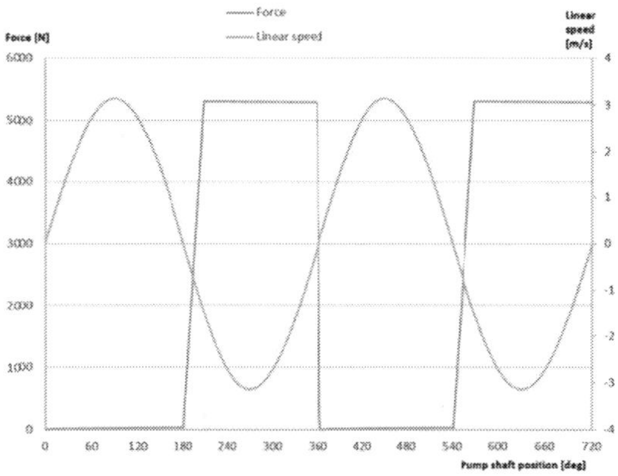

Fig. 13. Progression of force acting on the piston's face and piston's linear speed.

In order to determine power consumed by the pump, torque on the main shaft was determined. The value and progression of torque is presented in figure 14. The simulation accounted for friction losses between interoperating elements. The average torque required to drive the pump was calculated for the obtained curve. Based on this torque, knowing that the pump shaft was revolving at 1500 rpm, power consumption was determined to be 14.5 kW. This parameter seems unfavorable compared to conventional solutions, however the fact that the discharge per shaft revolution is much higher, as demonstrated in the simulation, must be accounted for. It seems that it is much more representative to relate the volume of pumped fluid to the pump's power. Very similar indicators of both pumps' work were obtained as a result of comparing the given parameters. Therefore, it is possible to achieve favorable working parameters of an engine using an injection pump with hypocycloid drive. Further optimization of the pump based on the results of workstation tests will go in the direction of reducing internal friction, which will additionally improve the entire system's efficiency.

Table 1. Comparision between hypocycloid and CP4 pump.

	Bosch CP4	Hypocycloid pump
Input power [kW]	3.95	14.51
Unit discharge [cm^3]	0.3 (100% filling)	1.15 (75% filling)
Discharge to power ratio	0.076	0.079

Figure 14 also presents the progression of lateral force in the mandrel – working element pair. Low values within the range of 10–5 N were obtained due to the systems vibrations. In the cam solutions presented earlier, for a cam-plate friction coefficient of 0.01, this force reaches up to several hundred N. Reduction of this force also provides the possibility of applying cheaper materials, which work under elastohydrodynamic lubrication conditions in the mandrel – bearing pair of the working element, besides reducing the adverse impact on wear of elements. DLC (Diamond Like Carbon) coatings are currently applied to cams and pushers, which, due to their high hardness thanks to properties similar to diamond, is wear resistant. High production costs related to advanced technology and the need to employ additional machines are the disadvantages of applying such coatings. In addition, in the case of insufficient lubrication, the surface layer undergo abrasive wear, which leads to contact with the cam's core and seizing of interoperating pairs.

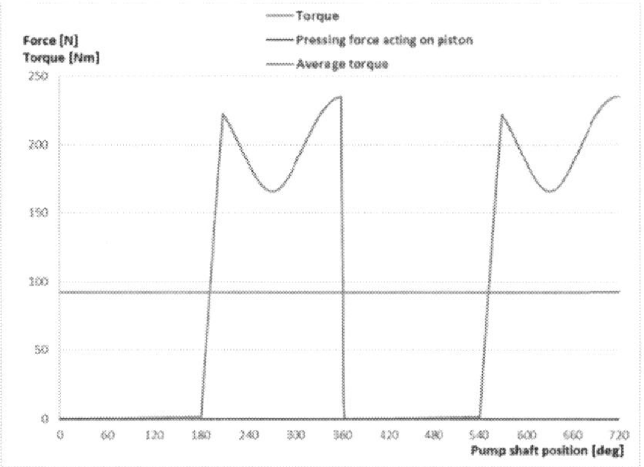

Fig. 14. Progression of pump's drive torque and lateral force acting on a piston and its linear speed.

4 Summary

Despite the dynamic development of spark ignition units, which are displacing Diesel engines on the passenger vehicles market as emissions standards become more strict, Diesel engines will continue to make up a large share of the total production of drive assemblies. Injection systems have a significant role to play in improving their operating parameters. Due to the high precision of injection process control, with high pressure of fuel compression, these systems are characterized by sensitivity to the quality of applied fuel due to the large forces acting on the system's elements. Numerous design solutions are susceptible to damage resulting from defective design of a given element, besides damage generated by fuel of insufficient quality. In the case of pump defects, leading to the creation of filings with diameters below several micrometers, other elements of the injection systems are also damaged very frequently, which increases repair costs significantly. The modern CR pump should be characterized by resistance to improper fuel quality, high compression pressure and low power consumption [11].

The authors have proposed in their article a hypocycloid drive fuel pump. The benefits presented by the drive system are the advantages of the presented solution both in terms of durability and achieved environmental parameters. The pump uses a hypocycloid mechanism instead of a conventional cam to drive the pumping section, which is lubricated using a separate lubricating oil system. The high stroke of the piston relative to the overall dimensions of the pump optimizes the flow, contributing to the compact size and the reduction in the number of pumping sections even in multi-cylinder engines. This pump is part of the latest trends in the design of high pressure injection systems. The lack of lateral force contributes to reducing the wear of the walls of pumping section. Due to the different design of the propulsion system, the problem of wear of the cam-piston knob caused by the resulting lateral force, also contributes to the complete damage to the pumps. In addition, the simulations showed that the torque curve does not deviate from the parameters of modern pumps. The large piston stroke of the pump gives the opportunity to create a system where the fuel tank is completely eliminated, which greatly simplifies the engine's layout and contributes to the reduction of harmful hydraulic volume. The direction of further work should be to optimize the flow of fuel in the canals and valve sections.

This article was financially supported within the project "New generation of common rail pumps" – Lider/015/273/L-5/13/NCBR/2014, implemented within the LIDER Programme, financed by the National Centre for Research and Development, Poland.

References

1. Materiały szkoleniowe Centrum Szkolenia Motoryzacji Autoelektronika Kędzia.
2. W. Karpiuk, G. Kinal, R. Smolec, Combustion Engines **162**, 988-995 (2015)
3. W. Urzedowska, Z. Stepien, Fuel Processing Technology **142**, 403-410 (2016), DOI: 10.1016/j.fuproc.2015.11.001

4. T. Osipowicz, T. Stoeck, Autobusy: technika, eksploatacja, systemy transportowe **16**, 178-181 (2015)
5. D. Pietras, Polish Maritime Research **23**, 58-63 (2016), DOI: 10.1515/pomr-2016-0032
6. R. Priyesh, Doctorate Thesis (2014)
7. G.M. Szymanski, F. Tomaszewski, *Proceedings of The 17th International Congress on Sound and Vibration* (Cairo, 2010)
8. www.2microntech.com (2017)
9. S.W. Kruczyński, B. Chrupek, Zeszyty Naukowe Instytutu Pojazdów **4**, 113-122 (2014)
10. J. Merkisz, J. Mizera, M. Bajerlein, L. Rymaniak, P. Maj, Experimental and Applied Mechanics **518**, 102-107 (2014), DOI: 10.4028/www.scientific.net/AMM.518.102
11. K. Szwedzka, J. Gruszka, P. Szafer, *IOP Conference Series-Materials Science and Engineering* **145**, 062007 (2016), DOI: 10.1088/1757-899X/145/6/062007

MATEC Web of Conferences **118**, 00021 (2017) DOI: 10.1051/matecconf/201711800021
VII International Congress on Combustion Engines

Practicability of passenger vehicle driving emission tests according to new European Union procedures

Jacek Pielecha[1,*], *Remigiusz* Jasiński[1], and *Agnieszka* Magdziak[1]

[1]Poznan University of Technology, Faculty of Machines and Transport, ul. Piotrowo 3, 60-965 Poznan, Poland

Abstract. The article compares driving test data using the latest legislative proposals applicable to passenger cars. Several measurements were performed on the same test route in accordance with the RDE test guidelines, which requires a number of criteria to be met. These criteria include: the length of the measuring segments, their overall test time share, and the dynamic characteristics of the drive. A mobile device for reading the EOBD System information was used to record the engine and vehicle operating parameters during tests. This allowed for the monitoring of parameters such as: load value, engine speed and vehicle velocity. The obtained results were then analyzed for their compatibility with the RDE procedure requirements. Despite the same research route, the obtained results were not the same. The analysis also uses the two-dimensional operating time share characteristics expressed in vehicle velocity and acceleration co-ordinates. As a result it was possible to compare the dynamic properties, share of operating time and, consequently, to check the validity of conducted drive tests in terms of their practicability and emission values.

1 Road driving emission tests requirements

The need to reduce negative impact on the environment has become a key driving force for the sustainable industry development. This has led legislators to adopt more stringent pollutant emission standards for the automotive sector [1-3]. The push towards this practice is still growing, as road transport is considered one of the major sources of pollution in the European Union. Currently, efforts are being made to introduce such emission testing procedures, which would extend to include the operating range of the engine used in real driving conditions [4, 5]. This is in contradiction to the previous testing procedures, where the exhaust emission values are measured in laboratory conditions (on chassis dynamometer for passenger cars) in the adopted type approval test [6-9]. Moreover, due to the global scope of production and sales of motor vehicles, there is a developing trend of type approval testing procedures unification [10, 11].

In January of 2011, the Commission had set up a working group composed of all stakeholders involved to develop a test procedure for real-time emissions that would better reflect the amount of pollutant emission measured in driving tests. To this end, following detailed technical discussions, the variant described in the Regulation (EC) No. 715/2007 [12] was adopted, i.e. the use of portable emission measurement systems (PEMS) and distance emission limit values. As agreed with the stakeholders in the CARS 2020 process, the RDE test procedures should be implemented in two stages: during the first transitional period, test procedures should only be used for emission monitoring purposes and then followed with binding quantitative RDE requirements for all new type approvals and new vehicles. RDE testing procedures have been introduced in the Commission Regulation (EU) 2016/427 [13]. Currently, quantitative RDE requirements have been established to limit emissions from exhaust systems under all operating conditions in accordance with the distance emission limits set out in Regulation (EC) No. 715/2007.

In order to allow manufacturers to gradually adapt to the RDE testing, the final quantitative RDE requirements should be introduced in two successive stages [13, 14]. In the first stage, which should come into force after 4 years from the mandatory application of the Euro 6 standard, a conformity factor of 2.1 should be used. The second stage should begin 1 year and 4 months after the first stage and should require full compliance with the NOx distance emission limit value of 80 mg/km as specified in Regulation (EC) No 715/2007 [10], plus a margin, taking into account additional measurement uncertainties resulting from the use of portable emission measurement systems (PEMS) (Table 1).

Table 1. RDE testing requirements in Europe [12, 15].

2015	2016	2017	2018	2019	2020	2021	2022
Euro 6b		Euro 6c				Euro 6d	
NEDC		WLTC					
Research and concept phase		Conformity Factor (CF)					
		$CF_{NOx} = 2.1$ $CF_{PN} = 1.5$				$CF_{NOx} = 1.5$ $CF_{PN} = 1.5$	

* Corresponding author: jacek.pielecha@put.poznan.pl

RDEs should cover all possible road situations, using a specific driving styles for the tested vehicles to intentionally alter the results to be more positive or negative that does not result from the technical characteristics of the vehicle, but from a very unusual driving style should be carefully avoided [16-18]. Therefore, complementary boundary conditions for RDE tests (Table 2) are introduced to prevent such situations. Owing to their nature, driving conditions occurring during individual PEMS drive tests cannot fully correspond to the "normal conditions of vehicle operation". The emission control intensity during such tests may therefore vary.

Table 2. Specific requirements for RDE tests [13, 14].

Parameter	Requirement
Ambient temperature (t_a)	• normal range: 0 °C ≤ t_a ≤ 30 °C • lower extended range: −7 °C ≤ t_a < 0 °C • upper extended range: 30 °C < t_a ≤ 35 °C
Driving test altitude (h)	• normal range: h ≤ 700 m a.s.l. • extended range: 700 m a.s.l. < h ≤ 1300 m a.s.l.
Impact evaluation of ambient weather and road conditions as well as the driving style	• total altitude increase: less than 1200 m / 100 km • relative positive acceleration (RPA): move than PRA_{min} (for all road conditions) • the product of velocity and acceleration ($V \cdot a_{pos}$): less than $V \cdot a_{pos\ max}$ (for all road conditions)
Vehicle technical state before the test	• engine cold start: cooling liquid temperature below 70 °C, time at least 300 s • cold start emission results not counted towards the RDE test results
Any vehicle stop	• no longer than 180 s
Vehicle aftertreatment systems operation	• a single regeneration of the particulate filter justifies repeating the RDE test; the occurrence of two filter regenerations is to be included in the results of the RDE test
Driving comfort systems operation	• regular use as intended by the manufacturer (for example: use of the air conditioning)
Vehicle load	• vehicle mass: driver (and passenger) along with the PEMS equipment; maximum load < 90% of the sum of the mass of the passengers and the vehicle curb weight
Test requirements	• duration 90 min – 120 min
Urban test portion requirements	• 29% – 44% share of the whole test time • distance: more than 16 km • vehicle velocity: up to 60 km/h • average velocity: 15 km/h – 40 km/h • vehicle stop: 6% – 30% of the urban portion of the test time
Rural test portion requirements	• 23%–43% share of the whole test time • distance: more than 16 km • vehicle velocity (V): 60 km/h < V ≤ 90 km/h
Motorway test portion requirements	• 23%–43% share of the whole test time • distance: more than 16 km • vehicle velocity: more than 90 km/h • vehicle velocity over 100 km/h for at least 5 min • vehicle velocity over 145 km/h no more than 3% of the test portion time

Therefore, and in order to account for the uncertainty of statistical and technical measurement procedures, in the future it may be possible to reflect in the distance emission limits, applicable to individual PEMS test drives, the characteristics of these drives, as determined by measurable parameters such as driving dynamics or engine load [19]. But in this case, it should not lead to an increased environmental impact and lowered effectiveness of RDE testing procedures, which should be demonstrated through peer-reviewed research tests. In addition, when assessing the emission control intensity in PEMS testing, only parameters that may be justified by objective scientific considerations, not just engine calibration, pollution control or emission control systems, should be considered [20].

2 Research methodology

2.1 Data collection

The Mobile EOBD Read System – TEXA OBD Log [21] (Fig. 1) was used for the test. This allowed the recording of engine parameters at 1 Hz directly after connecting it to the diagnostic socket. Values such as: sampling time, engine speed and load were monitored. In addition to the data from the motor controller, a GPS location signal was recorded. It is worth noting that during the tests the parameters recorded from the EOBD system [20] were stored in the memory of the TEXA OBD Log reader without the possibility of their continuous visualization.

Fig. 1. TEXA OBD Log mobile read system [21].

2.2 Exhaust emission measurement

The testing apparatus is presented in Fig. 2a. A portable Semtech DS analyzer [23] was used for the measurement of the exhaust emission from vehicles (Fig. 2b). It allowed measurements of carbon monoxide, carbon dioxide, hydrocarbons, non-methane hydrocarbons and nitrogen oxides exhaust emission.

A portable AVL MOVE PM (Micro Soot Sensor + gravimetric particulate measurement) analyzer was used to measure particle mass (Fig. 3a), while a PMP-compliant AVL condensation particle counter [24] was used to measure particle number (Fig. 3b; it is important to note that in accordance with the legislation, this device only measures solid particle). In terms of benchmarking and quality control, zero-span checks were performed before and after each measurement. A linearization of the equipment was carried out every three months. Post-processing plausibility checks were made on all data, focusing on CO_2, to ensure that the data collected was realistic; results from the Micro Soot Sensor were compared with PM and PN results to check their plausibility. The emission measurement equipment had a maximum mass of 116 kg (Gas PEMS – 25 kg; PM PEMS – 45 kg, PN Counter – 25 kg), together with an additional power supply (generator) – 21 kg. The total flow rate of the combined sample streams was approx. 20 dm^3/min. This was judged not to have any impact on the operation of the test vehicles or their engines, even when idle.

a) b)

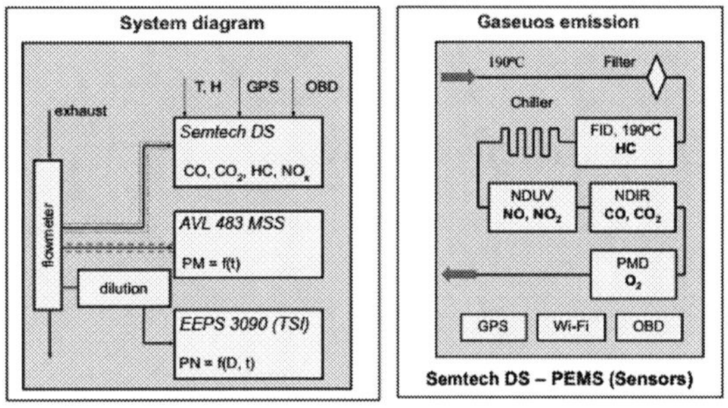

Fig. 2. The measuring systems used for testing under real driving conditions.

a) b)

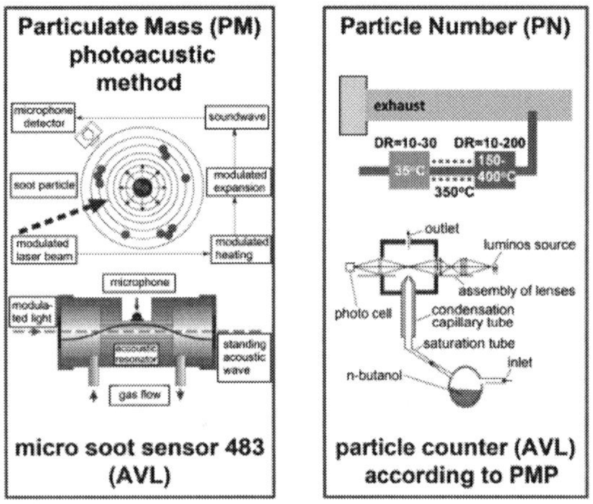

Fig. 3. Measurement system used for testing particle emission: a) analyser for measurement of the particle mass concentration – AVL MOVE (micro soot sensor + gravimetric), b) particle counter (AVL) for measurement of particle number.

2.3 Test route characteristics

Measurements were made 8 times conforming to the real driving conditions in urban, non-urban and motorways in Poznan and its surroundings. The average distance traveled was 80.2 km. Figure 4 shows an example of a route. The test route has been chosen so that it meets the requirements of the European Commission as described in the Regulation [13, 14], with particular attention paid to its topography. Table 3 shows the characteristics of the route in terms of terrain.

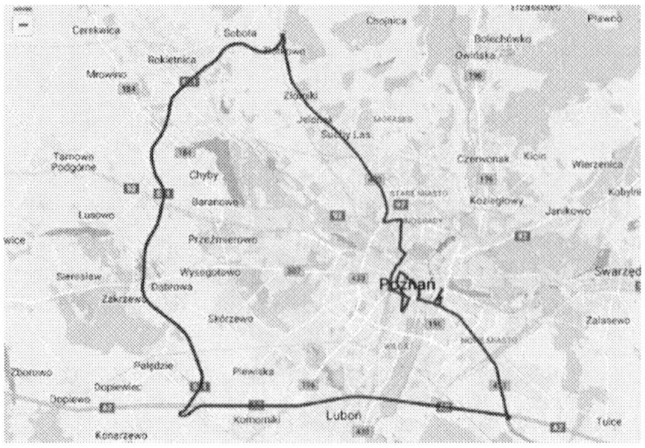

Fig. 4. Test route example.

Table 3. Topography characteristics of the test route.

Parameter	Test route
Altitude:	
minimum [m]	52
average [m]	82
maximum [m]	109
Maximum slope:	
increase [%]	4.1
decrease [%]	–3.7

The topography analysis of the test route confirms its compliance with the test requirement of the altitude difference between the start and end points of the test to not exceed 100 m. In addition, it was possible to distinguish between urban, non-urban and motorway sections.

3 Vehicle parameter analysis

In spite of traveling on a similar route, the results obtained varied. In Figure 5 results from all test drives are shown using the $V = f(S)$ relationship. It shows the division of the research route into 3 portions: urban (0 km/h - 60 km/h), rural (60 km/h - 90 km/h) and motorway (90 km/h). In spite of the similar research route, both the velocity results as well as its average values in the individual test portions are not the same. The driving parameters defined by acceleration, constant velocity, braking and stopping are similar. These parameters are systematized in Fig. 4a-d. The average values were: for acceleration – 24.6%, constant driving velocity – 34.7%, braking – 23.6%, and for parking – 16.9%.

Detailed requirements of the RDE road test standard are given below, where the verification of individual driving parameters (defined earlier) conformity has been demonstrated, and their values are compared also giving the permitted ranges (if required) and with mean values also determined.

The analysis of test distance data in the urban part showed that drive No. 2 was characterized by the longest distance traveled and the drive No. 4 with the shortest. However, the values of the distance covered for all test drives were within the permissible range, i.e. they were more than 16 km (Fig. 6a).

DOI: 10.1051/matecconf/201711800021

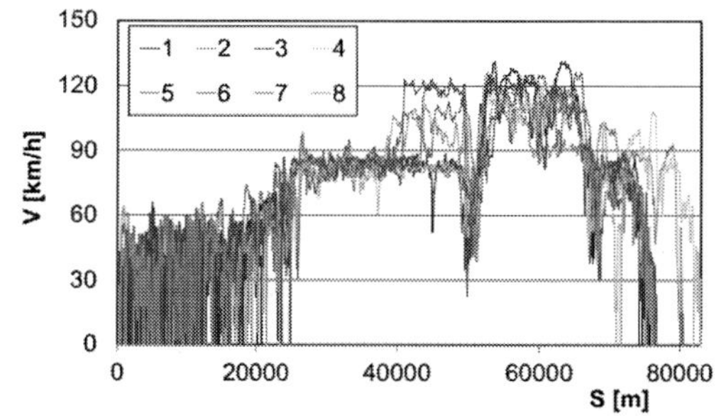

Fig. 5. Vehicle velocity for all RDE test drives.

Table 4. Drive parameters (acceleration, constant vehicle velocity, braking, stops) against the mean value.

Test drive number	Share [%]			
	$a > 0$	$V =$ const	$a < 0$	$V = 0$
Drive 1	20	39	20	21
Drive 2	20	51	19	10
Drive 3	22	41	20	17
Drive 4	21	41	20	18
Drive 5	23	34	23	19
Drive 6	26	34	26	14
Drive 7	33	21	30	16
Drive 8	32	17	31	20
Average	24.6	34.7	23.6	16.9

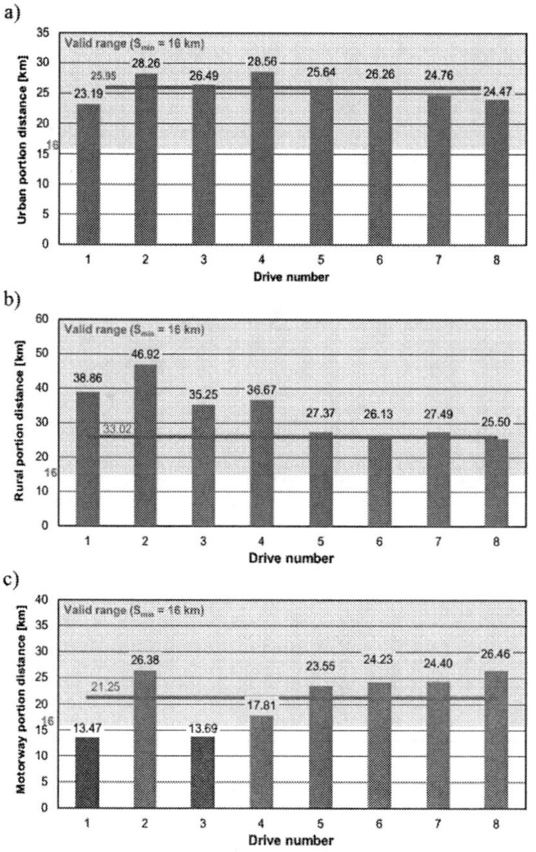

Fig. 6. Travel distance comparison of urban (a), rural (b) and motorway (c) test drives with the minimum distance (required) for all drives and the mean value.

Analysis of the travelled distance data in the rural portion showed that drive No. 2 was characterized by the longest distance traveled and the drive No. 8 with the shortest distance. The values of the distance covered for all test drives were within the permissible range, i.e. they were more than 16 km (Fig. 6b). The distance travelled on the motorway was the longest for the drive No. 8 and the shortest for the drive 1. The distance of the drives No. 1 and No. 3 were too small. Other drives were within the acceptable range, i.e. their distance was longer than 16 km and mean was 21.25 km (Fig. 6c).

The percentage time share of the urban portion in the drive tests was the largest for drive No. 3, and the smallest for drive No. 2. The percentages of urban share of all drives (except for drive No. 2) were within the permissible range of 29% – 44% (Fig. 7a). The percentage time share of the non-urban portion in the drive tests was the largest for the drive No. 1, and the lowest for the drive No. 8. Only the values obtained from the drives No. 5-8 were within the acceptable range of 23% - 43%. Others did not meet this requirement (Fig. 7b). The analysis of percentage time share data of the motorway section in the drive tests showed that the highest value was for the drive No. 8, and the smallest – for the drive No. 1. The shares of the motorway portion were within the permissible range of 23% - 43% (Fig. 7c) except for the values of drives No. 1, 3 and 4.

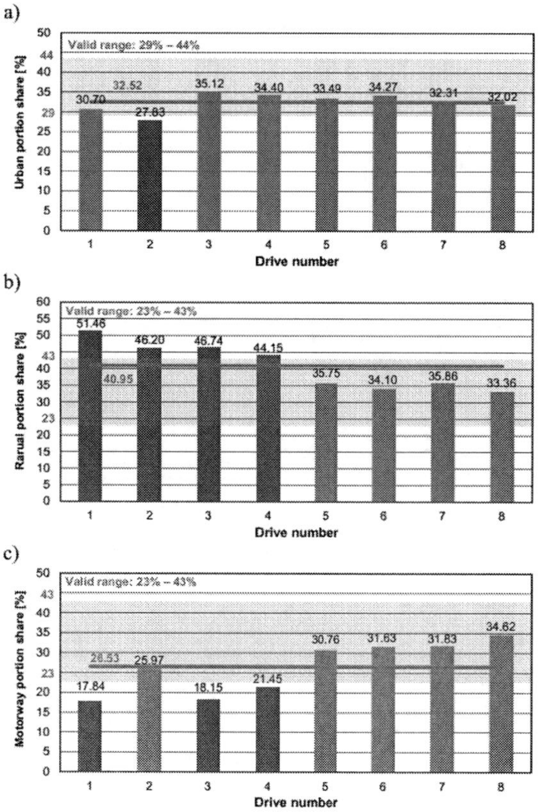

Fig. 7. Test time share comparison of urban (a), rural (b) and motorway (c) test drives with the minimum value (required) for all drives and the mean value.

Analysis of the average velocity data in urban test portion showed that the highest value was achieved in drive No. 2 and the smallest in drive No. 1. The average speed values of all drives are within the permissible range, i.e. between 15 km/h and 40 km/h (Fig. 8a).

Analysis of the average velocity data in non-urban test portion showed that the highest value was achieved in drive No. 5, and the smallest in drive No. 7. The average velocity for all drives were similar, although the legislator did not specify the permissible velocity range (Fig. 8b).

Analysis of the average velocity data in the motorway section showed that the highest value was achieved in the drive No. 1 and the lowest in the drive No. 2. The average velocity values did not have the permissible range defined (Fig. 8c).

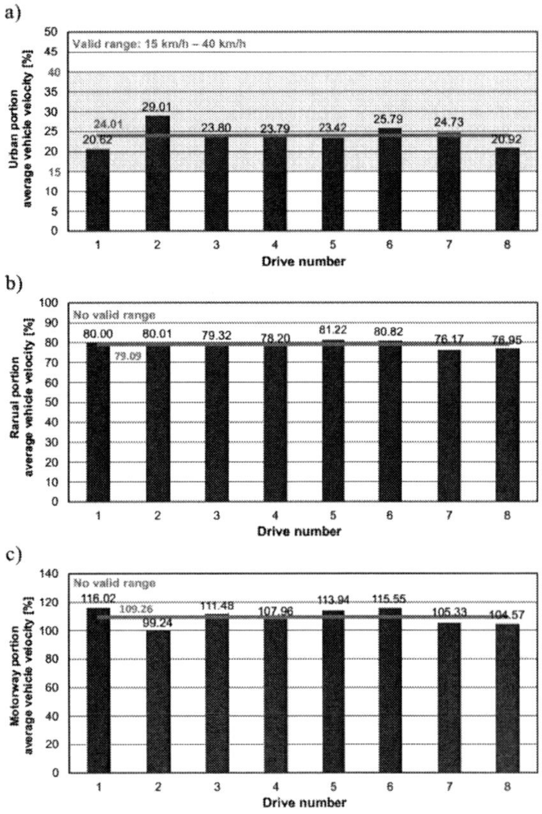

Fig. 8. Average vehicle velocity comparison of urban (a), rural (b) and motorway (c) test drives with the minimum velocity (required) for all drives and the mean value.

The duration of the test was the longest for drive No. 4 and the shortest for drive No. 6. All drives were carried out within an acceptable time range, i.e. 90 minutes to 120 minutes (Fig. 9).

Analysis of the drive test duration data at velocity above 100 km/h in the motorway section showed that the highest value was achieved by the drive No. 7 and the smallest in drive No. 2. In all drives the value is higher than the permissible range, i.e. more than 5 minutes (Fig. 10).

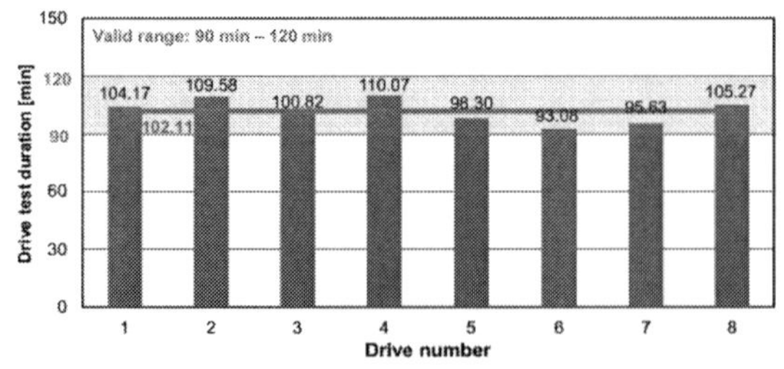

Fig. 9. Comparison of the drive test duration with the limit values (required) for all drives and with mean values shown.

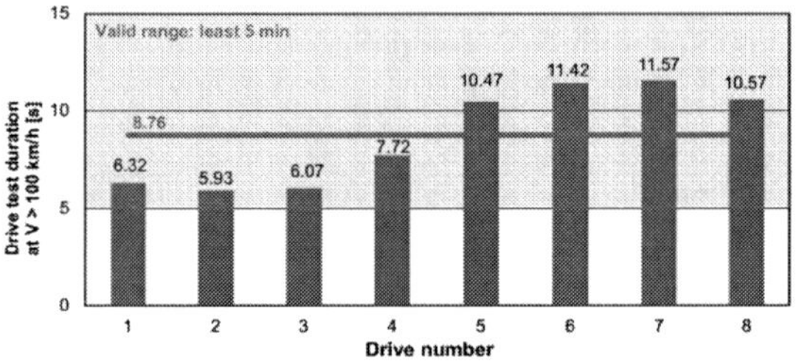

Fig. 10. Comparison of the drive duration with a velocity over 100 km/h on the motorway section of the test with the limit value (required) and the mean value shown for all drives.

Analysis of time share data of vehicle being immobile in the urban drive showed that the highest value was noted for drive No. 1 and the smallest for drive No. 2. All drives reached the stop time share within the permissible range, i.e. 6% - 30%. With the exception of drive No. 1 which was the only one that did not fulfill this requirement (Fig. 11).

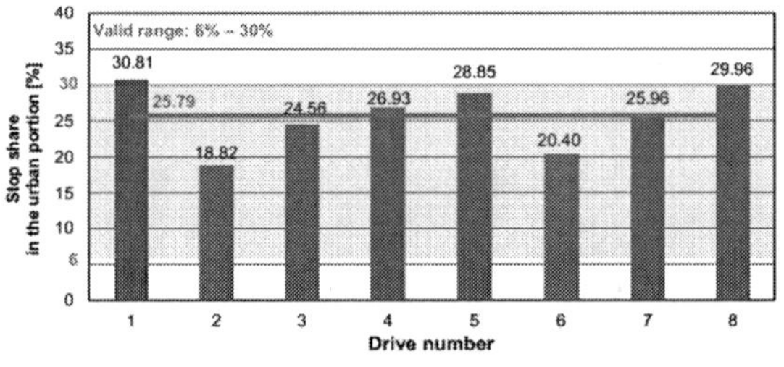

Fig. 11. Share of time spent stationary for the urban portion of the drive test with the limit values (required) for all drives and with mean values shown.

Analysis of the RPA dynamics data in the urban portion showed that the highest value was achieved in drive number No. 8 and smallest for drive No. 2. All drives reached a value greater than the permissible value based on the average velocity in the urban part (Fig. 12a).

The analysis of the RPA dynamics data in the non-urban portion showed that the highest value was obtained in the drive No. 8 and the smallest for drive No. 3. Drives number 1-3 did not exceed the minimum required value, i.e. they were incorrect, as opposed to the other drives (Fig. 12b).

Analysis of the RPA dynamics data in the motorway test portion showed that the highest value was achieved

 DOI: 10.1051/matecconf/201711800021

in drive No. 1 and smallest in drive No. 6. All the drives reached a value greater than the minimum required value dependent on the average velocity in the motorway section (Fig. 12c).

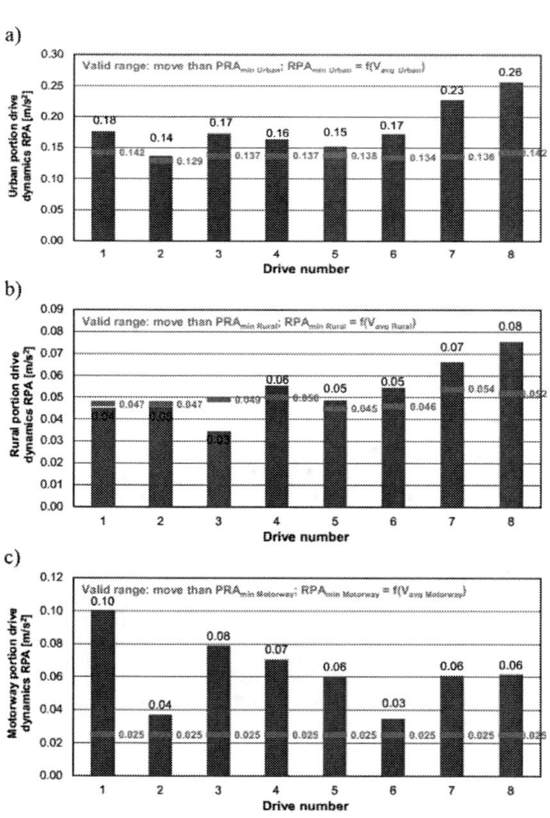

Fig. 12. Relative positive acceleration comparison of urban (a), rural (b) and motorway (c) test drives with the minimum value (required) for all drives and the mean value.

The analysis of the 95th percentile of the product of velocity and acceleration values for the urban portion of the test showed that the value was the highest for the 8th and the smallest for the 2nd drive. The values for all the drives were within the acceptable range, i.e. they were less than the maximum determined based on the average velocity in the urban part (Fig. 13a).

The analysis of the data for the 95th percentile of the product of velocity and acceleration in the non-urban portion of the test showed that the value was the highest for drive No. 6 and the lowest for drive No. 2. The values of all drives were within the acceptable range, based on the average velocity in the non-urban part (Fig. 13b).

The analysis of the data for the 95th percentile of the product of velocity and acceleration in the motorway portion of the test showed that the value was the highest for drive No. 6 and the lowest for drive No. 7. The value of all drives was within the acceptable range, i.e. they did not exceed the maximum value, determined by the average velocity in the motorway section (Fig. 13c).

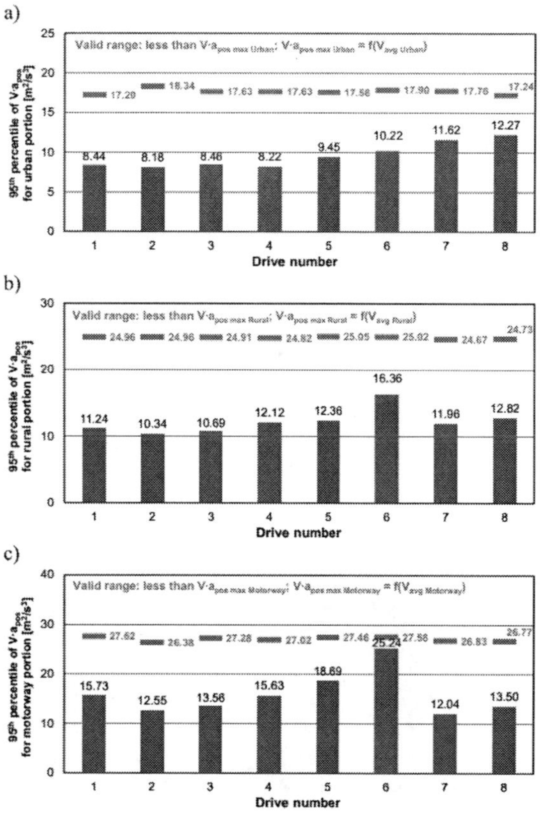

Fig. 13. Vehicle velocity and acceleration product 95th percentile comparison of urban (a), rural (b) and motorway (c) test drives with the maximum value (required) for all drives.

From the data presented in the Figures 6-12 it can be seen that the drives 1-4 only slightly differ from the guidelines for which the test should meet all requirements. In these cases, as non-compliant (the drives that fail to meet the criteria are shown on the graph in red) the following parameters can be listed: too short distance travelled on the motorway (drives No. 1 and 3), urban drive share too small (drive No. 2), too large share of the non-urban drive portion (drives No. 1, 2, 3 and 4), too small share of the motorway drive portion (drives No. 1, 3 and 4), too much time spent stationary in the urban portion (drive No. 1), too low drive dynamics in the non-urban portion (drives No. 1, 2 and 3).

Test drives number 5 to 8 meet all the requirements of the RDE procedure. In the remainder of the article the practicability study of the tests was performed based on the presented results for all drives, with only selected ones being presented.

4 Analysis of the results in the vehicle velocity-acceleration co-ordinates

The two-dimensional time share characteristics compiled in vehicle velocity-acceleration co-ordinates were used to compare the dynamic properties of the drive tests performed. The purpose of this characteristic is to replace the whole test drive with data points and to allow determining of the operating time share within the defined velocity and acceleration intervals. Although this characteristic does not take into account the dynamic properties of the vehicle, the published results of such an

a)

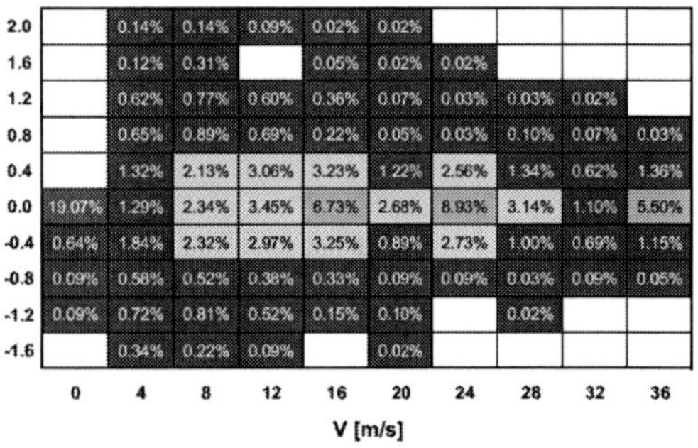

b)

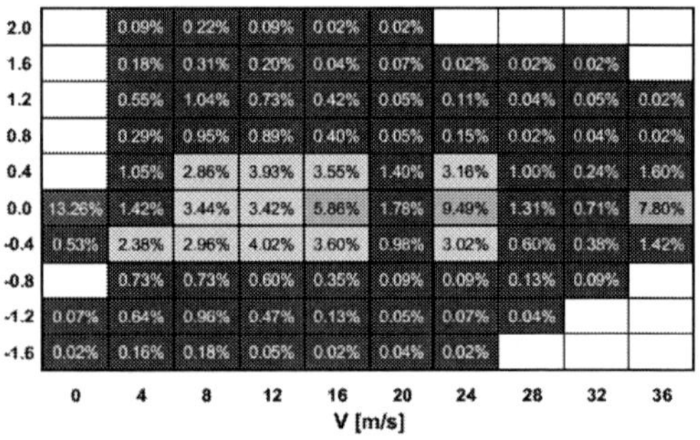

c)

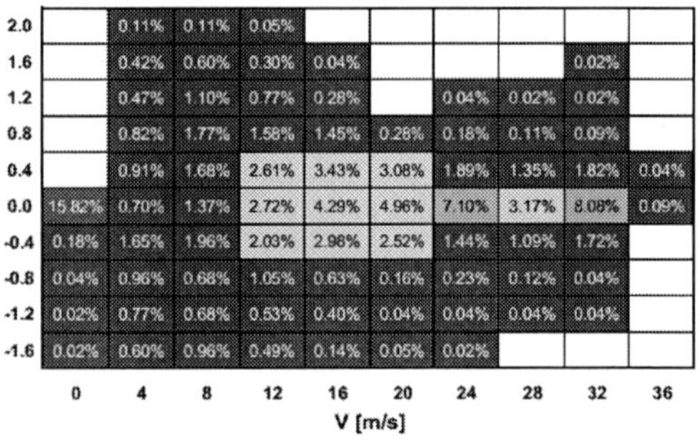

d)

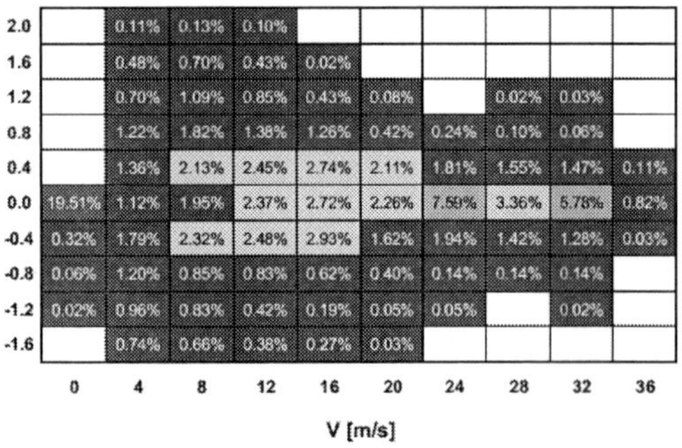

Fig. 14. Share of vehicle operating time in individual velocity and acceleration intervals: a) for drive No. 5, b) for drive No. 6, c) for drive No. 7, d) for drive No. 8.

optimization using this two-dimensional curve indicate that such a simplification is acceptable.

Figure 14 shows the two-dimensional characteristics of the vehicle's operating time for the drives that meet all the test requirements (drives No. 5-8), which can be characterized as follows: the highest operating time share of the vehicle in real driving test conditions is taken by the vehicle idle mode and the constant mean velocity values (V = 16 m/s – 36 m/s for a = 0 m/s^2).

For all drives, the largest share of vehicle operating time is in idle and the average velocity (V = 24 m/s for a = 0 m/s^2 except for drive No. 7). This is due to the constant driving velocity when travelling through in the non-urban area. In the case of drives that meet the test requirements, the extension of the vehicle's operating range into the variable acceleration values in the range of (-0.4 m/s^2, 0.4 m/s^2) can be seen. The acceptable drives were the ones numbered 5-8 and the average values for those drives were designated as "benchmark" values (Fig. 15).

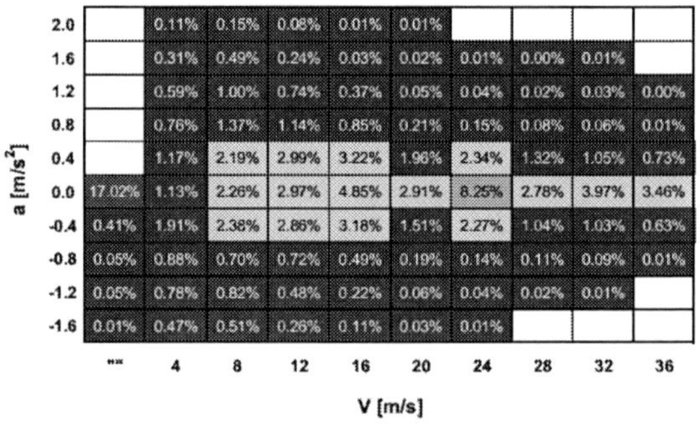

Fig. 15. Share of vehicle operating time in individual speed and acceleration ranges for the "benchmark" values.

5 Comparison of road emission tests practicability

The test cross-compliance comparison was performed by comparing the corresponding areas from the "benchmark" two-dimensional curve and the characteristic obtained from the drives No. 5-8 (for the remaining drives only the final values are given in Tab. 5).

Table 5. The values of the straight-line coefficient and the determination coefficient obtained from the comparison of the "benchmark" drives with all the drives No. 1-8.

Drive No.	Straight-line coefficient	Determination coefficient	Number of disqualifying parameters
1	0.752	0.852	5
2	0.695	0.558	3
3	0.776	0.805	4
4	0.800	0.868	2
5	0.895	0.956	0
6	0.964	0.854	0
7	0.960	0.890	0
8	0.924	0.941	0

The operating time share in the various vehicle velocity and acceleration ranges was compared resulting in a set of points for which the straight-line coefficient and the determination coefficient were calculated. These values determine the relationship between the researched variables (Fig. 16).

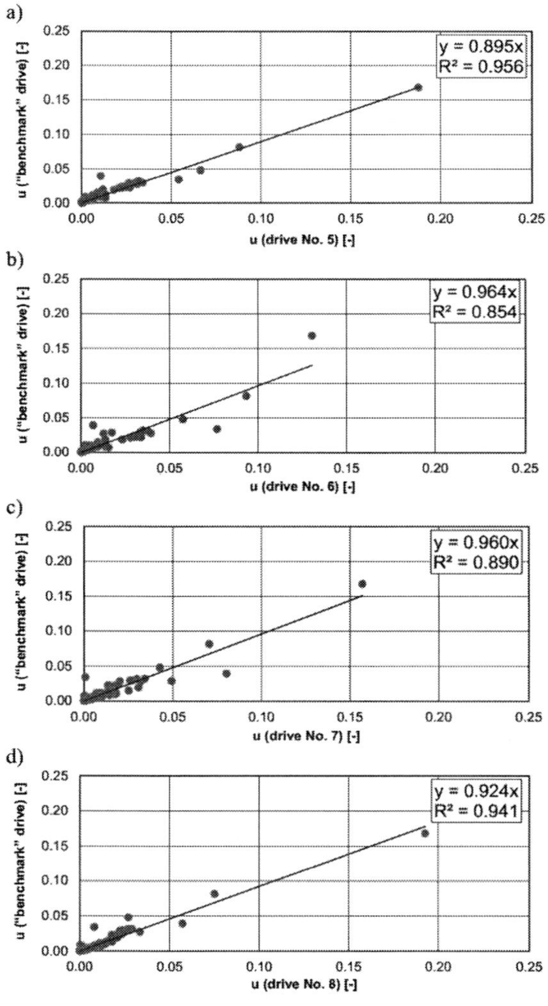

Fig. 16. Comparison of the vehicle operating time share in each velocity and acceleration compartment of the "benchmark" drive and: a) the drives No. 5, b) drive No. 6, c) drive No.7, d) drive No. 8.

Table 5 adds a parameter characterizing the number of disqualifying parameters, ones that to not fall within the permissible ranges (marked in red). The analysis of the data contained in Table 5 indicates that the discarded drives (No. 1-4) are characterized by the smallest straight line coefficient (in the range of 0.695-0.800), while for the acceptable drives the value of this coefficient ranges from 0.895 to 0.964. It should be noted that the accuracy of the test performed depends mainly on the straight-line coefficient value and much less on the value of the determination coefficient (Fig. 17).

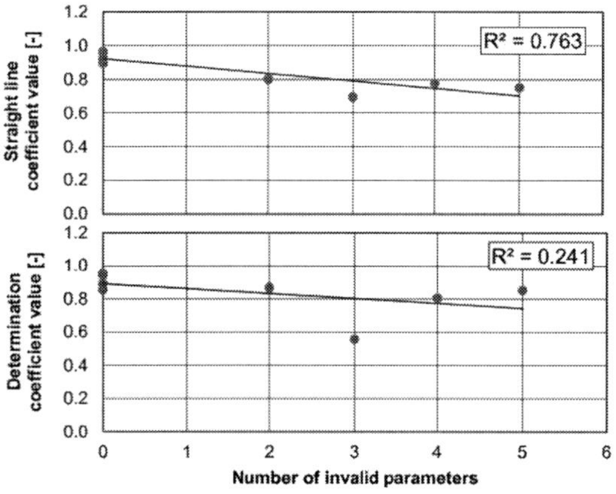

Fig. 17. The relationship between the straight-line coefficient and the determination coefficient from the number of disqualifying parameters.

6 Exhaust emission results

Previous studies with simultaneous emission measurement were used to determine the distance emission of pollutants. Due to the fact that the study was carried out for a passenger vehicle equipped with a Euro 6b gasoline engine (without a particulate filter – GPF), the emission values of carbon dioxide, carbon monoxide, hydrocarbons, nitrogen oxides and particle number were measured (Fig. 18). The results were presented for all eight test drives, indicating that the first four drives did not meet the formal requirements (marked in red), although this fact was not a disqualifying factor for determining distance emission. The disqualified drives were marked in red, whereas the range of values found in the acceptable drives (for drives No. 5-8) was marked in green.

The distance emission limits shown in Figure 18, labeled Euro 6d-Temp, refer to future emission standards (to be adopted from 1.09.2017), where the distance emission limit values are set as: for nitrogen oxides at 2.1, for particle number at 1.5 relative to the Euro 6b distance emission limits.

The analysis of the individual emission values of pollutants indicates the similar character of the obtained test data, which is discussed below.

The maximum value of carbon dioxide distance emission obtained in the test (Fig. 18a) for observed for drive number 8 (211.5 g/km), while the minimum value (185.1 g/km) for drive number 6 (for the acceptable drives). It should be noted that the drive number 8 is characterized by the greatest driving dynamics (relative positive acceleration, Fig. 12). However, considering all the drives, the minimum value of carbon dioxide emission obtained was 172.2 g/km, which was obtained for drive No. 2 – although this drive did not meet the RDE test requirements (which requires the lowest driving dynamics).

The maximum value of carbon monoxide distance emission (Fig. 18b) for the accepted drives was 356 mg/km (drive No. 8) and the smallest was 264 mg/km

DOI: 10.1051/matecconf/201711800021

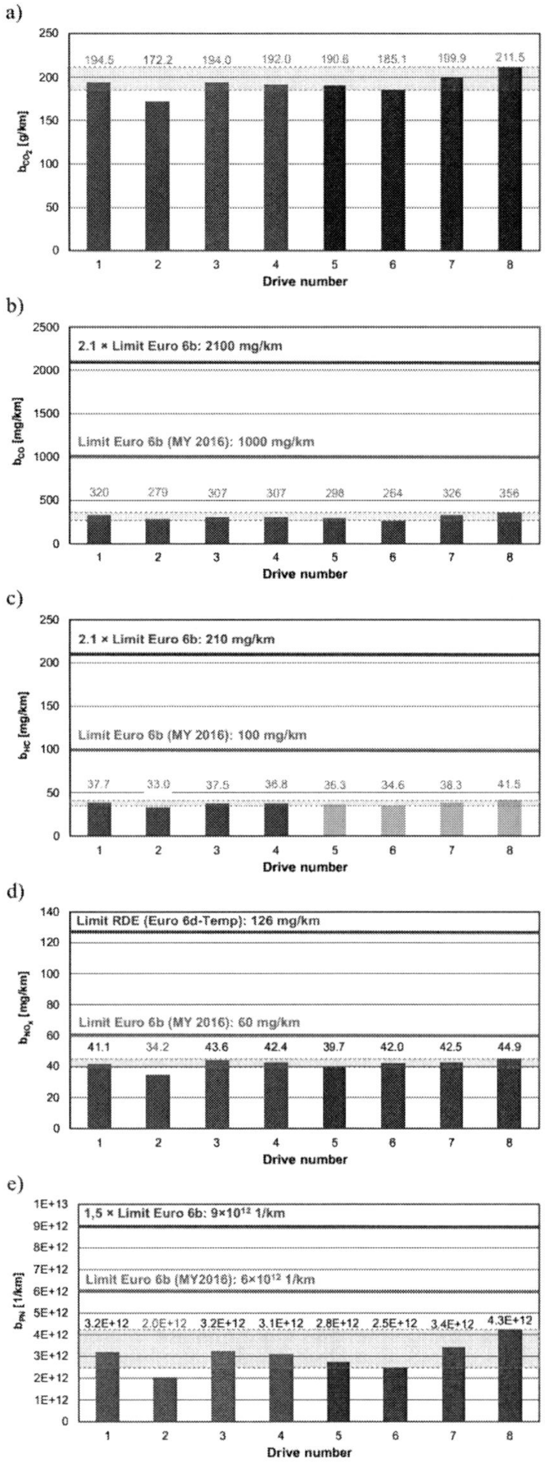

Fig. 18. Distance emission results for: a) carbon dioxide, b) carbon monoxide, c) hydrocarbons, d) nitrogen oxides, and e) particle number for all drives; the disqualified passages are shown in red, while the range of parameter value changes from valid drives (No. 5-8) are shown in green.

(drive No. 6). Test drive number 6 was characterized by a small RPA value, even relative to other accepted drives. The distance emission values in all the performed test drives did not exceed the Euro 6b limit values for passenger vehicles equipped with petrol engines (1000 mg/km).

The extreme hydrocarbons distance emission values (Fig. 18c) were obtained for the same drives as the ones for the distance emission of nitrogen oxides. The maximum value is 41.5 mg/km (drive No. 8) and the minimum is 34.6 mg/km (drive No. 6). The quoted values did not exceed the Euro 6b distance emission limit of 100 mg/km.

The distance emission of nitrogen oxides (Fig. 18d) varied between 39.7 mg/km (drive No. 5) and 44.9 mg/km (drive No. 8) for the acceptable drives. The value of distance emission did not exceed the distance emission limit, i.e. 60 mg/km, for any of the drives.

The nature of the particle number was similar to the previous results (Fig. 18e): the highest value was obtained for drive number 8 (4.3×10^{12} 1/km), and the lowest for drive number 6 (2.5×10^{12} 1/km).

In order to compare the distance emission change ranges for the individual exhaust gas components, a relative error of the calculated emission was determined for each component, in each drive, described by the equation:

$$\Delta b_j = \frac{b_{j,\mathrm{avg}(5\text{-}8)} - b_j}{b_j} \tag{1}$$

where:

$b_{j,\mathrm{avg}(5\text{-}8)}$ – average distance emission value from the drives numbered 5–8,

b_j – specific distance emission value of the given substance.

An analysis of the individual relative error values allows for the following observations to be made:

- The relative error value for carbon dioxide distance emission (Fig. 19a) indicates that it does not exceed 8% for the correctly conducted RDE test that meet all the requirements; with the exception of one drive – all were within the designated range. It cannot be clearly stated that the disqualified RDE tests (drives No. 1-4, which have some disqualified parameters) are characterized by lower carbon dioxide distance emission (however, it was a value of only a few percent). Only the value obtained during drive No. 2 deviated from the average value of carbon dioxide distance emission (was smaller) by about 12.5%.
- The relative error value for carbon monoxide distance emission between individual drives does not exceed 15% (Fig. 19b); all drives were within this range (also the non-RDE compliant). In this case the highest values of relative error (extreme values, of 14.5% and -15%) were obtained for the accepted RDE test drives. The relative error of carbon monoxide distance emission for the disqualified test drives was lower than that of the accepted tests, i.e. the carbon monoxide distance emission measured during non-RDE compliant tests were mostly close to

each other – as opposed to the values obtained in the tests that were in line with the RDE test procedure requirements.

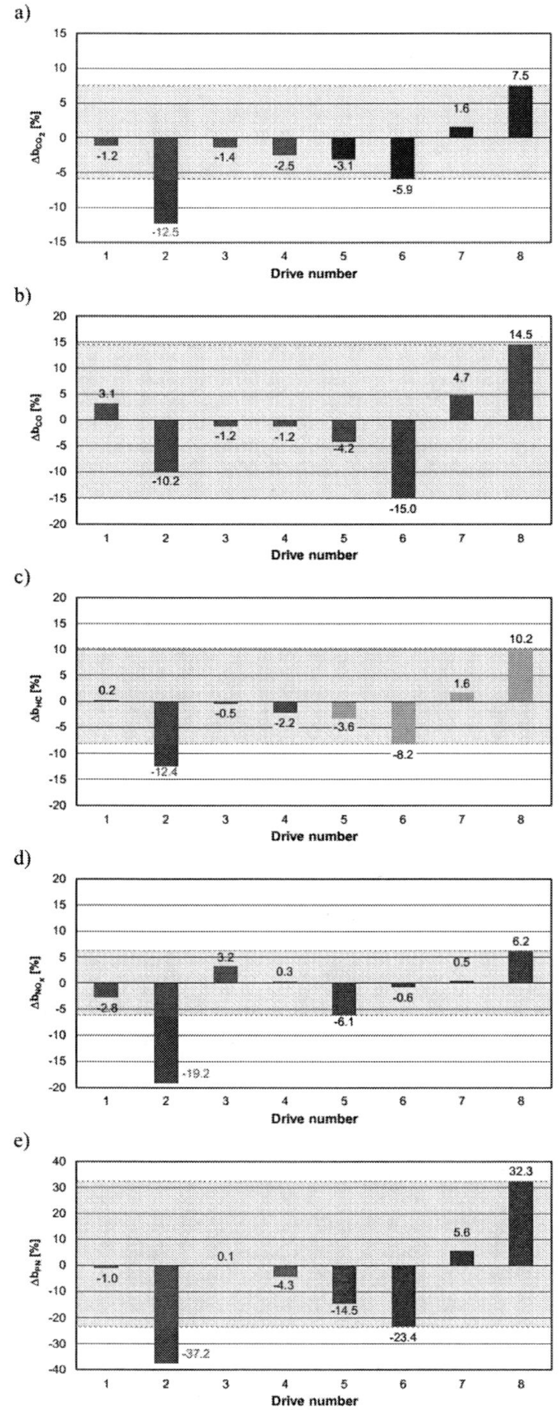

Fig. 19. Relative error for distance emission of: a) carbon dioxide, b) carbon monoxide, c) hydrocarbons, d) nitrogen oxides, e) particle number for all drives; unqualifying drives are marked with red, while the range of parameters changes from the valid test drives (No. 5-8) is marked in green.

- Hydrocarbons distance emission relative error values had a similar character as the values obtained for carbon monoxide (Fig. 19c); all drives – except for drive No. 2 again – were in the relative error range of ±10%, regardless of whether they were valid drives or not.
- The relative error value of the nitrogen oxide distance emission did not exceed 6% between individual drives (Fig. 19d); all drives were within this range – except for drive No. 2 (not meeting the requirements of the RDE test procedure), whose result was about 20% less than the average value of the average drive. At the same time, there is no clear trend in relative error values for the disqualified drives.
- The nature of the relative error of the particle number confirms the previous results; the relative error between each acceptable drive is the greatest among all measured emission and exceeds more than 30% (Fig. 19e); all drives were within that range – except for drive No. 2.

7 Summary

As of 1.09.2017, new testing procedures for determining the specific distance emission of pollutants for motor vehicles will be put in place. The European Commission has been working on them for several years, and the work is still ongoing, so the changes will be implemented in two steps. In the first phase (by 2020), the RDE tests will be used exclusively for monitoring and in the second phase (after 2020) they will be required as a part of the type approval tests. According to the new research procedure (RDE), determining pollutant exhaust emission will take place in real driving conditions. Thanks to this solution, the discrepancy between the results obtained in laboratory tests and the results obtained in road tests will be reduced. The new pollutant testing procedure has not yet been completely regulated (unlike trucks for which such conditions already exist), but there are proposals currently being examined by major research centers in Europe. These proposals differ based on the method of determining the emission values (moving average windows and power binning methods) and the test methodology including, for example, the measurement of hydrocarbon distance emission.

The article compares distance emission test data using the latest legislative proposals for passenger cars. The tests were carried out in accordance with the RDE guidelines for which a minimum test duration of 90 minutes - 120 minutes is required. The results obtained were analyzed in terms of their compliance with the requirements of the RDE procedure. Despite the similar research route, both the results of the velocity profile as well as the average velocity values in each part of the test were not the same. The driving parameters defined by acceleration, constant velocity, braking and stopping were similar. In road tests, it was shown that using a similar route led to obtaining different results in each drive, although the values were fairly close. With regard to the accuracy of the measurements in actual operation, it should be noted that the final result depends on the operating conditions of the vehicle. However, these conditions (such as traffic and congestion, driver's predisposition and driving style, as well as random events occurring during the drive) are unpredictable. Two-dimensional characteristics of the vehicle operating time, compiled in the vehicle velocity-acceleration coordinates, were also used in the analysis. This made it possible to compare dynamic properties, operating time shares and, consequently, to check the validity of carried out test drives.

The analysis of the individual requirements listed in Regulations 427/2016 [11] and 646/2016 [12] confirmed the validity of the RDE test procedure for four of the eight performed test drives. The invalid drives were characterized, among others, by: too short travel distance, too little operating time share of particular parts of the test, too long time spent stationary in the urban part and too little driving dynamics in the non-urban part. The analysis of the test results does not allow for unambiguous determination of the exhaust emission results inaccuracies with the assumed number of errors committed during road tests. It should be noted, however, that the highest values of distance pollutant emission were recorded for drives where a small part of the urban portion of the test was involved and a larger share of the motorway portion. This demonstrates the significant impact of the high engine load with the greatest pollutant emission intensity parts of the RDE test on the distance pollutant emission results (efficiency of the exhaust aftertreatment system is not at its peak).

The result of the study is the distance pollutant emission results value range for RDE tests carried out in accordance with the European Union procedures with the following required test parameters (for vehicles equipped with gasoline engines with direct injection systems compliant with the Euro 6b norms):

- for carbon dioxide distance emission – the emission value range was ±8%,
- for carbon monoxide distance emission – the emission value range was ±15%,
- for hydrocarbons distance emission – the emission value range was ±10%,
- for nitrogen oxides distance emission – the emission value range was ±6%,
- for particle number – the value range was ±32%.

The article does not, however, conclusively answer the issue of what has the greater impact on the exhaust gas pollutant emission: the share of the RDE test portion or the vehicle driving dynamics (defined by the relative positive acceleration or the product of velocity and acceleration). Determining these relations will be the subject of subsequent research using passenger cars powered using different fuels as well as hybrid vehicles.

The research was funded by project co-financed by the European Regional Development Fund in the Regional Program – Lubuskie 2020 (contract No. RPLB.01.01.00-08-0055/16-00).

References

1. P. Lijewski, J. Merkisz, P. Fuc, A. Ziolkowski, L. Rymaniak, W. Kusiak, European Journal of Forest Research **136**, 153-160 (2017), DOI: 10.1007/s10342-016-1015-2
2. J. Merkisz, W. Gis, *WIT Transactions on the Built Environment* **101**, 335-344 (2008), DOI: 10.2495/UT080331
3. A. Suchecki, J. Nowakowski, T. Knefel, *Transport Means – Proceedings of the International Conference*, 194-197 (2014)
4. K. Prainowski, J. Mamala, *IOP Conference Series-Materials Science and Engineering* **148**, UNSP 012022 (2016)
5. A. Bieniek, J. Jantos, J. Mamala, Transport **22**, 247-251 (2007)
6. P. Bielaczyc, J. Woodburn, D. Klimkiewicz, P. Pajdowski, A. Szczotka, Fuel Processing Technology **107**, 50-63 (2013), DOI: 10.1016/j.fuproc.2012.07.030
7. N. Ligterink, G. Kadijk, P. Van Mensch, S. Hausberger, M. Rexeis, TNO Report (2013)
8. W. Stütz, *International AVL Conference „Engine & Environment"* (2014)
9. T. Vlachos, P. Bonnel, A. Perujo, M. Weiss, et al. SAE Int. J. Commer. Veh. **7**, 199-215 (2014)
10. I. Khalek, T. Bougher, J. Jetter, SAE International Journal of Fuels and Lubricants **3**, 623-637 (2010)
11. P. Fuc, L. Rymaniak, A. Ziolkowski, *WIT Transactions on Ecology and the Environment* **174**, 207-218 (2013), DOI: 10.2495/AIR130181
12. Commission Regulation (EC) 715/2007 of the European Parliament and of the Council of 20 June 2007 on type approval of motor vehicles with respect to emissions from light passenger and commercial vehicles (Euro 5 and Euro 6) and on access to vehicle repair and maintenance information, European Commission (EC), Official J. European Union, L 171 (2007)
13. Commission Regulation (EU) 2016/427 of 10 March 2016 amending Regulation (EC) No 692/2008 as regards emissions from light passenger and commercial vehicles (Euro 6), Verifying Real Driving Emissions, Official J. European Union, L 82 (2016)
14. Commission Regulation (EU) 2016/646 of 20 April 2016 amending Regulation (EC) No 692/2008 as regards emissions from light passenger and commercial vehicles (Euro 6), Verifying Real Driving Emissions, Official J. European Union, L 109 (2016)
15. Commission Regulation (EC) 692/2008 of 18 July 2008 implementing and amending Regulation (EC) No 715/2007 of the European Parliament and of the Council on type-approval of motor vehicles with respect to emissions from light passenger and commercial vehicles (Euro 5 and Euro 6) and on

access to vehicle repair and maintenance information, European Commission (EC), Official J. European Union, L 199 (2008)

16. V. Franco, M. Kousoulidou, M. Muntean, L. Ntziachristos, S. Hausberger, P. Dilara, Atmospheric Environment **70**, 84-97 (2013) DOI: 10.1016/j.atmosenv.2013.01.006
17. T. Henrichs, *AECC Technical Seminar on Real-Driving Emissions of Particles (RDE PN)*, Diamant Conference Centre (2016)
18. J. Merkisz, S. Tarkowski, Eksploatacja i Niezawodnosc – Maintenance and Reliability **2**, 50-58 (2011)
19. M. Sobieszczanski, K. Brzozowski, J. Nowakowski, M. Tekielak, Eksploatacja i Niezawodnosc – Maintenance and Reliability **3**, 53-61 (2008)
20. J. Andersson, *AECC Technical Seminar on Real-Driving Emissions of Particles (RDE PN)*, Diamant Conference Centre (2016)
21. http://www.diag-tools.com/texa-obd-log-mobilny-czytnik-normy-eobd-p-478.html (11.3.2017)
22. F. Ortenzi, M.A. Costagliola, SAE Technical Paper 2010-01-1289 (2010)
23. Semtech-DS On-board vehicle emissions analyzer (2010)
24. Particle instruments: AVL Particle Counter (2010)

The influence of quantitative parameters of lubricant oil's on the damaged of car turbochargers – Event Tree Analysis (ETA) method

Maciej Bieliński[1], *Mateusz* Bor[1], *Tomasz* Borowczyk[1], *Marek* Idzior[1,*], *Wojciech* Karpiuk[1], and *Rafał* Smolec[1]

[1]Poznan University of Technology, Faculty of Machines and Transport, ul. Piotrowo 3, 60-965 Poznan, Poland

Abstract. The Event Tree Analysis (ETA) is a tool for assessing system and process safety. It facilitates detecting potential risks, as well as relations existing between actions or events. This method explores the path from the initiating event to the outcome (most often representing the immobilization of the tested object), with particular emphasis on intermediate stages affecting the technical condition of the object. Unlike the FTA analyses, which have a deductive nature, the ETA analysis is inductive. The aim of the work was to estimate the probability of occurrence of unwanted events concerning the operation of a car turbocharger. Based on the performed studies, the impact of the standard safety systems on the operation of the component was checked. Afterwards, more effective safety solutions were proposed. Initiating event was distinguished in the system. The occurrence of such an event could result in a complete immobilization of the component. To accomplish the task, it was necessary to use safety barriers aimed at reducing negative effects of initiating events. The calculation of probabilities of occurrences of various effects in each of the cases helped quantify the selected damages to the turbocharger.

1 Introduction and purpose of the study

To perform the analysis of the system in relation to its damage, it is essential to use unified procedures. The classic approach to determining reliability – as the probability of occurrence of suitability of a system within a given period of time – is replaced by an analysis of possibilities of implementing tasks imposed on the system [1]. Furthermore, in other publications, authors propose to assess the reliability of the process, and not only a single device.

In this paper, to evaluate the possibility of occurrence of the damage, the Event Tree Analysis was used. The damage concerned a component of the engine – a turbocharger, which is a rotary machine and an element of the forced induction of a combustion engine. Due to the operating principle and design features, the device works with many engine systems – intake, exhaust, lubrication systems and also, increasingly more often, with the cooling system.

A multitude of relations, complexity and working conditions are reasons due to which the device is sensitive to disturbances of quality parameters of the working media. Working conditions of turbochargers are highly variable. Such devices, arranged in exhaust systems of combustion engines, are exposed to the following factors (among others):

- high temperatures,
- fluctuations of exhaust gases,
- vibrations caused by the operating engine and the movement of the vehicle point,
- thermal shock caused by fluctuating temperature,
- impact loads.

Continuous improvement of environmental performance of combustion engines is connected with modifications of intake, fuelling, exhaust aftertreatment systems etc. The complexity and the number of moving parts, as well as small fitting tolerances in such systems mean that together with precision devices such as high-pressure pumps, injectors or turbochargers, the systems are particularly vulnerable to disturbances of the designed operating parameters of the working media. Devices operate properly as long as the quality and quantity parameters of lubricating oil, combustion air and the cooling liquid are maintained.

Damage to turbochargers in modern combustion engines are a prevalent phenomenon and their sources should be sought in dysfunctions of the following systems: engine lubrication, intake or exhaust systems (exhaust aftertreatment system). The authors' practical expertise proved that, many times, repairs ultimately constist in replacing the turbocharger (without solving the actual problem which initiated the damage). Subsequently, the damage re-occurs, often after a few or a dozen minutes of operation.

Given the large scale of the phenomenon, the aim of the study was specified to determine the probabilities of occurrence of unwanted event, in case of disorders of quality parameters of the working media of the motor

* Corresponding author: marek.idzior@put.poznan.pl

and turbocharger system. Based on the performed bench tests, the impact of safety systems on the work of the component was studied, and subsequently more effective solutions were proposed. A number of initiating events was distinguished in the system. The occurrence of such events would result in a complete immobilization of the component.

An Event Tree was created for one of such case. To accomplish the task, safety barriers were used. The purpose of the barriers was to reduce negative effects of occurrences of initiating events. The calculation of probability of occurrences of various effects helped quantify the selected damage to the turbocharger.

2 Research methodology, the applied equipment and the ETA method outlined

Measurement tests were conducted at a bench designed for testing car turbochargers. The bench includes a combustion engine for generating exhaust gases. Only a stream of hot gases is used to drive the turbocharger. Apart from that the turbocharger operates in an environment isolated from the engine. The bearings are lubricated using an independent lubrication system, including an electric pump, which makes it possible to adjust the whole range of pressure and output. The oil tank ensures that a specific temperature of the medium can be maintained. Compressor air is not used in any way. The system is not connected to the engine intake system. The bench was designed in a way to make it possible to simulate the phenomena affecting the disturbance in quality or quantity parameters of the working media of the engine.

a)

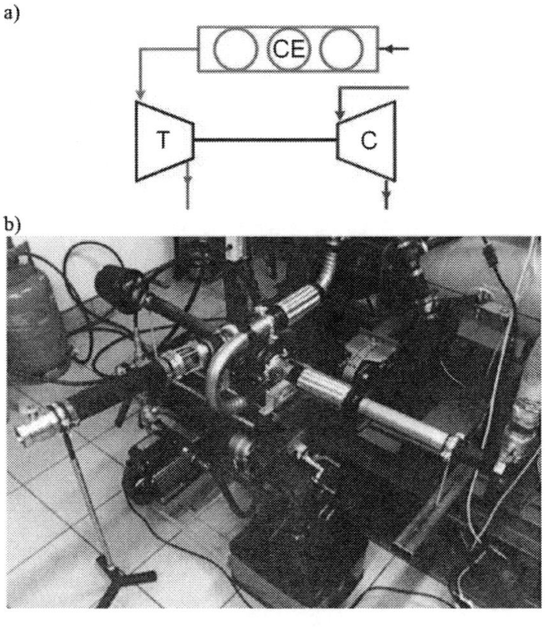

b)

Fig. 1. Test bench with a combustion engine as a gas generator: a) block diagram: CE – combustion engine, T – turbine, C – compressor, b) real view.

The main advantage of the bench is that the test conditions are as close as possible to the real conditions of a turbocharger coupled with an engine. Temperature, pressure and pressure fluctuations are similar to those existing in the exhaust manifold of a supercharged combustion engine. For checking and recording parameters of factors in various branches of the system, temperature sensors and pressure converters were used.

A disadvantage of the system using an engine as the exhaust gases generator is the fact that the energy stream of exhaust gases is strictly connected to the load and the rotational speed, resulting in dissipation of energy in the brake and in overcoming own resistance of the engine.

The bench is equipped with a four-cylinder, naturally aspirated, 16V FIAT SI (Spark Ignition) engine with a displacement of 1.2 dm^3, with the trademark of MPFI (Fig. 1). The obtainable power is 52 kW/6000 min^{-1} and the torque is 102 Nm/3250 min^{-1}. The used test stand is a multi-directional operation AMX-210/100 with a maximum load power of 100 kW and a maximum rotational speed of 10,000 min^{-1}. The maximum torque is 240 Nm. The engine is equipped with a complete set of instruments for continuous operation, as well as power and torque regulation, making it possible to produce a controlled stream of exhaust gases at given parameters [2].

For the purpose of the study we assumed that the operation of most turbochargers, their operational parameters and their design are not very different between manufacturers or models. The essence of reliability of turbochargers is to take up the problem in a systematic way. It is particularly noteworthy that during the identification and analysis of the damages, these machines should not be considered separately.

Due to the difficulty of modifying connectors for other turbochargers and the need of performing tests on one component, we refrained from using the classical method of assessing reliability and from determining empirical indices of reliability. Instead, we focused on other methods, such as ETA, FTA and FMEA. This article presents an analysis of damage, carried out using the ETA method, which is a part of a doctoral thesis of one of the authors.

The ETA method is a technique for identifying and assessing a sequence of events resulting from an initiating event. The purpose of this analysis is to determine the distribution of potential severity of the risk connected with the initiating event, as well as identifying the impact of the effectiveness of instruments responding to the risk. There are two forms of ETA: pre-occurrence and post-occurrence. The pre-occurrence ETA makes it possible to test the effectiveness of neutralizing the severity of a risk. The post-occurrence ETA is used to analyze the materialized risk and opportunities of managing the level of its severity [3, 4].

According to Oprychał, for the Event Tree method to be applied, it should be assumed that a major malfunction is a result of not just a single occurrence of an initiating event, but the confluence of several events or a series of events. The appearance of a single dysfunction does not generate a chain reaction, nor a serious malfunction. It is not until a sequence of adverse

DOI: 10.1051/matecconf/201711800022

conditions appears, causing subsequent damages, that may result in serious malfunctions.

Using the ETA method, one can graphically present a sequence of events and barriers, beginning with initiating a state likely to cause anomalies and resulting in effects arising at each stage of the analysis. Most often, the barriers include safety systems (e.g. an air filter, a CO_2 sensor or a fuse), as well as the human factor, which can directly minimize the effects of an occurrence of a malfunction (e.g. a CNC operator or a production line quality inspector). In this method it is assumed that each event in a sequence is a success or a failure (no intermediate states).The probabilities assigned to individual branches in Event Trees are conditional probabilities [5].

When conducting the analysis at each stage, one needs to ask the question "what if?" The operating algorithm is based on analyzing possibilities of evolving of the initiating event.

It is important to accurately determine safety barriers which effectively limit negative effects of the initial event. The ETA method is a logical, divalent tree with the initiating event at the base and a root leading to the event with the most serious effect. Each barrier creates a branch which is also a result of the considered logic state: a success (yes) or a failure (no) depending on whether the given barrier is effective or not. Events arranged in the branches are combinations of sequences of consecutive events.

The result of the tree analysis is the systematization of all possible sequences of events concerning the adopted classification of effects (in case of studying a turbocharger it is the degree of damages, their extent and consequences possibly caused by the faulty component). The ETA technique also makes it possible to calculate the probabilities of occurrence of each group of sequences.

In order to perform the analysis, it is necessary to estimate the probability of occurrence of the initiating events and the probabilities of occurrence of events at specific barriers, obtained from the bench tests. For the case discussed in this article, this step was performed based on one of the most common fault occurring in turbochargers, which was provided to facilities specialized in regenerating components. The additionally applied estimates take into account community interviews with people leading the above facilities. The selected case is presented and discussed in this article.

The ETA analysis was performed with the following steps:

1. Identifying initiating event which could lead to a complete immobilization of the turbocharger – the so-called state of unfitness (in extreme cases of the entire engine),
2. Suggesting alternative solutions for securing the system from critical effects,
3. Creating an Event Tree, Bullets,
4. Estimating the probabilities of occurrence of the initiating event and the occurrence of events at specific barriers,
5. Describing the malfunctions and determining their effects,
6. Calculating the probabilities of occurrence of a given type of effects for the existing and the proposed solutions,
7. Determining existing systems preventing the occurrence of an event or effects leading to the state of unfitness (so called safety barriers).

3 Results of the study

Based on the selected cause of damage of turbochargers, we focused on the following initiating event concerning the bearings lubrication system: clogged oil drains out of the center housing.

The selected initiating event is not random. According to information obtained in turbocharger regeneration facilities, it is the most common causes of damage of these types of devices. Therefore, it was decided to analyze the above case with the previously described test bench. Among others, we studied the impact of oil pressure increase in the turbocharger housing on the rotational speed of its shaft and the oil temperature. Tests were performed at the previously described bench. The change of passability (and thus a pressure increase) was simulated with a ball valve located directly behind the oil drain. Using appropriate sensors we recorded the rotational speed of the turbocharger shaft, pressure and oil temperature. We also studied the chemical composition of exhaust gases. Figure 2 presents significant correlations noted during the tests.

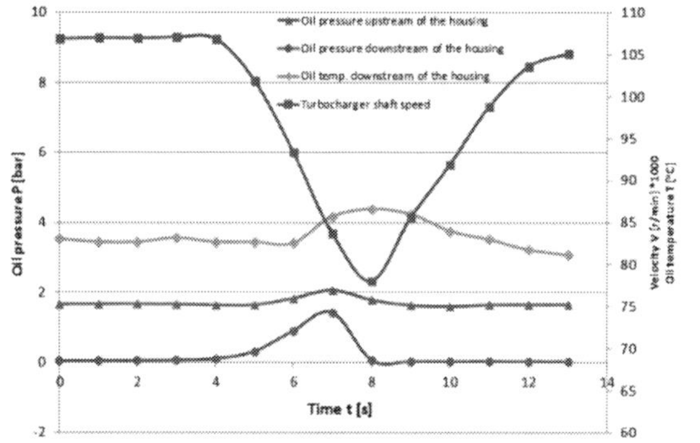

Fig. 2. The impact of oil pressure on the rotational speed of the turbocharger shaft.

The characteristics allows to conclude that oil, after flowing through bearings, is subject to expansion to ambient pressure, and then flows, gravitationally, to the oil sump. Oil flow throttling results in immediate increase in pressure. The pressurized medium disrupts the labyrinth flowing into the turbine chamber, where it is burned, and then flows into the exhaust system. For seal break to happen, it is necessary to produce force which creates resistance also in the system bearings. Moreover, it was noted that oil flow throttling also affects heat dissipation from the bearings in the housing, which certainly has a negative impact on the proper functioning of the device.

The first analyzed case of a turbocharger damage is a problem of an obstructed oil drain from the housing – event A, briefly defined as the "clogged center housing oil drains".

This phenomenon takes place in the oil drain channels, often located at parts of the engine exhaust system which heat up significantly. Some of the most vulnerable elements are large radius bends, narrow passages and mesh screens mounted inside the channels. The Event Tree caused by the reason is shown in Fig. 3.

The engine lubrication system contains the following safety barriers:

B – oil filter,

C – actions taken by a service technician.

Effects of different S events have the following interpretation:

S_1 – The oil filter stopped substances hindering the flow in oil channels. The intervention is to replace the filter.

S_2 – A serviceman performed oil change within the period recommended by the manufacturer (or before the end of this period). For the barrier to be effective, it must be assumed that the used oil has parameters suitable for the given engine, and the engine operating parameters, as well as the driving conditions of the car, were compliant with the manufacturer's guidelines [6].

S_X – This is a critical effect which causes the immobilization of the turbocharger and may affect the proper operation of the engine. In case of effect SX – repair or replacement of the turbocharger is necessary. Oil drain channels should be unclogged (sometimes aftermarket parts manufacturers make modifications involving the removal of the mesh screen from the tubing, which in extreme cases, in original parts, can cause blocking of the oil drain). In addition, the inside of the engine should be flushed and oil should be replaced (together with the filter).

Quantitative analysis of the success branch (to avoid a critical event) is described as $P(S_{1,2,3 \text{ or } 4})$ and the failure branch $1 - P(S_{1,2,3 \text{ or } 4})$. Graphical values are shown in Fig. 3.

The probabilities of occurrence of a given type of effects in quantitative terms are as follows:

$$P(S1) = P(A) \cdot P(B) \quad (1)$$

$$P(S2) = P(A) \cdot [1 - P(B)] \cdot P(C) \quad (2)$$

$$P(SX) = P(A) \cdot [1 - P(B)] \cdot [1 - P(C)] \quad (3)$$

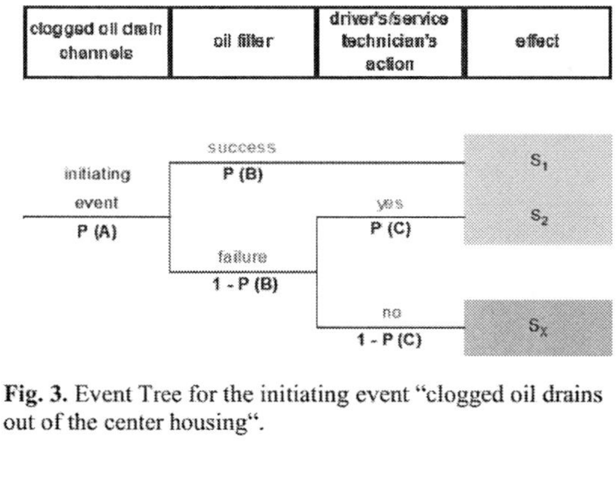

Fig. 3. Event Tree for the initiating event "clogged oil drains out of the center housing".

The probabilities of occurrence of initiating event P(A) was determined based on expertise and research. Clogged oil drains from the turbocharger housing are important for proper functioning of the device. As the oil is drained into the oil sump at atmospheric pressure, fluctuation of flow parameters is not uncommon. A well operating crankcase ventilation is particularly important in this case (commonly referred to as "a breather"). Moreover, the practiced solution of placing a mesh screen inside mounting screws (Fig. 4) or oil drain pipes effectively impede the liquid flow.

Fig. 4. Screw fixing the drainage hose oil from the turbocharger.

Flow disturbance results in impaired heat transfer and lubrication, and may also cause oil leakage through shaft seals of the turbocharger. During assigning values, taking into account the statistical analysis of the occurrence of the given cause, we assumed the range from "0" to "1", wherein "0" means no occurrence of the event and "1" a 100% probability of occurrence. The probability value P(A) was therefore defined at the level of 0.08. This means that in 8% of cases out of the group of all damaged turbocharger, a disrupt in oil flow inside the oil drain channel was noted. This faulty condition, without taking appropriate actions, can lead to the immobilization of the device.

Assigning values to barriers was done similarly to the initiating event. "0" means that the barrier does not work, does not affect or does not exist. The value of "1" is equivalent with a 100% effectiveness of the barrier. The probability values of the event at barrier P(B) were defined as 0.1. The oil filter is a barrier as it stops some part of contaminants, yet the phenomenon associated with blocking of flow is also caused by coking of oil (high temperature in the area of oil drain channels) and the aforementioned mesh screens. Oil quality parameters and the percentage of washing additives are also very important. Driver's/service technician's action as barrier C – relates to the quality and quantity of the lubricating medium in the system. Thus defines checking of the state of oil, caring about intervals between replacing parts and ensuring a proper oil viscosity and lubrication class. Taking into account all functions and capabilities, the parameter was estimated at 0.5.

Numerical values of probabilities of occurrence of a given type of effects in quantitative terms are as follows: $P(S_1) = 0.008$; $P(S_2) = 0.036$. The probability of occurrence of a critical event is equal to: $P(S_X) = 0.036$.When conducting the analysis at each stage, one needs to ask the question "what if?" The operating algorithm is based on analyzing possibilities of evolving of the initiating event.

DOI: 10.1051/matecconf/201711800022

4 Proposed solutions

In order to reduce the probability of occurrence of a critical event $P(S_X)$, we introduced barriers, which, in the research model, are specific technical solutions applied at the test bench. These were: an oil temperature sensor (indicated as barrier D) located in the turbocharger housing and an oil flow meter (indicated as barrier E) located in the oil drain channel evacuating the oil from the turbocharger housing to the engine oil sump.

The modified Event Trees are marked with the "prim" symbol. For the first described case, concerning the obstructed oil drain from the housing, an analysis shown in Figure 5 was performed.

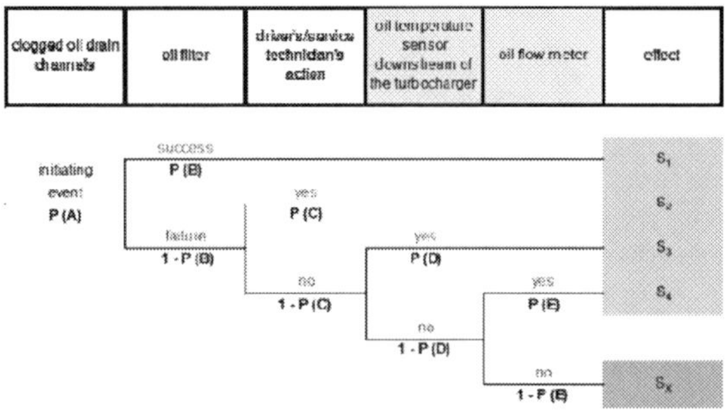

Fig. 5. Event Tree for the initiating event "clogged oil drains out of the center housing".

Probabilities in qualitative terms are analogous to the analyses carried out before introducing changes and for the initiating event equal P(A)' = 0.08; and for events at specific barriers P(B)' = 0.1; P(C)' = 0.5. For the proposed, new barriers, they equal P(D)' = 0.9; P(E)' = = 0.9. The values were taken at a 90% efficiency level, assuming that no solution gives a 100% certainty. Numerical values of probabilities of occurrence of a given type of effects in quantitative terms are as follows: $P(S_1)' = 0.008$; $P(S_2)' = 0.036$; $P(S_3)' = 0.0324$ i $P(S_4)' = = 0.00324$. The probability of occurrence of a critical event is equal to: P(SX)' = 0.00036.Typing the paper in this format is for the purpose of facilitating the editing process. The forms with text and figures will be printed as delivered. Only in very few exceptional cases will other forms be accepted. the Bibliography (in alphabetical order), e.g.: [1].

5 Summary and discussion of the results

The performed analysis, preceded with bench tests, proved that introducing additional safety barriers in the form of sensors, brings measurable benefits. For economic reasons, in mass production, the cost of applying additional sensors and including them in monitoring systems may adversely affect the price of the turbocharger/engine. However, taking into account the benefits, e.g. ones associated with meeting environmental requirements, the proposal presented in the article seems to be justified. In the total balance it makes it possible to significantly reduce the environmental cost (the impact of emissions of harmful substances from exhaust gases during operation) caused by allowing a critical event to occur, while reducing the cost of replacing devices with new ones, as well as the need to look for the cause [7, 8]. The model proved that the probability of occurrence of a critical event was significantly reduced.

Although the introduced indicator of improving reliability of the system using the same barriers for each case differs significantly, even a difference 90-times higher significantly affects the life of the component. The authors attempted to analyze damage to turbochargers with certain expert assumptions, which can be freely modified, taking input data from research or service repair statistics. As the estimated values may be subject to discussion, the proposed monitoring solutions show clear relations, having a positive reliability and durability impact [9]. The ETA method makes it possible, in a transparent way, to define interrelations occurring between the operation and unfitness of safety systems. Such systems should be taken into account at the stage of designing devices. The paper should be as informative as possible. The research results have to be described in an explicit way. The applied research apparatus and methodology of full value should be introduced precisely. Very extensive literature surveys are not recommended, provided they are not the main target of the paper.

This article was financially supported within the project "New generation of common rail pumps" – Lider/015/273/L-5/13/NCBR/2014, implemented within the LIDER Programme, financed by the National Centre for Research and Development, Poland.

References

1. D. Bobrowski, *Metody sieciowe w inżynierii niezawodności* **I** (1999)
2. M. Idzior, W. Karpiuk, M. Bieliński, T. Borowczyk, P. Daszkiewicz, P. Stobnicki, Technika Transportu Szynowego, 3965-3977 (2012)
3. J. Merkisz, J. Mizera, M. Bajerlein, L. Rymaniak, P. Maj, Experimental and Applied Mechanics **518**, 102-107 (2014), DOI: 10.4028/www.scientific.net/AMM.518.102
4. blogoryzyku.blogspot.com (2016)
5. K. Szwedzka, J. Gruszka, P. Szafer, *IOP Conference Series-Materials Science and Engineering* **145**, 062007 (2016), DOI: 10.1088/1757-899X/145/6/062007
6. M. Idzior, M. Bieliński, T. Borowczyk, W. Karpiuk, Logistyka 1129-1139 (2010)
7. M. Idzior, W. Karpiuk, M. Bieliński, T. Borowczyk, Problemy recyklingu 201-207 (2012)

8. L. Oprychał, K. Fiedler, W. Jankowski, A. Mazurczyk, *Metody Komputerowe w Projektowaniu i Analizie Konstrukcji Hydrotechnicznych* (1994)
9. M. Orczyk, F. Tomaszewski, B. Czechyra, G.M. Szymanski, *Proceedings of the International Congress on Sound and Vibration* (Athens, 2016)

MATEC Web of Conferences **118**, 00023 (2017) DOI: 10.1051/matecconf/201711800023
VII International Congress on Combustion Engines

Number and mass analysis of particles emitted by aircraft engine

Remigiusz Jasiński[1,*]

[1]Poznan University of Technology, Faculty of Machines and Transport, ul. Piotrowo 3, 60-965 Poznan, Poland

Abstract. Exhaust emissions from aircraft is a complex issue because of the limited possibility of measurements in flight conditions. Most of the studies on this subject were performed on the basis of stationary test. Engine certification data is used to calculate total emissions generated by air transport. However, it doesnt provide any information about the local effects of air traffic. The main threat to local communities is particulate matter emissions, which adversely affects human health. Emissions from air transport affect air quality, particularly in the vicinity of the airports; it also contributes to the greenhouse effect. The article presents the measurement results of the concentration and size distribution of particles emitted during aircraft landing operation. Measurements were carried out during the landings of aircraft at a civilian airport. It was found that a single landing operation causes particle number concentration value increase of several ten-fold in a short period of time. Using aircraft engine certification data, the methodology for determination of the total number of particles emitted during a single landing operation was introduced.

1 Introduction

Air transport is part of the global transport system due to the high speed and intercontinental range of transport operations. For many years, there has been a strong correlation between the rate of growth in major world economies and the development of air transport [1]. Comparing the growth of world GDP (Gross World Product) and air traffic, it can be noted that the growth rate of air is greater than the growth of the world economy. According to the analysis of the largest aircraft manufacturers (Airbus and Boeing) this is due to a very dynamic development of low-cost transportation, increase in the aircraft capacity, fleet modernization, and the dynamic growth in emerging markets such as China, India and Japan. The Far East market, according to forecasts, will generate the highest growth of 31% of world traffic to year 2029. The rapid development of the Asian market will mean that 27% of the entire global air transport will be taking place in Asia [2]. In the last decade, a twofold increase in global passenger air transport was reported (Fig. 1), while the global cargo increased by 50% (Fig. 2).

The development of air transport is associated with increased external environmental costs which, in accordance with the definition of the European Union includes the impact of the air transport related to: air pollution, climate change and noise among others [3, 4]. Next to the nitrogen oxides, the most dangerous toxic compound of exhaust are particles. Particles emitted by aircraft are responsible for heart disease, lung disease, and early mortality. The above fact was the subject of many research in which it was found that the particles may penetrate deep into the respiratory system [5-8]. Most of studies investigate the effect of the particulate matter mass on humans, without consideration the particle origin [9]. The most dangerous are particles with diameters less than one micron, which mainly come from combustion processes and are referred to as ultrafine particles [10-13].

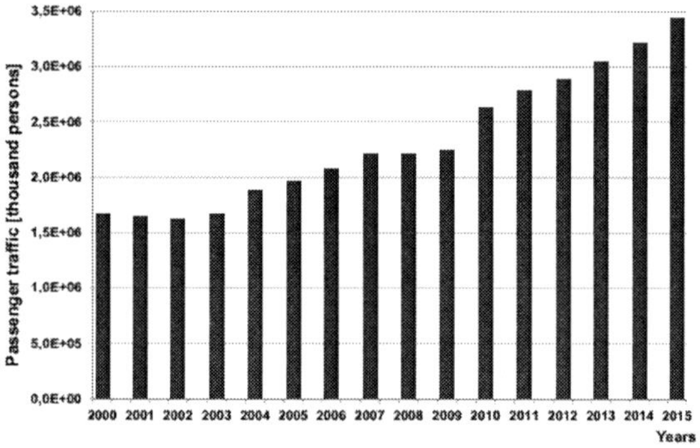

Fig. 1. World passenger air transport in the years 2000-2015 (Central Statistical Office of Poland).

Most of the studies aimed at determining the particle size distribution of the aircraft engines were made within a few hundred meters of the airport. In many studies it has been found that the particle number (PN) emission factors are higher for the operating conditions of engines with low thrust values [14-19]. Therefore, special attention is required for such operations as landing and

* Corresponding author: remigiusz.w.jasinski@doctorate.put.poznan.pl

taxiing of aircraft, due to engine conditions that favor high particulate emissions with relatively long periods of time. Under these conditions dominates PN with dimensions of 20–30 nm, resulting from nucleation mode. Besides nucleation mode, there are many processes of forming particles such as: "Aitken mode" with particles in the size range between 25 and 100 nm , "accumulation mode" with particles between 30 and 300 nm, condensation/evaporation, coagulation, dry deposition and dilution [20-25]. In terms of combustion engines the highest particle number concentrations are in the nucleation mode, while most of the mass is found in the accumulation mode [26, 27]. Nucleation particles are particularly formed by condensation within a few seconds after emissions in ambient air. Particles in the accumulation mode are primarily formed in the combustion chamber [20, 28]. Research shows that ultrafine particles may be associated with cardiovascular diseases, while the mass of particles is more important for respiratory effects [29].

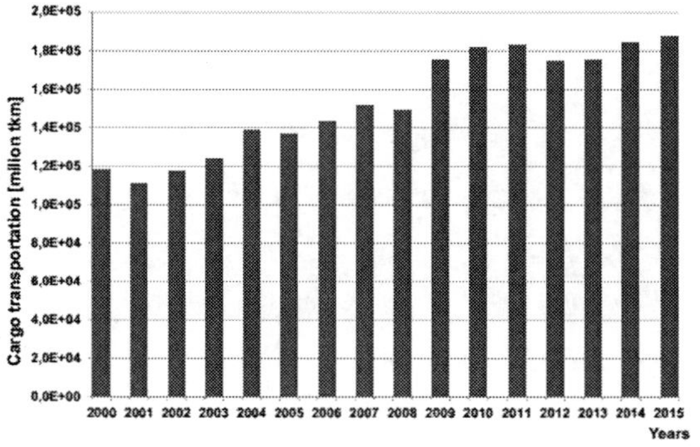

Fig. 2. Global cargo air transport in the years 2000-2015 (Central Statistical Office of Poland).

One of the most important source of nanoparticles (submicron) are airports [30-32]. More and more research is being conducted to evaluate the impact of airport operations on air quality in areas adjacent to airports [33]. In the case of airports, engines are not the only source of particulate matter. Other sources of particulate matter are aircraft tires, brakes, ground handling equipment and others vehicles. In view of the above, it is essential that, in the case of an airport ecological assessment, all sources of particulate matter should be included.

More and more research is about areas adjacent to the airports. In this case, it is necessary to make an effort to model the dynamics of the aircraft plume and the dispersion of pollutants [34]. Publications indicate that nanoparticles from aircraft engines dominate in the radius of 10 km from the airport, and their concentration is up to 4 times higher than in the other areas [35]. This fact confirms the belief that air quality standards based on mass concentrations of particles do not explicitly answer the question of the degree of threat to human health. Air quality standards should be extended by measuring the number of particles. In addition, the LTO (Landing and Take-off) test, which is the basis of aircraft certification, does not really include particulate matter testing. It includes only Smoke Number which is more a measure of exhaust smoke. The conventional approach to toxic emissions from aircraft engines assumes its control only in the phases of take-off, landing and taxiing of the aircraft (LTO). Considered air operations, which are subject to analyzes of the impact of air transport on air quality are held at altitude of up to 3000 feet. This is evidenced by the provisions of the aircraft engines certification (ICAO Annex 16), made with the aim to reduce emissions in areas adjacent to airports. Current research indicates, however, that the issue at cruising altitudes (above 3000 feet) can provide a significant portion of the total emissions affecting human health [4, 36-38]. It may turn out that further research and future assessment of the impact of aviation on air quality will include the full emissions to take account of the overall impact of aircraft on the environment.

Current world trends in the field of environmental assessment of the engines lead to measurements in real operating conditions of the means of transport. This type of measurements are difficult in the case of aircraft because of the size of the measuring apparatus but also for safety reasons. For this reason, attempts of assessment the impact of airports on air quality in areas adjacent take into account the dependencies resulting from measurements of exhaust emissions in stationary conditions. For a deeper analysis of airport operations and their impact on environmental pollution, it is advisable to look for the relationship between measurements made under stationary but also real operating conditions. Due to the technical difficulties associated with the placement of equipment on an aircraft, it is possible to make the assessment of the aircraft operation under specific conditions (such as landing operation) and looking for relationship with results obtained under stationary conditions.

2 Methodology

2.1 Stationary research

Stationary tests were performed on the engine dynamometer during the service test. The research was conducted in 31st Tactical Aviation Base in Krzesiny in western Poland. Due to the fact that the research was conducted at military facility some details of the test object are confidential.

The research object was Pratt & Whitney engine, F100-PW-229 (Fig. 3). This engine is a drive of fighter aircraft F-16. Its maximum values of thrust are 79.13 kN – without the use of afterburner, and 128.91 kN with afterburner. It is a turbofan, twin-shaft engine with hydraulically adjustable nozzle. It is equipped with three-stage low pressure compressor and a ten-stage high pressure compressor. The combustion chamber is annular.

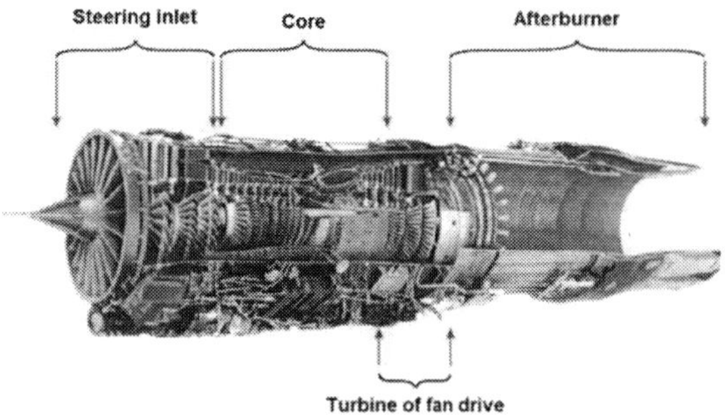

Fig. 3. Cross section of the engine Pratt & Whitney F100-PW-229 [12].

The engine is fitted with anti-icing system, wherein the heating of the engine inlet is managed by air taken from the relief valve at the compressor fifth stage (Fig. 4). Technical parameters of the engine are as follows:

- Maximum diameter: 1080 mm,
- Length: 4855 mm,
- Engine weight: 1370 kg,
- Specific fuel consumption: 0.693 kg/(kG h),
- Specific fuel consumption with afterburner: up to 2.6 kg/(kG h).

Fig. 4. The view of Pratt & Whitney F100-PW-229.

Measurement of particle diameters was performed with a EEPS 3090 (engine exhaust particle sizer™ spectrometer) mass spectrometer. It enabled the measurement of a discrete range of particle diameters (from 5.6 nm to 560 nm) on the basis of their differing speeds. The degree of electric mobility of particulate matter is changed exponentially, and measurement of their size is carried out at a frequency of 10 Hz. The exhaust gases are routed through a dilution system and to the mass spectrometer while maintaining at the desired temperature. The initial filter retains particles with a diameter greater than 1 micron, which are outside of the measuring range of the device. After passing through the neutralizer the particles are directed to the charging electrode; after getting electrically charged they can be classed by their size. The particles deflected by the high-voltage electrode go to an annular slit, which is the space between the two cylinders. The gap is surrounded by a stream of clean air supplied from outside. Exhaust cylinder is built in a stack of sensitive electrodes isolated from one another and arranged in a ring. The electric field present between the cylinders causes the repulsion of particles from the positively charged electrode; then the particles are collected on the outer electrodes. When striking the electrodes, the particles generate an electric current, which is read by a processing circuit [39].

2.2 Research under real operating conditions

Studies at the military airport were conducted in 31st Tactical Aviation Base in Krzesiny in western Poland, approximately 1.5 km from the aircraft touchdown point. The aircraft used in the tests was F-16 Fighting Falcon Multirole Fighter (Fig. 5). Meteorological conditions on the day of the measurements were characterized by lack of precipitation and wind speed of 10 km/h (Fig. 6). The measuring apparatus was set up in the approach axis for the landing of passenger aircraft (Fig. 7).

Fig. 5. The view of F-16 Fighting Falcon.

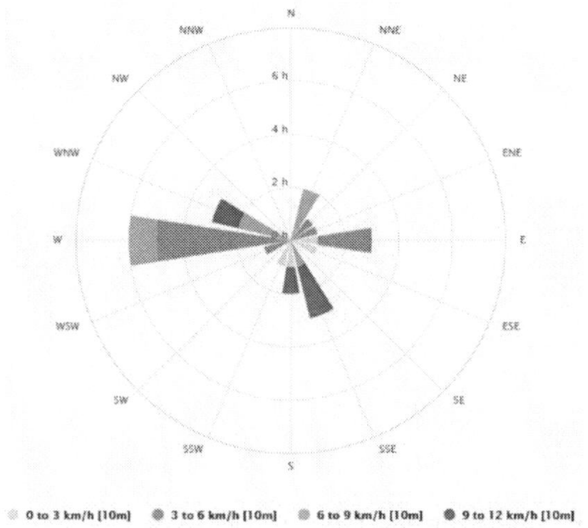

Fig. 6. Wind rose of the research day [meteoblue.com].

The measurements were divided into three phases: pre-landing measurement – to determine the measurement background; measurement during the landing – to determine changes in the concentration of particles during the landing operation; measurement after landing – to determine the maximum concentration of particles.

Fig. 7. The view of F-16 Fighting Falcon.

Measurements of the concentration of particles in the ambient air during the landing of the fighter were made 13 times. Landings were made at 15% throttle opening. The same parameters of the engine were set up during engine dynamometer measurements. The descent angle was about the standard 3°.

3 Results and analysis

3.1 Stationary research

The purpose of the research on the engine dynamometer was to determine the concentration of the particles (PN) in the jet engine exhaust gases, operating at the parameters corresponding to the landing of the F-16 fighter. In the initial phase of the measurements, the engine was working at the thrust corresponding to the idling. After engine stabilization (and stabilization of PN concentration), the engine parameters have been changed so that they correspond to the landing (Fig. 8). This was associated with an increase in the value of thrust. According to the literature [14, 15], the increase of thrust value was associated with a decrease in PN concentration directly behind the confusor.

Particles size distribution obtained from the engine during engine dynamometer test is shown in the Figure 9. Particles with diameters of 5–15 nm were dominated. The characteristic value of the diameter of the particles (the most common) obtained distribution was 9 nm. Particles with diameters greater than 20 nm were negligible part of the total concentration of PN. Particles emitted by the jet engine were of the nanoparticle type, mainly formed in nucleation mode.

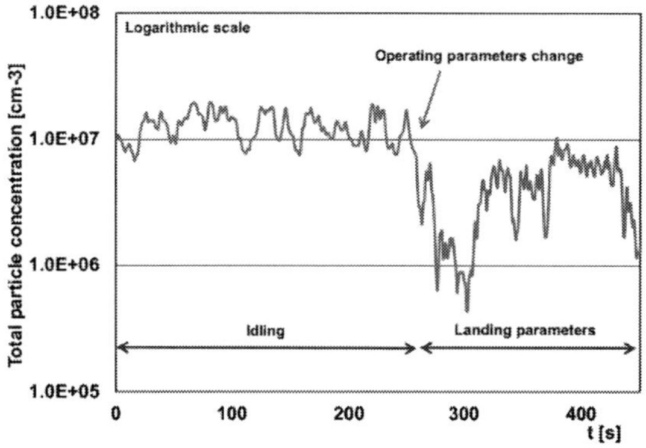

Fig. 8. Total concentration of particles obtained during stationary test.

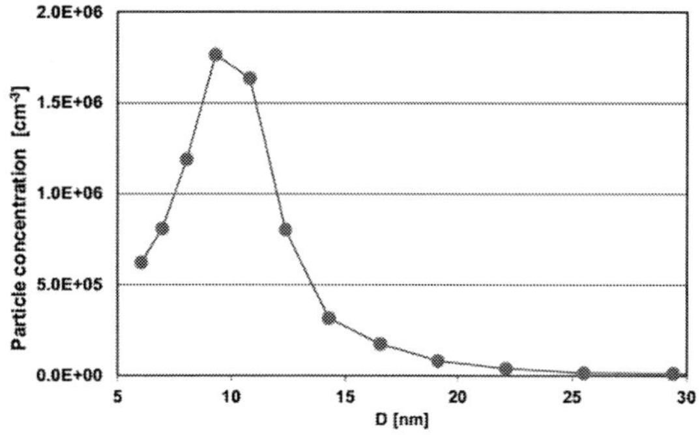

Fig. 9. Size distribution of particles obtained during stationary test.

3.2 Research under real operating conditions

Research of the particles concentration in the air during landing of the aircraft were conducted before, during and after the landings. The measurements were made during the landing of 13 planes, all in the same configuration. The measurements were carried out for few hours. Recorded results are shown in Figure 10. Presented data were reduced of periods in which the concentration of particles was close to the measuring background (long before or after the landing operation).

A detailed record of the change in PN concentration during landing and the process of return to the measured background value is shown in Figure 11. The PN concentration specified as the measuring background was 10^4 cm^{-3}. The landing operation resulted in a 100-fold increase in the concentration of PN which reached 10^6 cm^{-3}. Approximate to maximum PN concentration persisted for about 20 seconds. Then another landing took place, which led to the return to the maximum PN concentration. After about a minute, the particles emitted by the landed aircraft dispersed, which appeared as a return to the measured background value.

 DOI: 10.1051/matecconf/201711800023

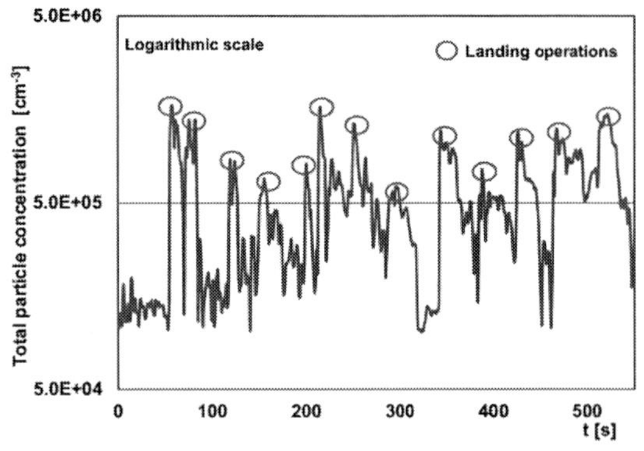

Fig. 10. The record of the PN concentration during research.

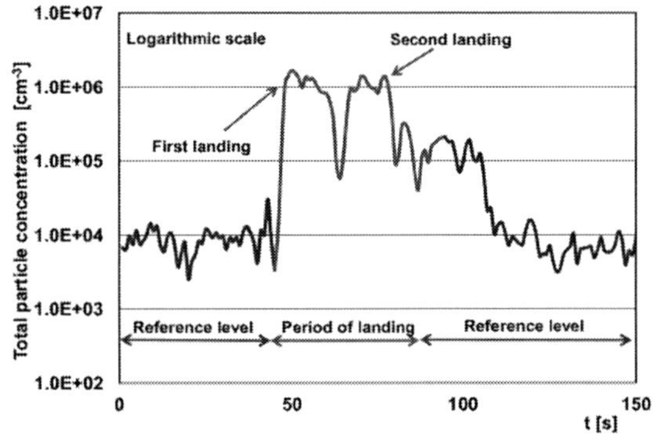

Fig. 11. The process of changes in the concentration of PN during landing.

In the process of analyzing the received measurement results, the maximum PN concentration values during the thirteen landing operations were selected. The data are shown in Figure 12, while the average maximum PN concentration of all tests was 1.13×10^{6} cm^{-3}. Received value reduced by measuring background was 8 times lower than the PN concentration in stationary tests, just behind the engine exit.

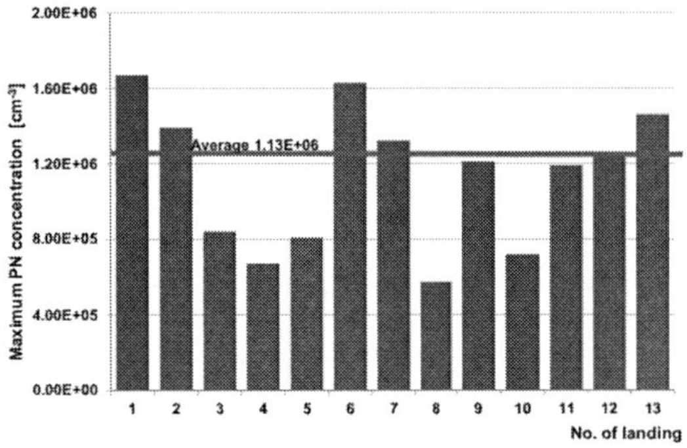

Fig. 12. The process of changes in the concentration of PN during landing.

Particle size distributions obtained during all thirteen landings are shown in Figures 13–15. The resulting distributions are similar to the distribution obtained during stationary measurements.

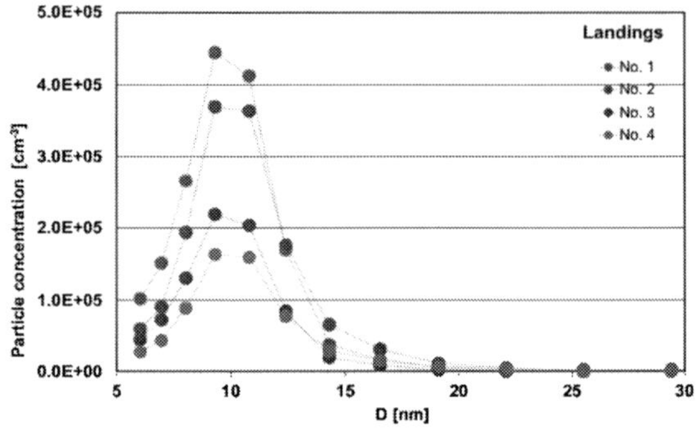

Fig. 13. Size distributions obtained during landings (No. 1–4).

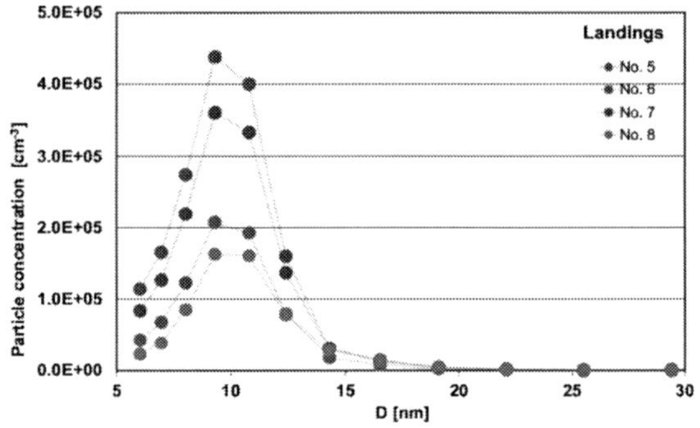

Fig. 14. Size distributions obtained during landings (No. 5–8).

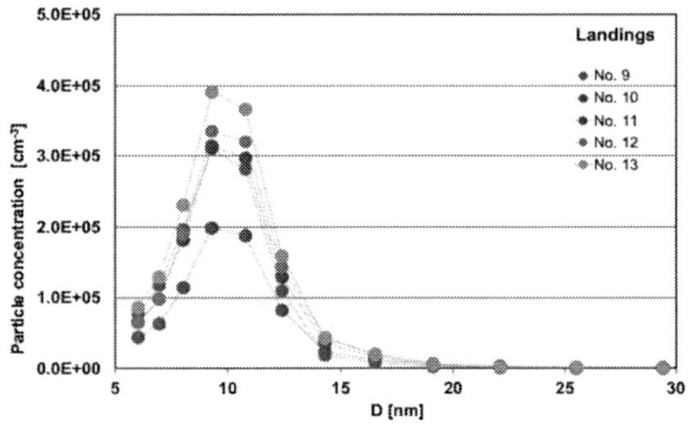

Fig. 15. Size distributions obtained during landings (No. 9–13).

All particle size distributions obtained from the engine during measurements in area adjacent to the airport are close to each other. This proves that the results are characterized by high repeatability. Particles with diameters of 5–15 nm were dominated. The characteristic value of the diameter of the particles was 9 nm (same as in distribution obtained during stationary tests). Particles with diameters greater than 20 nm were negligible part of the total concentration of PN.

4 Conclusions

Landing aircrafts have a significant impact on the concentration of particulate matter in the area adjacent to the airport. A single landing operation causes an increase of the value of particle number concentration tens of times. In addition, even up to 100 times higher concentration of PN in air and the dominance of particulates characteristic for jet engine emissivity, proves that approximately one minute after aircraft landing the air has (in terms of PN) a similar composition as exhaust gases of the internal combustion engine. Obtained particle size distributions indicate that particles emitted by jet engines are characterized by a set of dimensions in the range 5–15 nm.

Such a large number of particles with very small diameters has a negligible effect on the mass concentration of the particles in the air. Due to their small size, however, they represent a very serious threat to human health. This fact confirms the belief that air quality standards based on mass concentrations of particles do not explicitly answer the question of the degree of threat to human health. Air quality standards should be extended by measuring the number of particles.

The study presented in this article was performed within the statutory research (contract No. 05/52/DSMK/0265).

References

1. D. Tłoczyński, Logistyka **3** (2015)
2. Boeing, Business & Market Environment (2015)
3. S. Huderek-Glapska, Logistyka **2** (2014)
4. J. Merkisz, J. Pielecha, S. Radzimirski, *Springer Tracts on Transportation and Traffic* **4**, 1-170 (2014), DOI: 10.1007/978-3-319-02705-0
5. G. Hoek, B. Brunekreef, S. Goldbohm, P. Fisch-Er, P.A. Van Den Brandt, Lancet **360**, 1203-1209 (2002)
6. L. Bayer-Oglesby, C. Schindler, M.E. Hazenkamp-Von Arx, C. Braun-Fahrlander, D. Keidel, R. Rapp, American Journal of Epidemiology **164**, 190-1198 (2006)
7. N. Künzli, R. Kaiser, S. Medina, M. Studnicka, O. Chanel, P. Filliger, Lancet **356**, 795-801 (2000)
8. B. Brunekreef, R. Beelen, G. Hoek, L. Schou-Ten, S. Bausch-Goldbohm, P. Fischer, B. Armstrong, E. Hughes, M. Jerrett, P. Van Den Brandt, HEI Research Report (2009)
9. H.R. Anderson, Atmospheric Environment **43**, 142-152 (2009)
10. J. Merkisz, J. Pielecha, *Springer Tracts on Transportation and Traffic* **8**, 9-17 (2015), DOI: 10.1007/978-3-319-15928-7_2
11. J. Merkisz, J. Pielecha, *Springer Tracts on Transportation and Traffic* **8**, 69-92 (2015), DOI: 10.1007/978-3-319-15928-7_7
12. C. Sioutas, R.J, Definno, M. Singh, Environmental Health Perspectives **113**, 947-955 (2005)
13. D. Taler, A. Cebula, Inzynieria Chemiczna i Procesowa **25**, 2331-2342 (2004)
14. S.C. Herndon, J.T. Jayne, P. Lobo, T.B. Onasch, G. Fleming, D.E. Hagen, P.D. Whitefield, R.C. Miake-Lye, Environmental Science & Technology **42**, 1877-1883 (2008)
15. S.C. Herndon, T.B. Onasch, B.P. Frank, L.C. Marr, J.T. Jayne, M.R. Canagaratna, J. Grygas, T. Lanni, B.E. Anderson, D. Worsnop, R.C. Miake-Lye, Aerosol Science and Technology **39**, 799-809 (2005)
16. M.T. Timko, E. Fortner, J. Franklin, Z. Yu, H.W. Wong, T.B. Onasch, R.C. Miake-Lye, S.C. Herndon, Environmental Science & Technology **47**, 3513-3520 (2013)
17. C.C. Wey, B.E. Anderson, C. Wey, R.C. Miake-Lye, P.D. Whitefield, R. Howard, Journal of Propulsion and Power **23**, 898-905 (2007)
18. J. Merkisz, J. Markowski, J. Pielecha, *WIT Transactions on Ecology and the Environment* **174**, 219-230 (2013), DOI: 10.2495/AIR130191
19. J. Merkisz, J. Pielecha, S. Radzimirski, Springer Tracts on Transportation and Traffic **4**, 67-90 (2014), DOI: 10.1007/978-3-319-02705-0_4
20. M.M. Maricq, Aerosol Science **38**, 1079-1118 (2007)
21. M. Ketzel, R. Berkowicz, Atmospheric Environment **38**, 2639-2652 (2004)
22. P. Kumar, M. Ketzel, S. Vardoulakis, L. Pirjola, R. Britter, Journal of Aerosol Science **42**, 580-603 (2011)
23. J. Merkisz, J. Pielecha, *Springer Tracts on Transportation and Traffic* **8**, 27-46 (2015), DOI: 10.1007/978-3-319-15928-7_4
24. J. Merkisz, J. Pielecha, *Springer Tracts on Transportation and Traffic* **8**, 1-139 (2015), DOI: 10.1007/978-3-319-15928-7
25. J. Merkisz, J. Pielecha, P. Lijewski, A. Merkisz-Guranowska, M. Nowak, *WIT Transactions on Ecology and the Environment* **174**, 27-38 (2013), DOI: 10.2495/AIR130031
26. M.P. Keuken, J.S. Henzing, P. Zandveld, S. Van Den Elshout, M. Karl, Atmospheric Environment **54**, 320-327 (2012)
27. J. Merkisz, J. Pielecha, *Springer Tracts on Transportation and Traffic* **8**, 131-139 (2015), DOI: 10.1007/978-3-319-15928-7_10
28. Z. Stelmasiak, D. Pietras, *IOP Conference Series Materials Science and Engineering* **148** (2016), DOI: 10.1088/1757-899X/148/1/012087
29. R.W. Atkinson, G.W. Fuller, H.R. Anderson, R.M. Harrison, B. Armstrong, Epidemiology **21**, 501-511 (2010)
30. J.S. Kinsey, Y. Dong, D.C. Williams, R. Logan, Atmospheric Environment **44**, 2147-2156 (2010)

31. M. Mazaheri, G.R. Johnson, L. Morawska, Atmospheric Environment **45**, 3500-3507 (2011)
32. J. Merkisz, J. Markowski, J. Pielecha, *IEEE Vehicle Power and Propulsion Conference VPPC*, 897-902 (2012)
33. D. Westerdahl, S.A. Fruin, P.L. Fine, C. Sioutas, Atmospheric Environment **42**, 3143-3155 (2008)
34. R.H. Barrett, R.E. Britter, I.A. Waitz, Atmospheric Environment **74**, 247-258 (2013)
35. N. Hudda, T. Gould, K. Hartin, T. Larson, S. Fruin, Environmental Science & Technology **48**, 6628-6635 (2014)
36. US EPA, Integrated science assessment for oxides of nitrogen – health criteria (2008)
37. US EPA, Integrated science assessment for sulphur oxides – health criteria (2008)
38. US FAA, Environmental tool suite frequently asked questions.
39. J. Merkisz, J. Markowski, J. Pielecha, *Selected issues in exhaust emissions from aviation engines*, (Nova Science Publishers, 2014)

MATEC Web of Conferences **118**, 00024 (2017)
DOI: 10.1051/matecconf/201711800024
VII International Congress on Combustion Engines

Automotive Thermoelectric Generator impact on the efficiency of a drive system with a combustion engine

Andrzej Ziolkowski[1,*]

[1]Poznan University of Technology, Faculty of Working Machines and Transportation, ul. Piotrowo 3, 60-965 Poznan, Poland

Abstract. Increasing the combustion engine drive systems efficiency is currently being achieved by structural changes in internal combustion engines and its equipment, which are geared towards limiting mechanical, thermal and outlet losses. For this reason, downsizing. In addition to these changes, all manner of exhaust gas energy recovery systems are being investigated and implemented, including turbocompound, turbogenerators and thermoelectric generators. The article presents the author's idea of a thermoelectric generator system of automotive applications ATEG (Automotive Thermoelectric Generator) and the study of the recovery of exhaust gas energy stream. The ATEG consists of a heat exchanger, thermoelectric modules and a cooling system. In this solution, 24 commercial thermoelectric modules based on Bi_2Te_3 (bismuth telluride) were used. Measurements were made at two engine test sites on which SI and CI engines were installed. The exhaust gas parameters (temperature and mass flow rate), fuel consumption and operating parameters of the ATEG – the intensity and the voltage generated by the thermoelectric modules and the temperature on the walls of the heat exchanger – were all measured in the experiments. Based on the obtained results, the exhaust gas energy flow and the power of the ATEG were determined as well as its effect on the diesel engine drive system efficiency.

1 Introduction

According to the sustainable development policy implemented within the framework of the Europe 2020 strategy it is necessary to reduce the consumption of conventional energy sources (oil, gas and coal) in order to build a competitive low-carbon economy. This policy is implemented by a number of legal norms that imply limiting emissions of GHG (greenhouse gases) especially CO_2 [1-4].

In 2007, the EC (European Commission) proposed a 30% reduction of greenhouse gas emissions in developed countries by the year 2020, and suggested that the European Union itself should make a firm independent commitment to achieve at least a 20% reduction in greenhouse gas emissions [5-6]. For the automotive industry road emissions standards were introduced for either CO_2 emission or fuel consumption [7-9]. According to Arena, et al. [10], CO_2 emissions from road transport sector constitutes 16.4% of global emissions. Powering vehicles is currently responsible for 38% of the annual global demand for oil. Road transport is also responsible for emissions of CO, HC, NO_x, PM [11-14].

Most of the chemical energy contained in the fuel is drained through both the cooling system, as well as in the form of heat and pressure in exhaust gases [15]. There is a large potential for heat energy recovery from exhaust gases, as its share in the energy balance for SI engines can constitute as much as 45%. For CI engines, it is assumed that the share of thermal energy loss in the exhaust gas is a maximum of 35% [16]. For this reason, systems are developed and implemented that work in the Clausius-Rankine cycle as well as thermoelectric generators for automotive applications ATEG [17-21]. At a conference organized by German Institute of Automotive Engineering IAV in 2014, Iriyama [22] the representative of Toyota Motor Corporation, presented the company's strategy for reduction of fuel consumption in order to comply with the increasingly stringent CO_2 emission limits. The strategy is based on reducing the heat loss in vehicles – primarily through the use of automotive thermoelectric generators (ATEG). A prototype TEGS (Thermoelectric Generation System) based on an ATEG was presented. Cleary et al. [23] presented two prototypes of ATEGs dedicated to passenger vehicle and heavy-duty vehicle (HDV) engines. In the first case, the ATEG was presented with a power rating of 200 W dedicated to Honda Accord fitted with an SI engine. The second generator was dedicated to a tracked combat vehicle (Bradley) used mainly by the US army. In addition to the above solutions, prototypes developed by international research and development centers have been presented in literature [24-29]. The determining factor in the development of ATEGs is to develop thermoelectric materials with a higher ZT index. Intense research is therefore carried out to implement thermoelectric modules based on materials for mass production capable of generating greater power [30-33] and working within a broader temperature range compared to conventional Bi_2Te_3 based applications.

* Corresponding author: andrzej.j.ziolkowski@put.poznan.pl

2 Characteristic of ATEG generator

ATEG is a modular generator and it consists of three main elements: a heat exchanger, commercial thermoelectric modules and cooling elements (Fig. 1, Tab. 1). The main part of the heat exchanger, onto which thermoelectric modules are arranged has a rectangular shape. Its width is almost twice its height. In order to ensure a uniform temperature distribution along the heat exchanger a ribbing of increasing cross-section was mounted inside. It was calculated that four modules would be arranged longitudinally on each side of the heat exchanger. There were two rows of modules on the top and bottom sides of the heat exchanger and one row on each side. This gives a total of 24 modules. There was one dedicated cooling panel for each row of modules, which was attached to the heat exchanger by the means of a screw connection. This allowed the use of thermoelectric modules of different thickness. A special silicate thermal grease was applied between the modules and the walls of the heat exchanger, which facilitated the transfer of heat from the walls of the heat exchanger to the hot side of the module and helped reduce possible heat leaks arising from the surface roughness of the heat exchanger and other material imperfections and faults.

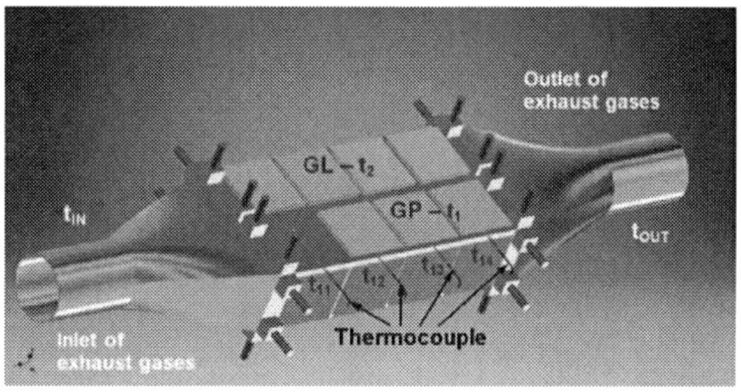

Fig. 1. View of ATEG generator.

Table 1. Characteristics of the ATEG generator.

Parameter	Value
Length	671 mm
Width	200 mm
Height	135 mm
Inlet diameter	60 mm
Material	Aluminum alloy with thickness of 2 mm
Number of modules	24
Module arrangement	Top and bottom walls – 2 rows with 4 modules each, Side walls – 1 row with 4 modules
Type of modules used	TMG-241-1.4-1.2 by Ferrotec NORD
Number of cooling panels	6
Generated nominal power	225.6 W
Generated nominal voltage without load	300 V
Generated nominal voltage with load	151.2 V

3 Research methodology

3.1 Measurement conditions in the engine test beds

Measurements of the recovery of waste energy from the exhaust gas using an ATEG were carried out on a static using a 1.2 TSI (turbo stratified injection) and on dynamic engine dynamometer using a 1.3 SDE (small diesel engine). In order to determine the engine power band for the test bed of heat recovery from exhaust gases tests were made under real operating conditions in accordance with the RDE methodology. The study selected a two vehicles equipped with a 1.2 TSI and 1.3 SDE engines, which is also located on an engine test beds (Tab. 2).

Table 2. Characteristics of the engines used for test.

Parameter	Value	
	1.2 TSI	1.3 SDE
Type of engine	Spark ignition	Compression ignition
Displacement	1.2 dm^3	1.3 dm^3
Number of cylinders	4	4
Power output	77 kW @ 5000 rpm	66 kW @ 4000 rpm
Torque	175 Nm @ 1550–4100 rpm	200 Nm @ 1750 rpm
Emissions standard	Euro 5	Euro 4
Exhaust gas aftertreatment	TWC with lambda control	DOC

In order to best mimic the registered working conditions of the first vehicle's drive unit when testing it on a static engine test bed, it was decided that the research will be carried out at three different load characteristics – constant speed value of the engine crankshaft with the load changing at 20 N·m intervals (Fig. 2 – marked with pink dots) first characteristics has been defined for 1800 rpm (six operating points), the second for 2200 rpm (five operating points) and third to 2800 rpm (five operating points). The study also included the engine idling state (800 rpm). This gives a total of 17 operating points of the engine, where the tests were carried out under laboratory conditions.

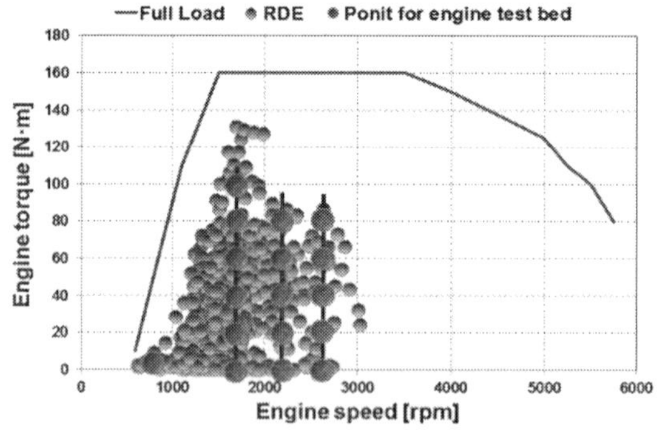

Fig. 2. Operating points for the engine of the test vehicle achieved in a test carried out in accordance with the RDE methods with its full power characteristic.

DOI: 10.1051/matecconf/201711800024

Using the Step Sequence Editor, the recorded driving profile of the vehicle expressed as a function $V = f(t)$ along with the road gradient was uploaded to the ISAC 400 software in dynamic test bed. Additionally, the vehicle speed profile was integrated with the gear ratios of the transmission in the simulation software, which was necessary for a proper operation on the test stand. A similarity of the speed profile recorded in the road test to that obtained on the test stand was observed (Fig. 3). Only in individual points of work, differences were spotted. This may result from the characteristics of the transmission and clutch operation. In the road test, the use of clutch by the driver is practically unrepeatable, while on the test stand the characteristics of the clutch and transmission operation is preset and repeatable for each gear change. This, however, does not significantly influence the difference between the average speeds in the road test and those reproduced on the test stand – they were 31.8 and 31.7 km/h respectively.

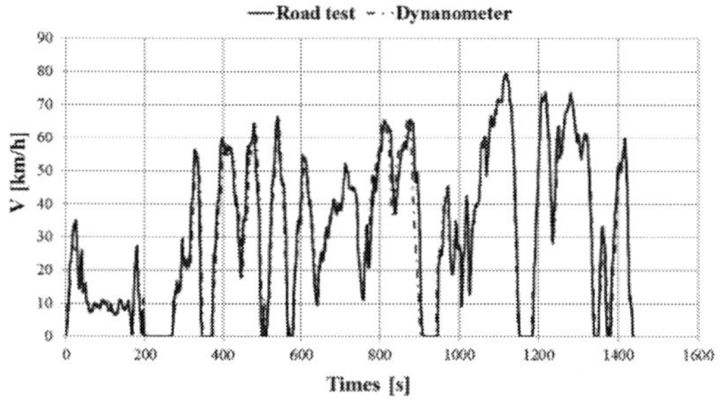

Fig. 3. Comparison of the speed profiles in the road test and on the test stand.

The fact of successful reproduction of the actual driving profile of a vehicle on an engine test bed is confirmed by the linear relation between the actual and the reproduced profiles, whose coefficient of determination is $R^2 = 0.93$ (Fig. 4). From the statistical analysis, we know that obtaining a coefficient in the range of 0.9–1.0 confirms a very good compatibility of the model with the exogenous variable.

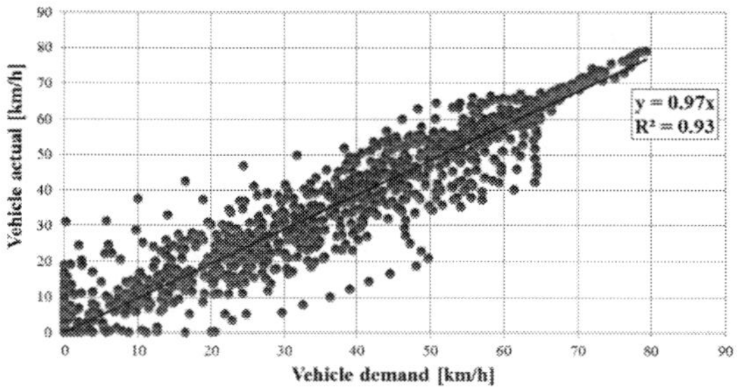

Fig. 4. Correlation of the speed in the road test and that obtained on the test stand.

During the tests of heat recovery from exhaust gases performed on a static and dynamic engine test bed the following parameters were measured:

- momentary fuel consumption G [g/s] – mass fuel gauge from Automex and AVL company,
- crankshaft rotational speed S [rpm] and engine torque T [N·m] – stationary engine test bed made by Automex company, dynamic engine test bed made by AVL company and equipped with an eddy current brake with an inductive engine speed sensor,
- exhaust mass flow rate EFM [kg/h] – with Semtech-Ecostar,
- exhaust gas temperature in measurement points [°C] – thermocouple,
- the temperature of the hot side of the thermoelectric modules [°C] (first and last module in each row) – thermocouple,
- the mass flow m_{cool}.[dm^3/h] and temperature t_{cool}. [°C] of the coolant,
- voltage U [V] and current I [A] generated by the modules – a custom-made measurement system of author design.

3.2 Equipment used for tests

The first instrument used for research was the Semtech-Ecostar. This is another device from the Setmtech® series – a successor to the Semtech DS analyzer. Example of use Semtech DS in research are presented in [34-39]. The main difference between these devices is that Semtech-Ecostar is made up of separate modules (Fig. 5):

- FEM module for measuring the concentration of CO_2, CO, HC (NDIR analyzer),
- NO_x module for the measurement of NO_x (NO and NO_2) concentration using the NDUV analyzer,
- FID module to measure THC with a FID analyzer,
- PDM module to power supply system for each component.

Fig. 5. Semtech-Ecostar instrument used to measure emissions in accordance with the RDE methodology.

In the case when of all the modules are used (of the gas measurement section) they must be connected together by a CAN communication wire (green lines in Fig. 6). The FEM module is connected via a special cable with a flow meter for measuring thermodynamic parameters of exhaust gases (mass flow, temperature and pressure). This line is supplied a sample of exhaust gas that passed through the NDIR analyzer (measuring CO_2, CO and HC) is directed to the NO_x module with three connectors, wherein the measurement of NO and NO_2 takes place.

Apart from the exhaust gas sample a special conduit transmits the signal from the vehicle global positioning system (GPS) and weather sensor used to measure the atmospheric conditions (temperature, pressure, humidity), which are connected to the flow meter. A computer for system control and a signal converter from the vehicle's diagnostic system are connected using the AUX2 connector to the PDM, FEM and NO_x modules. Each module is equipped with one AUX1 and one AUX2 connection point. The signal from the power supply module for the heated wire is sent to the FID module. Semtech-Ecostar device can be powered by a PDM module, to which voltage is supplied from the grid or from the car battery. It is also possible to directly connect the modules to the mains. Due to the modular design of the instrument it can perform measurement of all gaseous compounds simultaneously, or separately using a single module.

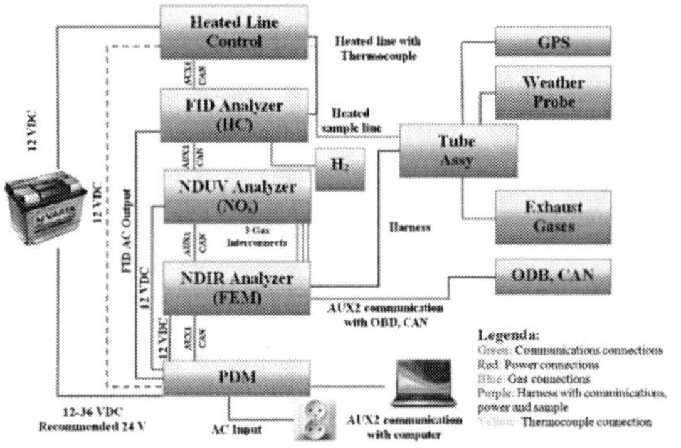

Fig. 6. Diagram of the Semtech-Ecostar system used for testing – section responsible for the measurement of gaseous compounds.

To register the temperatures and pressures two IOTECH Personal Daq/3000 signal converters were used. The measurement module was equipped with a USB interface and high-speed A/D converter (1 MHz/16 bit). The transmitter has 16 single-ended type of analog inputs (8 differential inputs), 4 analog outputs, 24 digital input/output lines, and it is possible to program the machine in 7 ranges from ±100 mV to ±10 V. Information from the sensor is transmitted to a computer which records the data of a given frequency.

The custom measurement system to measure the voltage and current generated by the modules consists of: the main module (which includes, among others a USB-600 measuring card manufactured by National Instruments having 8 analog inputs and 4 digital inputs), two converters for voltage and current measurement fitted on a PCB, a slide resistor for setting the load of the system as well as control software. The main module is supplied with a voltage of 220 V. The software allows for monitoring of the generated voltage and current in real time (in numerical and graphic form) as well as data recording with a frequency of 1 Hz. It uses algorithms to calculate the generated power in Watts and the energy expressed in kWh. A measuring device records the total value of the voltage and current of all modules, but also has the ability to measure these values for individual modules.

3.3 Method for placing ATEG generator in the exhaust system of engines tested

Both test stands were equipped with complete exhaust systems such as those used in motor vehicles. In order to indicate the location of the ATEG in the system a set of preliminary measurements of the exhaust gases thermodynamic parameters were carried out as according to the diagrams in Fig. 7–8.

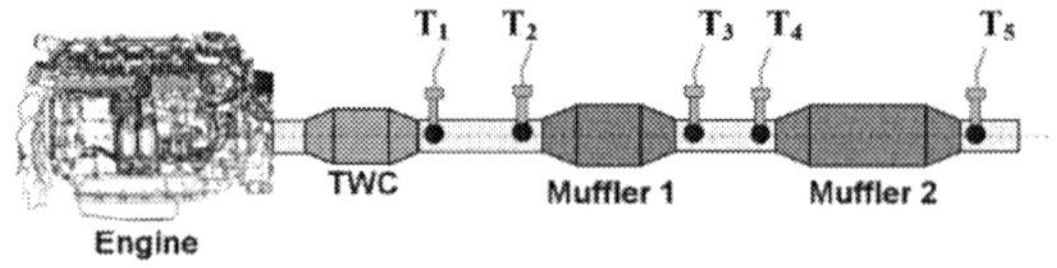

Fig. 7. The distribution of temperature and pressure measurement points in the 1.2 TSI engine exhaust system.

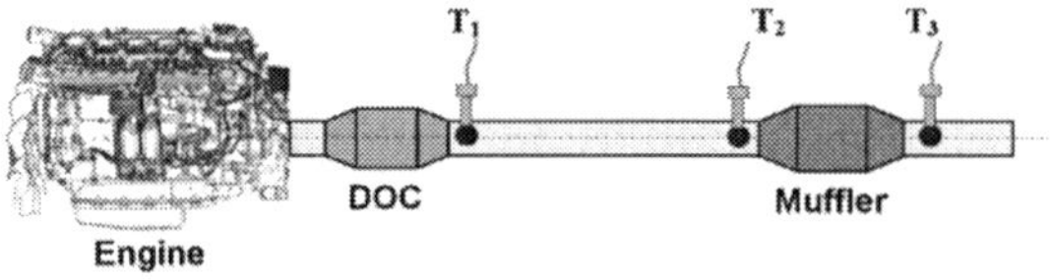

Fig. 8. The distribution of temperature and pressure measurement points in the 1.3 SDE engine exhaust system.

At all points of the 1.2 TSI engine operation constant temperature characteristics have obtained in the exhaust system. Therefore, not all research results have been presented. On the first engine characteristics at $S = 1800$ rpm and $T = 60$ N·m, the highest exhaust gas temperature was recorded after the TWC catalytic reactor, which was $T_1 = 580.1$ °C. The resulting temperature characteristics were used to determine the location of the ATEG in the 1.2 TSI engine exhaust system. From the initial tests, locating the generator directly behind the TWC catalytic reactor was excluded as a possibility, because in most of the analyzed engine operating points the exhaust gas temperature in that area exceeded 600 °C. This was mainly related to loads over 40 N·m. The design assumptions are that the exhaust gas temperature at the entrance to the ATEG will not exceed 500 °C. This was related both to the materials used for the heat exchanger construction as well as to the maximum operating temperature of the hot side of the used TEM modules. The second potential solution was to locate the energy recovery system at the first muffler location. But this was only possible for a limited engine operating range. For this reason, it was decided that the ATEG in the energy recovery study of the waste energy stream would be placed at a position slightly away from the second muffler (Fig. 9). It has also been assumed that the first muffler will be removed from the exhaust system because it causes a significant drop in exhaust gas temperature (40–80 °C).

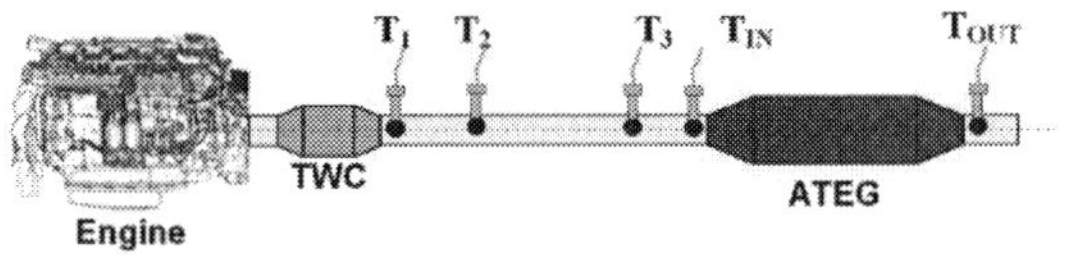

Fig. 9. Distribution of temperature measurement points in the 1.2 TSI engine exhaust system, with the ATEG.

For the 1.3 SDE engine, the exhaust gas temperature was relatively low, which resulted in the placement of the generator directly behind the catalytic converter DOC (Fig. 10).

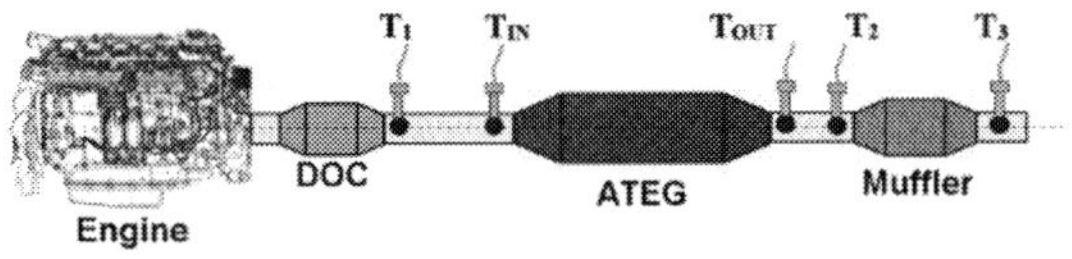

Fig. 10. Distribution of temperature measurement points in the 1.3 SDE engine exhaust system, with the ATEG.

4 Energy balance of engine exhaust system with ATEG

4.1 Energy balance of engine 1.2 TSI and 1.3 SDE engine

The choice of the 1.2 TSI engine operating points for the analysis was primarily related to the steady-state temperature characteristics, as was the case for measurements made for the exhaust system without the ATEG. Primarily, it was shown that the exhaust gas temperature was significantly reduced in the muffler. This was confirmed by the recorded temperature characteristics at points T_2 and T_3 for the considered operating area of the combustion engine. When a muffler was installed in the exhaust system, the temperature difference between T_2 and T_3 ranged between 40–80 °C. Removing the muffler resulted in a decrease in that range. For individual engine load characteristics at $T = 60$ N·m it reached a value of:

- $S = 1800$ rpm: $T_2 - T_3 = 17.5$ °C,
- $S = 2200$ rpm: $T_2 - T_3 = 10.7$ °C,
- $S = 2600$ rpm: $T_2 - T_3 = 6.2$ °C.

The exhaust gas temperature at the entrance of the ATEG was similar to that recorded before the second muffler. For $S = 1800$ rpm and $T = 60$ N·m it was 390 °C (Fig. 11).

For the measurements performed on the dynamic engine test bed with the 1.3 SDE, the temperature at point 1, behind the diesel oxidation catalyst, did not exceed 250 °C (Fig. 12). At the entry to the ATEG it reached a maximum of 200 °C. It was also observed that the measured temperature increased with time. This was related to the thermal stabilization of the exhaust system, which was reached after 400 s of operation. The maximum temperature drop in both the ATEG and the muffler was 53 and 31 °C.

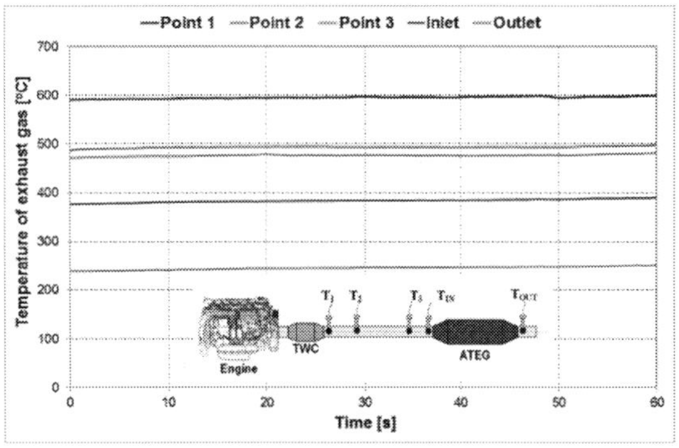

Fig. 11. The exhaust system temperature of a 1.2 TSI engine equipped with an ATEG on a static engine test bed.

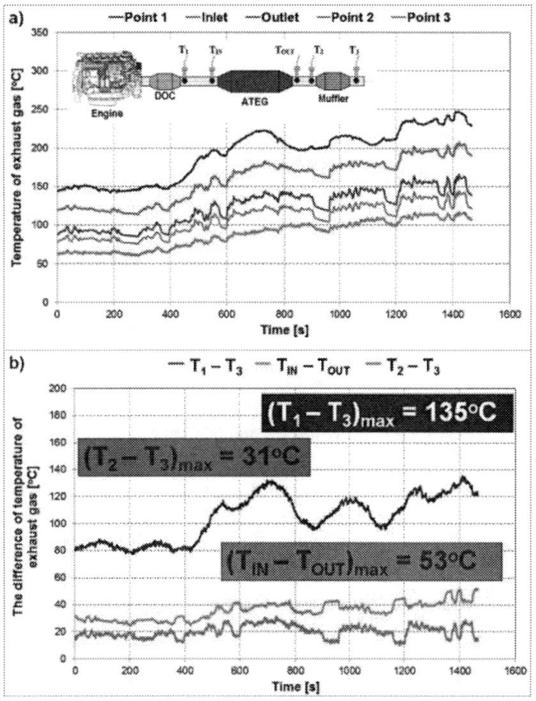

Fig. 12. The exhaust system temperature of a 1.3 SDE engine equipped with an ATEG on a dynamic engine test bed in an actual vehicle driving cycle: a) at points T_1–T_3 and T_{IN}–T_{OUT}, b) differences $T_1 - T_3$, $T_{IN} - T_{OUT}$ and $T_2 - T_3$.

4.2 The temperature distribution on the walls of the ATEG

An important factor influencing the obtained ATEG power was the temperature distribution on its walls. Due to the nature of the TEM modules operation the distribution must be practically uniform on each of them. The developed design uses 24 commercial TEM units, which are combined in series into a single system. In this combination, the current flows through all the elements, maintaining a constant value of intensity I, while the electrical voltage U is additive. In the case of TEM modules its value is closely linked to the temperature difference between the hot and the cold side. If temperature differences occur between the individual TEM modules when connected in series, the total value of the voltage and current that the system generates will change. This will translate into power generated by the ATEG. Therefore, the temperature distribution on the individual walls of the ATEG during the tests was presented. The generator consists of four walls, which for the purposes of this analysis have been indicated as: top, bottom, left and right walls. The top and bottom walls provided two rows of four TEM modules, while the lateral (left and right) walls one row each. One sensor for measuring the temperature was placed on each hot side. Temperature measurement points on the ATEG walls were defined in the following manner (Fig. 13):

- upper right wall UR – T_{11}, T_{12}, T_{13}, T_{14},
- upper left wall UL – T_{21}, T_{22}, T_{23}, T_{24},
- right wall R – T_{31}, T_{32}, T_{33}, T_{34},
- left wall L – T_{41}, T_{42}, T_{43}, T_{44},
- lower right wall LR – T_{51}, T_{52}, T_{53}, T_{54},
- lower left wall LL – T_{61}, T_{62}, T_{63}, T_{64}.

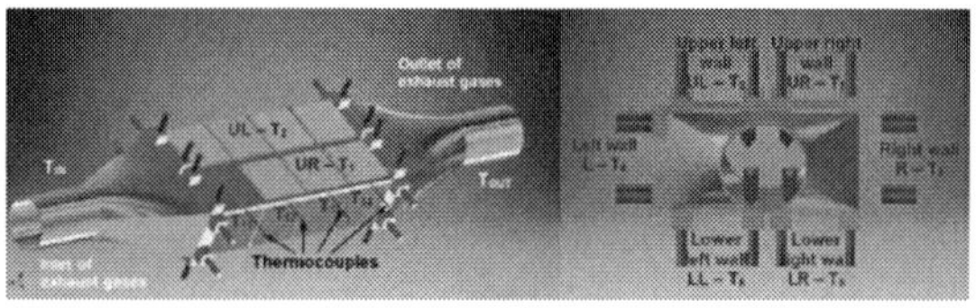

Fig. 13. Location of temperature measuring points on the walls of the ATEG.

Temperature in ATEG was measured in 24 points. The analysis of the results revealed that the temperature differences between the various modules were not significant, and demonstrated a linear relationship. Therefore, only the data for the first and last TEM unit in the row will be presented.

The test driving cycle performed on a dynamic engine dynamometer with the 1.3 SDE similar temperature distributions on each wall were obtained. The study was carried out for a thermally stable engine – the coolant temperature was 90 °C. The engine was heated up before each measurement, and then left for the temperature of the exchangers walls to stabilize. The temperature difference between the first and the last TEM module on each panel was calculated for the mean value obtained throughout the test cycle:

- $\Delta T_1 = T_{14} - T_{11}$,
- $\Delta T_2 = T_{24} - T_{21}$,
- $\Delta T_3 = T_{34} - T_{31}$,
- $\Delta T_4 = T_{44} - T_{41}$,
- $\Delta T_5 = T_{54} - T_{51}$,
- $\Delta T_6 = T_{64} - T_{61}$.

On the top and bottom walls of the heat exchanger the biggest drop in temperature was recorded for ΔT_1, which was 4.9 °C and ΔT_2 = 4.9 °C (Fig. 14a). On the side walls the differences were slightly lower ΔT_3 = = 4.5 °C and ΔT_4 = 4.6 °C (Fig. 14b). Recorded temperature distributions on the walls of the heat exchanger resulted in the obtained values of voltage and current generated by the TEM modules in a given row. Comparing the temperatures of the upper and lower walls to the side walls resulted in greater differences than the change registered between the far modules in a row.

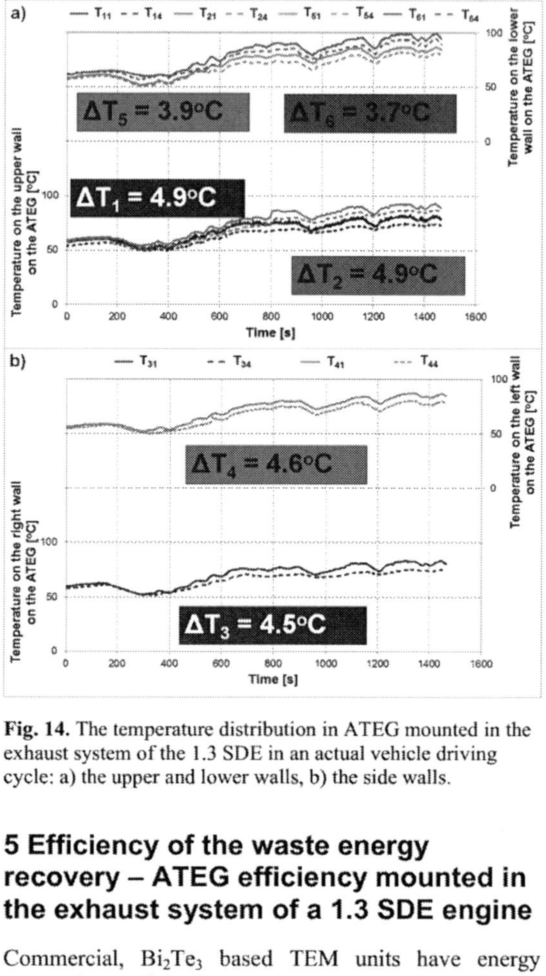

Fig. 14. The temperature distribution in ATEG mounted in the exhaust system of the 1.3 SDE in an actual vehicle driving cycle: a) the upper and lower walls, b) the side walls.

5 Efficiency of the waste energy recovery – ATEG efficiency mounted in the exhaust system of a 1.3 SDE engine

Commercial, Bi_2Te_3 based TEM units have energy conversion efficiencies not exceeding 3%. For the purpose of the research, commercial TMG-241-1.4-1.2 modules made by Ferrotec NORD were used with a power of 9.4 W each, for a total of 225.6 W. In order to determine the ATEG exhaust gas energy recovery potential, the efficiency of waste energy recovery (tantamount to ATEG efficiency) was defined, which was expressed as the ratio of power generated by the TEM modules to the amount of energy present in the exhaust gas heat exchanger fitted with the ATEG:

$$\eta_{\mathrm{ATEG}} = \frac{P_{\mathrm{ATEG}}}{Q_{\mathrm{EX}}} = \frac{P_{\mathrm{ATEG}}}{\dot{m}_{\mathrm{EX}} \cdot c_p \cdot (T_{\mathrm{EX}} - T_{\mathrm{A}})} \tag{1}$$

where: η_{ATEG} – ATEG efficiency [–],
P_{ATEG} – power generated by the TEM modules [kW],
Q_{EX} – energy rate in the exhaust gas [kW],
$\dot{m}_{EX}$ – exhaust mass flow rate [kg/s],
T_{EX} – temperature of exhaust gas [°C],
T_A – temperature of ambient air [°C].

The power of the TEM module is determined based on the voltage and current recorded by the developed measurement system. The amount of energy in the exhaust gas is calculated using equation (1). The exhaust gas temperature T_{EX} is construed as the average ATEG temperature:

$$T_{EX} = \frac{T_{IN} - T_{OUT}}{2} \tag{2}$$

where:
T_{EX} – average temperature inside the ATEG [°C],
T_{IN} – temperature at the inlet to the ATEG [°C],
T_{OUT} – temperature at the outlet from ATEG [°C].

In the measurements carried out on the 1.3 SDE engine on a dynamic test stand, relatively low values of power generated by the ATEG were achieved. Therefore, only the waste heat recovery efficiency for the module coolant temperature of $T_{cooling}$ = 10°C was presented (Fig. 15). The highest value occurred at deceleration (simulated vehicle braking) and zero acceleration (vehicle stationary). This was mainly due to the advantage of a constant exhaust gas mass flow resulting in a uniform temperature distribution on the walls of the heat exchanger and the ATEG. This reduced the pressure pulsations in the engine exhaust system. With the increasing speed (acceleration of the vehicle) there was a cyclical decline in the waste energy recovery efficiency due to sudden changes in the exhaust gas thermodynamic parameters. The key, in this case, was the relatively low temperature of the exhaust gas – the value on the walls of the heat exchanger did not exceed 100 °C in the tests.

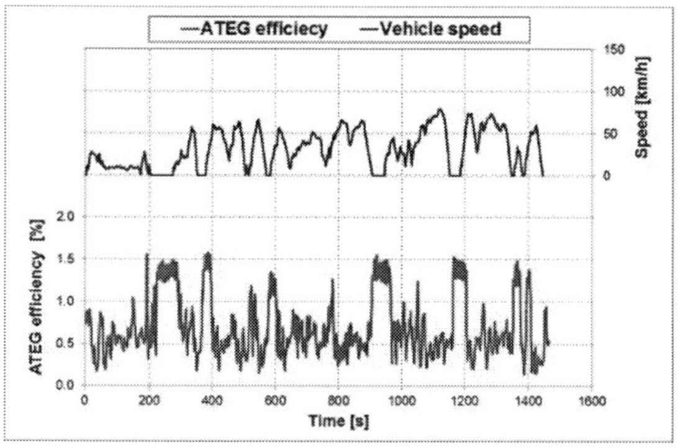

Fig. 15. The efficiency of the ATEG fitted in the exhaust system of a 1.3 SDE engine in an actual vehicle driving cycle.

In addition to determining the ATEG efficiency, it is also important to determine the potential exhaust gas energy recovery (waste energy) that is transferred to the heat exchanger walls. In this case, the temperature drop ΔT_{ATEG} plays a key role. Some of the designated waste energy is lost on the walls of the diffuser and the confusor of the heat exchanger as well as following the internal flow resistance. For this analysis, it was assumed that it equaled half of the total ATEG calculated exhaust gas energy flow. In the test driving cycle, the average value of the exhaust gas energy flow in the ATEG was Q_{EX} = 0.67 kW, and reached a maximum of 4.05 kW (Fig. 16).

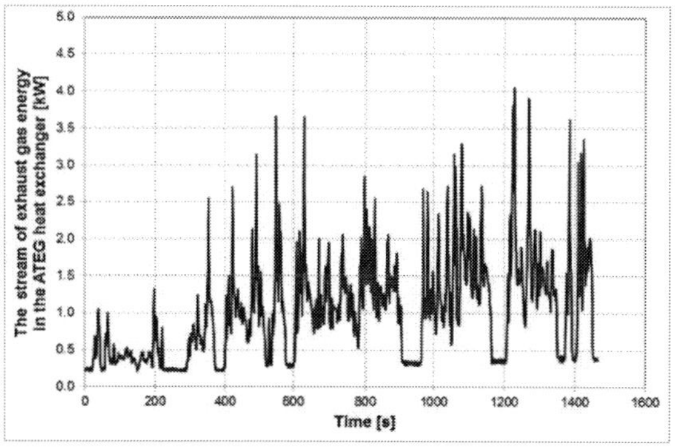

Fig. 16. The exhaust gas energy flow in the ATEG heat exchanger mounted in the exhaust system of a 1.3 SDE engine in an actual vehicle driving cycle.

6 Drive system efficiency with the 1.2 TSI and 1.3 SDE engine including the ATEG

The overall efficiency of an internal combustion engine is a measure of the use of energy contained in the fuel. This provides a view of the efficiency of conversion of the heat energy contained in the fuel into mechanical energy of the engine. This efficiency is also considered the inverse of specific fuel consumption and can be expressed as:

$$\eta_o = \frac{P_e}{G \cdot GCV} = \frac{1}{BSFC \cdot GCV} \tag{3}$$

where:
η_o – overall engine efficiency,
P_e – engine power output,
G – second-by-second fuel consumption,
GCV – gross calorific value,
BSFC – brake specific fuel consumption.

In order to determine the efficiency of a drive system η_{DSE} fitted with an ATEG, formula (3) should be used taking into consideration the power generated by the generator P_{ATEG}. Adding the ATEG power P_{ATEG} (using the values of voltage and current obtained using the developed measurement system) to the output of the engine P_e is justified, because the generator produces electrical power from the exhaust gas energy, which, in internal combustion engines, is treated as a loss (failure to use this waste energy generated by the combustion of the air-fuel mixture). This increases the efficiency of the drive system because the engine produces a greater amount of energy with the same amount of fuel. The thermal energy loss of the exhaust gas is also limited. In order to determine the increase of the powertrain efficiency through the application of an ATEG in the exhaust system, the overall efficiency of the engine η_o must be deducted from the efficiency of the powertrain η_{DSE} (4).

$$\Delta\eta_{DSE} = \eta_{DSE} - \eta_o = \frac{(U \cdot I)_{ATEG}}{G \cdot GCV} \tag{4}$$

where:
η_{DSE}– the efficiency of the drive system including the ATEG,
U_{ATEG} – voltage generated by the TEM modules,
I_{ATEG} – current generated by the TEM modules.

Brake specific fuel consumption in the studies conducted for 1.2 TSI engine was in the range of 261.3–482.5 g/kWh (Tab. 3). The smallest value indicates the greatest overall efficiency of the engine, which was recorded for S = 2600 rpm and loads of 80 N·m. It amounted to 32.8% efficiency respectively. For S = 1800 rpm and T = 100 Nm overall efficiency was 29.4% and for S = 2200 rpm and T = 80 N·m amounted to 32.3%. For idling (the lowest possible speed of the crankshaft without load) and neutral (increased speed of the crankshaft without load) BSFC and efficiency are not calculated, because the engine does not generate any power output that would be transmitted to the receiver. It only generates power as needed to overcome the resistance of its operation. Due to the existence of these points in real road driving conditions a decision was made to also include these points in the analysis. In these cases, only the value of power generated by ATEG will be presented without determining its effect on the overall efficiency of the engine.

Table 3. Comparison of specific fuel consumption and overall efficiency of the tested engine.

Engine speed [rpm]	Torque [N·m]	BSFC [g/kWh]	Efficiency [%]
900	0	–	–
1800	0	–	–
	20	482.5	17.8
	40	370.2	23.1
	60	324.8	26.3
	80	293.2	29.2
	100	291.4	29.4
2200	0	–	–
	20	476.8	18.0
	40	345.9	24.8
	60	286.6	29.9
	80	265.1	32.3
2600	0	–	–
	20	492.7	17.4
	40	367.9	23.2
	60	267.9	31.2
	80	261.3	32.8

In all load characteristics for 1.2 TSI engine a trend of increased ATEG power generation with increasing engine torque (Fig. 17) could be observed. This is logical, because at higher loads the engine produces more exhaust gases at a higher temperature. The highest values of P_{ATEG} were as follows:

- 1800 rpm – 90.1 W,
- 2200 rpm – 98.8 W,
- 2600 rpm – 189.3 W.

For idling of the engine (S = 900 rpm & T = 0) P_{ATEG} did not exceed 1 W. The obtained test results did not reach the ATEG nominal power, which is equal to 225.6 W. The system generated at most 84% of its maximum nominal power.

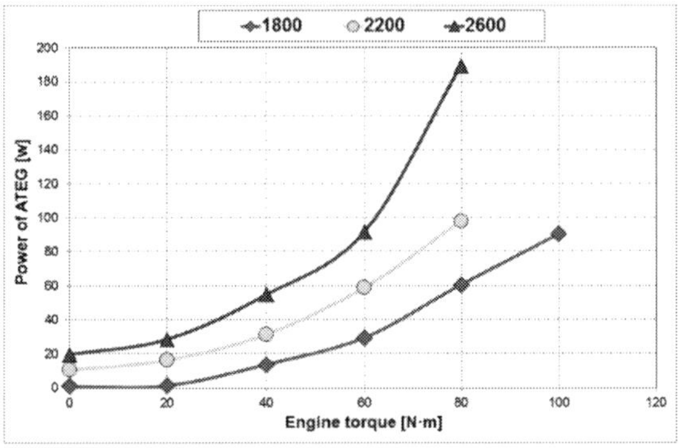

Fig. 17. The calculated power generated by the thermoelectric generator modules (ATEG) during tests.

The efficiency of the drive system including, the energy generated by the ATEG was determined using equation (4) for both engines. According to this assessment, the largest growth of efficiency of the drive system with 1.2 TSI engine was obtained for S = 2600 rpm (Fig. 18. The energy generated by the ATEG over the tested range of engine operation does not constitute a significant share of the total energy. This was mainly the result of the failure to achieve the nominal power of the generator. The above analysis has not taken into account the power loss occurring in the ATEG associated with resistance to the flow of exhaust gases.

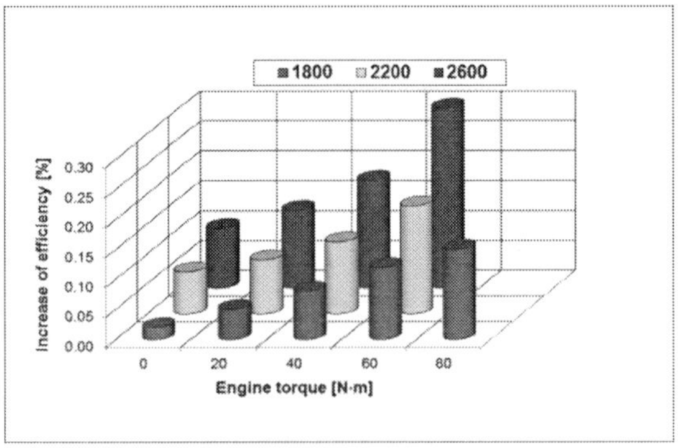

Fig. 18. Increase in overall efficiency of the test engine, through the application of ATEG in the exhaust system.

The increase in the efficiency of the drive system fitted with the 1.3 SDE engine by utilizing the power generated by the ATEG did not exceed 0.27% (Fig. 19) and the average value throughout the experiment was 0.06%. This was mainly due to the low exhaust gas temperature, which was a consequence of low energy consumption in the test driving cycle. The greatest increase in the value of powertrain efficiency was recorded at maximum torque, which was reflected in the temperature of the exhaust gas – the efficiency was the highest in these operating points. These, however, were only single points recorded in the measurement during high accelerations.

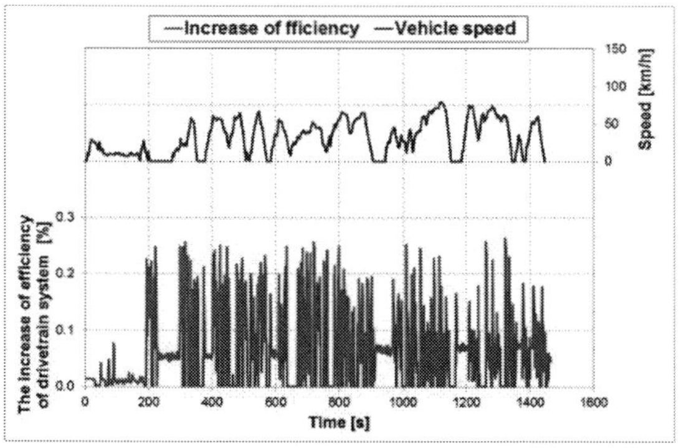

Fig. 19. The increase in the efficiency of the drive system fitted with the 1.3 SDE engine including the ATEG in an actual vehicle driving cycle.

7 Conclusion

The paper presents a study of the impact of a developed ATEG prototype the efficiency of the drive system fitted with the modern spark ignition engine designed in accordance with the principles of downsizing and compression ignition engine. These studies were performed on a static and dynamic engine test bed. In order to determine the engine power band and vehicle speed profile for the test bed of heat recovery from exhaust gases tests were made under real operating conditions in accordance with the RDE methodology.

The developed design of the ATEG was characterized by a nearly uniform distribution of temperature on each row of the TEM modules. The greatest differences of temperature between the first and the last module did not exceed 5 °C. The temperature differences among the walls were slightly higher. Such a distribution of temperatures results in similar values of voltage and current generated by the TEM modules. The study also determined the exhaust gas energy recovery potential in the ATEG heat exchanger. The average value of waste energy for 1.3 SDE engine, taking into account the loss, was 0.67 kW and the maximum was 4.05 kW. The efficiency of the ATEG varied in the range of 0.14–1.6%. The increase in the efficiency of the drive system fitted with the 1.2 TSI engine by utilizing the power generated by the ATEG did not exceed 0.29%. For 1.3 SDE this value did not exceed 0.27%. In further works, aimed at improving the efficiency of the ATEG, a change in the ribs of the internal heat exchanger is planned in order to improve the transfer of the energy flow from the exhaust gas to the generator walls.

The research was funded by the National Centre for Research and Development (Narodowe Centrum Badań i Rozwoju) – research project within the Applied Research Programme (contract No. PBS3/A6/25/2015).

References

1. Z. Juda, M. Noga, *IOP Conference Series – Materials Science and Engineering* **148**, UNSP 012042 (2016), DOI: 10.1088/1757-899X/148/1/012042
2. Z. Korczewski, J. Rudnicki, *Computational Methods in Marine Engineering VI*, 490-501 (2015)
3. J. Kromulski, T. Pawlowski, J. Szczepaniak, W. Tanas, A. Wojtyla, M. Szymanek, J. Tanas, W. Izdebski, Annals of Agricultural and Environmental Medicine **23**, 373-376 (2016), DOI: 10.5604/12321966.1203908
4. J. Szczepaniak, T. Pawlowski, R. Grzechowiak, J. Kromulski, J. Rutkowski, Applied Engineering in Agriculture **31**, 709-716 (2015)
5. M. Andrzejewski, A. Merkisz-Guranowska, Journal of KONES Powertrain and Transport **22**, 257-261 (2015)
6. J. Merkisz, M. Andrzejewski, P. Fuc, P Lijewski, Journal Acta Naturalis Scientia **1**, 11-16 (2014)
7. M. Bajerlein, L. Rymaniak, Experimental and Applied Mechanics **518**, 96-101 (2014), DOI: 10.4028/www.scientific.net/AMM.518.96
8. P. Lijewski, J. Merkisz, P. Fuc, Croatian Journal of Forest Engineering **34**, 113-122 (2013)
9. P. Lijewski, J. Merkisz, P. Fuc, M. Kozak, L. Rymaniak, Applied Mechanics and Materials **390**, 313-319 (2013), DOI: 10.4028/www.scientific.net/AMM.390.313
10. F. Arena, L. Mezzana, A. Doyon, H. Suzuki, et al., Arthur D. Little (2014)
11. J. Nowakowski, K. Brzozowski, T. Knefel, *Transport Means – Proceedings of the International Conference* 333-336 (2015)
12. J. Merkisz, M. Bajerlein, P. Daszkiewicz, *Lecture Notes in Information Technology* **13**, 106-113 (2012)
13. J. Merkisz, P. Lijewski, P. Fuc, S. Weymann, Applied Engineering in Agriculture **31**, 875-879 (2015), DOI: 10.13031/aea.31.11225
14. J. Merkisz, J. Pielecha, P. Lijewski, A. Merkisz-Guranowska, M. Nowak, *WIT Transactions on Ecology and the Environment* **174**, 27-38 (2013), DOI: 10.2495/AIR130031
15. A. Elalem, M.S. El-Bourawi, International Journal of Material Forming **3**, 663-666 (2010), DOI: 10.1007/s12289-010-0857-2
16. K. Brzozowski, J. Nowakowski, Eksploatacja i Niezawodnosc – Maintenance and Reliability **16**, 407-414 (2014)

17. P. Bombarda, C.M. Invernizzi, C. Pietra, Applied Thermal Engineering **30**, 212-219 (2010), DOI: 10.1016/j.applthermaleng.2009.08.006
18. J. Dong, Y.J. Wang, R. Zhang, B. Wang, *Lecture Notes in Electrical Engineering* 139-147 (2014), DOI: 10.1007/978-3-662-45043-7_15
19. N. Esponisa, M. Lazard, L. Aixala, H. Scherrer, Journal of Electronic Materials **39**, 1447-1455 (2010), DOI: 10.1007/s11664-010-1305-2
20. S. Glover, D.L. Glover, G. Mccullough, S. Mckenna, International Journal of Automotive Technology **16**, 399-409 (2015), DOI: 10.1007/s12239-015-0041-2
21. K. Shiho, P. Soonseo, K. Sunkook, R. Seok-Ho, Journal of Electronic Materials **40**, 812-816 (2011), DOI: 10.1007/s11664-011-1580-6
22. Y. Iriyama, *4th Thermoelectric Conference Utilizing Waste Heat in Transport and Industry* (Berlin, 2014)
23. M. Cleary, Y. Zhang, A. Mazzotta, H. Young, et al., *4th Thermoelectric Conference Utilizing Waste Heat in Transport and Industry* (Berlin, 2014)
24. S. Kim, S. Park, S. Kim, S. Rhi, Journal of Electronic Materials **40**, 812-816 (2011), DOI: 10.1007/s11664-011-1580-6
25. S. Kumar, S.D. Heister, X. Xu, J.R. Salvador, et al., Journal of Electronic Material **42**, 665-674 (2013), DOI: 10.1007/s11664-013-2471-9
26. S. Kumar, S.D. Heister, X. Xu, J.R. Salvador, et al., Journal of Electronic Materials **42**, 944-955 (2013), DOI: 10.1007/s11664-013-2472-8
27. R. Kühn, O. Koeppen, J. Kitte, Journal of Electronic Materials **43**, 1521-1526 (2014), DOI: 10.1007/s11664-013-2769-7
28. D. Tatarninov, M. Koppers, G. Bastian, D. Schramm, Journal of Electronic Materials **42**, 2274-2281 (2013), DOI: 10.1007/s11664-013-2642-8
29. D. Tatarninov, D. Walling, G. Bastian, Journal of Electronic Materials **41**, 1706-1712 (2012), DOI: 10.1007/s11664-012-2040-7
30. R. Amatya, R.J. Ram, Journal of Electronic Materials **41**, 1521-1526 (2012), DOI: 10.1007/s11664-011-1839-y
31. P. Gao, I. Berkun, R.D. Schmidt, M.F. Luzenski, X. Lu, et al., Journal of Electronic Materials **43**, 1790-1803 (2014), DOI: 10.1007/s11664-013-2865-8
32. J.Y. Jang, Y.C. Tsai, Applied Thermal Engineering **51**, 677–689 (2010), DOI: 10.1016/j.applthermaleng.2012.10.024
33. K.T. Wojciechowski, M. Schmidt, R. Zybala, J. Merkisz, P. Fuc, P. Lijewski, Journal of Electronic Materials **39**, 2034-2038 (2010), DOI: 10.1007/s11664-009-1010-1
34. J. Merkisz, M. Andrzejewski, A. Merkisz-Guranowska, I. Jacyna-Gołda, Journal of KONES Powertrain and Transport **21**, 219-226 (2014)
35. J. Merkisz, M. Idzior, J. Pielecha, W. Gis, *WIT Transactions on the Built Environment* **111**, 181-189 (2010), DOI: 10.2495/UT100171
36. J. Merkisz, P. Lijewski, P. Fuc, S. Weymann, Eksploatacja i Niezawodnosc – Maintenance and Reliability **15**, 364-368 (2013)
37. J. Merkisz, J. Pielecha, *IOP Conference Series-Materials Science and Engineering* **148**, UNSP 012078 (2016), DOI: 10.1088/1757-899X/148/1/012078
38. J. Pielecha, J. Merkisz, J. Markowski, R. Jasinski, *E3S Web of Conferences* **10**, UNSP 00073 (2016), DOI: 10.1051/e3sconf/20161000073
39. J. Merkisz, J. Pielecha, W. Gis, *Proceedings of the Ninth Asia-Pacific International Symposium on Combustion and Energy Utilization* 477-482 (2008)

Particle number and particulate mass emissions of heavy duty vehicles in real operating conditions

Lukasz Rymaniak[1,*], *Andrzej* Ziolkowski[1], and *Dawid* Gallas[1]

[1]Poznan University of Technology, Faculty of Machines and Transport, ul. Piotrowo 3, 60-965 Poznan, Poland

Abstract. The article investigates the issue of PM emissions from HDV vehicles. The theoretical part discusses the problem of emission of this toxic compound in terms of particle structure taking into account the mass and dimensions of PM. Next, the methodology of the research and the results of the measurements performed under the conditions of actual operation were presented. The test drive routes were chosen in accordance with the operational purpose of the selected test vehicles. Two heavy vehicles were used for the study: a tractor with trailer and an eighteen meter long city bus. The test vehicles complied with the Euro V standard, with the second vehicle additionally complying with the EEV standard and being equipped with a DPF. The analysis of the research results was performed in the aspect of determining the operating time densities of vehicles and their drive systems as well as defining their emission characteristics and ecological indicators. PM and PN emissions were measured in the tests and particle size distribution was determined. It was shown that the exhaust gas aftertreatment system used in the city bus had a positive influence on the ecological indicators and had contributed to the reduction of PN emissions for heavier particles.

1 Introduction

The automotive industry represents a significant share of the global CO_2 emissions. All internal combustion engines manufactured for vehicles must meet specific emissions requirements. Depending on the region, there are different standards for the type approval and limit values of emissions of particular harmful substances [1]. With the introduction of the Euro 5b norm for passenger cars (PC) and Euro VI for heavy-duty vehicles (HDV), the control of particulate matter (PM) emissions has been extended to include their numbers – indication: particle number (PN) [2]. PM emissions are a very serious problem due to their composition and toxicity. For example, in terms of size, it was found that particles with a diameter greater than 10 μm accumulate in the upper respiratory tract, whereas particles with a diameter of 2.5 μm enter the deeper parts of the lungs, where they are difficult to excrete and begin to accumulate [3-5]. It is therefore imperative to use construction and design solutions for vehicles that will have a beneficial effect on reducing the emissions of this compound both in terms of mass and number [6].

In the European Union the Euro standards apply to both PC and HDV. In the case of city buses, the standard expects engine dynamometer tests to be performed for the European Transient Cycle (ETC), European Load Response (ELR) and European Stationary Cycle (ESC) for Euro III–V norm and the World Harmonized Transient Cycle (WHTC) and World Harmonized Stationary Cycle (WHSC) for Euro VI. These tests relate to heavy duty vehicles, but the measurement procedures are performed for the internal combustion engines themselves in laboratories. It should be noted, however, that the results of laboratory tests may vary considerably from the actual emissions from a HDV, especially if the vehicle is equipped with a hybrid drive, for example. Hence, research in real operating conditions is being used increasingly more often. In this type of measurement it is possible to evaluate the emission of pollutants in a much larger range of engine operating parameters variability [7-10]. For HDVs, road tests are already necessary in terms of checking the emissions compliance of vehicles already in service [11]. On the other hand, for the PC type vehicles, real driving emissions (RDE) will appear in the type approval procedures starting on 1st of September 2017 [12, 13].

2 Causes of PM emissions

The process of PM formation is complicated and consists of many complex chemical and physical reactions, often occurring simultaneously. Decomposition of fuel into aromatic hydrocarbons or acetylene is assumed to be one of the main processes involved in PM formation. The basic building block of PM is the soot, which is formed from said fuel decomposition products. Among the main causes of soot formation in the engine combustion chamber are: locally too rich fuel mix ($\lambda < 1$), incorrect fuel injection parameters, low combustion temperature and cold engine start. The direct causes are: the temperature distribution, the concentration of substrates and combustion products, the pressure values, and the kinetics of the chemical reactions. Pure carbon can also

* Corresponding author: lukasz.rymaniak@put.poznan.pl

be formed in the cylinder of an internal combustion engine by endothermic CO decay.

The soot's outer structure, its dielectric properties, and the radical properties favor the absorption of various kinds of particles/impurities. In the field of internal combustion engine research PM_{TOTAL} is defined as the organic or inorganic products collected on a filter through which the stream of diluted exhaust gas passes at 52 ±3 °C [14]. PM_{TOTAL} is divided based on the process of its formation into two basic fractions: PM_{SOF} and PM_{INSOL}. The first is defined as a soluble fraction, while the second one is insoluble, the primary component of which is carbon resembling the structure of graphite, PM_{INSOL} also contains water associated with sulphates, nitrates, metals and other carbon containing compounds. In addition, metals produced by the engine abrasive effects may appear in their composition. Metal particles from the exhaust aftertreatment system – catalyst particles, can be particularly dangerous. The PM_{SOF} soluble fraction consists of compounds derived from fuel PM_{FUEL}, formed by the combustion of an overly rich mixture, flame extinguishing process and lowering the combustion temperature due to expansion. This fraction also contains the PM_{LUBE}, that refer to components derived from the lubricating oil.

All of these causes of PM formation directly or indirectly affect the shape, structure, and properties of the created PM. The use of different technical solutions in modern compression ignition (CI) engines, the chemical composition of fuel and lubricants, power systems, supercharging and turbulence in the cylinder led to the formation of different particulate structures [15]. In terms of modern solutions used in the construction of internal combustion engines and vehicles one should consider their morphology in two aspects. The first one covers the most commonly considered physical parameters, i.e. the mass and the number (Fig. 1). On the other hand, the second aspect concerns the chemical parameters – the structure of the particle and its composition.

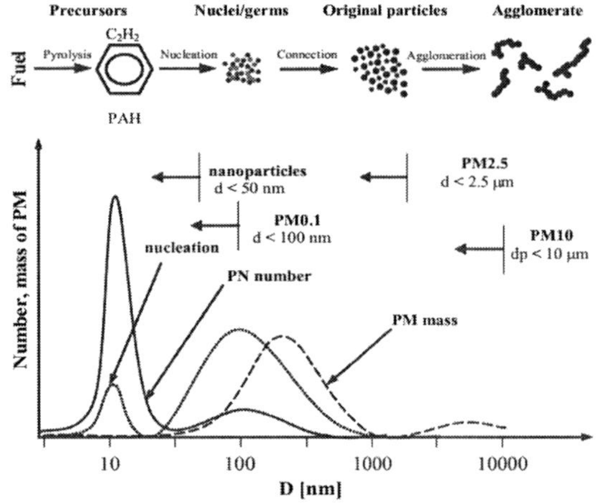

Fig. 1. The assumed pattern of particle formation along with size distribution [16, 17].

3 Structure of PM

Divergence in the number of particles can be analyzed mainly in terms of engine performance and vehicle performance. The mass of particulates is a parameter whose analysis is more difficult to perform. This is mainly due to the differences in particle structures that occur directly affecting their density and mass (Fig. 2). Measurement equipment usually assumes density of 1 g/cm^3 as a default value and a spherical shape of particulate matter. It is known, however, that these parameters can significantly differ in reality – apart from the size (diameter), solid particles are characterized by their volume and mass. For particles of spherical shapes, the physical significance of these parameters is known and is defined as the volume or mass of the material from which the particle is made, with and without considering internal voids.

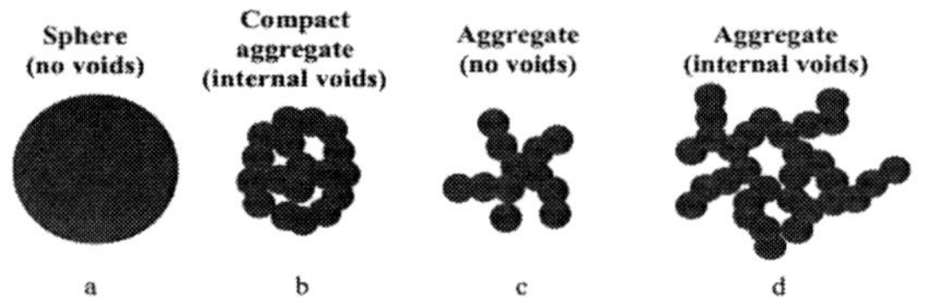

Fig. 2. Solid particle structures resulting from physicochemical formation characteristics and simplifications taken for measurement [18].

The volume of spherical particles and PM agglomerates without voids is defined with the formula (Fig. 2a, c) [18]:

$$V_{cs} = \frac{\pi}{6} \cdot d_{cs}^3 \tag{1}$$

where: V_{cs} – spherical volume of PM or agglomerate concentrated without voids, d_{cs} – diameter of solid particle without voids.

For spherical particles and agglomerates with voids, in order to determine the actual volume of PM (Fig. 2b, d), the volume of the voids should be taken into account [18]:

$$V_{pz} = \frac{\pi}{6} \cdot d_{pp}^3 = V_{cs} + V_{pp} \tag{2}$$

where: V_{pz} – spherical volume of PM or agglomerate with voids, V_{pp} – total volume of void spaces in PM, d_{pp} – PM diameter taking into consideration internal voids.

The mass m_{cs} describing PM can be considered according to the accepted definition of density. The first is the density of solid matter ρ_{cs} without voids:

$$m_{cs} = V_{cs} \cdot \rho_{cs} = \frac{\pi}{6} \cdot d_{cs}^3 \cdot \rho_{cs} \tag{3}$$

where: ρ_{cs} is density considering the solid and liquid phase:

DOI: 10.1051/matecconf/201711800025

$$\rho_{cs} = \frac{\rho_a \cdot V_a + \rho_b \cdot V_b + \ldots}{V_a + V_b + \ldots} \tag{4}$$

where: $\rho_a \cdot V_a + \rho_a \cdot V_a + \ldots$ – PM solid and liquid phase characterizing parameters.

The second definition uses density with voids ρ_{pz}:

$$m_{cs} = V_{pz} \cdot \rho_{pz} = \frac{\pi}{6} \cdot d_{pp}^3 \cdot \rho_{pz} \tag{5}$$

where: ρ_{pz} – PM density with voids:

$$\rho_{cs} = \frac{\rho_a \cdot V_a + \rho_b \cdot V_b + \ldots}{V_a + V_b + \ldots} \tag{6}$$

Assuming that the PM does not contain empty spaces in its structure, then the values are equal $\rho_{cs} = \rho_{pz}$. In the paper by Baron at al. [19] assumed density inequality $\rho_{cs} > \rho_{pz}$ and determined the fraction of PM volume taken up by void spaces:

$$\beta = \left(\frac{\rho_{cs}}{\rho_{pz}} \right)^{1/3} \tag{7}$$

A defined Baron coefficient allows to describe the relation between spherical or agglomerate diameters with and without empty spaces:

$$d_{pp} = \beta \cdot d_{cs} \tag{8}$$

In the research field of PM distribution, the value is described using the diameter d_a defined as the diameter corresponding to the electrical charge of particle. For spherical particles and concentrated agglomerates it is described by:

$$d_a = d_{cs} = d_{pp} \tag{9}$$

while the diameter d_a for PM in the form of dispersed agglomerates is described by:

$$d_a > d_{pp} \tag{10}$$

The size of the initial particles, their formation and the amount of voids all determine the PM structures. This has a direct effect on the diameter distribution and this also applies to density differences:

$$\rho_{cs} > \rho_{pz} \tag{11}$$

Such relations directly affect the mass and the course of PM decomposition. The biggest differences in measurements concern the expanded particle agglomerates with internal voids, isolated from the atmosphere of the medium in which the particle is located. In this case, the measuring apparatus identifies the compounds to be measured as large diameter particles but having a small mass. This results in different shapes of the characteristics of raw exhaust gas particle distribution curve and the distribution curve after chemical treatment in exhaust gas aftertreatment systems. It should be noted that sintering of particles has a significant effect on the PM agglomerate mass distribution after chemical treatment in the exhaust gas aftertreatment system. This especially applies to agglomerates with a dispersed structure, leading to a reduction in the volume ratio of PM to its density. Another phenomenon is agglomeration and the reshaping of agglomerates as a result of collisions. This process occurs both in the exhaust system and in the ducts and open pores of the particulate filter walls. This is another reason for the decrease of the PM volume to mass ratio.

In literature, the soot emitted from CI engines is mainly analyzed according to three criteria: the type of the engine, the emission standards of the exhaust gas components that the engine meets and the parameters of its operation (Fig. 3). The soot breakdown according to the second mentioned criterion is mainly attributable to a change in the version of direct fuel injection systems. This is about changing the parameters of injection pressure, droplet size and dose distribution. It also highlights the optimization of the combustion process as well as the use of modern exhaust gas treatment systems.

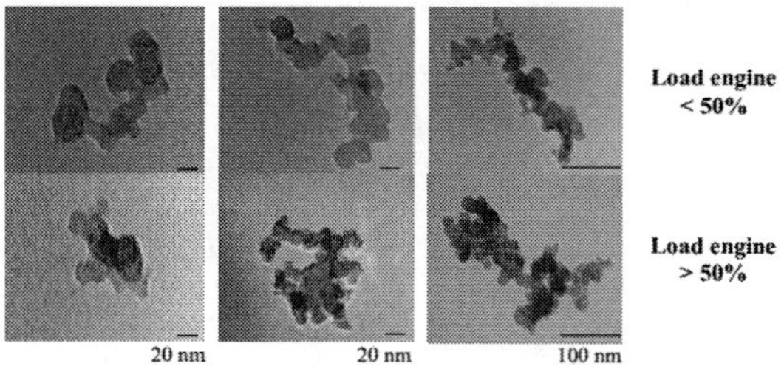

Fig. 3. TEM images of various soot aggregates emitted at different loads from a heavy duty vehicle engine [20].

3 Research methodology

3.1 Tested vehicles

Two HDVs were used for the measurements. The first one – designated as vehicle A – was an articulated lorry designed for long haulage of goods with a maximum authorized mass of over 16,000 kg (Tab. 1). It was equipped with a V8 engine with displacement of 15.6 dm^3. It generated a maximum output power of $P =$ $= 412$ kW (560 hp) at 1900 rpm and maximum torque of $T = 2700$ N·m in the engine speed range of 1100–1400 rpm. The engine complied with the Euro V emission standard and was equipped with an selective catalytic reduction (SCR) system. During the measurements, vehicle A was loaded with a load of 25,000 kg. The vehicle was also fitted with a driver monitoring system. The system, thanks to a continuous analysis of the signals from a series of sensors, provides real time hints and, at the end of the trip, generates a report on the driving style. The hints and the evaluations are presented and there are 4 categories: driving uphill, predicting, braking and gear shifts. The idea behind the system is to improve the driving skills in terms of fuel consumption and proper use of modern solutions such as: automatic transmission, retarder or the electronic braking system (EBS).

The second vehicle (B) was an 18 meter long city bus. As a source of propulsion, it used an 6-cylinder in-line internal combustion engine with a displacement of 9.2 dm^3 generating a maximum power of P = 231 kW at 1900 rpm. It complied with the EEV (enhanced environmentally friendly vehicle) emission standard and was equipped with a continuous regeneration trap (CRT) and SCR catalytic system. The first of these is the combination of a diesel particulate filter (DPF) and an oxidation catalyst (DOC). The vehicle was loaded during the measurements with a load corresponding to the maximum number of passengers it can carry.

Table 1. Tested vehicles technical parameters.

Vehicle A – tractor with trailer	
	Displacement: 15.6 dm^3 Maximum power: 412 kW @ 1900 rpm Maximum torque: 2700 N·m @ 1000–1400 rpm Emission standard: Euro V (SCR)
Vehicle B – urban bus	
	Displacement: 9.2 dm^3 Maximum power: 265 kW @ 1900 rpm Maximum torque: 1450 N·m @ 1100–1700 rpm Emission standard: EEV (CRT and SCR)

3.2 Test routes

For the vehicle A authors selected a road portion of the length of 27 km (Tab. 2). The chosen route characterizes the operation of vehicles accurately of the GVW (gross vehicle weight) exceeding 16,000 kg (long haulage) in the area where the measurements were carried out. The test route started and ended in the industrial zone (point A) where a production facility is located at which approximately 50 heavy-duty vehicles are handled daily. The test route can be divided into two portions: a drive on the municipal roads (portion A–B) and national and regional roads. The drive on the national regional/roads depends on the driving direction from/to the entrance to the A2 expressway (Koło) (point D). In the case of driving to the 'Koło' expressway entrance to the route went through the sections B–C and C–D. In the reverse situation i.e. exiting the expressway and driving to the production facility via bypasses: sections D–C and C-B (on the D–C road portion heavy duty trucks exceeding GVW of 7000 kg are not allowed.) The above route can be deemed representative of the national transport and logistic infrastructure – representing the road infrastructure and the distribution of production facilities in small and medium-sized towns. For vehicle B, the route was dedicated to public transport buses and was a regular bus route in Poznan. In the hours when the study was conducted, it was one of the busiest lines serviced by the municipal communication company in Poznań. It had 46 stops and started in the northern part of town (Sobieskiego). Then ran through the center and ended in the southern part of town (Dębina) near the A2 motorway. Its total length was 16 km.

Table 2. Test routes characteristics [21].

Vehicle A test route	
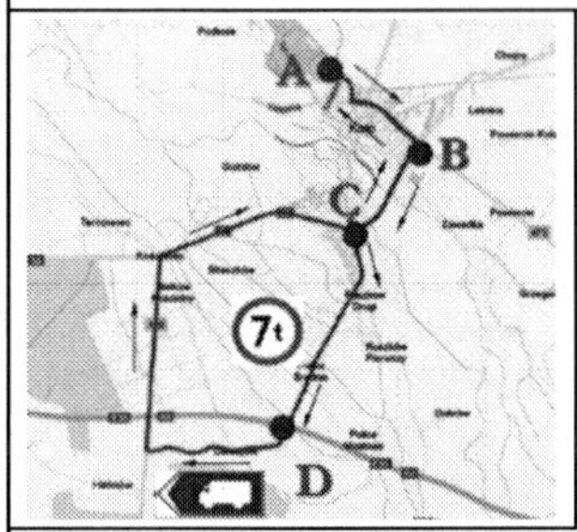	Distance: 27 km Average speed: 45 km/h Top speed: 90 km/h Time idle: 5–10% Characteristics: urban, non-urban, motorway driving
Vehicle B test route	
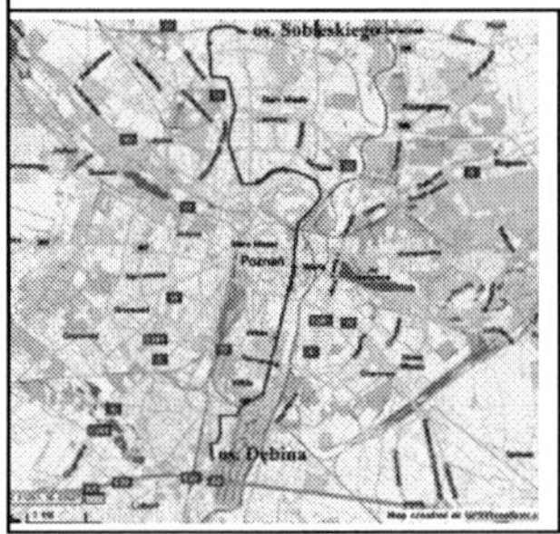	Distance: 16 km Average speed: 18 km/h Top speed: 60 km/h Time idle: 25–35% Number of stops: 46 Characteristics: urban center and main commuting routes

3.3 Measurement equipment

The purpose of the research was to analyze the mass, number and size distribution of emitted particles. For this reason, three instruments from the PEMS group (portable emission measurement system) were used: Semtech DS from Sensors Inc., Micro Soot Sensor from AVL and Engine Exhaust Particulate Sizer from TSI (Fig. 4). The first of these is dedicated to gas measurement of exhaust gas compounds. However, in this research measurement, the Semtech DS was used to measure the exhaust flow rate with the use of a Pitot tube. Also measured were: the speed and position of the vehicle (GPS), engine operating parameters (crankshaft speed and load) and the atmospheric conditions. The Micro Soot Sensor was used to measure the concentration of particulate matter, which was based on the determination of the carbon content of the molecule using the photoacoustic method. The exhaust gas sample is subjected to radiation with modulated light, which results in alternating heating and cooling of the sample, thereby changing its size periodically and generating vibration of the medium. Thanks to the use of sensitive microphones tuned to a certain amplitude and frequency, a sound wave is recorded. However, it occurs only when there is a particulate in the sample to be tested – then the impulse is directly proportional to the particulate

DOI: 10.1051/matecconf/201711800025

concentration. Otherwise the microphones will not register any signal. For correct operation of the Micro Soot Sensor, it is necessary to dilute the flue gas sample to prevent condensation of the solids and to maintain their proper temperature.

Measurement of the number and dimensional distribution of particle diameters was done using an Engine Exhaust Particulate Sizer, which measures particles in the range of 5.6–560 nm. First, the exhaust sample is filtered from particulate matter larger than 1 μm in diameter. It is then directed to the charging electrode where the ions are generated. This combination of particles and electrical charge allows to classify particles by their size. The electrode is located at the very top of the device. The particles then enter the ring gap, which is the space between the two concentric cylinders. The gap is surrounded by a stream of clean air from the outside. The outlet cylinder is built in the form of a stack of very sensitive electrodes arranged in a ring in such a way that they are isolated from one another. The electrodes on one side are connected to a sensitive amplifier and grounded on the other. Such a connection creates an electric field between the cylinders made up of the electrodes. A mixture of fresh air and solid particles, which are charged positively, are repulsed from the high voltage electrode and passed between the sensitive electrodes. Particles striking the electrodes generate a current that is read out in real time by a driver with ten readings of different particle sizes per second. Example of use PEMS in research are presented in [2, 7, 9, 10, 13, 22].

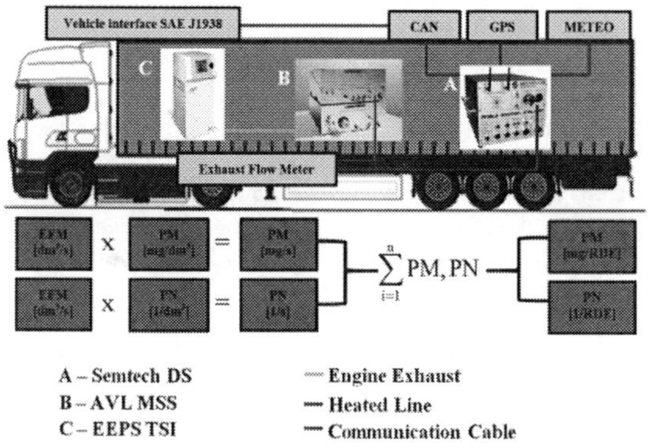

Fig. 4. PEMS type equipment used for the research.

As shown in the diagram depicted in Fig. 4, the values of the mass and number concentration of particles were multiplied by the exhaust gas flow, thereby obtaining the streams of the their volumes. After adding them, the mass and number of particles from the entire RDE test were obtained. By referring these values to the road traveled by the vehicle and the work done by the engine, the road and specific particulate matter mass and particle number were obtained. Particle size distribution is presented as the sum of the number of particles in the range of 5.6–560 nm.

4 Research results

4.1 Parameters of researched vehicles in road tests

The time-density characteristics (TD) are commonly used in the construction and optimization work done on internal combustion engines and entire propulsion systems. Using their assumptions, it is possible to determine, among others, the emission intensity of harmful exhaust compounds or the internal combustion engine time density in the crankshaft speed ranges and loads in a given measurement tests cycle. Therefore, it is possible to describe the motion of a vehicle as a discrete function. They also allow to describe the test drive characteristics for different speed ranges and acceleration values. The characteristics relating to the engine and vehicle operating parameters were performed for half-closed intervals. This resulted in more accurate analysis. In addition, consideration of traffic conditions is given to the vehicle being idle (speed and acceleration equal to 0), which is very important due to the specificity of the operation of city buses – vehicle B.

Registered data from the GPS positioning system and on-board diagnostic systems were used to develop the internal combustion engines and vehicles operating parameters in road tests (Fig. 5). The tested vehicles are classified as heavy vehicles and are similarly type approved in terms of exhaust emissions. However, the presented relations indicate that their typical operating conditions vary considerably. The heavy-duty vehicle A used for long-distance transport obtained a significant operating time density at constant speeds ($a = 0$ m/s^2). In the vehicle speed interval (16 m/s; 24 m/s⟩ the operating time density of 25.3% was calculated for the entire test. For the idle interval, the operating time density was 2.4%. The operating conditions contributed to achieving similar acceleration value ranges for acceleration greater and smaller than 0 m/s^2. For the city bus, a different characteristic distribution was obtained. The greatest operating time density being 24.3% for vehicle idle interval, which was a consequence of the way the vehicle was operated (i.e. the service of bus stops). For the interval of constant speed (8 m/s; 12 m/s⟩ the time density value of 9% was obtained. The vehicle drive system dynamics had the greatest influence on the distribution of acceleration greater than 0 m/s2. For acceleration in the intervals ⟨–0.8 m/s^2; 0 m/s^2) and ⟨0 m/s^2; 0.8 m/s^2) in the vehicle speed interval (0 m/s;10 m/s⟩ the operating time density was 16% and 23.7% respectively. The maximum speeds achieved by the vehicles in the test were: for vehicle A – 23.3 m/s and for vehicle B – 15.6 m/s.

Data obtained from the researched vehicles was analyzed with consideration given to the internal combustion engines operating parameters. For vehicle A the largest operating time density was 43.2% and 25.9% in two main intervals. The first of those occurred in the interval described by the speed parameters (1000 rpm; 1200 rpm⟩ and engine load of ⟨0%; 80%⟩. The second one was recorded at the lowest engine loads in the

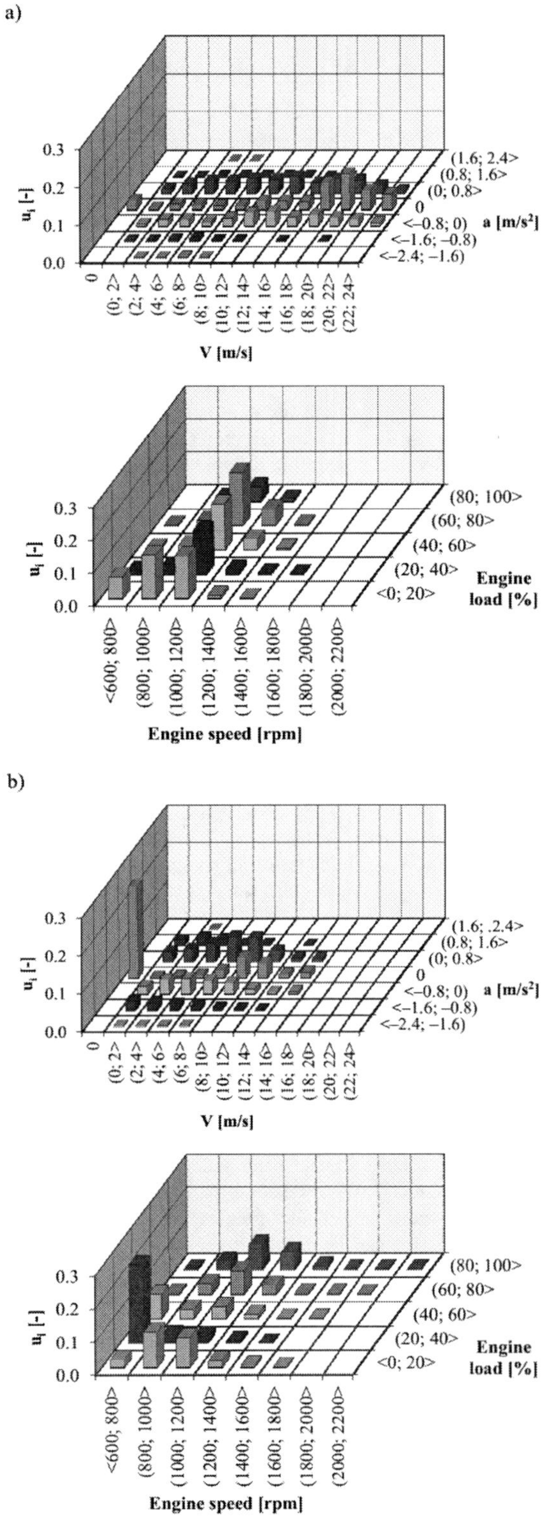

Fig. 5. Operating parameters of the tested vehicles in intervals of: speed – acceleration of vehicle, and crankshaft speed – engine load: a) vehicle A, b) vehicle B.

engine speed interval (800 rpm; 1200 rpm⟩. In terms of idle engine speed and minimum load, the obtained time density was 6.5%. The remainder of the individual intervals yielded no more than 4%. The internal combustion engine of vehicle B had a much greater variation in its operating parameters, with the exception of one of the intervals characterized by the lowest crankshaft rotational speed and load interval of (20%; 40%⟩.

The time density in the aforementioned interval is 24%. This value is due to the operating nature of the city bus – the high proportion of idle operation connected with the servicing of bus stops and traffic congestions led to a constant load on the engine. It should be noted that this is highly dependent on the number of functional systems used in the vehicle (such as: air conditioning, heating, pneumatic handling of doors, vehicle suspension, etc.). In other single intervals the time density did not exceed 5%.

4.2 Particle emissions in mass, number and size distribution in road tests

PM emissions from modern combustion engines are a significant problem. In both CI and SI engines with direct injection systems and devices are being used increasingly more often to reduce the emission of this toxic compound in terms of mass and number. Among the most important of these solutions are high-pressure fuel injection with dose distribution and particulate filters. The first of these is a structural design that directly affects the combustion process. Very small drops evaporate more rapidly, resulting in a reduction of soot formation. The second solution allows the effective reduction of the PM emissions generated in the combustion process. However, the use of filters requires the ability to regenerate them and in such a way that will be as little a nuisance to the vehicle user as possible. This is a difficult issue that requires taking many aspects into account (the quality of the regeneration has a significant effect on the filter life).

The summary of the PM emissions per second from the road tests as a function of the engine performance is shown in Fig. 6. For vehicle A, a close dependence of the PM emissions on the engine load is evident. The effect of engine speed is only relevant for the first load range up to 20%. For the most common operating parameter range, described by the following parameters: (1000 rpm; 1200 rpm⟩ and ⟨0%; 80%⟩, PM emissions average 4.4 mg/s. This is a small result compared to other intervals. For the highest load interval, the average PM emission is 13 mg/s. Vehicle A was not equipped with a DPF system and therefore a higher average PM emission of 6.5 mg/s was achieved for the entire road test than for vehicle B.

The DPF filter used in the city bus contributed to an effective mass reduction of PM emissions, the average emission was 0.5 mg/s. In this case, there is another tendency of the second order dependence on the operating parameters of the internal combustion engine. The emission value mainly depends on the crankshaft

DOI: 10.1051/matecconf/201711800025

rotational speed and to a lesser extent on the engine load. For the maximum speed interval and the load interval of (80%; 100%⟩ PM emission was determined at 1.6 mg/s. For the highest time density interval recorded, the per second emission of the compound was 0.1 mg/s.

a)

b)

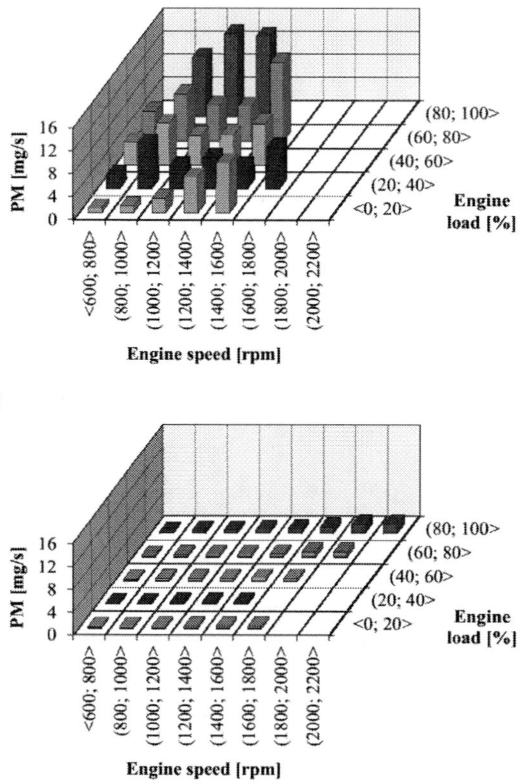

Fig. 6. PM emission intensity in the speed and torque intervals during road tests: a) vehicle A, b) vehicle B.

The road and specific PM emissions of tested vehicles were determined for the carried out drive tests (Fig. 7). The area of the engine operating parameters variability for vehicle A was smaller compared to the city bus (vehicle B). This is advantageous in terms of PM emissions because it largely depends on the contribution from the dynamic operating conditions – frequent changes in engine speed and load. However, the lack of an exhaust aftertreatment system has resulted in higher emissions than for vehicle B as a result. For long-distance heavy vehicle, the road PM emission was 0.39 g/km, while the specific PM emission was 0.15 g/(kW·h). In the tests of the urban bus, the values were lower by 87.2% and 86.6%, respectively. The vehicle B's internal combustion engine worked in a greater range of operating parameters variability, which was due to its operating conditions. However, for the interval where the highest time density occurred, the PM emission rate was characterized by low values. This translated into the obtained results of the road and specific PM emissions.

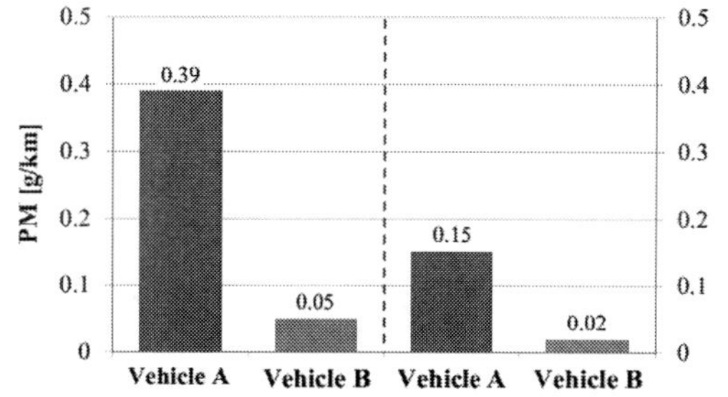

Fig. 7. Comparison of road and specific PM emissions based on road tests.

The vehicle A PN emission rate depended on both the crankshaft rotation speed and the engine load (Fig. 8). Within the interval of the highest time density, the average emission was 2.5E+10 particles per second. The maximum emission rate of PN of 7.3E+10 per second was recorded in the engine speed interval (1400 rpm; 1600rpm⟩ and load (20%; 40%⟩. In the maximum load interval, the average emission rate was 3.4E+10 per second and for the entire operating range reached 2.7E+10 per second. The comparison of the obtained emission characteristic with the mass PM emission rate (Fig 6a) indicates that there is no unambiguous relationship between the PM mass and the number emitted. This conclusion is based on vehicle A in which the PM emission is limited to only through the applied structural engineering in the internal combustion engine itself.

Different relations were obtained using the distribution of the characteristics measured for the city bus. For vehicle B, a similar emission rate of PN and PM was obtained. This means that the performance of the exhaust aftertreatment system used depends primarily on the crankshaft rotational speed. At the operating point with the highest time density, the PN emission rate was 4.2E+8 per second. For the entire road test, the average value was 3.9E+8 per second, and in the highest load interval it was 4.1E+8 per second. The highest PN emission rate 7E+9 per second occurred in two single engine speed intervals of (1400 rpm; 1800rpm⟩ for engine load of (40%; 60%⟩.

Due to the toxic effect of PM on living organisms, it is necessary to be aware of their emission from vehicles including for their size distribution. Nanoparticles are carcinogenic and, according to the environmental protection initiatives, their numbers should be reduced. Fig. 9 shows the total PN emissions with respect to measured particle diameters. However, it should be noted that the test vehicles have travelled different routes, so it is not possible to compare the values themselves, but only the characteristics of the obtained size distributions. For vehicle A the highest PN emission (above 1.9E+12 1/RDE) occurred in the diameter intervals ⟨6 nm, 10.8 nm⟩ and ⟨29.4 nm; 52.3 nm⟩ (Fig. 9). The emission totals in these intervals were: 2.8E+13 1/RDE and 1.4E+13 1/RDE respectively. For the interval between the indicated range, a lower PN emission value was recorded, on average 7.4E+11 1/RDE. For single diameters above 254 nm, the obtained

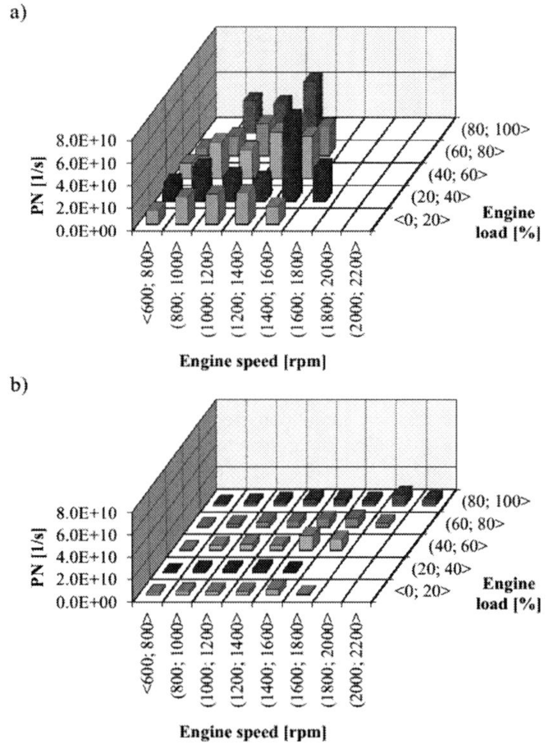

Fig. 8. PN emissions intensity in rotational speed and torque intervals during road tests: a) vehicle A, b) vehicle B.

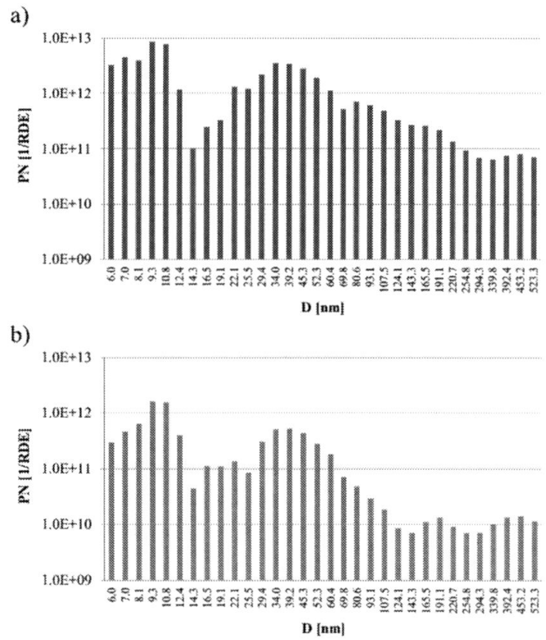

Fig. 9. Summary of PN emission as a function of particle diameters during road tests: a) vehicle A, b) vehicle B.

results were not greater than 7.9E+10. Smaller PN emission values were obtained for each diameter in case of the city bus, however, the characteristic of the size distribution remains very similar. It can therefore be assumed that the equipped filter limits the PN emissions for all the measured particle diameters. For particle diameters up to 10.8 nm over the whole test route, the emission value of 4.6E+12 1/RDE was obtained. The PN emission did not exceed 1.4E+10 1/RDE for diameters greater than 124 nm.

In order to compare the obtained PN emissions for the tested vehicles, road and specific indicators have been identified (Fig. 10). As with the PM mass comparison, the city bus was characterized by more favorable environmental indicators. This was also the result of using a DPF system that effectively reduced the number of particles in the exhaust gases. The road PN emission of vehicle A was 1.9E+12 per km, while the specific PN emission was 7.45E+11 1/(g/kW·h). The second test vehicle achieved results respectively 27% and 24.7% lower than vehicle A. The discrepancies are lower than in the case of mass indicators. This means that the filter used in the city bus effectively reduces the PM emission by weight, so according to the accepted research method, by limiting the larger diameter particles.

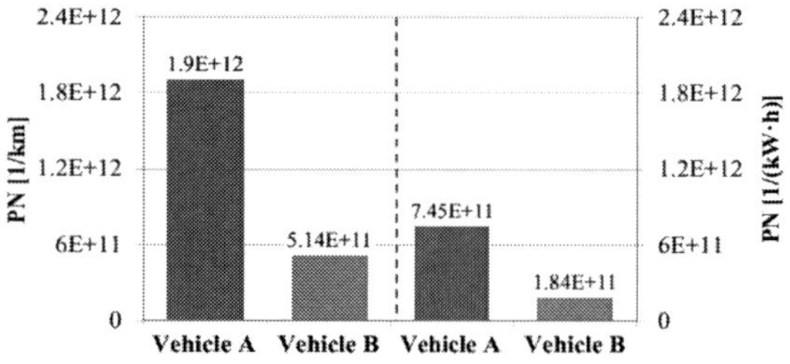

Fig. 10. Comparison of road and specific PN emission based on road tests.

5 Conclusions

The restriction of regulations describing the emission standards for toxic compounds is a major factor in the continuous development of internal combustion engines and exhaust gas aftertreatment systems. In these activities, particular emphasis is placed on limiting NO_x, PM, emissions, as well as CO_2 emission, which is directly related to the vehicle fuel consumption. Within the scope of PM regulation, not only limits on emitted mass are introduced but also on emitted numbers. Effective research on the environmental performance of vehicles is increasingly being used in real driving conditions. They allow for a comprehensive assessment of the pollutant emissions of a given test vehicle in a wide and variable range of engine operating parameters.

The article discusses the causes of PM generation and emission in internal combustion engines. The results of road tests on PM emissions in terms of mass, number and size distribution were also presented. The tested vehicles were both heavy vehicles: a tractor with a trailer and an eighteen meter city bus. The test conditions have been selected in such a way so that they as closely as possible reflect the vehicles real operating conditions.

For this reason, vehicle A was working in a much smaller range of variability than vehicle B. The analysis of the environmental indicators obtained implies that the second test vehicle has obtained more favorable final values. This was due to the use of an exhaust aftertreatment system in the form of a DPF. The road and specific PM emissions for this vehicle were respectively 87.2% and 86.6% lower than for the other vehicle, while the PN emissions were 27% and 24.7% lower. This means that the exhaust aftertreatment system effectively reduces the emission of higher mass particles, those with a larger diameter (according to the accepted methodology). Particle size distribution analysis for PN indicates that the obtained results have similar characteristics for both test vehicles. The determined relations, supplemented with data of the test conditions, can be very useful for carrying out various types of further research and development work. This includes the modification of the exhaust aftertreatment systems components as well as the control of combustion engines operation.

The research was funded by the National Centre for Research and Development (Narodowe Centrum Badań i Rozwoju) – research project within the INNOTECH Programme (contract No. INNOTECH-K2/IN2/36/182269/NCBR/12).

References

1. J. Merkisz, J. Pielecha, S. Radzimirski, *Springer Tracts on Transportation and Traffic* **4**, 1-170 (2014), DOI: 10.1007/978-3-319-02705-0
2. J. Merkisz, J. Pielecha, *Springer Tracts on Transportation and Traffic* **8**, 131-139 (2015), DOI: 10.1007/978-3-319-15928-7_10
3. Z. Stepien, J. Czerwinski, P. Comte, S. Oleksiak, Energy & Fuels **30**, 7268-7276 (2016), DOI: 10.1021/acs.energyfuels.6b00583
4. M. Pośniak, Medycyna Pracy **54**, 389-393 (2003)
5. R. Jasinski, J. Pielecha, J. Markowski, E3S *Web of Conferences* **10**, UNSP 00030 (2016), DOI: 10.1051/e3sconf/20161000030
6. A. Suchecki, J. Nowakowski, T. Knefel, *Transport Means – Proceedings of the International Conference*, 194-197 (2014)
7. J. Merkisz, J. Pielecha, P. Lijewski, A. Merkisz-Guranowska, M. Nowak, *WIT Transactions on Ecology and the Environment* **174**, 27-38 (2013), DOI: 10.2495/AIR130031
8. J. Szczepaniak, T. Pawlowski, R. Grzechowiak, J. Kromulski, J. Rutkowski, Applied Engineering in Agriculture **31**, 709-716 (2015)
9. J. Merkisz, J. Pielecha, *Springer Tracts on Transportation and Traffic* **8**, 69-92 (2015), DOI: 10.1007/978-3-319-15928-7_7
10. J. Merkisz, P. Lijewski, P. Fuc, S. Weymann, Eksploatacja i Niezawodnosc – Maintenance and Reliability **15**, 364-368 (2013)
11. Commission Regulation (EU) No 582/2011 of 25 May 2011 implementing and amending Regulation (EC) No 595/2009 of the European Parliament and of the Council with respect to emissions from heavy duty vehicles (Euro VI) and amending Annexes I and III to Directive 2007/46/EC
12. Commission Regulation (EU) 2016/427 of 10 March 2016 amending Regulation (EC) No 692/2008 as regards emissions from light passenger and commercial vehicle (Euro 6)
13. J. Pielecha, J. Merkisz, J. Markowski, R. Jasinski, *E3S Web of Conferences* **10**, UNSP 00073 (2016), DOI: 10.1051/e3sconf/20161000073
14. O.R. Ignatov, A.P. Polivyanchuk, Combustion Engines **148**, 48-52 (2012)
15. J.O. Müller, D.S. Su, R.E. Jentoft, J. Kröhnert, F.C. Jentoft, R. Schlögl, Catalysis Today **102**, 259-265 (2005)
16. P. Fuc, L. Rymaniak, A. Ziolkowski, *WIT Transactions on Ecology and the Environment* **174**, 207-218 (2013), DOI: 10.2495/AIR130181
17. H. Richter, J.B. Howard, Progress in Energy and Combustion Science **26**, 565-608 (2000)
18. P.F. Decarlo, J.G. Slowik, R. Douglas, P. Davidovits, J.L. Jimenez, Aerosol Science and Technology **38**, 1185-1205 (2004)
19. T.L. Barone, A.A. Lall, J.M.E. Storey, G.W. Mulholland, V.Y. Prikhodko, J.H. Frankland, J.E. Parks, M.R. Zachariah, Energy & Fuels **25**, 1978-1988 (2011)
20. A. Soewono, S. Rogak, Aerosol Science and Technology **45**, 1206-1216 (2011)
21. http://gpsvisualizer.com (access: 24.04.2017)
22. P. Lijewski, J. Merkisz, P. Fuc, Croatian Journal of Forest Engineering **34**, 113-122 (2013)

MATEC Web of Conferences **118**, 00026 (2017) DOI: 10.1051/matecconf/201711800026
VII International Congress on Combustion Engines

Comparison of exhaust emission on the basis of Real Driving Emissions measurements and simulations

Mateusz Nowak[1,*], and *Jacek* Pielecha[1]

[1]Poznan University of Technology, Faculty of Machines and Transport, ul. Piotrowo 3, 60-965 Poznan, Poland

Abstract. Designing of modern transport systems involves the need to meet a large number of requirements. The influence of designed road infrastructure on the environment is very wide and important. The most valid aspect in this case is the reduction of emissions of harmful compounds by increasing the fluency of vehicles flow and building collision free road intersections. But it should be started from establishing the initial emission level of harmful compounds. This paper presents a methodology for determining exhaust emissions from vehicles moving on the national road no. 50 in area of Zyrardow. Modern measuring tools such as the PEMS and the microscopic road simulation software, using the application to determine exhaust emissions, were used for this purpose.

1 Introduction

Important issues in case of the impact of automotive on the environment are traffic congestion, its fluency and the share of heavy vehicles. In 2013, motor vehicles in Poland had a significant share in pollutant emissions [1]:

- 32% of nitrogen oxides,
- 22% of hydrocarbons,
- 20% of carbon monooxide,
- 19% particulate matter.

Air pollution caused by fuel combustion occurs mainly in large urban agglomerations where there is heavy traffic and there are a lot of buildings which prevents good ventilation. This problem may rise, as it is expected, that between 2015 and 2020 the traffic volume of vehicles will increase about 15%. This problem can be solved, for example, by introducing vehicles that meet stringent emission standards. However, sudden change of the entire fleet of vehicles is impossible. For this reason, many solutions need to be introduced at the same time, which will enlarge the possibility to achieve the stated goal. The directions of development of road infrastructure and transport systems have been included in the relevant documents: European Transport Policy [2, 3] and Transport Policy of the Country 2006-2025 [4].

Despite the fact that in Poland in 2007-2015 more than 200% of the increase in motorway and expressway networks was registered, national road infrastructure still requires development and modernization expenditures. New ventures significantly increase the capacity of national roads and meet citizen' expectations for comfortable, fast and secure communication. The main objective of the National Road Construction Program for 2014-2023 (with a view to 2025) adopted in 2015 is to build a coherent and modern national road system to ensure the efficient functioning of road and passenger transport. The program is diagnosing the state of the current road sector, defining both the objectives to be achieved and the key areas constituting the so called "Bottlenecks" in passenger and freight transport. The document assumes the completion of the construction of expressways and motorways and the construction of dozens of bypasses in national roadways. In addition, the Program includes a list of investment tasks related to the improvement of traffic safety so called: Dangerous Places Elimination Program.

The implementation of the projects planned in the above mentioned documents allows many places to shift transit traffic from city centers to newly built ring roads and improve the smoothness of car communication. Investments in the construction of bypasses allow about 90% of all transit traffic of cars and trucks outside the city to be delivered in large cities. Traffic in cities without ring roads represents a very serious burden on road infrastructure, while at the same time being a source of significant noise and air pollution. However, the construction of the ring road is not limited to the construction of a new single-lane or two-lane road, for example. It consists also the construction of viaducts over the ring road, intersections and access roads, as well as solutions that improve traffic safety and provide better environmental protection. In Poland in 2014-2018, 12 new bypasses are planned to be built in the most sensitive transit points (Fig. 1).

To accurately assess the exhaust emission of a single vehicle and the entire traffic stream, existing and new designed road infrastructure, there should be the appropriate method selected. The accuracy of the chosen method affects many aspects. Most important from the perspective of reliability of the assessment are two issues: the degree of simplicity adopted in emission modeling and sources of emission data. In the second

* Corresponding author: mateusz.s.nowak@put.poznan.pl

case, they may be: homologation data, emission factors as a function of the mass of consumed fuel, road measurements and simulation evaluations of road infrastructure [5, 6]. Increasingly popular are the Real Driving Emissions measurements of vehicles in which actual emissions of harmful emissions are known. The performance of the vehicle's driving units in bench tests is often understated and does not correspond to those used in road conditions, which contributes to the false image of vehicle exhaust emission. Computer simulations are also gaining popularity, which has several advantages, including the ability to change arbitrary the traffic density and parameters of the road infrastructure, limited only by the functionality of the software, and time savings compared to traditional methods.

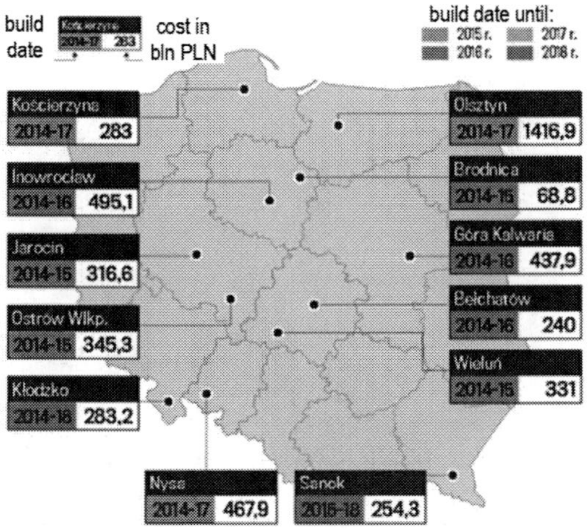

Fig. 1. Location of new bypasses in Poland in 2014-2018 [7].

This article compares the exhaust emission factors for the selected road infrastructure section obtained during RDE (Real Driving Emissions) and computer simulation using specialized software. Until now the authors obtained results of gaseous compounds. It is well known that the particulate matter also is very harmful. In future the authors will determine the influence of road infrastructure on particulate matter emission. This aspect needs to be considered very precisely, because there are some important cases that must be taken into consideration, as: mass and number evaluation or particulate filter regeneration [8-10].

2 Research methodology

The issue described later in the article concerns National Road No. 50 within Zyrardow (Fig. 2). At present, DK 50 is run by the Zyrardow bypass, but all of the performed evaluations have been applied to the situation when all transit traffic was carried out by the town. Due to the high daily traffic (about 17,000 vehicles per day) and the urban nature of the road, the traffic was intermittent often by stops. This situation is confirmed by pictures from the camera recording the image in front of the research vehicle (Fig. 3).

In order to reduce the impact of the driver's driving style on the obtained results, the study was conducted in the form of driving behind the preceding vehicle, trying to maintain its speed, acceleration, thus adopting its driving style. Such a methodology is commonly used and is called driving "behind the leader". The beginning and the end of registration of the measured values fell to the point of connection of the bypass and the national road no. 50. A lot of attention was paid to repetition of the measurements, so 10 road trips were performed during the road tests (5 passes in both directions). Time of measurements: 10:58-14:55. The ambient temperature was 23-29 °C.

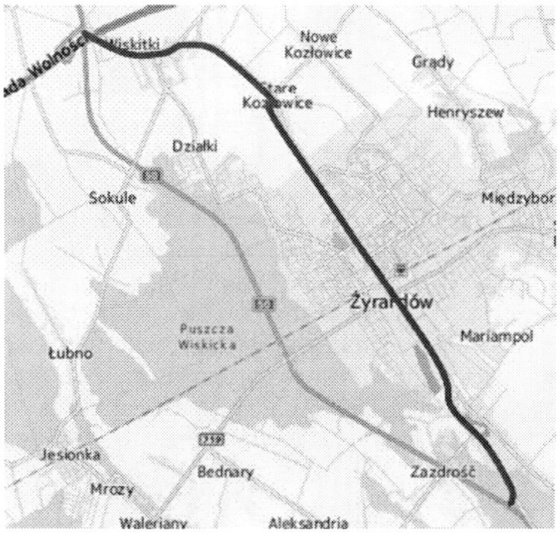

Fig. 2. Research route (using [11]).

Fig. 3. Situation on the road during RDE measurements.

Passes during road tests were carried out by a passenger car, driven by a 1.8 liter supercharged SI engine (Table 1). This vehicle complied the Euro 4 emission standard. The exhaust aftertreatment system is typical for SI engines with multi-point injection, a three-way catalytic converter.

Table 1. Research vehicle.

Vehicle type	Passenger car
Engine type	SI, 4-cyl.
Emission standard	Euro 4
Max. power/max. torque	130 kW/250 Nm
Aftertreatment system	three-way catalytic converter
Mileage	80,000 km
Curb weight	1430 kg
Load	300 kg

Exhaust emission measurements were conducted using the Semtech DS mobile exhaust gas analyzer system (Fig. 4). This instrument has been designed to measure exhaust emissions in real traffic conditions and meets the requirements of the relevant standards [12-15]. For the concentration of harmful compounds in the exhaust can be refer to the driving characteristics and conditions of the drive system, the Semtech DS device enables communication with the vehicle diagnostic system. It is possible to read data from systems supported by different data protocols, making it compatible with PC, LDV and HDV vehicles. Also a GPS module and a meteorological system for measuring pressure and ambient temperature is included. All of these are recorded at a frequency of 1 Hz, which allows the observation of dynamic changes in the concentration of substances contained in the exhaust gases.

For the Semtech DS, but also for other types of systems used to measure exhaust emissions of gaseous pollutants [16, 17], collected exhaust gases are supplied with a heated line (Fig. 4). This element is maintained at 191 °C to prevent the hydrocarbons condensation in the measuring line. The next step is to filter out the mechanical pollutants, then the exhaust gas is conducted directly into the FID analyzer where the hydrocarbon concentration is measured. Subsequently, the analyzed gases are cooled to 4 °C, followed by nitrogen, carbon monoxide, carbon dioxide and oxygen measurement. After passing the whole circuit, exhaust gases are discharged into the environment. Signals from all analyzers and GPS modules, weather stations, and on-board diagnostic systems go to the control unit.

The computer simulations of the passage through the analyzed road infrastructure were made using PTV Vissim software [18], which is a tool for traffic simulation in the microscopic scale [19-21]. The advantage of this system is the ability to accurately design the geometry and parameters of the modeled road section and traffic [22]. The programmed software has extensive databases, including traffic models for different vehicle groups such as passenger cars, trucks, buses, motorcycles. In addition, the expanded version contains data to simulate the exhaust emission of pollutants from SI and CI vehicles that meet current and previous emission standards. The software has implemented a VERTIS+ emission models [23-25] developed jointly with TNO (Netherlands Organization for Applied Science Research) from the Netherlands. This kit is used to forecast emission indicators and energy usage rates that are representative for vehicle fleets in different countries. Emission factors vary by vehicle type and traffic situation. The VERSIT+ emission database is based on information from 12,000 different driving cycles including road test results. They are generalized for vehicles and may vary from specific vehicle groups.

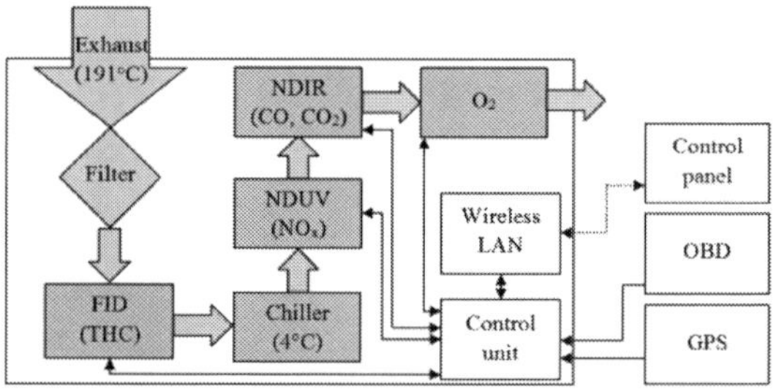

Fig. 4. Functional diagram of the Semtech DS exhaust emission device.

During the analysis in PTV Vissim software, it is possible to view vehicles in the network, for example, from the perspective "from bird's eye view" or from the perspective of the selected vehicle on the network (Fig. 5). At any time of simulation, it is also possible to preview selected parameters of any vehicle, such as speed, position in the network or temporary exhaust emissions. In addition, current data on vehicle traffic on the analyzed road infrastructure was used in the simulations [26].

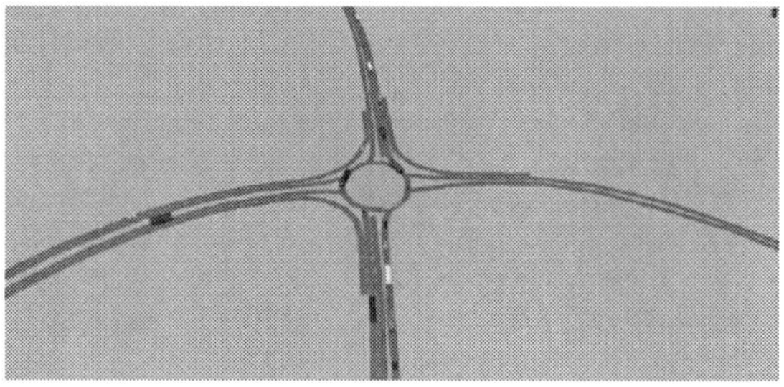

Fig. 5. View of selected part of modeled infrastructure in PTV Vissim.

3 Research results

3.1 Road tests

The characteristics of rides by analyzed section of road infrastructure (in the north-south direction – referred to as 1-x and in the south-north direction – designated as 2-x): time, distance and average speed of each passage are shown in Fig. 6–8. Values of these parameters were recorded at 1 Hz using GPS data, then summed and calculated to determine the average speed. From the presented data it follows that the travel time from north to south was from 1086 s to 1209 s (at an average of 1137 s, so the scatter was from 51 s less to 72 s more), whereas in the opposite direction time was definitely longer (for ex. due to the higher traffic volume) and was from 1461 s to 1594 s. The distance between the research trail in successive passes was about 13.7 km. In this respect, high repetition was also achieved, because for urban journeys distance differences are only 0.39 km, which is about 2% of the smallest recorded value. The average speed of passing through the city was 38.5 km/h.

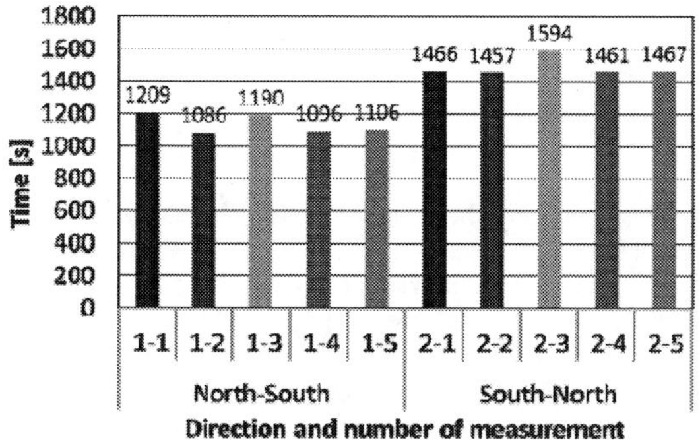

Fig. 6. Drive time during RDE measurements.

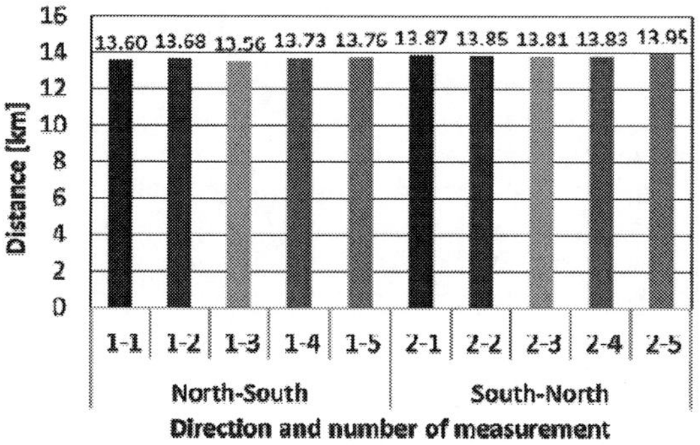

Fig. 7. Travelled distance during RDE measurements.

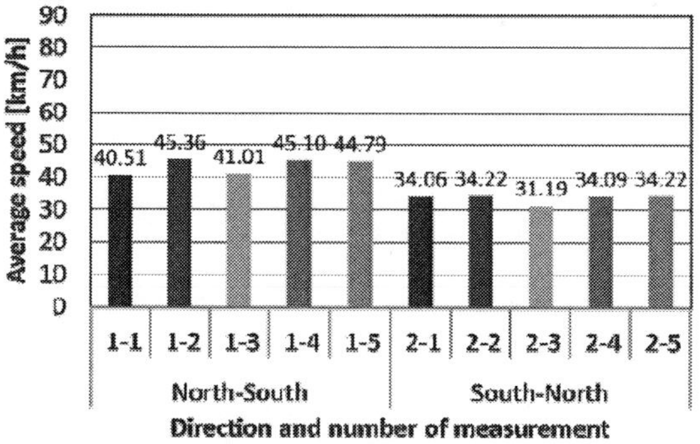

Fig. 8. Average speed during RDE measurements.

At the next stage of the analysis, the course of the masses of harmful compounds was presented in growing way. This was intended to demonstrate not only the final value, but also the changes in individual journeys. The intensity of emissions of particular harmful compounds, summed up every 1 s, allowed for the mass of the harmful compound emitted on the measuring section. The following diagrams show cumulative emission values – the maximum values represent the total mass of the harmful compound determined in each run. Because of space constraints, only values recorded during north-south passes are shown.

The carbon monoxide masses (cumulatively) generated during the passage through the city are characterized by a specific increase in this value in the middle of the route (Fig. 9). This has to do with intersections that occurred on this part of road - and this would require multiple stops and acceleration to a steady speed. All recorded mileage during the measurements was characterized by considerable repetition. The resulting end values largely relied on dynamic traffic conditions.

The mass of the hydrocarbon emitted for all journeys through the center of Zyrardow marked with the symbols 1-x is within 150-200 mg (Fig. 10). The mass distribution of hydrocarbons was small and the graphs were similar to those of carbon monoxide masses.

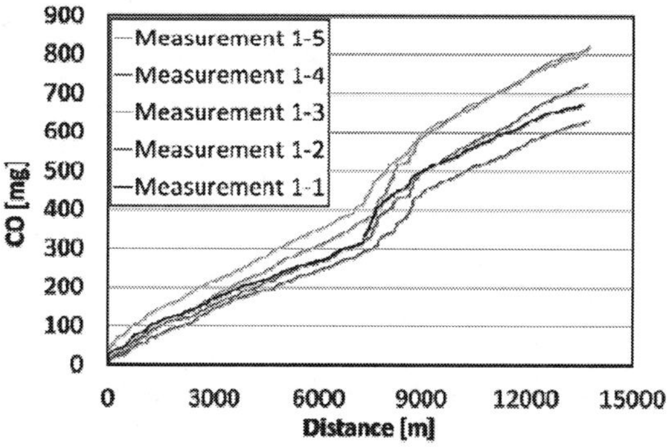

Fig. 9. Cumulative mass of carbon monoxide during RDE measurements for the north-south direction.

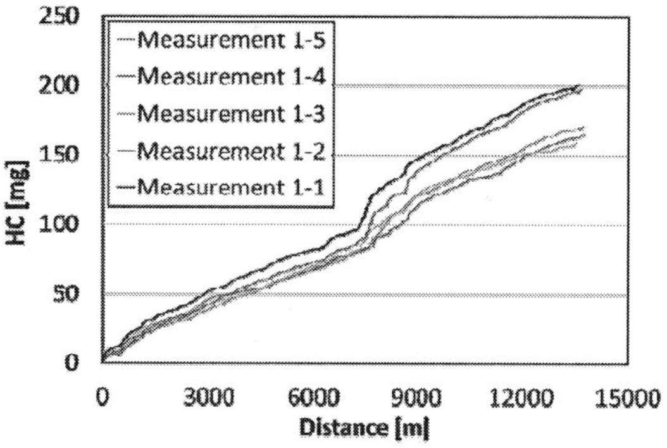

Fig. 10. Cumulative mass of hydrocarbons during RDE measurements for the north-south direction.

The course – of cumulative defined mass of nitrogen oxides during passing through the city was very repetitive (Fig. 11). The final values obtained are in range from 170 mg to 220 mg. For the same direction of study scatter results were not greater than 50 mg. Such a small dispersion of nitrogen oxides can be seen in similar operating conditions of the combustion engine due to the high volume of traffic. The lack of dynamic acceleration of the vehicle contributed to the low emissions of this compound.

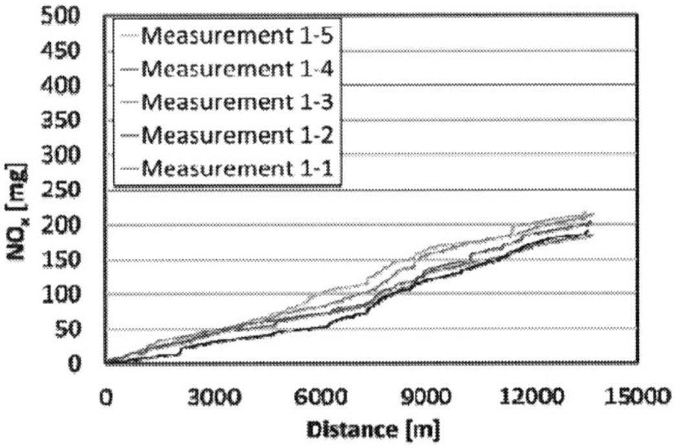

Fig. 11. Cumulative mass of nitrogen oxides during RDE measurements for the north-south direction.

Cumulative carbon dioxide mass is characterized by the smallest scatter (regardless of the direction of the study). The spread of these values during urban journeys does not exceed 200 g (which is less than 10% of the obtained results – Fig. 12). Carbon dioxide emissions can be directly attributed to fuel consumption, Also, for each test, similar fuel consumption values were achieved.

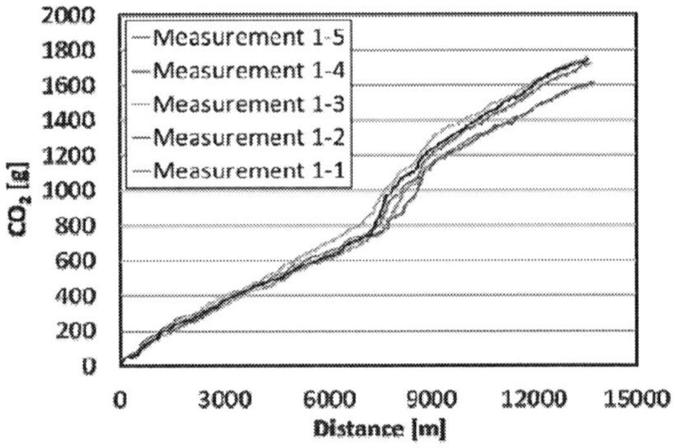

Fig. 12. Cumulative mass of carbon dioxide during RDE measurements for the north-south direction.

3.2 Simulations

A route model (Fig. 13) was created for the purpose of simulation of the vehicle passing through the city, consisting of the characteristic sections of the road (road mark 50 – before the bypass opening). These were, for example, straight ahead driving and crossing through crossections with traffic lights, located in the middle of the town (in different configurations.) The initial phase of the modeling was the selection of part of the road, followed by the geometry of the model based on the Google satellite image. The end stage was the parameterization of the modeled sections of the road infrastructure, which consisted mostly of setting the routes of individual vehicle groups and description of the type and emission structure of vehicles moving along the road.

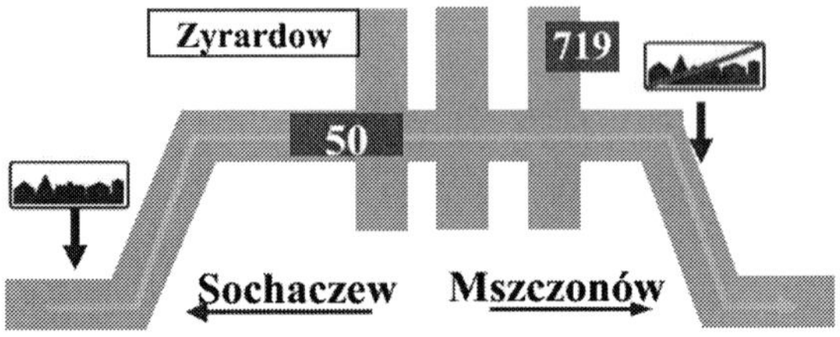

Fig. 13. Scheme of the road infrastructure model created in PTV Vissim.

For the assumed parameters, simulations of the modeled section of the road were carried out. The parameters of the journey (time of travel, distance and average speed) are shown in Fig. 14. The analysis of this drawing and its comparison with parameters obtained during road tests are the basis for the claim that the modeling of the analyzed road infrastructure was performed correctly. Travelling time during the simulation was 1193 s, which, compared to the average of 1313 s, is less than 120 seconds, or 2 minutes. At such a long distance, it should be considered this result as rewarding. In the case of a section length of 13.4 km, we managed to obtain almost the same result as the average from road tests (13.8 km). The distance and time parameter, the average speed is 40.3 km/h and its value is about 5% higher than the average value obtained during the real tests.

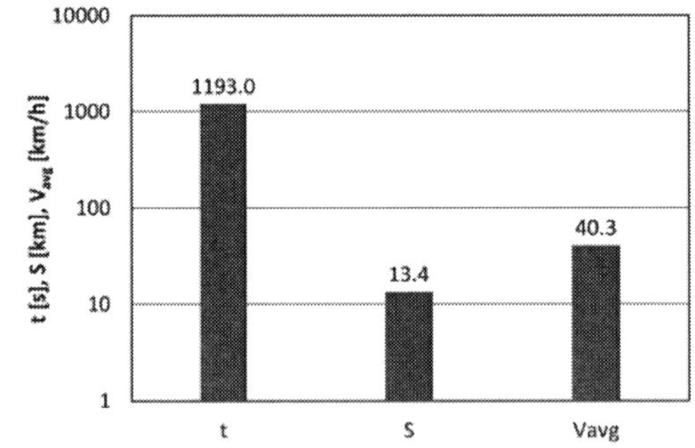

Fig. 14. Trip parameters of Euro 4 passenger car during simulation in PTV Vissim software.

Emission values obtained during the simulation are shown in Fig. 15. The Euro 4 vehicle emits 1701 mg of carbon monoxide, 210 mg of hydrocarbons, 407 mg of nitrogen oxides and 1987 g of carbon dioxide under the simulator. Not all of these values are consistent with road measurements, as will be discussed later in this chapter. The comparison of the mass of carbon monoxide emitted during the passage of the National Road 50 within Zyrardow (Fig. 16) shows that the results obtained during the simulation are much higher than the results obtained from the measurements. Also pay attention to the characteristic values of mass at arrival at intersections, passing through three intersections and reaching the end point of measurements. The sum of the masses from all the sections and the comparison with the results of the measurements shows a significant difference, which is 133% higher than the average from RDE tests. This proves that the results of the simulation are significantly overestimated in relation to the road value.

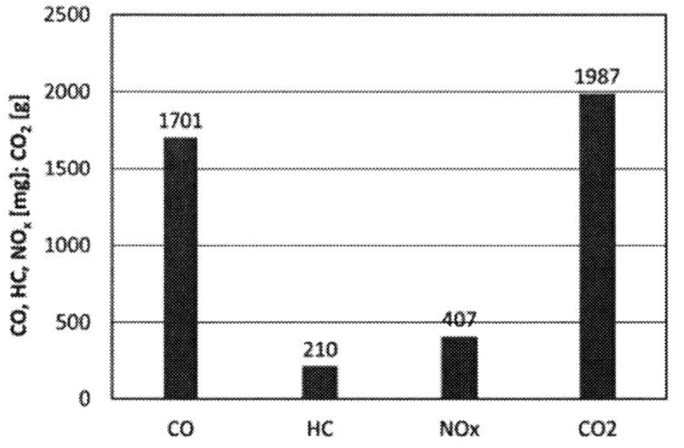

Fig. 15. Emission parameters of Euro 4 passenger car during simulation in PTV Vissim software

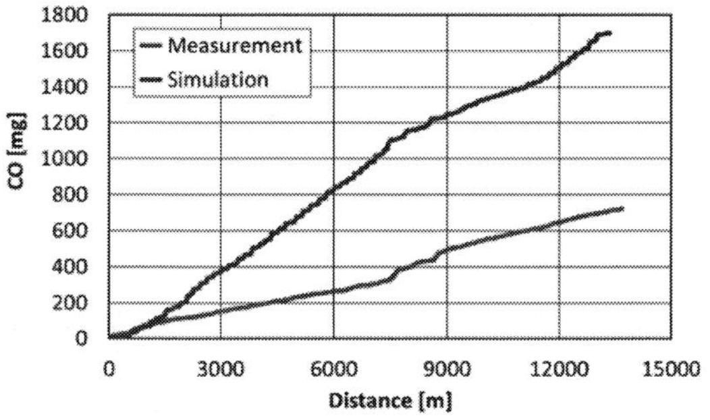

Fig. 16. The mass of carbon monoxide during RDE measurements and simulation.

Different character of the results (real and simulated) was obtained with respect to the comparison of the mass of hydrocarbons emitted during the passage through the city. In the first part of the route (up to about 3 km), the higher values were obtained from the measurements in real traffic conditions, while in the rest of the route the simulation results were higher (Fig. 17). For the whole run, the simulation results are only 17% higher than the measured values.

The comparison of the mass values of nitrogen oxides determined from the tests and during the simulation shows, as in the case of carbon monoxide, the incompatibility of the obtained values (Fig. 18). They note the increased mass values of this compound in all the parts of the urban route examined, which are twice as the value obtained from the measurements.

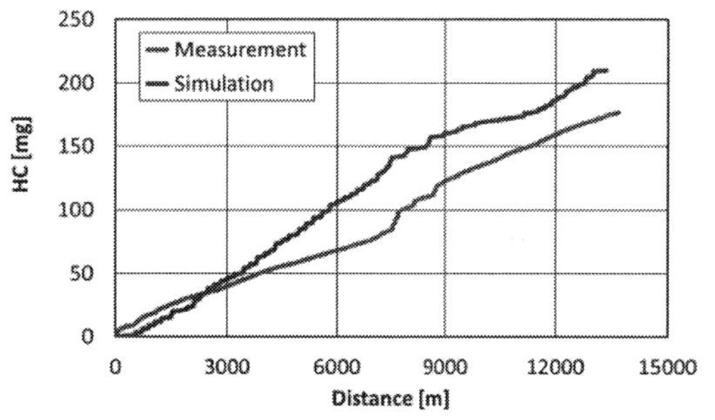

Fig. 17. The mass of hydrocarbons during RDE measurements and simulation.

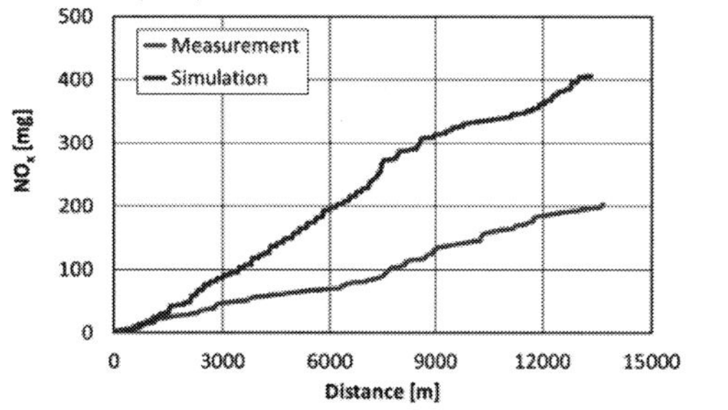

Fig. 18. The mass of nitrogen oxides during RDE measurements and simulation.

The highest compliance of mass results from measurements and simulation was recorded for carbon dioxide (Fig. 19). In the case of characteristic sections, where the compatibility of the results was significant (except the approach to the intersection), the similarity of the results achieved was the same across the entire urban route. The results obtained from traffic measurements differed by approximately 17% in relation to the determined value obtained using the Vissim software.

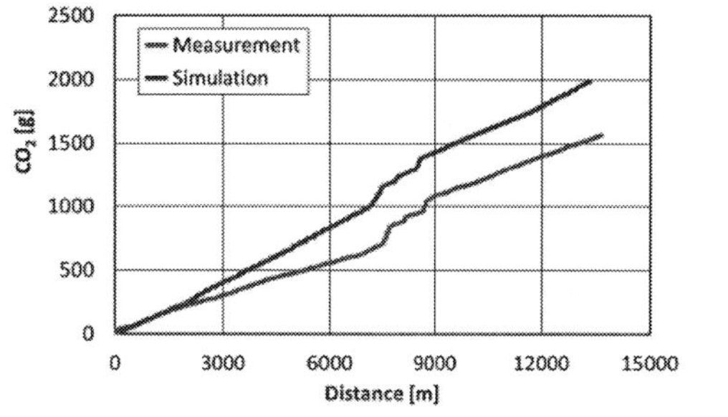

Fig. 19. The mass of carbon dioxide during RDE measurements and simulation.

It was assumed that the incompatibility in the mass of emitted harmful compounds during the simulation results from improper emission values of in the simulation software. For this reason, the emission values have been changed based on the data obtained during the RDE emission test. In the further part of the article the modified emission data is used.

3.3 Entire vehicles stream simulation

Using modified emission data and urban traffic patterns and vehicle structure information, simulations of the average speed and emissions of the entire vehicle stream were made (taking into account the differentiation of vehicles in terms of emission type and class) under such conditions after the analyzed road. In addition, the analysis of these parameters was performed for 50% less and 50% greater traffic density than the standard value. The analysis of Fig. 20 shows that, with the increase in vehicle traffic, the average speed of vehicles in the modeled road network decreases. For traffic density 50% smaller than standard, the average vehicle speed was 42.7 km/h. Increasing the traffic density to 100% and 150% resulted with reduction in average speeds to 40.3 km/h and 27.2 km/h, respectively.

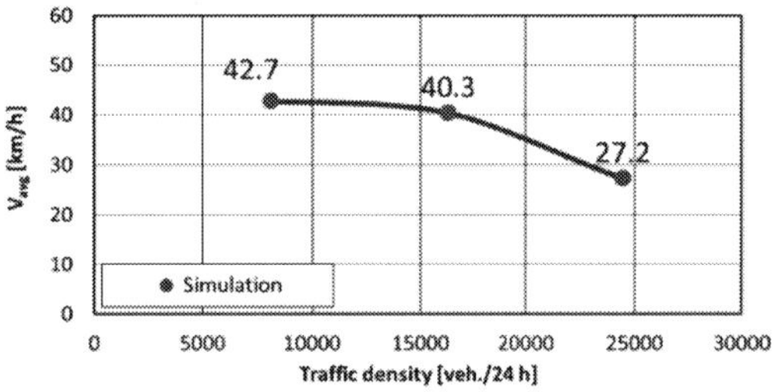

Fig. 20. Average speed of vehicles in the road network during the simulation of different traffic density on selected road.

Inversely look the graphs of the mass of harmful emissions from vehicle streams emitted over a 24-hour period. In this case, the increase in traffic intensity increases the emission of all analyzed components of the exhaust gases. For standard traffic density, the daily carbon monoxide emission is 390 kg/day (Fig. 21).

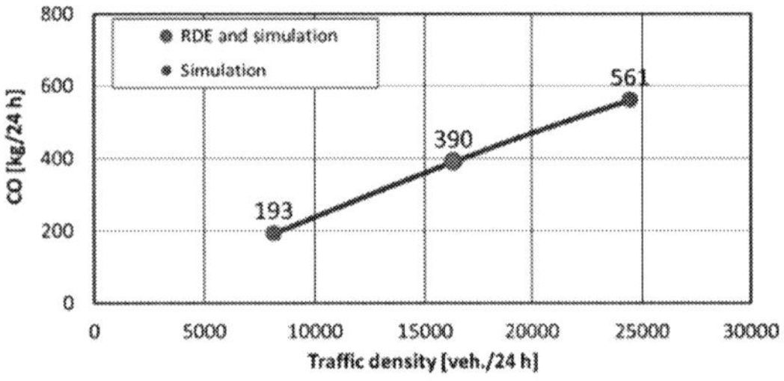

Fig. 21. Daily carbon monoxide emission from vehicles in the road network during the simulation of different traffic density on selected road.

A 50% reduction in the number of vehicles in the network resulted in a 50% reduction in carbon monoxide emissions, while a 50% increase in vehicle traffic led to a 44% increase in carbon monoxide emissions. Also, the mass of hydrocarbons increases with the increase in vehicle traffic in the modeled road network (Fig. 22). It shows that the mass of hydrocarbons emitted during the passage of the 50% smaller vehicle stream is 47% less than the passage of the whole vehicle stream through the city, and for much greater traffic – 46% greater. The mass of emitted nitrogen oxides during the passage of 50% less vehicle stream than that resulting from the General Measurement of Movement was 132.5 kg/day. For standard vehicle traffic, this result was approximately 52% higher. On the other hand, 50% more traffic from standard vehicles contributes to a 50% increase in nitrogen oxide emissions (Fig. 23).

Also similar in character to previous ones is the mass of carbon dioxide: for low traffic, the passing of the entire vehicle stream across the city generates 52% less mass than the standard traffic, and for higher traffic density, respectively 45% more carbon dioxide (Fig. 24).

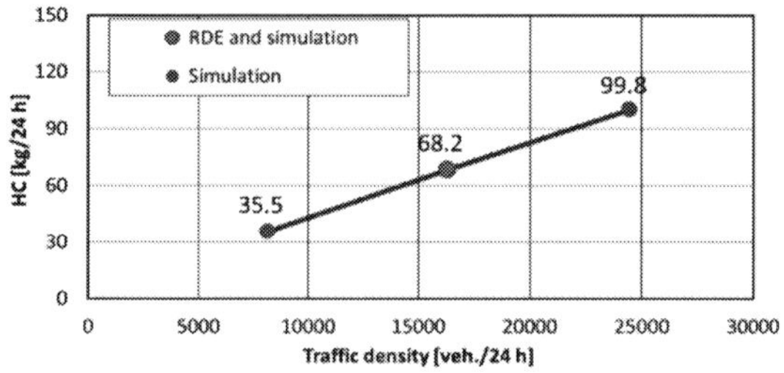

Fig. 22. Daily hydrocarbons emission from vehicles in the road network during the simulation of different traffic density on selected road.

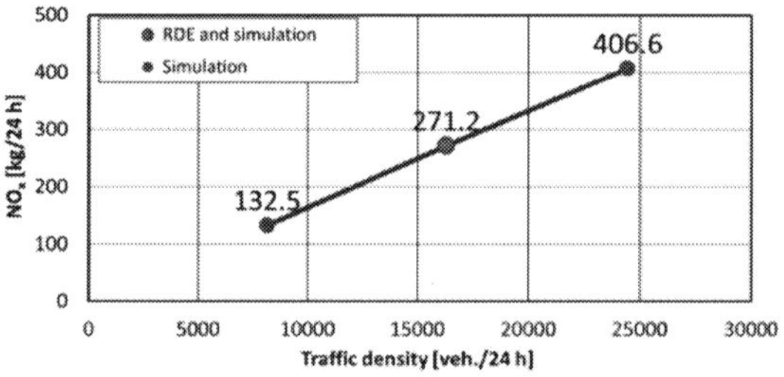

Fig. 23. Daily nitrogen oxides emission from vehicles in the road network during the simulation of different traffic density on selected road.

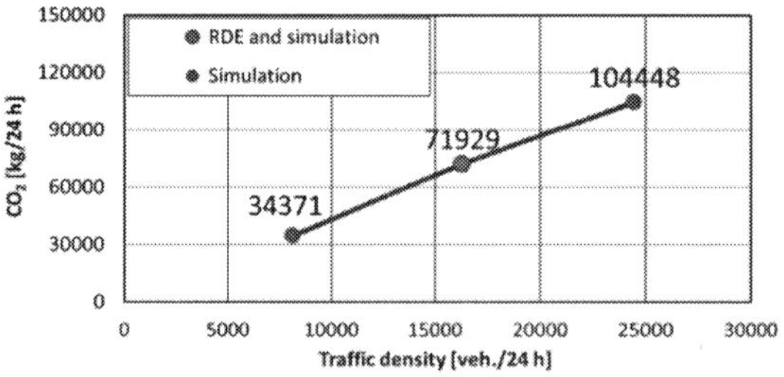

Fig. 24. Daily carbon dioxide emission from vehicles in the road network during the simulation of different traffic density on selected road.

4 Conclusions

Reducing the negative impact of transport on the environment can take place in many ways. Most often these actions are directed towards changing vehicles moving on the road to more environmentally friendly. But another aspect that can also contribute to reducing environmental pollution is a well-designed road infrastructure that ensures the smoothest possible flow.

During design of new roads, it is necessary to assess the existing infrastructure in terms of ecology and to carry out an analysis of the environmental impact of the new road infrastructure. For this purpose, specialized computer software is useful for analyzing not only emissions from a single vehicle but also the emission of the entire traffic stream. Another advantage of this approach is the ability to evaluate multiple variants of the course and configuration of the new road and almost any change in the number of vehicles moving in the modeled road infrastructure. However, in order to present the results in a reliable manner, attention should be paid to the appropriate validation of the model, which should include, among other things, determining the direction of travel, selecting the appropriate structure of vehicles in terms of categories and emission standards, and applying appropriate emission values. Particularly advantageous in the last aspect is the use of data obtained during emission measurements in real driving conditions.

The research was funded by project co-financed by the European Regional Development Fund in the Regional Program – Lubuskie 2020 (contract No. RPLB.01.01.00-08-0055/16-00).

References

1. Ochrona środowiska 2015 (GUS, 2015)
2. Commission of the European Communities: Communication from the Commission to the Council and the European Parliament: Keep Europe Moving – sustainable mobility for our continent. Mid-term review of the EC's 2001 Transport White Paper, Brussels 22.06.2006 (COM (2006) 314 final)
3. Commission of the European Communities. White Paper: European transport policy for 2010: time to decide, Brussels 12.09.2001 (COM (2001) 370 final)
4. Assessment of greenhouse gas emissions abatement potential in Poland by 2030 (McKinsey&Company, 2010)
5. K. Prainowski, J. Mamala, *IOP Conference Series-Materials Science and Engineering* **148**, UNSP 012022 (2016)
6. J. Mamala, S. Brol, G. Graba, *International Symposium on Electrodynamic and Mechatronic System*, MAG 033, 41-48 (2013)
7. A. Kublik, Gazeta Wyborcza (2013)
8. P. Fuc, L. Rymaniak, A. Ziolkowski, *WIT Transactions on Ecology and the Environment* **174**, 207-218 (2013), DOI: 10.2495/AIR130181
9. A. Suchecki, J. Nowakowski, T. Knefel, *Transport Means – Proceedings of the International Conference*, 194-197 (2014)
10. Z. Stepien, L. Ziemianski, G. Zak, M. Wojtasik, L. Jeczmionek, Z. Burnus, Fuel **161**, 278-286 (2015), DOI: 10.1016/j.fuel.2015.08.071
11. www.wego.here.com (2017)
12. P. Bonnel, M. Weiss, A. Provenza, *8th Annual SUN Conference* (Ann Arbor, 2011)
13. I. Khalek, *5th Annual SUN Conference* (Ann Arbor, 2008)
14. T. Korniski, C. Gierczak, T. Wallington, *4th Annual SUN Conference* (Ann Arbor, 2007)
15. F. Ortenzi, M.A. Costagliola, SAE Technical Paper 2010-01-1289 (2010)
16. AVL GAS PEMS (AVL, 2012)
17. AVL M.O.V.E System Control™ (AVL, 2013)
18. PTV Vissim, Traffic Simulation (2012)
19. M. Maciejewski, Transport Problems **4**, 27-38 (2010)
20. M. Maciejewski, T. Maciejewski, Uwarunkowanie makroskopowych modeli ruchu drogowego. Transport **95**, 309-319 (2013)
21. M. Maciejewski, B. Piątkowski, W. Walerjańczyk, Przegląd Komunikacyjny **2**, 27-31 (2014)
22. PTV Vissim – Modules, Emissions Modelling (2015)
23. N.E. Ligterink, R. De Lange, *17th Transport and Air Pollution Symposium and 3rd Environment and Transport Symposium* (Toulouse, 2009)
24. R. Smit, R. Smokers, E. Rabe, Transport and Environment **12**, 414-422 (2007)
25. VERSIT+ TNO state-of-the art road traffic emission model, TNO Science and Industry (2015)
26. Generalny Pomiar Ruchu w 2015 roku (Warszawa, 2016)

Specific emissions analysis for a combustion engine in dynamometer operation in relation to the thermal state of the exhaust gas aftertreatment systems in a modified NRSC test

Jerzy Merkisz[1,*], and *Maciej* Siedlecki[1]

[1]Poznan University of Technology, Faculty of Machines and Transport, ul. Piotrowo 3, 60-965 Poznan, Poland

Abstract. Exhaust gas aftertreatment systems have been present in motor vehicles for decades and have contributed to reducing their impact on the environment and people. Most of them for oxidation or reduction of harmful emissions of particulates and fumes require a certain temperature to be reached that changes with the exhaust temperature, i.e. the points of engine operation. The article describes the effect of oxidation reactor and particulate filter temperatures on specific emissions of gaseous compounds and particulate matter during the modified NRSC engine test. Before the first measurement cycle, the engine was idling, before the second measurement cycle, the exhaust system was heated with exhaust gases at full engine load until passive regeneration of the particle filter occurred (noticeable decrease in instantaneous particle concentration).

1 Introduction

The process of burning fuel in engines causes the formation of a number of harmful compounds. The products of complete and total combustion include carbon dioxide and water vapor; however, nitrogen oxides, carbon monoxide, hydrocarbons, solid particles and others are also produced by imperfections in this process. For motor vehicles, they have been limited by the Euro emission standards since 1992 [1]. The legal emission limits are further reduced with the introduction of subsequent emission standards. In spite of the technological advances in engine construction and better control of the combustion process in the internal combustion engine by using advanced electronic fuel injection systems, it is not possible to meet these standards without the use of in-engine and exhaust aftertreatment systems [2, 3].

The latter have been used in vehicles for decades, practically in all vehicles with an internal combustion engine. The basic element for SI engines is a three-way catalytic reactor, and for CI engines, due to different air excess coefficient values, oxidation reactor (DOC) [4]. Currently, diesel particulate filters (DPFs) and selective catalytic reduction are common in vehicles with CI engines [5, 6]. This is also the first solution currently being applied to vehicles with spark ignition engines due to the introduction of further limits for specific emission of harmful exhaust gases [7, 8].

These systems require a certain temperature to operate correctly, depending on the construction and materials used as the substrate, the intermediate layer and the catalytic layer. An enriched fuel blend is usually used in the engine during the cold start, which further enhances its emissivity in order to warm up both the engine and the exhaust system faster. This temperature mainly depends on the operating point of the engine, and more specifically its load [9]. This means, however, that the first few moments after engine start the emissions from the engine are significant and continue until the set temperature has been reached by the exhaust aftertreatment systems and by the internal combustion engine itself [10, 11].

At present, the engine design companies are looking for the most environmentally friendly solutions. Virtually every manufacturer of passenger vehicles currently offers hybrid vehicles combining the advantages of an internal combustion engine and a large group also offer full electric vehicles. Despite the gradual replacement of the classic diesel engine, it still remains the main source of propulsion for the vehicles currently in operation, and this will clearly not change in the coming years [12, 13]. Only further improvement in fuel consumption reduction and associated emissions of harmful compounds is left to the engine designers.

In order to counteract the effects of the under-heated propulsion system operation in today's vehicles, heating devices are used, which accelerate the process of reaching the operating temperature of both the exhaust gas aftertreatment systems and other systems like the lambda probe [14].

This article examines the effect of pre-heating the engine exhaust aftertreatment system on the emission of toxic compounds scaled to the unit of generated energy.

* Corresponding author: jerzy.merkisz@put.poznan.pl

DOI: 10.1051/matecconf/201711800027

2 Research methodology

The tested CI engine had a displacement of 1.3 dm^3 (Fig. 1). The engine was approved with the EURO 4 standard, but for the requirements of the test the factory standard DOC was removed, and the EGR valve was sealed shut. The basic engine technical data has been listed in Tab. 1.

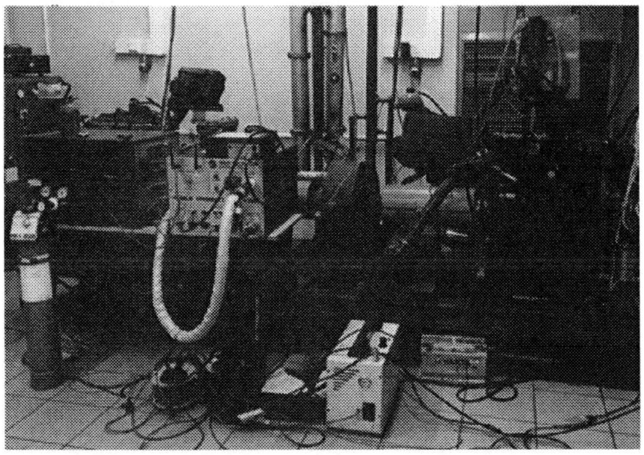

Fig. 1. The tested engine mounted on an engine dynamometer.

The tests were performed on the AUTOMEX engine dynamometer system using an electric brake. Brake information is listed in tab. 2. The engine operating point is set with a high accuracy on the control cabinet (Fig. 2) in each measurement case.

Table 1. Basic information about combustion engine.

Number of cylinders / displacement	4, in-line/1.3 dm^3
Bore / stroke	70 mm / 86 mm
Maximum power	52 kW at 4000 rpm
Maximum torque	180 Nm at 1750 rpm
Compression ratio	18,0:1
Supercharging	Turbocharger (FGT)
Exhaust emission standard	Euro IV

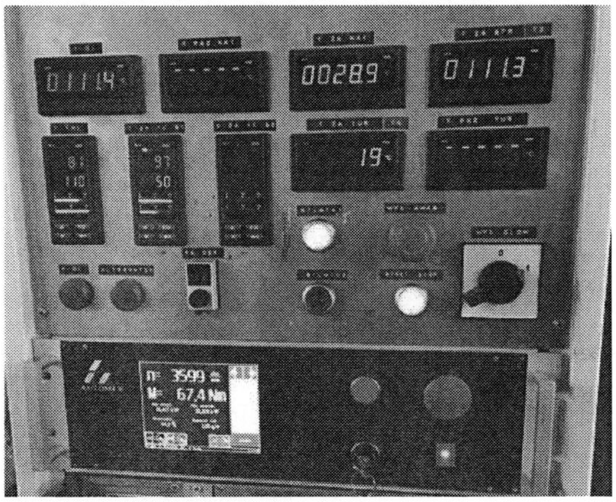

Fig. 2. Control cabinet used in research.

Table 2. Engine dynamometer technical data.

Brake type	AMX-210/100
Maximum power	100 kW
Maximum engine speed	10,000 rpm
Maximum torque	240 Nm
Direction of rotation	Any

The exhaust system was retrofitted in DOC and DPF (Fig. 3). These elements are placed exactly in the exhaust gas flow axis and additionally sealed with a high temperature paste and then screwed together. Their airtightness was checked before the start of each test cycle.

Fig. 3. The exhaust aftertreatment system used in the engine tests (DOC + DPF).

The elements were brand new before the measurement, both of which used metallic substrates in the form of folded sheet metal with applied catalytic layers. The DOC and DPF systems are presented in Fig. 4 and 5.

The assessment of the exhaust emissions level was performed with the PEMS equipment measuring the concentration of the exhaust gas components. In the work described in the paper, SEMTECH DS – a portable exhaust emissions analyzer manufactured by Sensors Inc. was used (Fig. 6).

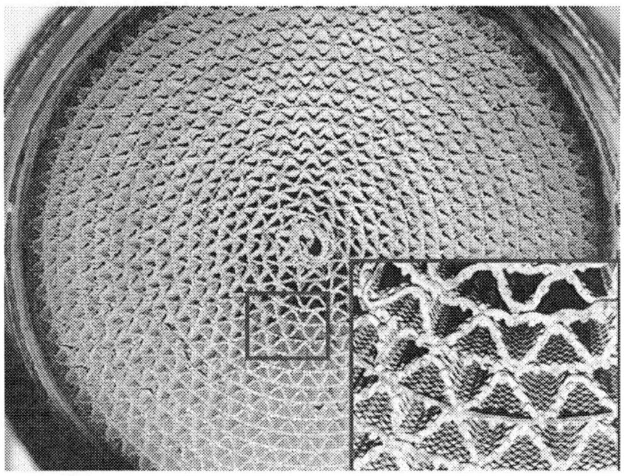

Fig. 4. The inside of the DOC.

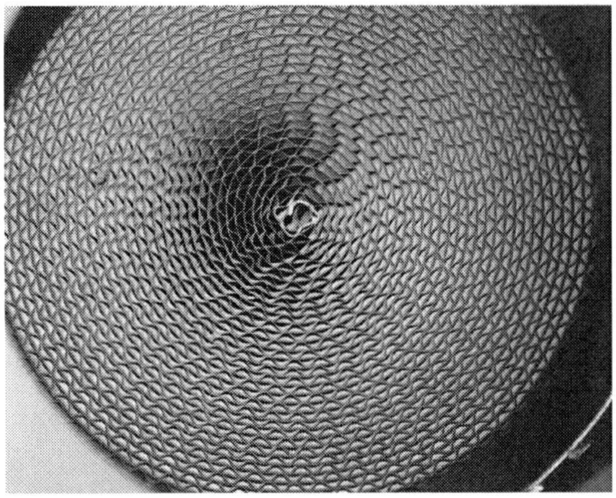

Fig. 5. The inside of the DPF.

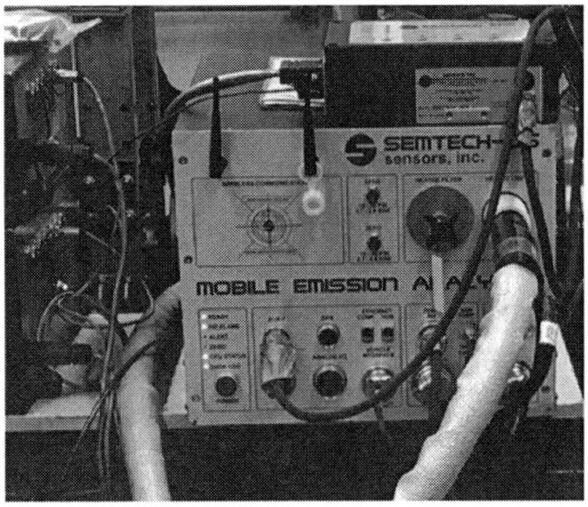

Fig. 6. SEMTECH DS analyzer by Sensors Inc.

The entire volume of the exhaust gas from the exhaust system was sent to the mass flow meter and then through a measurement probe (maintaining the temperature of 191 °C) to the analyzer. The device filtered the exhaust gas to separate the particulate matter (PM). In the next step, the system measured the concentration of hydrocarbons in a FID (Flame Ionization Detector) (Fig. 7). The exhaust gas was then chilled to the temperature of 4 °C and the concentrations of nitrogen oxides (NDUV, Non-Dispersive Ultra-Violet), carbon monoxide/carbon dioxide (NDIR, Non-Dispersive Infrared) and oxygen (electrochemical analyzer) were measured. The device is compatible with the vehicle's on-board diagnostic system (recording of the operating parameters – engine speed and load) and GPS (latitude and longitude for determining the vehicle speed), which was not used in this research.

The authors used MSS (*Micro Soot Sensor*) by AVL (Fig. 8a) for the measurement of particulate matter. The device uses laser light dispersion triggered by particulate matter contained in the exhaust gas. AVL MSS can determine the real time concentration of PM in the exhaust gas.

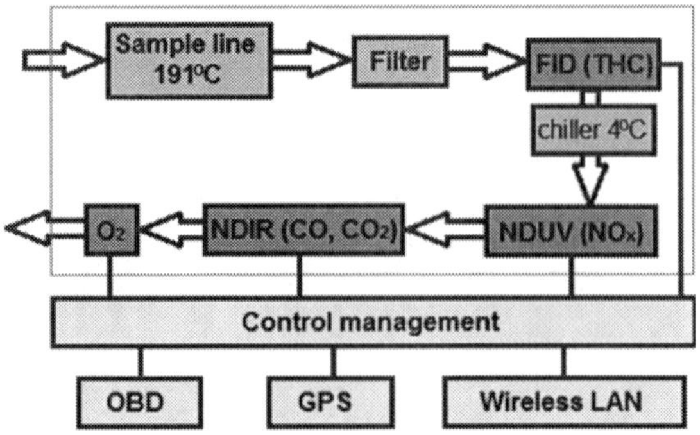

Fig. 7. Diagram of the portable exhaust emissions analyzer (SEMTECH DS) with the exhaust gas flow (═) and electrical connections (---) marked [15].

The number of the particles were measured by EEPS TSI (Fig 8b). In this device, the solid particles get an electrical charge and then they reach the electrodes, the larger the particle the further the electrode reached.

a) b)

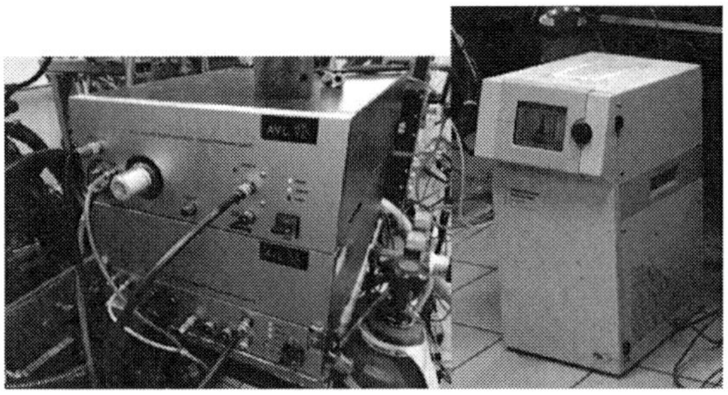

Fig. 8. View of the: a) AVL Micro Soot Sensor, b) TSI EEPS.

At the start of the measurements, the analyzers were calibrated with reference gases according to the manufacturer's guidelines. This process consists of delivering a strictly defined gas composition to the gas analyzers. This operation is repeated for several gas canisters, thereby restoring the original operating characteristics of the device.

In addition, before the measurements were made, the ambient air was used to determine the data offset. The device did not report any operational errors when performing measurements.

Engine operating points were determined in accordance with a modified NRSC test. This is a stationary test introduced with Directive 97/68/EC in 1997. This directive has had numerous updates, but the test is still used today to measure the emissions of harmful gases contained in exhaust gases in non-road vehicle tests. This choice was made because it is a stationary test that can be performed on a test bench. Modification compared to the original test consisted of changing the operating points to make them more in-line with the real conditions of non-road vehicles operation [16]. The changes involved reducing the maximum engine load while increasing the load at points where it was low. This change was made on the basis of a comprehensive analysis of operating points of NRMM vehicle engines in actual operation. The weight of each point's contribution to the total value was also determined based on the time that the engine spent operating in given point while in actual operating conditions. The operating points are shown in table 3 and have been graphically represented in Fig. 9.

Table 3. Operating points measured for the tested engine [16].

Operating point	Crankshaft rotational speed [rpm]	Load [Nm]	Weight average
1	2400	148,5	0.15
2	2400	107	0.2
3	2400	66	0.2
4	3600	97,5	0.1
5	3600	67	0.1
6	3600	36,6	0.1
7	770	3,3	0.15

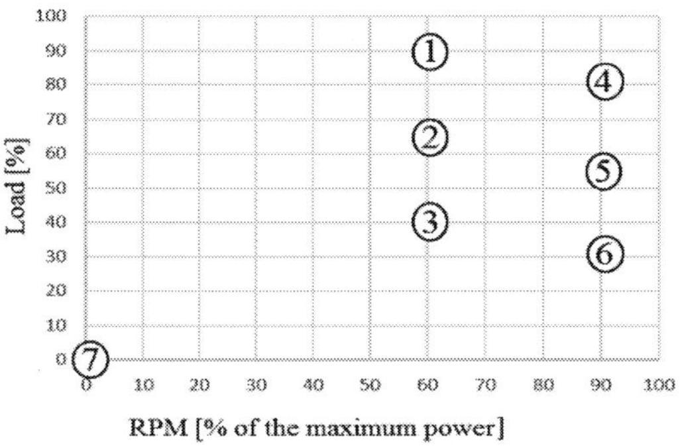

Fig. 9. Operating points for the tested engine [16].

The engine worked for 60 seconds at each operating point, and the shift between the points was dependent on the filter inlet and outlet temperatures. Measurement was taken when the temperature has not changed by more than 1 °C for 10 seconds. The entire cycle was performed three times in each case to eliminate any random errors. The warm up of the system consisted in maintaining the engine operating point so that the filter temperature was above 400 °C for a period of 10 minutes and during that time a passive regeneration of the DPF occurred. This process involves the use of nitrogen dioxide which, at elevated temperatures after contact with the solids, causes them to oxidize and as a result, purge the particles that accumulated in the filter. The system was considered cold when the engine worked for 10 minutes at idle before measurements. In this case, the approximate temperature of the exhaust aftertreatment system was 140 degrees centigrade.

3 Results and analysis

The measured concentrations were used in calculations on a spreadsheet to determine the value of specific emissions. Both emission concentration and instantaneous power measurements were averaged over the entire measuring cycle lasting 60 seconds according to the weight of the individual operating points. The results of the emissivity of the individual compounds were presented in table and mean values were presented graphically. Due to the tests performed on the engine dynamometer, all the values reported were calculated as per unit work expressed in kWh. Taking into account the output value, i.e. the lack of exhaust aftertreatment systems as a benchmark, the result was divided by these values and converted into percentages.

Carbon monoxide emission research results are presented in Tab. 4 and the average of the test in Fig. 10. All values are given with a background reference, which is 100%.

Table 4. Results of the carbon monoxide emission at specific points of engine operation.

Point no.	Crankshaft rotational speed [rpm]	Load [Nm]	W/o after-treatment [g/kWh]	Cold after-treatment [g/kWh]	Hot after-treatment [g/kWh]
1	2400	148.5	0.18	0.18	0.14
2	2400	107	0.28	0.31	0.03
3	2400	66	0.47	0.39	0.37
4	3600	97.5	0.29	0.21	0.21
5	3600	67	0.38	0.36	0.35
6	3600	36.6	1.25	0.53	0.57
7	770	3.3	5.64	3.38	3.38

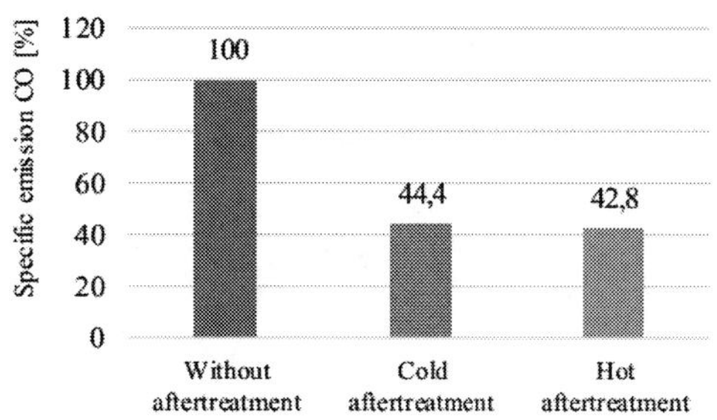

Fig. 10. Relative specific CO emissions for the cold and hot exhaust gas treatment system.

For carbon monoxide there is a noticeable reduction in emissivity for both cold and hot systems. The carbon monoxide emissions is almost halved and demonstrates high efficiency of the aftertreatment system despite various engine loads during the test. For the hot exhaust system, a slightly worse result was obtained due to the high temperature of the DOC mounted before the DPF. No significant differences in emissivity for cold and hot systems were found in the results between the individual operating points.

The reduction in emissions was mainly due to the DOC. The greatest differences between systems at different temperature states can be seen for the first measurement point, which makes it possible to consider the heating of the system as an effective solution for reducing emissions. At the later operating points, the relative differences were very small. Hydrocarbon emissions are listed in Tab. 5 and shown in Fig. 11.

Table 5. Results of specific hydrocarbon emissions at different points of engine operation.

Point no.	Crankshaft rotational speed [rpm]	Load [Nm]	W/o after-treatment [g/kWh]	Cold after-treatment [g/kWh]	Hot after-treatment [g/kWh]
1	2400	148.5	0.18	0.18	0.02
2	2400	107	0.28	0.31	0.03
3	2400	66	0.47	0.39	0.06
4	3600	97.5	0.29	0.21	0.21
5	3600	67	0.38	0.36	0.35
6	3600	36.6	1.25	0.53	0.57
7	770	3.3	5.64	3.38	3.38

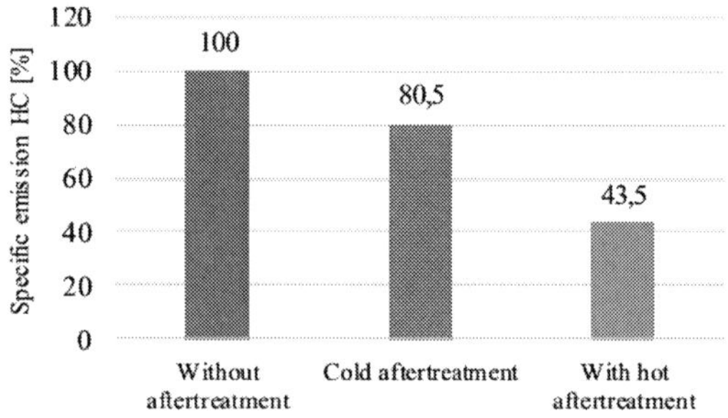

Fig. 11. Relative specific HC emissions for the cold and hot exhaust system.

In the case of hydrocarbons, the cold system reduces their emissions by 20% over the measurement without exhaust gas aftertreatment systems. After warming up, this value is increased almost twice, to 43% of the initial value. Unlike in the case of carbon monoxide, there can be a significant positive influence of heating the system on its effectiveness. Again, the impact of the particle filter application is rather small and the oxidation reactor has significant impact on the change in emissivity. The biggest differences are seen for the first three operating points where the success of the hot system is 90%. Results of nitrogen oxide emission tests are presented in Tab. 6 and the mean of the measurements in Fig. 12.

Table 6. Specific emission results for nitrogen oxides at individual engine operating points.

Point no.	Crankshaft rotational speed [rpm]	Load [Nm]	W/o after-treatment [g/kWh]	Cold after-treatment [g/kWh]	Hot after-treatment [g/kWh]
1	2400	148.5	7.18	7.20	7.80
2	2400	107	8.04	8.26	10.86
3	2400	66	8.38	16.20	14.30
4	3600	97.5	9.66	10.01	9.55
5	3600	67	11.33	13.02	12.88
6	3600	36.6	14.21	16.16	15.86
7	770	3.3	45.87	49.26	43.62

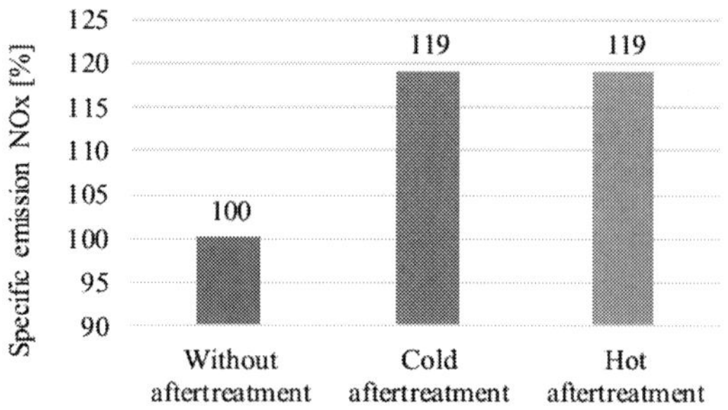

Fig. 12. Relative specific NO_x emissions for the cold and hot exhaust system.

Emissions of nitrogen oxides are increased when using exhaust aftertreatment systems. The increase in both cold and hot cases amounted to approximately 19%. This increase was most apparent in the medium load range (operating points 3 and 6). This can be due to the high temperature inside the DOC despite the temperature of the exhaust gas. Again the increased emissions are affected by the use of an oxidation reactor where NO is converted to NO_2 for example, which is used for passive filter regeneration. For this reason, it was decided that the increase in NO_2 emissions must be analyzed separately (Tab. 7 and Fig. 13).

Table 7. Results of the specific emission of nitrogen dioxide at individual points of engine operation.

Point no.	Crankshaft rotational speed [rpm]	Load [Nm]	W/o after-treatment [g/kWh]	Cold after-treatment [g/kWh]	Hot after-treatment [g/kWh]
1	2400	148.5	0.18	2.66	1.79
2	2400	107	0.20	3.99	2.55
3	2400	66	0.30	6.00	5.44
4	3600	97.5	0.48	3.18	2.96
5	3600	67	0.71	4.59	4.28
6	3600	36.6	0.65	4.46	3.68
7	770	3.3	8.65	11.13	12.70

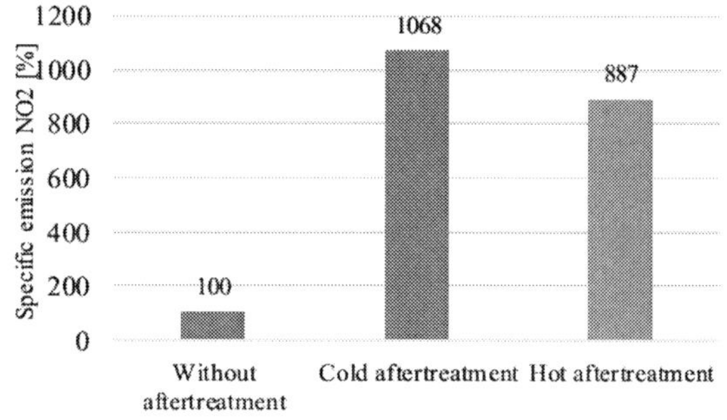

Fig. 13. Relative specific NO_2 emissions for the cold and hot exhaust system.

The results clearly show a high NO oxidation share in NO_2 generation. Especially at maximum loads, the exhaust aftertreatment system increases NO_2 emissions by as much as twenty times. The average of the entire test is more than a tenfold increase in emissions for the cold system and nearly nine fold for the hot system. The cold system produces much more nitrogen oxide IV for the first three measurement points. Emission results for PM are found in Tab. 8 and Fig. 14.

Table 8. Specific PM emission results at individual engine operating points.

Point no	Crankshaft rotational speed [rpm]	Load [Nm]	W/o after-treatment [g/kWh]	Cold after-treatment [g/kWh]	Hot after-treatment [g/kWh]
1	2400	148.5	0.76	0.53	0.56
2	2400	107	0.15	0.13	0.19
3	2400	66	0.12	0.11	0.13
4	3600	97.5	0.91	0.64	0.72
5	3600	67	0.40	0.25	0.27
6	3600	36.6	0.41	0.28	0.31
7	770	3.3	2.54	2.51	2.62

In the case of particulate matter, the use of additional exhaust aftertreatment systems contributed to a 16% and a 10% decrease in particle mass emissions, for the hot and cold systems respectively. This difference was quite small, probably due to the considerable distance of the filter from the exhaust manifold and consequently its temperature. The second reason may be the lack of the control system adjustment to the presence of a particulate filter. The cold system proved to be more effective in oxidizing the particles.

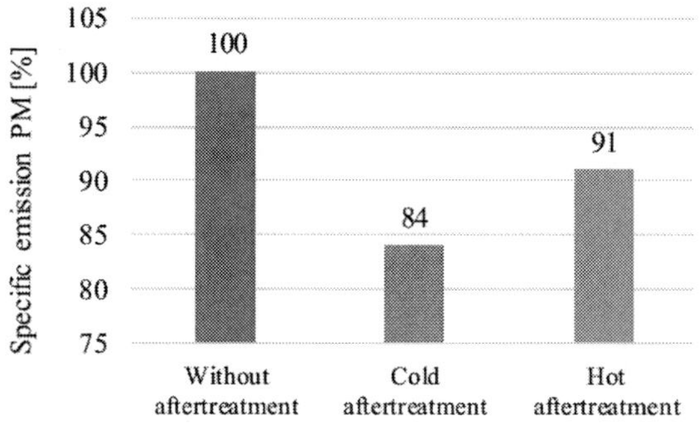

Fig. 14. Relative specific emission of PM for the cold and hot exhaust gas aftertreatment system.

This could be due to the increased amount of nitrogen oxide IV required for passive regeneration. The biggest differences were observed for the highest engine loads. For points with little or no load, the differences are minimal, which indicates a very low efficiency of the aftertreatment system. The cold system results in reduced PM emissions at all measuring points. The final component tested was the number of particles, the results of which are shown in Tab. 9 and Fig 15.

Table 9. Specific PN emission results at individual engine operating points.

Point no.	Crankshaft rotational speed [rpm]	Load [Nm]	W/o after-treatment [g/kWh]	Cold after-treatment [g/kWh]	Hot after-treatment [g/kWh]
1	2400	148.5	6.73E+11	5.03E+11	7.15E+11
2	2400	107	3.59E+12	1.03E+12	1.27E+12
3	2400	66	2.58E+12	1.26E+12	1.11E+12
4	3600	97.5	1.35E+12	1.23E+12	1.05E+12
5	3600	67	4.47E+12	1.26E+12	1.40E+12
6	3600	36.6	5.89E+12	1.59E+12	1.69E+12
7	770	3.3	6.67E+12	5.40E+12	5.67E+12

Unlike small differences in particulate mass, the particle number is reduced by nearly half due to the presence of the exhausr gas aftertreatment system, which would indicate that the DPF has a high trapping efficiency for the smallest sized particles. Just like in the PM evaluation, the cold system is more effective, although the difference is less significant and mainly affects the first two operating points. This would indicate that this system worked with lower efficiency after being heated up before use in terms of particle number emissions.

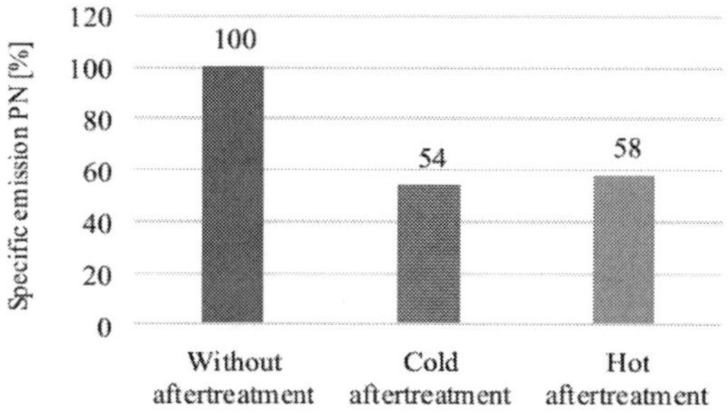

Fig. 15. Relative specific PN emission for the cold and hot exhaust gas aftertreatment system.

4 Summary

The exhaust aftertreatment systems are nowadays indispensable for virtually any vehicle with an internal combustion engine. Without it, strict type approval standards simply cannot be met. However, the test conditions in the laboratory often differ from what happens in real life. Frequent starts of the car engine cause the systems to be underactive and prevents them from operating with satisfactory efficacy in reducing toxicity.

The undertaken research efforts described in the article were concerned with the assessment of the exhaust aftertreatment systems pre-heating influence on the internal combustion engine emissivity, based on the exhaust gas measurements. The use of a Euro IV approved engine in the exhaust emission tests greatly influenced the overall emissivity values. Carbon monoxide emission was significantly reduced. A slight reduction in the carbon dioxide was observed as well as in the mass and number of emitted particles. The emission of nitrogen oxides, especially NO_2, has been significantly increased. Self-warming of the system has a clear positive effect on the carbon monoxide and hydrocarbon emissions. Nitrogen oxides are not much different, but differences appear in the analysis of the nitrogen oxide VI. More of this compound is emitted from a cold engine. In the case of particles, it is surprising that the cold system reduces both the number and the mass of the particles more effectively than when heating up.

The research was funded by project co-financed by the European Regional Development Fund in the Regional Program – Lubuskie 2020 (contract No. RPLB.01.01.00-08-0055/16-00).

References

1. M. Bajerlein, P. Fuc, P. Lijewski, L. Rymaniak, A. Ziolkowski, M. Dobrzynski, Combustion Engines **154**, 708-715 (2013)
2. J. Merkisz, P. Fuc, P. Lijewski, M. Siedlecki, A. Ziolkowski, *IOP Conference Series-Materials Science and Engineering* **148**, UNSP 012077 (2016), DOI: 10.1088/1757-899X/148/1/012077
3. M. Bajerlein, L. Rymaniak, P. Swiatek, A. Ziolkowski, P. Daszkiewicz, M. Dobrzynski, Experimental and Applied Mechanics **518**, 108-113 (2014), DOI: 10.4028/www.scientific.net/AMM.518.108
4. P. Fuc, P. Lijewski, A. Ziolkowski, M. Dobrzynski, Journal of Electronic Materials **46**, 3145-3155 (2017)
5. M. Bajerlein, L. Rymaniak, Technika Transportu Szynowego **10**, 2773-2782 (2013)
6. P. Fuc, L. Rymaniak, A. Ziółkowski, *WIT Transactions on Ecology and the Environment* **174**, 207-218 (2013)
7. W. Burget, *9th AVL International Commercial Powertrain Conference* (2017)
8. R. Jasinski, J. Pielecha, J. Markowski, *E3S Web of Conferences* **10**, UNSP 00030 (2016), DOI: 10.1051/e3sconf/20161000030
9. P. Fuc, J. Merkisz, P. Lijewski, M. Bajerlein, A. Ziolkowski, L. Rymaniak, M. Dobrzynski, Combustion Engines **154**, 701-707 (2013)
10. J. Markowski, J. Pielecha, R. Jasinski, T. Kniaziewicz, P. Wirkowski, *E3S Web of Conferences* **10**, UNSP 00140 (2016), DOI: 10.1051/e3sconf/20161000140
11. J. Merkisz, J. Markowski, J. Pielecha, *WIT Transactions on Ecology and the Environment* **174**, 219-230 (2013), DOI: 10.2495/AIR130191
12. J. Merkisz, J. Markowski, J. Pielecha, *IEEE Vehicle Power and Propulsion Conference VPPC*, 897-902 (2012)
13. J. Merkisz, P. Fuc, P. Lijewski, J. Pielecha, Transportation Research Procedia **14**, 3070-3078 (2016), DOI: 10.1016/j.trpro.2016.05.452
14. M. Idzior, M. Bajerlein, P. Lijewski, P. Fuc, *Global Science and Technology Forum* (Singapore, 2013)
15. Sensors Inc. Emissions Measurement Solutions (Erkrath, 2010)
16. P. Lijewski, Postdoctoral thesis (Poznan, 2013)

Analysis of tractor particulate emissions in a modified NRSC test after implementing a particulate filter in the exhaust system

Maciej Siedlecki[1], *Piotr* Lijewski[1,*], and *Sylwester* Weymann[2]

[1]Poznan University of Technology, Faculty of Machines and Transport, ul. Piotrowo 3, 60-965 Poznan, Poland
[2]Industrial Institute of Agricultural Engineering, ul. Starolecka 31, 60-963 Poznan, Poland

Abstract. Retrofitting, which means retrofitting old generation engine systems with modern exhaust aftertreatment systems, is becoming increasingly popular, which allow vehicles to adhere to the newer and more stringent emission norms. This can save the operators of such vehicles money using older engineered designs without the need to design a new unit or buy an expensive new machine or vehicle. At present, there is a growing interest in emissions from off-road vehicles and the introduction of minimum limits for older vehicles that must be met in order to be able to allow for their operation. For the purposes of this article, the Stage IIIA farm tractor has been fitted with a particulate filter in the exhaust system. The study investigated the impact of the use of exhaust aftertreatment systems on particle emissions in terms of mass, size distribution and number using PEMS analyzers in the modified NRSC stationary test by engine loading, using a mobile engine dynamometer and comparison of test results.

1 Introduction

Agricultural vehicles are the largest subgroup of machines belonging to the NRMM category. This is due to a significant amount of area used for agriculture in Poland [1]. For decades of crop use, the work has been heavily mechanized and the average yield per hectare has increased significantly [2]. The use of an internal combustion engine, however, is associated with the emission of harmful compounds into the atmosphere. This is countered by the increasingly recent emissions standards for new vehicles, developed over several years [3].

Despite attempting to apply the hybrid system or even the total electrification of NRMMs, manufacturers have still not achieved satisfactory results in terms of powertrain size and its potential range [4]. Both the use of fuel cells and electrochemical batteries cannot currently compete with the energy density and power of fuels based on oil processing, as well as the time it takes to refuel. NRMM vehicles, despite their large size, usually do not have enough room to accommodate enough batteries to provide a range similar to that obtained on liquid fossil fuel [5, 6].

Recuperation, that is typical for modern vehicles, cannot be used in this case either. The driving speeds obtained in this case are small, which results in low kinetic energy, and the braking of the vehicle through the resistance of the work done causes it to stop very quickly [7]. Because of that, the energy consumption and the emission of the vehicle are not reduced as much as in the case of a personal cars [1, 8].

Current technology for the construction of internal combustion engines, however, does not allow for a drastic reduction of their environmental impact [9, 10]. For this reason, one of the current actions to protect the planet is to reduce emissions of older vehicles. This trend is called retrofitting, and is based on the use of modern exhaust gas aftertreatment systems [2]. Today's exhaust aftertreatment systems are usually small in size, and the advancement of the current technology allows the choice of materials and, above all, choosing the intermediate and active layers in such a way as to allow passive filter regeneration. It involves the use of nitrogen oxide (IV) to reduce the particle mass and size. Importantly, it does not require any changes in construction or control of the combustion engine. It is particularly desirable to reduce particle emissions in terms of number and mass as well as nitrogen oxides, emissions of which are reduced by the particle filter and selective catalytic reduction respectively [11].

These systems are currently standard issue in vehicles that meet the current Stage IV emission standard and will be present in Stage V [12] vehicles. Numerous studies have shown the efficacy of emission reduction systems in reducing emissions, especially in terms of particle mass and number [13-16]. Particularly, the latter value is important because it has been proven that nanoparticles are the most dangerous to humans, and those are precisely the particles most commonly emitted from diesel engines [17-20], its emission value for passenger vehicles and HDVs is also subject to limits.

Therefore, the chances of reducing the environmental impact are reflected in the retrofitting process. Quite a large reduction in limits, especially particulate matter,

* Corresponding author: piotr.lijewski@put.poznan.pl

DOI: 10.1051/matecconf/201711800028

occurred in the Stage IIIB standard, which entered into force at the end of 2009-2011, depending on the engine power [21, 22]. This change was noticed by the German authorities and currently vehicles that do not meet at least Stage IIIB cannot participate in the construction work in the capital of Berlin. From the economics point of view, it is much better to use a modern exhaust aftertreatment system than to replace a fleet of vehicles whose value will be further reduced [2, 23].

This article presents the effect of the use of a particulate filter in the exhaust system of an agricultural tractor on the emissivity of particulate matter in terms of mass and number. The tests were carried out in a hall where the tractor was subjected to load with a mobile brake through the power take-off shaft.

2 Test vehicle

A Deutz-Fahr Agrotron X720 tractor was used for the tests. During the test, it was in the hall, where it was connected by a shaft from one side to the mobile chassis dynamometer (Fig. 1). The vehicle was checked for possible faults and was technically sound before testing. The tractor engine was heated to its operating temperature by the engine oil and the cooling liquid prior to testing. The basic data of the tractor internal combustion engine is presented in Tab. 1.

Fig. 1. Image of the tractor during test.

Table 1. Deutz-Fahr Agrotron X720 tractor technical parameters.

Engine type, number and arrangement of cylinders, number of valves	6-cylinder, 4 valves per cylinder, CI engine
Injection system	Common Rail
Displacement	7.1 dm^3
Cylinder diameter/piston stroke	115×149 mm
Maximum power	198 kW at 2300 rpm
Maximum torque	1050 Nm at 1400 rpm
Turbocharging	VGT turbocharger with intercooler
Aftertreatment systems	EGR, DOC, custom DPF filter
Emission standard	Tier 3/Stage IIIA

The Eggers-Dynamometer PT 301 MES mobile dynamometer was used to load the engine. This is an air cooled dynamometer based on the work of electric motors. The basic brake parameters are shown in Table 2. During the tests, a reduction ratio (2.1:1) relative to the engine speed was introduced to the brake software, which enabled the actual torque generated by the internal combustion engine to be used later to calculate the unit emissions.

Table 2. Basic technical parameters of the dynamometer used for the test.

Braking system	retarder
Coolant	air
Top engine speed	3600 rpm
Maximum torque	7200 Nm
Rotation direction	left/right

The particle filter cartridge with which the tractor was retrofitted was made of metal. Filters of this type are built by spiral coiling of previously catalytically coated sheets (Fig. 2), and then wrapping the insulating layer and dipping in a metal casing. The external dimensions of the filter used for the test are 870 mm length, maximum diameter of 400 mm, and the inlet and outlet with diameters of 100 mm. These dimensions were selected according to manufacturer's recommendations based on the engine displacement. The filter was incorporated into the outlet system with steel pipes and sealed with high temperature tape and steel clamps (Fig. 3). It was thus precisely embedded in the exhaust gas flow axis.

Fig. 2. View of the DPF carrier used.

Fig. 3. DPF mounted in the exhaust system of the tested tractor.

The standard tractor exhaust manifold (Fig. 4) has been replaced by an element that directs the exhaust gas in parallel to the substrate (Fig. 5). The authors have not decided to modify the standard exhaust system by altering the engine exhaust system, due to its large complexity (the intake manifold is integrated with the turbine which also contains the DOC), and by limiting the structural modifications of the tractor to a minimum. The diameter of the pipe has been chosen to match the standard exhaust system without damaging the exhaust.

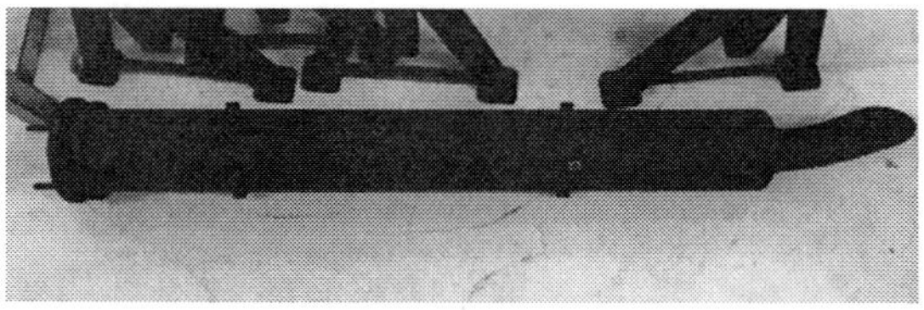

Fig. 4. Standard engine exhaust system.

Fig. 5. Exhaust system modifications for the test.

3 Apparatus used

The assessment of the exhaust emissions level was performed with the PEMS equipment measuring the concentration of the exhaust gas components. In the research described in the article, Semtech DS – a portable exhaust emissions analyzer manufactured by Sensors Inc. was used (Fig. 6).

Fig. 6. SEMTECH DS and AVL Micro Soot Sensor analyzers during tests.

The entire volume of the exhaust gas from the exhaust system was sent to the mass flow meter (Figure 7) and then through a measurement probe (maintaining the temperature of 191 °C) to the analyzer [24]. The device filtered the exhaust gas to separate the particulate matter (PM). In the next step, the system measured the concentration of hydrocarbons in a FID (Flame Ionization Detector). The exhaust gas was then chilled to the temperature of 4 °C and the concentrations of nitrogen oxides (NDUV, Non-Dispersive Ultra-Violet), carbon monoxide/carbon dioxide (NDIR, Non-Dispersive Infrared) and oxygen (electrochemical analyzer) were measured. The device is compatible with the vehicle's on-board diagnostic system (recording of the operating parameters – engine speed and load) and GPS (latitude and longitude for determining the vehicle speed).

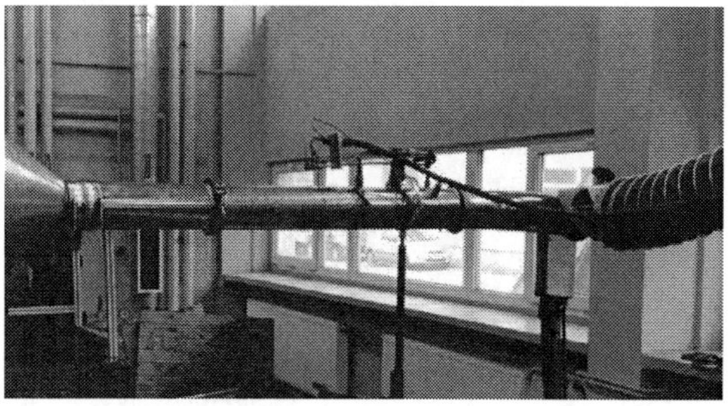

Fig. 7. View of the flow meter.

The authors used MSS (*Micro Soot Sensor*) by AVL (Fig. 6) for the measurement of particulate matter. The device uses laser light dispersion triggered by particulate matter contained in the exhaust gas. AVL MSS can determine the real time concentration of PM in the exhaust gas. The number of the particles was measured by EEPS TSI (Fig. 8).

In this device, the solid particles get an electrical charge and then they reach the electrodes, the larger the particle the further the electrode reached.

Fig. 8. TSI EEPS device during tests.

Before starting the measurements, the analyzers were calibrated with reference gases according to the manufacturer's guidelines. This calibration consists of providing a gas with a precise composition to individual analyzers. Each analyzer is provided with 5 gas canisters having different ratio of ingredients to each other and containing gases of the highest possible purity. This allows to restore the original characteristics of individual devices and guarantee measurement of the actual value of individual toxic components.

These operations were done every single cycle that was performed to make sure the results are comparable. In addition, the offset value was determined using outside air before the measurements. This action alleviates the impact of the environment in which the measurements are made on the results. None of the analyzers reported errors in operation during tests.

4 Research methodology

The tractor was tested in operating points according to the modified NRSC type approval test. This test was changed by prof. Lijewski based on the study of the operating points of the internal combustion engine of NNRM in real operation [25]. Various field work performed by agricultural tractors, combine harvesters and construction machinery was investigated. Based on the obtained results the engine's operating points have been changed compared to the NRSC test, which only partially maps the actual operation. Based on the research, it has been found that the range of the engine's operating points is much narrower than the test specifies. The points with the highest loads were shifted towards the lower loads and the points with the low loads of the engine were shifted upwards. No external load at the specified engine speeds has been added to the test to verify the behavior of the filter with low temperature exhaust and significant air excess ratio.

The differences relate to the load intervals used, the engine rotational speeds are selected in the same manner as in the norm, respectively 60 and 90% rotational speed at the maximum engine power. Such a test allows for better mapping of real conditions during stationary tests The operating points of combustion engine used in the studies are presented in Tab. 3 and Fig. 9.

The authors did not use the characteristics of the engine provided by the manufacturer. For the purpose of the tests, the actual external characteristics of the engine were drawn up and the torque at the given crankshaft rotational speed was determined.

Table 3. Engine operating points [25].

Point No.	Engine speed [% of the speed at maximum engine power]	Engine speed [rpm]	Load [%]	Load [Nm]	Weight
1	60	1410	90	878	0.15
2	60	1410	65	633	0.2
3	60	1410	40	390	0.2
4	60	1410	0	0	0.1
5	90	2115	80	672	0.1
6	90	2115	55	462	0.1
7	90	2115	30	252	0.15
8	90	2115	0	0	0.15
9	idle	850	–	–	0.2

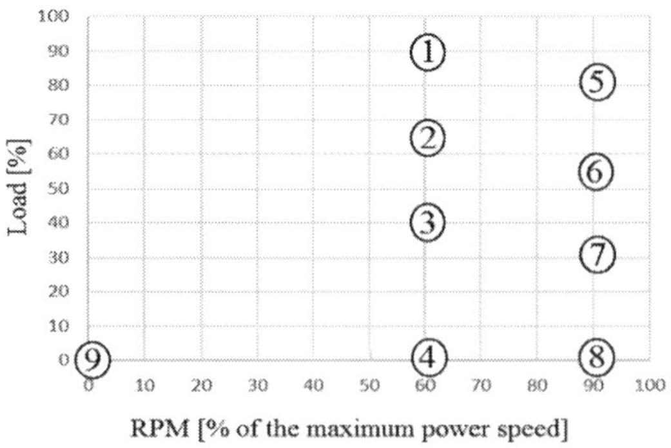

Fig. 9. The operating points of combustion engine [25].

At each operating point the engine worked for 60 second and the transition time between points was determined by the inlet and outlet exhaust gas temperature. The measurement was taken when the temperature has not changed more than 1 °C for a period of 10 seconds. The entire cycle was performed three times, and the presented results are the weighted average values of the obtained results.

5 Result analysis

Based on the concentration readings on the analyzers and the mass flow rate of the exhaust gas and the instantaneous engine power, the specific PM emissions were calculated (Tab. 4). In an similar manner, the specific particle number emissions were calculated using the volumetric concentration of particulate matter (Tab. 5).

Table 4. Results of specific PM emission with and without DPF.

Point No.	Engine speed [rpm]	Load [Nm]	Without DPF [g/kWh]	With DPF [g/kWh]
1	1410	878	0.43	0.06
2	1410	633	0.42	0.13
3	1410	390	0.64	0.33
4	1410	0	1.12	0.65
5	2115	840	0.73	0.15
6	2115	672	0.55	0.08
7	2115	462	1.10	0.25
8	2115	0	1.92	0.92
9	850	157	0.35	0.43
Weighted average	–	–	0.73	0.34

There is a noticeable decrease in particle mass at each point of the engine operation, which is particularly high for points where the engine is operating under heavy load (seven fold difference for the first operating point). It can be argued that the filter is more effective at higher exhaust temperatures where passive regeneration using nitrogen dioxide for oxidation of solid particles occurs. As the load decreases, the efficiency decreases. Increased emissions at idle may be caused by increased exhaust gas flow resistance at low exhaust gas

DOI: 10.1051/matecconf/201711800028

temperatures preventing oxidation of the particles. Finally, the average test result for the system without a filter is 0.73 and the system with filter 0.34 g/kWh (Fig. 10). The obtained values of PM emission are much higher than limits even for Stage IIIA.

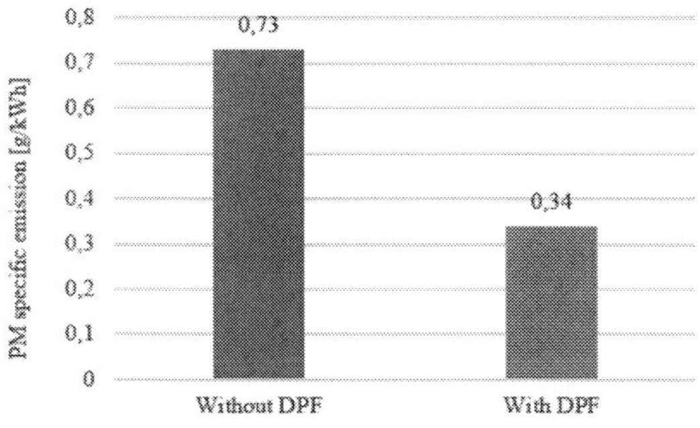

Fig. 10. Weighted average of the specific PM emissions for the system with and without the DPF.

For a number of particles, a similar relationship can be observed as for the mass, i.e. the most effective filter operation takes place at the highest load settings (operating points 1 and 5), where the number of particles decreases more than four times (Tab. 5). For points with a low load, the filter performance is similar to or greater than that of the standard exhaust system, again due to low exhaust temperature and no passive regeneration. As a result, the number of particles decreases almost twice from 1.81E+12 to 9.22E+11 (Fig. 11).

Table 5. Results of PN emission with and without DPF.

Point No.	Engine speed [rpm]	Load [Nm]	Without DPF [1/kWh]	With DPF [1/kWh]
1	1410	878	1.43E+12	3.27E+11
2	1410	633	9.79E+11	4.48E+11
3	1410	390	5.25E+11	8.79E+11
4	1410	0	3.71E+11	3.34E+11
5	2115	840	4.02E+12	8.35E+11
6	2115	672	4.04E+12	9.54E+11
7	2115	462	3.05E+12	2.63E+12
8	2115	0	4.75E+11	6.47E+11
9	850	157	2.94E+11	3.2E+11
Weighted average	–	–	1.81E+12	9.22E+11

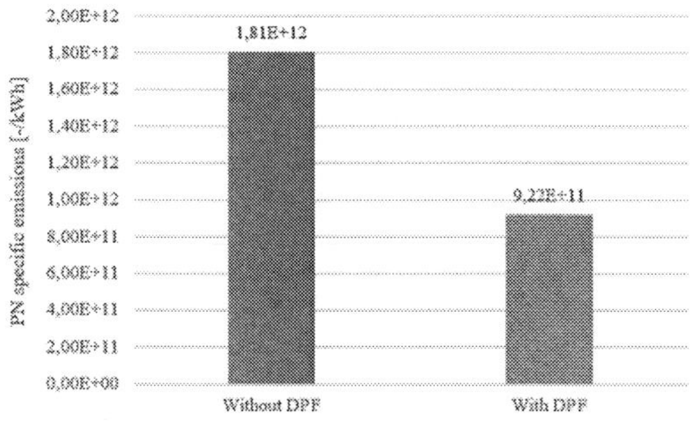

Fig. 11. Weighted average of the specific PN emissions for the system with and without the DPF.

The particle size distribution was also analyzed. This is particularly important because as mentioned the smallest particles are the most dangerous for humans, and the number itself does not provide the information on what particle size is permanently oxidized by the exhaust aftertreatment systems. The average particle size distribution for the standard system is shown in Fig. 12, and after using the DPF in the exhaust system is shown in Fig. 13.

It is noteworthy that the size distribution is very similar in both cases. There are fairly equal values for the smallest particles and for particles with the size of several dozen nanometers, the distribution is similar to a Gaussian curve. What differentiates these distributions is the scale used. After using the filter, a considerable reduction in the number of medium and large particulates in the exhaust gas can be observed. The values for the smallest particles are practically unchanged.

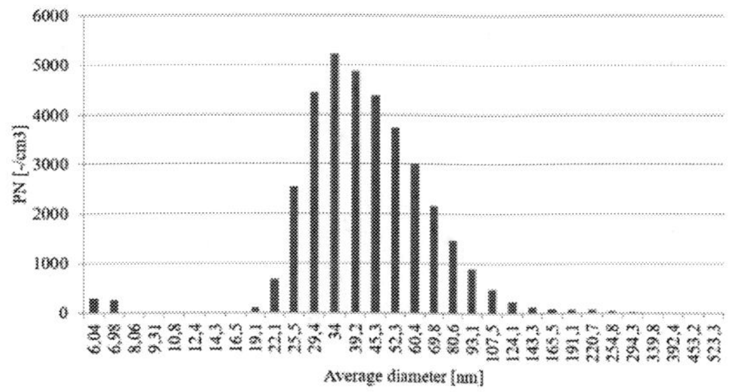

Fig. 12. The average particle size distribution of the standard exhaust system without DPF.

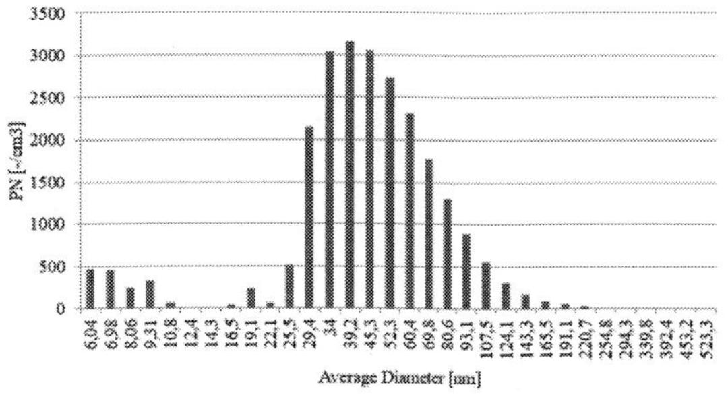

Fig. 13. The average particle size distribution of the standard exhaust system with DPF.

This demonstrates the efficiency of the filter in stopping and oxidation of large particles with little to no effect on small particles. In addition, the catalytic layer used could affect the reduction of large particles without complete oxidation.

6 Summary

Current trends in combustion engines aim to reduce their negative impact on the environment by combining it with other engine types (hybrids) or the use of alternative fuels (fuel cells, methane, LPG). The use of these solutions in NRMM vehicles causes a number of complications that make it as if for the next few years the CI internal combustion engine dominates the propulsion system of this group. It is characterized primarily by the availability of fuel and the ability to quickly refuel. The use of modern exhaust aftertreatment systems will allow to achieve the emission values that are not yet possible for several years, especially for nitrogen oxides and particulate matter. However, this applies to new vehicles, while older vehicles, whose emissions may be several times higher, are still in use.

For this reason, the popularity of downsizing, which means retrofitting vehicles with modern exhaust gas aftertreatment systems. This solution is relatively inexpensive, and the theoretical method allows for significant reduction of emissions especially PM, PN and nitrogen oxides.

The research carried out in the article concerned the retrofitting of a farm tractor with a universal particle filter with the size chosen relative to the engine displacement. Placing the filter in the exhaust system has measurable effects on particle emissions both in terms of number and mass of particulate matter. The use of the filter made it possible to reduce the emitted mass and number of solid particles from the exhaust system by a half.

The trend, which means retrofitting older tractors with newer exhaust aftertreatment systems, can generate positive environmental effects far exceeding the gains obtained through improving new constructions where the resulting effects are much smaller. In addition to the number of used vehicles, the new vehicles represent only a few percent of vehicles currently in use. The use of the modified NRSC test allowed the engine to work in conditions similar to those occurring during field operations. The reduction in the emissivity of the tested tractor would probably have been even greater if the filter was closer to the exhaust system (higher temperature) and if it had been optimized for that particular engine.

The research was funded by project co-financed by the European Regional Development Fund in the Regional Program – Lubuskie 2020 (contract No. RPLB.01.01.00-08-0055/16-00).

References

1. J. Merkisz, P. Lijewski, P. Fuc, M. Siedlecki, S. Weymann, Applied Engineering in Agriculture **31**, 875-879 (2015)
2. H. Adam, *9th AVL International Commercial Powertrain Conference* 48-54 (2017)
3. P. Fuc, J. Merkisz, P. Lijewski, M. Bajerlein, A. Ziolkowski, L. Rymaniak, M. Dobrzynski, Combustion Engines **154**, 701-707 (2013)
4. R. Borsboom, *9th AVL International Commercial Powertrain Conference* 9-13 (2017)
5. W. Burget, *9th AVL International Commercial Powertrain Conference* 15-18 (2017)
6. F. Rahe, R. Resch, *9th AVL International Commercial Powertrain Conference*, 114-120 (2017)
7. J. Merkisz, P. Lijewski, P. Fuc, M. Siedlecki, A. Ziolkowski, *IOP Conference Series-Materials Science and Engineering* **148**, UNSP 012077 (2016), DOI: 10.1088/1757-899X/148/1/012077
8. M. Bajerlein, L. Rymaniak, Technika Transportu Szynowego **10**, 2773-2782 (2013)
9. M. Idzior, M. Bajerlein, P. Lijewski, P. Fuc, *Annual Global Science and Technology Forum* 227-233 (2013)
10. J. Merkisz, P. Fuc, P. Lijewski, J. Pielecha, Transportation Research Procedia **14**, 3070-3078 (2016), DOI: 10.1016/j.trpro.2016.05.452
11. M. Bajerlein, L. Rymaniak, A. Ziółkowski, P. Świątek, Technika Transportu Szynowego **10**, 2763-2772 (2013)
12. W. Eckerle, V. Sujan, G. Salemme, *9th AVL International Commercial Powertrain Conference* 19-47 (2017)
13. M. Bajerlein, P. Fuc, P. Lijewski, L. Rymaniak, A. Ziolkowski, M. Dobrzynski, Combustion Engines **154**, 708-715 (2013)
14. J. Markowski, J. Pielecha, R. Jasinski, T. Kniaziewicz, P. Wirkowski, *E3S Web of Conferences* **10**, UNSP 00140 (2016), DOI: 10.1051/e3sconf/20161000140
15. J. Merkisz, J. Markowski, J. Pielecha, *WIT Transactions on Ecology and the Environment* **174**, 219-230 (2013), DOI: 10.2495/AIR130191
16. J. Pielecha, J. Merkisz, J. Markowski, R. Jasinski, *E3S Web of Conferences* **10**, UNSP 00073 (2016), DOI: 10.1051/e3sconf/20161000073
17. P. Fuc, P. Lijewski, A. Ziolkowski, *IOP Conference Series-Materials Science and Engineering* **148**, UNSP 012065 (2016), DOI: 10.1088/1757-899X/148/1/012065
18. P. Fuc, L. Rymaniak, A. Ziolkowski, *WIT Transactions on Ecology and the Environment* **174**, 207-218 (2013), DOI: 10.2495/AIR130181
19. J. Merkisz, J. Markowski, J. Pielecha, *IEEE Vehicle Power and Propulsion Conference VPPC*, 897-902 (2012
20. J. Merkisz, J. Pielecha, *Springer Tracts on Transportation and Traffic* **8**, 131-139 (2015), DOI: 10.1007/978-3-319-15928-7_10
21. P. Lijewski, J. Merkisz, P. Fuc, SAE Technical Paper 2012-01-1965 (2012)
22. J. Kromulski, T. Pawlowski, J. Szczepaniak, W. Tanas, A. Wojtyla, M. Szymanek, J. Tanas, W. Izdebski, Annals of Agricultural and Environmental Medicine **23**, 373-376 (2016), DOI: 10.5604/12321966.1203908

23. J. Szczepaniak, W. Tanas, T. Pawlowski, J. Kromulski, Annals of Agricultural and Environmental Medicine **21**, 403-406 (2014), DOI: 10.5604/1232-1966.1108613
24. Sensors Inc. Emissions Measurement Solutions (2010)
25. P. Lijewski, Post-doctoral thesis (Poznan, 2013)

Laser diagnostics for urea-water solution spray characterization

Łukasz Jan Kapusta[1,*], and *Andrzej* Teodorczyk[1]

[1]Warsaw University of Technology, Faculty of Power and Aeronautical Engineering, ul. Nowowiejska 21/25, 00-665 Warszawa, Poland

Abstract. In this study, we summarize the laser techniques used for urea-water solution (UWS) spray characterization at the Institute of Heat Engineering (ITC), Faculty of Power and Aeronautical Engineering at Warsaw University of Technology. In presented studies several techniques for both, global and local spray characterization were used. Shadowgraphy-based long distance microscopy was used to visualize individual droplets and primary breakup. High speed imaging of Mie scattering (scattering on the gas-liquid interface) signal was used for global spray characterization. Combination of LIF (Laser Induced Fluorescence) and Mie scattering allowed to determine qualitative droplet size distribution across the whole spray cloud. The structured illumination technique used to modulate laser light sheet allowed to minimize the effects of multiple scattering in detection of Mie signal, what indicated huge potential of this technique in characterization of UWS sprays. The results presented here prove the importance of laser diagnostics in SCR systems development.

1 Introduction

Laser diagnostics for spray characterization has been well established in the area of fuel injection and mixture formation in both CI (Compression Ignition) and SI (Spark Ignition) engines. The application of sprays in CI engines due to implementation of SCR (Selective Catalytic Reduction) systems has extended. Moreover, due to development of close-coupled SCR system designs, and therefore decreased distance for water evaporation and urea decomposition the importance of spray in proper NOx reduction increases. NOx reduction efficiency strongly depends on the quality of mixing UWS with surrounding exhaust gases. UWS-surrounding gas mixing process is the main factor determining NH3 distribution before the inlet to the SCR catalyst where NOx reduction takes place.

The accurate measurement methods for fast and reliable UWS sprays characterization are of high importance in terms of spray pattern and injection process optimization. This is especially important for system optimization using CFD (Computational Fluid Mechanics) methods. It is of crucial importance to provide accurate input parameters such as droplet size distribution, spray angle, and spray pattern geometrical parameters.

In this paper, we summarize the laser techniques used to determine these parameters. The UWS spray characterization was done in the Laser Diagnostics Laboratory at Warsaw University of Technology. We used several techniques which allowed both, global and local spray characterization: shadowgraphy-based long distance microscopy, high speed imaging of Mie scattering (scattering on the gas-liquid interface), LIF/Mie (Laser Induced Fluorescence together with Mie scattering).

Due to relatively large droplets and relatively dilute sprays generated by SCR system injectors the droplet diameters may be determined by direct visualization using high magnification optics. As shown by Grout et al. [1], even direct Mie scattering imaging can be used to determine the size of individual droplets. They used this approach to determine temporal change of the droplet size and calculate the evaporation rate according to D-square evaporation law. Commonly used method for individual droplets visualization is backlight imaging (shadowgraphy) [1, 2]. This method coupled with long distance microscopy allows to visualize individual droplets of diameter lower than 10 μm provided that the spray is not optically dense. This feature is important in terms of proper droplet size distribution determination especially when the range of droplet sizes is relatively wide. Therefore, in this study we used shadowgraphy-based long distance microscopy. This technique allows also to precisely visualize primary breakup what was done in this study as well.

High speed imaging of Mie scattering signal is a convenient technique for fast global spray characterization. Simple global illumination (e.g. halogen lights) can be used for spray angle and spray tip penetration determination [3]. Its great advantage is that it can be used for determination of initial velocity of a liquid emerging the nozzle just after injector opening provided the frame rate is high.

LIF/Mie technique was used for qualitative analysis of droplet size distribution across the whole spray cloud. This technique was used for many years for gasoline spray characterization. It is based on simultaneous LIF and Mie signal measurements. Since under assumption

* Corresponding author: lukasz.kapusta@itc.pw.edu.pl

of spherical droplet LIF signal is proportional to the volume of the droplet and Mie signal to its surface the LIF/Mie ratio represents the SMD (Sauter Mean Diameter). In order to determine quantitative SMD distribution calibration factor for LIF/Mie ratio shall be first determined. The calibration procedure was not included in the study and the results presented here are intended to give qualitative information on the SMD distribution across the spray. As explained earlier, in order to determine SMD quantitatively shadowgraphy-based long distance microscopy was used.

Recent findings of Berrocal [4] and Mishra [5] show that even in fairly dilute sprays effects of multiple scattering. Therefore, additionally Mie signals were captured using structured illumination. The results were compared with conventional Mie signals in order to conclude on the effects of multiple scattering on detected Mie signal and recognize the potential of this technique in LIF/Mie droplet sizing in the area of SCR sprays, what is planned to be done in near future.

2 Experimental setup

2.1 Shadowgraphy-based long distance microscopy

In all experiments the same injector was used (commercial 3-hole Bosch injector for SCR systems no. 0 280 158 720). UWS in this study was replaced by pure water. Only in LIF/Mie droplet sizing study water was doped with eosin Y for the purpose of LIF signal generation. Injection pressure was of 5 bar. Water was injected into ambient conditions. The environment and water temperature was of 293 K.

The setup for shadowgraphy included several components among which the most important was the Nd:YAG laser Quanta-Ray Pro-230. The 2nd harmonic of 532 nm wavelength was used to illuminate dye in the diffuser. The diffuser changed the wavelength to 574-580 nm. The light emitted by the diffuser was used to illuminate the spray. The camera was equipped with long distance microscope, which was adjusted to visualize area of 88x85 mm of the spray. The center of the image was located at 30 mm from the injector tip, in the axis of the injector. The schematic diagram of the experimental setup is shown in Fig. 1.

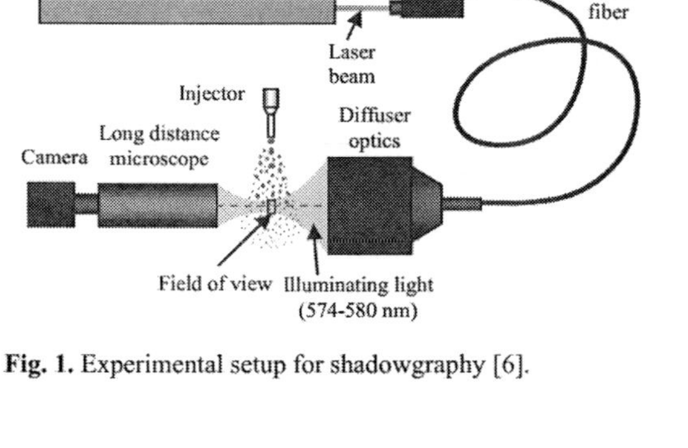

Fig. 1. Experimental setup for shadowgraphy [6].

2.2 High speed imaging of Mie scattering

The setup for high speed imaging of Mie scattering included the high speed camera and the light source. In this setup, global illumination was used. Two halogen lights of power 500 W each were used to illuminate the spray cloud. Photron SA1.1 camera with Nikon f 2.8 50 mm lens was used. The frame rate was of 10 000 fps. The resolution was of 1024×450 pix. The setup is shown in Fig. 2.

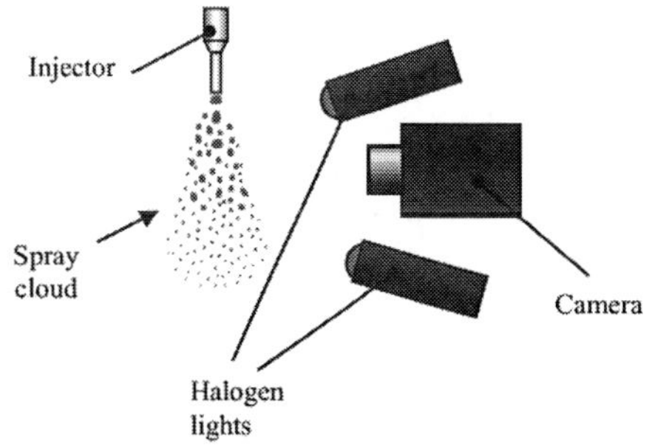

Fig. 2. Experimental setup for high speed imaging of Mie scattering.

2.3 LIF/Mie droplet sizing

In order to capture both signals LIF and Mie generated by the same droplets in the setup for LIF/Mie droplet sizing two cameras were used at the same time. Otherwise the number of images used for averaging would be significantly higher. The cameras were equipped with different filters to capture pure LIF and Mie signal. The cameras were located perpendicular to the light sheet – each camera was located at different side of the laser light sheet. As a light source second harmonic beam generated by Nd:YAG pulse laser was used. Note that injector used here was 3-hole injector and the spray cloud generated by the injector was formed by 3 single sprays. The light sheet was crossing one jet directly in its axis while going between the other two. The laser light sheet was of 60 mm wide and started 5 mm from injector outlet, since at the exit of the nozzle liquid ligaments and strongly deformed droplets are present. The experimental setup is shown in Fig. 3.

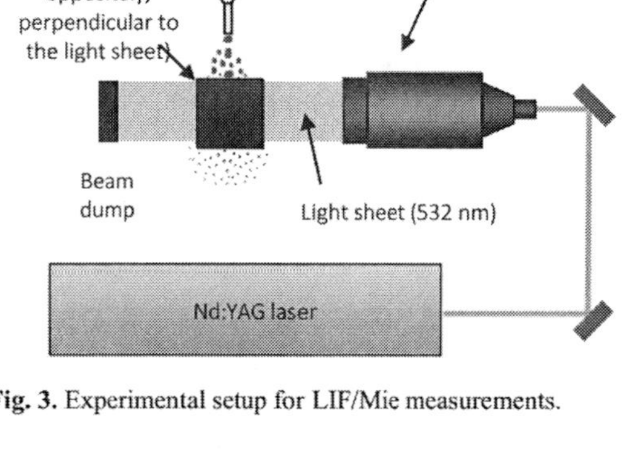

Fig. 3. Experimental setup for LIF/Mie measurements.

2.4 SLIPI

The structured illumination technique referred to as SLIPI (Structured Illumination Planar Imaging) was developed for sprays by Berrocal et al. [7] and Kristenson et al. [8] and commercially implemented by LaVision [9]. The experimental setup including the SLIPI optics for structured light generation is shown in Fig. 4.

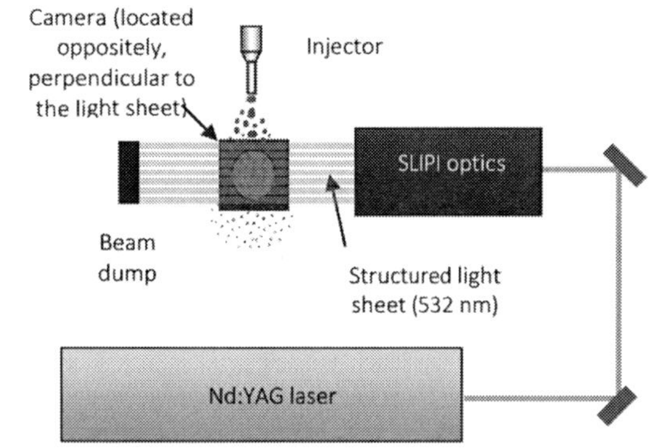

Fig. 4. Experimental setup for LIF/Mie measurement using structured illumination [10].

The modulated light sheet used in SLIPI visualization was entering the spray the same way as in LIF/Mie experiments, i.e. crossing one jet directly in its axis while going between the other two.

The SLIPI processing is performed by recording several modulated images where the modulation is vertically shifted between each recording [11]. The studies using structured illumination shown here were done only for Mie signal. The camera was located perpendicularly to the light sheet. As a light source again second harmonic beam generated by Nd:YAG pulse laser was used. The conventional images were reconstructed from three series of modulated images. The structured light sheet was created by SLIPI optics delivered by LaVision. The schematic setup of SLIPI optics is shown in Fig. 5.

In order to obtain LIF signal injected water was doped with Eosin Y.

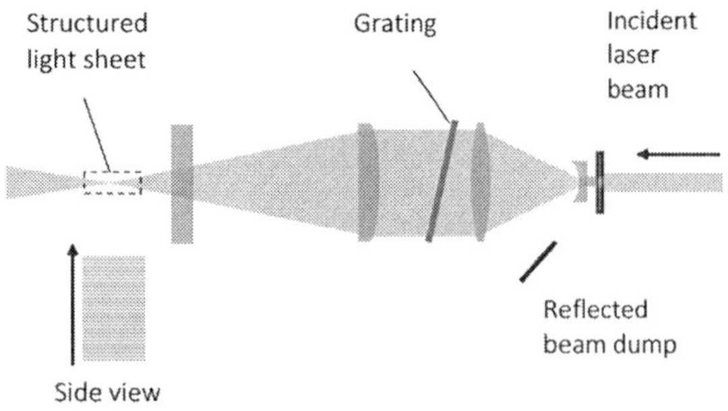

Fig. 5. Schematic setup of complete SLIPI optics - top view [10].

3 Results

3.1 Shadowgraphy-based long distance microscopy

Shadowgraphy-based long distance microscopy was used to visualize primary breakup (Fig. 6a) and determine droplet size distribution for the purpose of spray model calibration done in [12]. Due to frequency of the laser limited to 10 Hz only one image per injection was taken. The single image with indicated recognized particles is shown in Fig. 6b.

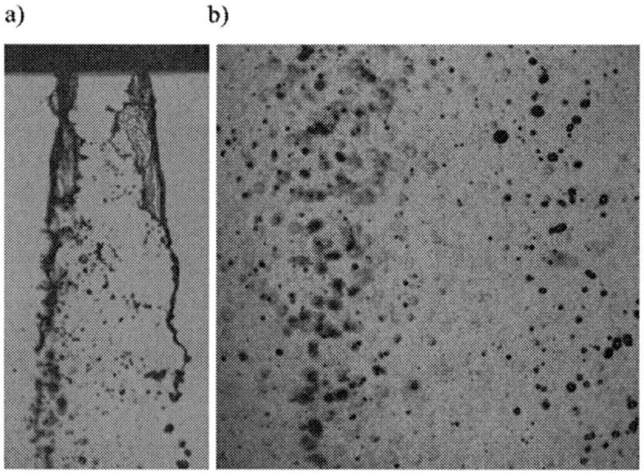

Fig. 6. Instantaneous shadowgraphy images; a) primary breakup (field of view: 39 × 85 mm), b) indicated recognized particles (field of view: 85 × 88 mm).

In order to perform accurate statistical analysis 100 injections were studied. At 100 images 6376 droplets in total were recognized. The detected droplet sizes were in the rage from 0 µm to 300 µm. This range was divided into 30 subranges each of 10 µm width. This data was used to construct droplet size distribution histogram (Fig. 7).

Basing on the acquired information on droplet sizes, SMD was calculated. The SMD was of 92.7 µm.

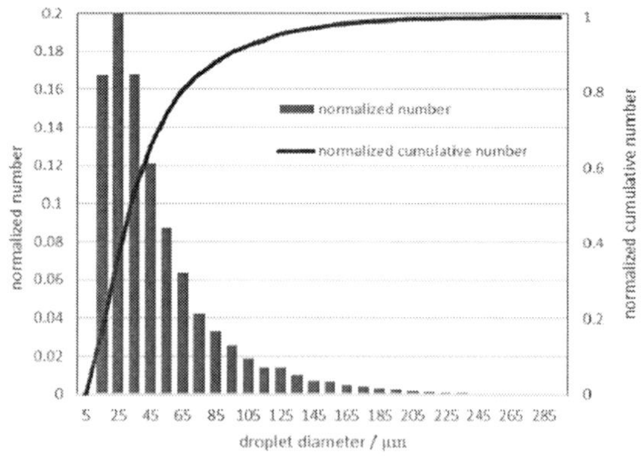

Fig. 7. Droplet size distribution.

3.2 High speed imaging of Mie scattering

High speed imaging was used to determine spray penetration and initial liquid jet velocity in order to use the data for spray model calibration [12] as well. The single frames captured 1 ms, 2 ms, 3 ms and 7 ms after start of injection are shown in Fig. 8 respectively from left.

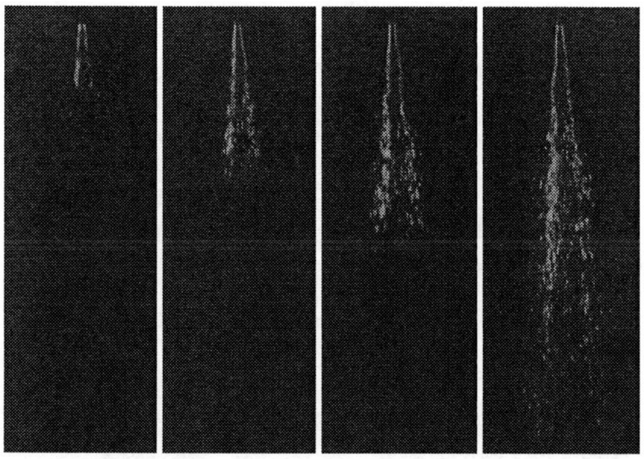

Fig. 8. Instantaneous Mie scattering high speed images 1 ms, 2 ms, 3 ms and 7 ms after start of injection, respectively from left.

Basing on the acquired images the spray penetration was determined using LaVision DaVis version 8.2. The spray penetration evolution is shown in Fig. 9. The initial velocity of the liquid jet emerging the nozzle after the injector opening determined from first frames where jet occurred was of 24 m/s. The image where the spray was already developed was used to determine the spray visualization angle. Spray visualization angle determined from image captured 7 ms after start of injection was of 16.54 deg.

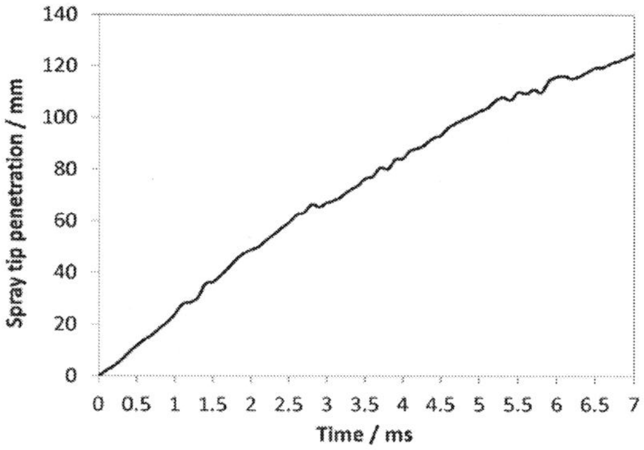

Fig. 9. Spray penetration evolution (adopted from [12]).

3.3 LIF/Mie droplet sizing

The procedure of image acquisition included obtaining 200 frames for each signal (LIF and Mie). Both signals were captured for the same laser pulse. Instantaneous LIF and Mie images for the same laser pulse are shown in Fig. 10. Note that the field of view at all single images presented in Figs 10-13 is of 60 mm high and 45 mm wide. Note that laser light sheet enters the spray from right on all LIF images and from left on all Mie images.

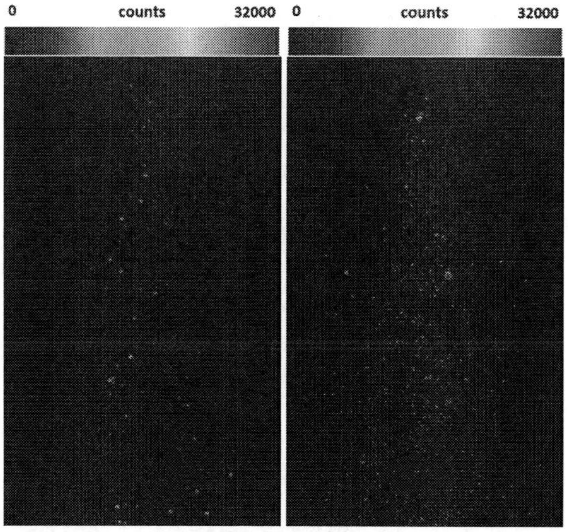

Fig. 10. Instantaneous LIF and Mie images, respectively from left.

Additionally to spray images the background image was recorded for each signal. The background images were recorded with laser pulse but without the spray. The instantaneous images shown in Fig. 10 were processed in order to subtract the background signal. Instantaneous LIF and Mie images for the same laser pulse after background subtraction are shown in Fig. 11. The 200 instantaneous images recorded for each signal after background subtraction were averaged. The averaged images of LIF and Mie signals are shown in Fig. 12. The averaged images shown in Fig. 12 were used to create LIF/Mie ratio image (shown in Fig. 13). Note that laser light sheet in Fig. 13 enters the spray from right.

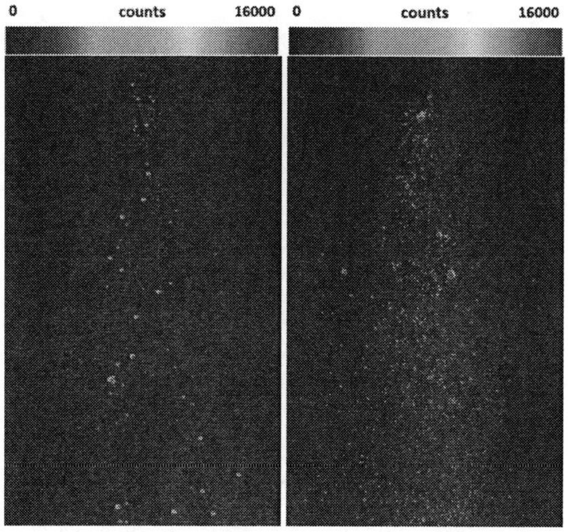

Fig. 11. Instantaneous LIF and Mie images after background subtraction, respectively from left.

DOI: 10.1051/matecconf/201711800029

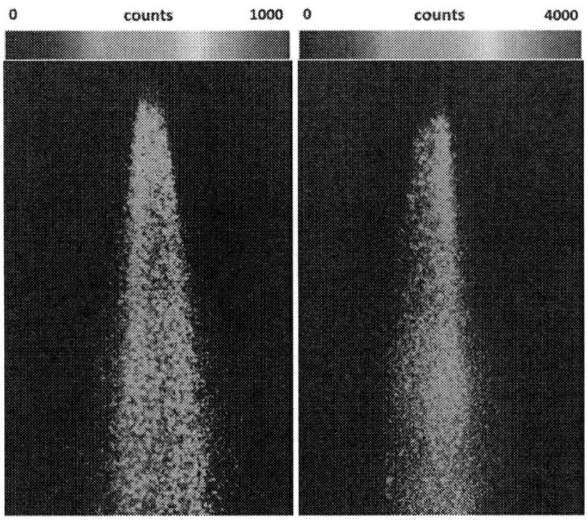

Fig. 12. Averaged LIF and Mie images, respectively from left.

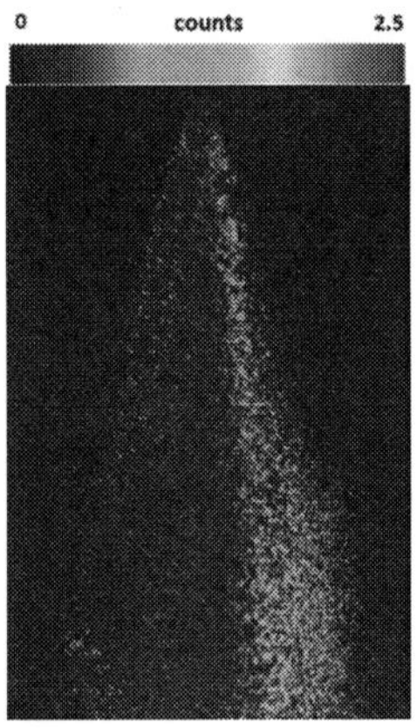

Fig. 13. LIF/Mie ratio.

LIF/Mie ratio in most of the visualized area remained unchanged, what suggests that SMD remains unchanged as well. One needs to be aware that at 90 deg detection angle the scattered light intensity with the droplet diameter is irregular [13]. Therefore, LIF/Mie ratio shall be treated as qualitative indicator, not as direct SMD result. Slightly stronger signal can be observed along right nozzle axis (the spray emerging that nozzle was illuminated directly in the middle). This seems to be in accordance with the shadowgraphy results shown in Fig. 6b. There, between the jets only small number of very small droplets can be observed, while on the right side the main stream of large droplets is present. Note that the visualization depth of shadowgraphy is much higher than in case of light sheet imaging. Even though in Fig. 6b large droplets are present also on right side of the image they are blurred. This indicates that they are located a bit closer or further than the focal point. Therefore, they are not detected by light sheet imaging. The other increase of LIF/Mie ratio can be observed at higher distance from the nozzle on the right side of the spray (Fig. 13), what suggests bigger droplets. This result is questionable since there is no reason to observe bigger droplets in that area. This might have come either from multiple scattering or other noise signal. Therefore the conventional imaging in case was compared with SLIPI imaging [10].

3.4 SLIPI

For the purpose of comparing conventional and SLIPI imaging, only Mie signal was used. The procedure of image acquisition included obtaining series of frames per one modulation phase. Full investigation into influence of number of frames on both SLIPI and conventional imaging results is shown in [10]. Here only results for case using 1000 frames per each phase are shown. Again, only one image per injection was captured. Recording series of 1000 images required 1000 separate injections. 1000 instantaneous images for each modulation phase were averaged. The average phase images are shown in Fig. 14.

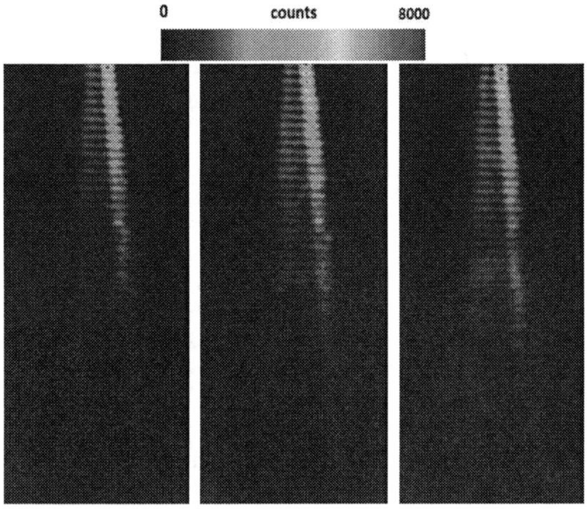

Fig. 14. Averaged Mie images for three modulation phases; the visualized area on single image is of 60 mm high and 22 mm wide.

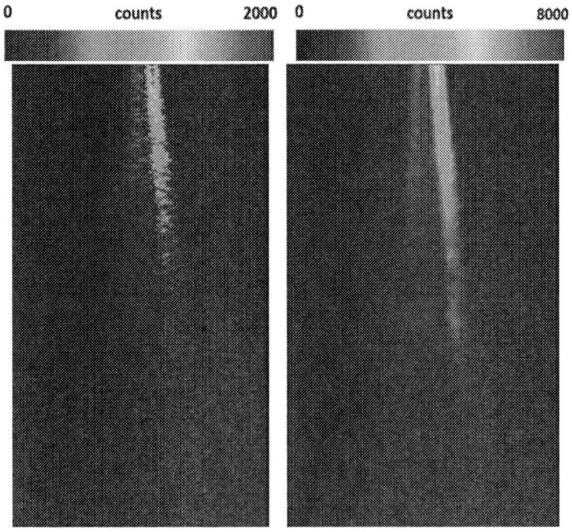

Fig. 15. SLIPI (left) and conventional (right) Mie scattering image [10]; the visualized area on single image is of 60 mm high and 35 mm wide.

The averaged images were used to create SLIPI image and reconstruct conventional images. Both SLIPI and conventional images are shown in Fig. 15.

In general, averaged Mie signal determined using SLIPI and conventional processing were very similar. The SLIPI technique was however performing much better in areas between the single spray jets. In conventional image Mie signal between the spray jets was of similar intensity as in the background, especially at higher distances from the nozzle exit. The areas where droplets were observed in SLIPI image are clearly visible and can be easily distinguished from the background. The difference is much higher when SLIPI Mie signal (Fig. 15-left) is compared with Mie signal obtained using flat light sheet (Fig. 12-right). However, it needs to be noted that the number of frames used to create these images was different. The observed difference between SLIPI and conventional images indicates huge potential of SLIPI technique in droplet size distribution determination for SCR system applications by means of LIF/Mie method, what is planned to be done in near future.

Additionally, the obtained conventional and SLIPI images shown in Fig. 14 were used to determine the visualization angle. The visualization angle for that case was very similar for conventional and SLIPI imaging, 17.58 deg and 17.7 deg respectively.

4 Conclusions

The techniques used in the study were capable to deliver all the information on the sprays generated by the SCR system injector needed for their proper characterization. Such a wide information on spay parameters can be successfully used to adjust and calibrate CFD calculation setup for full SCR system investigations. Obtaining these parameters required several techniques which were based on the same laser system. Shadowgraphy-based long distance microscopy was used to visualize individual droplets and primary breakup. High speed imaging of Mie scattering allowed to determine global spray parameters, specifically spray tip penetration, and spray angle. Combination of LIF and Mie scattering allowed to determine qualitative 2D droplet size distribution across the whole spray cloud.

The studies done with structured illumination showed huge potential of SLIPI technique in droplet size distribution determination for SCR system applications by means of LIF/Mie method. This is planned to be done in near future.

It needs to be noted that the visualization angle determined using global illumination was different from the angle determined using light sheet imaging (both, SLIPI and conventional). For global illumination, the angle was of 16.54 deg while for light sheet imaging was of 17.58 deg and 17.7 deg for conventional and SLIPI imaging, respectively. Imaging using global illumination is affected by the thickness of the spray cloud and therefore much stronger signal is detected in the center of the spray than in the peripheral region. Therefore, light sheet imaging result is more accurate.

The laser diagnostics provided all the data on SCR spray parameters critical for SCR systems development. In general, it may be concluded that the results presented in the study prove importance of laser imaging techniques in SCR systems development.

This work was supported by the European Smart Growth Operational Programme 2014-2020 through the project "Development of mixing and urea-water solution conversion unit in SCR systems in order to start production of exhaust system for compression ignition engine that meets the Euro 7 emission standards", grant number: POIR.04.01.04-00-0060/15-02.

References

1. S. Grout, J.B. Blaisot, K. Pajot, G. Osbat, Fuel **106**, 166-177 (2013), DOI:10.1016/j.fuel.2012.09.022
2. L. Postrioti, G. Brizi, C. Ungaro, M. Mosser, F. Bianconi, Fuel **150**, 548-557 (2015), DOI:10.1016/j.fuel.2015.02.067
3. I. Pielecha, International Journal of Automotive Technology **15**, 47-55 (2014), DOI: 10.1007/s12239-014-0005-y
4. E. Berrocal, *18th Int Symp Appl Laser Imaging Tech to Fluid Mech* (2016)
5. Y.N. Mishra, E. Kristensson, E. Berrocal, Opt Express **22**, 4480 (2014), DOI:10.1364/OE.22.004480
6. L.J. Kapusta, P. Jaworski, A. Teodorczyk, J. Kowalski, Journal of KONES Powertrain Transp **22**, 91-98 (2015), DOI:10.5604/12314005.1165976
7. E. Berrocal, E. Kristensson, M. Richter, M. Linne, M. Aldén, Opt Lett **16**, 17870-17881 (2008), DOI:10.1364/OE.16.017870
8. E. Kristensson, E. Berrocal, M. Richter, S. Pettersson, M. Aldén, Opt Lett **33**, 2752-2754 (2008), DOI:10.1364/OL.33.002752
9. P. Leick, R. Grzeszik, S. Arndt, S. Wissel, *24th Eur. Conf. Liq. At. Spray Syst.* 1-10 (Estoril, 2011)
10. L.J. Kapusta, *Proc. 8th Eur. Combust. Meet.* 1541-1546 (Dubrovnik, 2017)
11. E. Berrocal, E. Kristensson, P. Hottenbach, M. Aldén, G. Grünefeld, Appl Phys B **109**, 683-694 (2012), DOI:10.1007/s00340-012-5237-9
12. J. Bachanek, R. Rogóż, L.J. Kapusta, A. Teodorczyk, *Proc. 8th Eur. Combust. Meet.*, 996-1000 (Dubrovnik, 2017)

An impact assessment of functional systems in vehicles on CO_2 emissions and fuel consumption

Jerzy Merkisz[1,*], *Michal* Dobrzynski[1], and *Kamil* Kubiak[1]

[1]Poznan University of Technology, Faculty of Machines and Transport, ul. Piotrowo 3, 60-965 Poznan, Poland

Abstract. The number of facilities for a driver increasing the travel comfort in a PC vehicle is constantly increasing. Most of them are powered by electricity generated at a cost of additional engine load. Power is received directly from the crankshaft by an alternator or air conditioning compressor, which negatively affects fuel consumption. This paper presents the results of CO_2 emission testing in actual traffic conditions for vehicles with on and off functional systems responsible for driving comfort. In addition, an assessment of fuel consumption calculated using the carbon balance method was performed. Measurements were made for cold start and preheated combustion unit the research used the latest PEMS mobile apparatus.

1 Introduction

Among the many factors that negatively affect the quality of human life is motorization. Striving for a higher standard of living forces automotive companies to introduce new technological solutions, whose task is to improve the safety of travel, lowering the cost of travel, and increasing the comfort of driving and adjusting the vehicle to the tastes of the driver. Modern functional systems in the vehicle are directly associated with improved driving characteristics, as well as the comfort of traveling. There are many expressions of travel comfort, which are not precisely defined, and that is used intuitively with variable and often overlapping meaning: comfort, passenger comfort, driving quality, driving comfort [1]. Improvements in operating conditions can be defined on the basis of the concept definition "Comfort" in the article [2] presented, and it should be understood as follows "a set of devices and conditions that ensure comfort and fully meet the aesthetics". It can be stated that travel comfort refers to the "state of physical, mental and aesthetic satisfaction associated with the movement of a passenger on a given route". There is also a whole range of external conditions that ensure a person's sense of safety, characterized by a state of physical and mental fulfillment, and the elimination of unforeseen situations [3, 4].

Improvements of the usefulness of relevant material from many points of view. One of the aspects is the continual increase in the number of vehicles on the roads, as a consequence it leads to an ever-increasing transit time on the same stretch of road [5, 6]. Another source is the growing supply of car manufacturers in the world markets, which forces increasing competitiveness of the offered products. This is especially evident in the regions of North America and Europe where saturation of the car market can be observed – an insignificant increase in the number of vehicles. The only exceptions are developing countries, where the share of vehicles per capita is inconsiderable, and this increase is almost four times larger [8].

Undoubtedly the consequence of the development and the increasing number of systems improving the use of vehicles is to increase energy intensity, which translates into unit fuel consumption, and increased exhaust gas emissions. Current regulations concerning emissions testing from motor vehicles do not allow for the precise definition of the impact of using additional electric power receivers on the emissions in real operating conditions RDE. The progressing in recent years development and miniaturization of measuring apparatus in the PEMS group, makes it possible to perform more and more accurate vehicle environmental testing. Considering the current trends aimed at ever increasing dissemination and use of functional systems mounted in vehicles it is justified to undertake activities that influence the assessment of their ecological aspects [4, 8, 9].

2 Methodology of research

2.1 Apparatus used for measuring emissions of harmful substances in exhaust gases

Measurement of fuel consumption and emission of harmful exhaust gas components, carried out within the framework of the research, required the use of specialized exhaust gas toxicity measurement apparatus [7, 10]. The research performed (in the conditions of actual vehicles operation) used the mobile test apparatus type PEMS, namely SEMTECH DS of Sensors Inc. (Fig. 1), a unique combination of high measuring accuracy and the ability to record the following parameters:

* Corresponding author: jerzy.merkisz@put.poznan.pl

DOI: 10.1051/matecconf/201711800030

- concentrations of CO and CO_2 (analyzer NDIR), $NO_x = NO + NO_2$ (analyzer NDUV), HC (analyzer FID), O_2 (electrochemical sensor),
- thermodynamic exhaust parameters (mass flow, temperature, pressure) – a flow meter using Pitot tube,
- environmental conditions – atmospheric pressure, temperature, humidity,
- vehicle position and speed – GPS,
- data from vehicle diagnostic network – data transmission protocol CAN SAE J1939/J2284.

In the tests carried out the main unit, the SEMTECH DS device, was placed inside the vehicle (Fig. 2). On the roof of the vehicle was a GPS positioning system and weather sensor, which were connected to the main unit of the instrument. Connection to the diagnostic network was made by a special data module using a dedicated protocol for PC category vehicles. The sample of exhaust gas for analysis was taken from the flow meter and transported by a heated line to maintain the temperature of 190 °C. This is to protect against emission condensation on the walls of the cord. Then the sample, after passing through the filter, goes to the analyzer NDIR, which made it possible to register the CO_2 concentration. Controlling and monitoring the SEMTECH DS instrument was done using a laptop connected to the main unit via a wireless network.

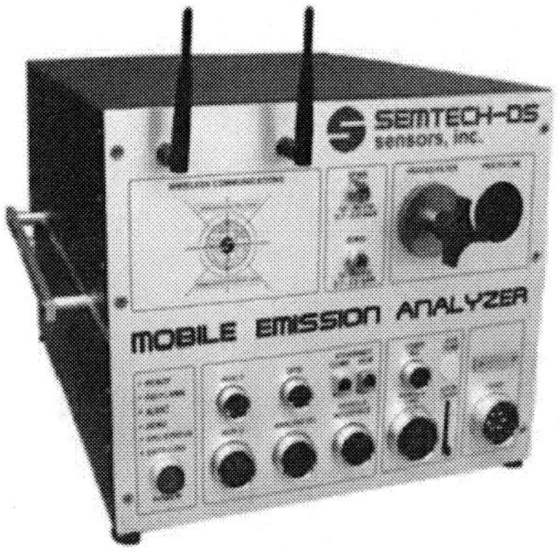

Fig. 1. Mobile exhaust gas analyzer SEMTECH DS.

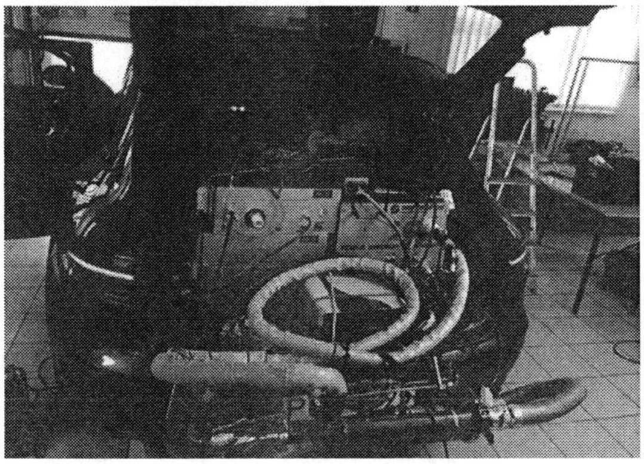

Fig. 2. SEMTECH DS measuring device during installation in a vehicle.

Measurement ranges of the apparatus allow for its application in a wide range. In addition to measuring the concentration of individual gaseous pollutants in the exhaust gases, it also allows measurement of mass flow rate (with the mentioned flowmeter). It is necessary to calculate the emission of these compounds (e.g. road, specific). It is important to note that due to the varying displacement of engines in vehicles, there is also a difference in the amount of exhaust gas. Therefore, it is necessary to select exhaust gas flow meters for testing that have an appropriate diameter (Fig 3). Equipment SEMTECH DS, when using the so-called carbon balance method, also allows to calculate the fuel consumption of the vehicle.

Currently this type of measurement is set to be integrated into the European legislative action (especially in the measurement of gas component emissions). For environmental assessment of vehicles the road tests are most desirable (in this case actual operating conditions), because only then it is possible to obtain information about real vehicle emissions. Tests are performed, for example, on stretches from several to dozens of kilometers in different road conditions. The biggest disadvantage, however, is the cost of measuring equipment and adaptation to the conditions in the vehicle. On the other hand, reliable measurement results are obtained, Impossible to obtain under simulation conditions or on a stationary dynamometer.

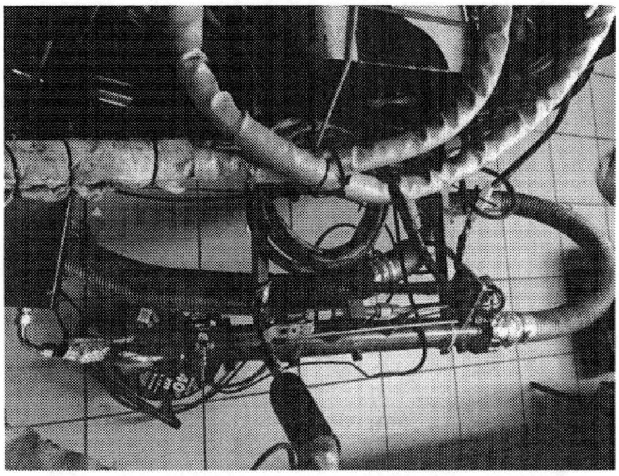

Fig. 3. Mass flow mass meter installed on exhaust system.

2.2 Research objects

The research used PC category vehicles equipped with a number of functional systems vehicle A (Fig. 4a) equipped with a spark-ignition engine powered by direct injection of gasoline. The vehicle was equipped with a three-way catalytic converter TWC, along with a lambda probe working in the feedback loop. The second vehicle marked B (Fig. 4b) powered by diesel fuel, which carries out a combustion cycle with compression ignition, equipped with an additional DPF in the engine exhaust system, and selective catalytic reduction in SCR. The Table 1 shown below contains the basic parameters in used propulsion units.

Table 1. Vehicle technical parameters.

Parameter	Tested vehicle	
	Vehicle A	Vehicle B
Engine type/fuel	4-stroke, SI/Gasoline	6-stroke, CI /Diesel
Displacement [dm^3]	2.0	3.0
Compression ratio	9.6	16.8
Max. power [kW]/[rpm]	185/6800	193/4000
Max. torque [N·m]/[rpm]	370/ 1600–4500	580/ 1750–2500
Aftertreatment system	TWC	DOC/DPF/SCR
Drive system	Start–stop	Start–stop
Vehicle curb weight [kg]	1845	2185

Presented objects as standard are equipped with automatic transmission gears, and their propulsion units meet the Euro 6 emission standard of exhaust gas [11]. Modern solutions used in this category of vehicles are activated directly by the user. Depending on the demand, these functions allow for making changes in the operating characteristics of the power unit, transmission ratios, suspension, thermal conditions inside the vehicle, and audio-visual elements. All implemented components require the proper functioning of the power supply generated by the internal combustion engine used to drive the vehicle, affecting its additional load during regular operation.

a)

b)

Fig. 4. Research objects used for research: a) vehicle A powered by gasoline, b) vehicle B powered diesel fuel.

2.3 Measurements carried out in real vehicle traffic conditions

As standard, type approval tests for passenger vehicles are carried out on chassis dynamometer, whose results are inadequate to the conditions that exist during actual operation. To make a real assessment of the pollutant emissions to determine the impact of the application non-essential vehicle assistance systems measurements were made based on a route representative of the category of vehicles tested [9, 12]. Road emission studies were carried out on the urban route in the Poznan agglomeration (Fig. 5). In some parts of the route there were a large number of vehicles causing road congestion, thanks to that, it was possible to reproduce conditions in the strict city center. In addition, on the selected route it was possible to obtain larger average speeds during the ride, which allowed to simulate driving through non-urban regions and on the main communication arteries. The total length of the measuring section was 12 km.

Fig. 5. Route used in RDE.

Passages were made to achieve the purpose of the research in the following configurations, and the adopted signs defined in the following way the work of the systems:

- "cold engine, no power receivers" – deactivated power receivers,
- "hot engine, no power receivers" – deactivated power receivers,
- "hot engine, active power receivers" – activated all circuits with energy consumption,
- "hot engine, active power receivers, sport mode" – change of operating characteristics of the drive unit, power transmission system, suspension system, activated all power systems.

The set of devices that are power receivers in the case in question are:

- interior air conditioning system,
- ventilated / heated seats,
- interior lighting,
- audio-visual system.

3 Research results

Scheduled route drive cycles have been repeated many times, each time making changes to the configuration implemented in the vehicle's functional systems. The first stage of the study refers to the registration of emissions from the vehicle without the use of additional power receivers, both in the conditions before reaching the thermal equilibrium of the drive unit, as well as after reaching the state of thermal equilibrium. These measurements made it possible to set a reference point for the intended purpose in the resulting work.

The first test vehicle identified as A was powered with Gasoline. On the basis of the recorded data CO_2 concentration was determined (Fig. 6). The results of the measurements are shown in (Tab. 2).

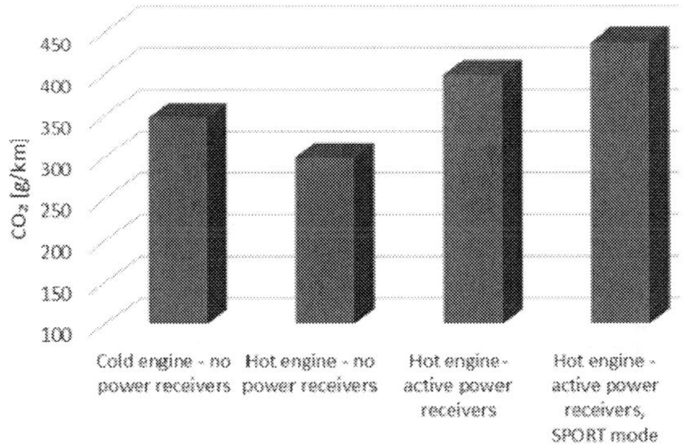

Fig. 6. Carbon monoxide emission in vehicle A.

Table 2. Results of measurements from vehicle B.

Parameter	Cold engine, no power receivers	Hot engine, no power receivers	Hot engine, active power receivers	Hot engine, active power receivers, sport mode
CO_2 [g/km]	347.2	299.3	398.6	437.8
Fuel consumption [dm^3/100 km]	14.4	12.4	16.6	19.5
Distance traveled [km]	12.1	12.2	12.3	12.1

Obtained results of road emission during the conducted tests allowed to determine the fuel consumption (Fig. 7) the carbon balance method, which uses the formula:

$$FC_w = \frac{1155}{\rho_{\text{fuel}}} \cdot [(0,866 \cdot \text{HC}) + (0,429 \times \text{CO}) + (0,273 \times \text{CO}_2)] \quad (2)$$

where: FC – fuel consumption [dm^3/100 km]; HC, CO, CO_2 – emission of ingredients harmful [g/km]; ρ_{fuel} – fuel density at temperature 15 °C [g/cm^3].

During the first passage the propulsion unit was struggling to achieve thermal equilibrium, the power receivers were inactive. Concentration of carbon dioxide emission in the test reached 347 g/km, and the fuel consumption was 14.5 dm^3/100 km. Another passage made after reaching the rated operating conditions has demonstrated a reduction in unit fuel consumption and emissions of CO_2. In the next stage of the study, "warm engine, active power receivers" the drive components worked in standard mode with all functional systems activated. Registered measurements showed the highest concentration CO_2, and an increase in fuel consumption of over 52%, reaching the value of 19 dm^3/100 km. Working conditions during the mobile test extended by the active work system suspension system, and the modified operating characteristics of the drive unit marked as "SPORT mode" showed the highest levels of measured parameters. The results obtained can be attributed to the unfavorable influence of drive control on the selected test section. During the vehicle speed test phase the gearbox mode controller has forced high speed values. Change gear ratio to the higher gear was in this case only after exceeding 60% of the range engine crankshaft rotational speed, thus increasing the share of working time during generated loads. Fuel consumption and carbon dioxide emissions have reached values, which significantly deviated from the previously performed tests.

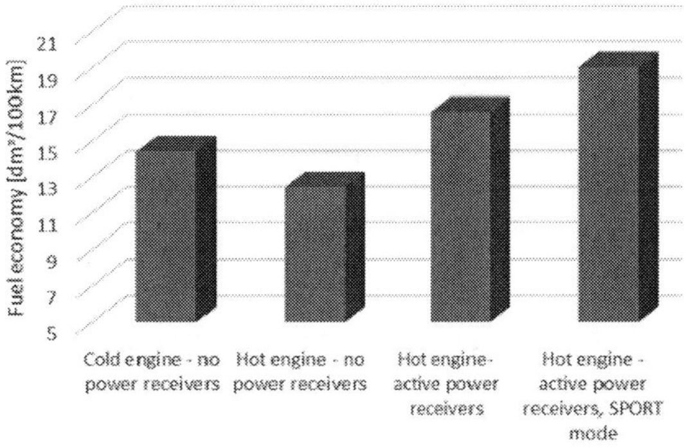

Fig. 7. Fuel consumption while traveling from vehicle A.

The second of the objects tested powered by diesel fuel exhibits similar tendencies of obtained characteristics relative to the first of the tested vehicles. Considering, however, for a different type of drive unit used smaller differences in the designated fuel consumption were recorded (Fig. 8) as well as the concentration of emissions CO_2 (Fig.9). The results are shown in Table 3.

Notably, road tests were carried out under similar ambient conditions. The concentration of carbon dioxide in the first case reached 218 g/km, at the same time fuel consumption ratio was 8 dm^3/100 km. Another attempt

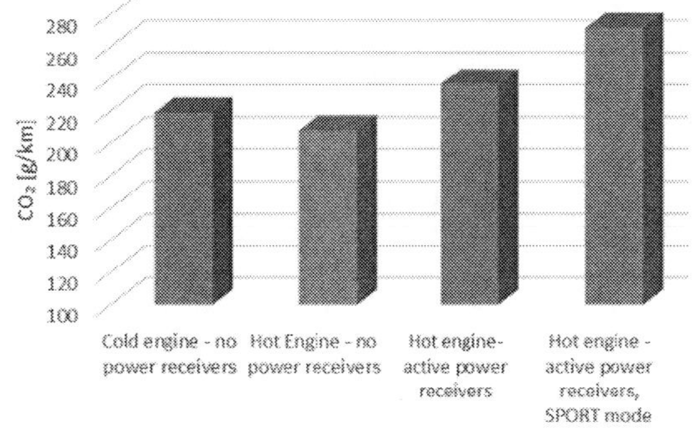

Fig. 8. Carbon monoxide emission in vehicle B.

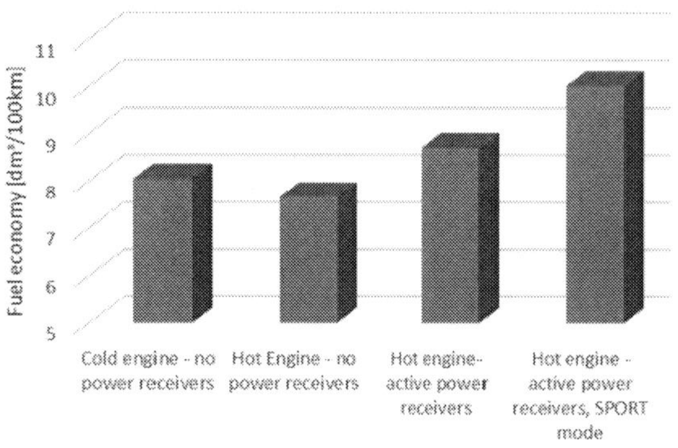

Fig. 9. Fuel consumption while traveling from vehicle B.

Table 3. Results of measurements from vehicle B.

Parameter	Cold engine, no power receivers	Hot engine, no power receivers	Hot engine, active power receivers	Hot engine, active power receivers, sport mode
CO_2 [g/km]	347.2	299.3	398.6	437.8
Fuel consumption [dm^3/100 km]	8.1	7.6	8.7	10.0
Distance traveled[km]	12.1	12.2	12.3	12.1

was made after stabilization of the engine operating conditions and an increase in CO_2 concentration by 13% was recorded. Under test conditions extended by the active suspension system, and the modified operating characteristics of the drive unit has occurred at the highest fuel consumption in the test. In the case of the vehicle B the differences referring to the adopted starting point in the performed analyzes have reached a maximum change of 30%, both for fuel consumption and for carbon dioxide emissions.

4 Summary

In this article a comparison of tests performed in actual operating conditions is drawn. The unified research route, and use for testing vehicles of one category It had a positive effect on the comparative analysis of the results obtained, because they have excluded any discrepancies associated with the route profile, physical characteristics of the tested vehicles, and drive components.

Based on the results obtained it can be seen that the use of additional solutions in the form of functional elements implemented in vehicles has a significant impact on the emissions, and also on the vehicle fuel consumption. The comparative analysis made recorded increases in value, which in the extreme cases have reached levels exceeding 50%, exposing the adverse effects on the ecological aspects, as well as on the economic considerations during operation.

In the context of the research done it is worth noting that the current research of exhaust emissions for previous Euro categories do not take into account the impact of high-involvement of additional components installed in vehicles on their ecological indicators. Use of the research methodology based on the application of PEMS type apparatus, as used in the realized research work, allows for more precise estimation of the exhaust emissions in real operating conditions from vehicles equipped with a series of increasingly accessible facilities. It is therefore reasonable to conduct further research based on the methodology presented in this article.

The research was funded by project co-financed by the European Regional Development Fund in the Regional Program – Lubuskie 2020 (contract No. RPLB.01.01.00-08-0055/16-00).

Fundusze Europejskie Program Regionalny Lubuskie Unia Europejska Europejski Fundusz Rozwoju Regionalnego

References

1. M. Andrzejewski, *PhD Thesis* (Poznan, 2013)
2. J. Merkisz, J. Pielecha, S. Tarkowski, Autobusy, technika, eksploatacja, systemy transportowe **13**, 300-305 (2012)
3. K. Brzozowski, J. Nowakowski, Eksploatacja i Niezawodnosc – Maintenance and Reliability **4**, 56-62 (2011)
4. Z. Stepien, L. Ziemianski, G. Zak, M. Wojtasik, L. Jeczmionek, Z. Burnus, Fuel **161**, 278-286 (2015), DOI: 10.1016/j.fuel.2015.08.071
5. J. Merkisz, I. Pielecha, *Alternatywne napędy pojazdów* (Poznań, 2006)
6. J. Merkisz, J. Pielecha, P. Lijewski, A. Merkisz-Guranowska, M. Nowak, *WIT Transactions on Ecology and the Environment* **174**, 27-38 (2013), DOI: 10.2495/AIR130031
7. J. Merkisz, M. Bajerlein, P. Daszkiewicz, *Lecture Notes in Information Technology* **13**, 106-113 (2012)
8. R. Jasinski, J. Pielecha, J. Markowski, *E3S Web of Conferences* **10**, UNSP 00030 (2016), DOI: 10.1051/e3sconf/20161000030
9. J. Merkisz, J. Pielecha, I. Pielecha, M. Szukalski, Transport Samochodowy (2010)
10. J. Pielecha, J. Merkisz, J. Markowski, R. Jasinski, *E3S Web of Conferences* **10**, UNSP 00073 (2016), DOI: 10.1051/e3sconf/20161000073
11. K. Engeljehringer, 2nd *International Exhaust Emissions Symposium* (2011)
12. P. Fuc, L. Rymaniak, A. Ziolkowski, *WIT Transactions on Ecology and the Environment* **174**, 207-218 (2013), DOI: 10.2495/AIR130181

MATEC Web of Conferences **118**, 00031 (2017) DOI: 10.1051/matecconf/201711800031
VII International Congress on Combustion Engines

Evaluation of turbine microjet engine operating parameters in conditions conducive to inlet freezing

Jaroslaw Markowski[1,*], *Damian* Olejniczak[1], and *Pawel* Wirkowski[2]

[1]Poznan University of Technology, Faculty of Machines and Transport, ul. Piotrowo 3, 60-965 Poznan, Poland
[2]Polish Naval University, Faculty of Mechanical and Electrical Engineering, ul. Smidowicza 69, 81-127 Gdynia, Poland

Abstract. The problem of turbine microjet engine operation is related to flight conditions of unmanned aircraft. These flights are often performed at low altitudes, where, in autumn and winter conditions, the air can be characterized by high humidity and low temperature. Such operating conditions may cause freezing the turbine engine inlet. In particular, this problem may be related to microengines, which most often are not equipped with a de-icing installation. Frosting of the inlet violates the air flow conditions at the engine inlet and may cause unstable operation and even outages, which eventually may lead to a loss of aircraft's stability and breakdown. Therefore, an attempt was made to evaluate the changes in operational parameters of the turbine microjet engine under conditions leading to the freezing of the inlet. The engine test was performed in stationary conditions and the analysis of the obtained results are presented in this article.

1 Introduction

The process of converting the state of water contained by the air in the form of steam into a solid form accumulating on the elements of the aircraft structure during flight is called icing. The occurrence of aircraft icing is a phenomenon that adversely affects the volatility of the aircraft. It changes the wing profile and increases the weight of aircraft [1-3]. Icing depends on many factors and occurs under favorable weather conditions such as high humidity and low temperature. The atmospheric conditions conducive to the icing of the inlet system of the turbine engine are:

- low dew point temperature,
- high air humidity,
- presence of water in the atmosphere in all forms, especially in the form of clouds, precipitation, overflowing water,
- air temperature less than 10 °C.

The icing of the turbine engine inlet system is particularly undesirable because of the negative effects on engine performance. It narrows the cross section of the engine intake duct, which translates into a decrease in the mass of the flowing medium, and consequently results in reduced engine thrust [4, 5]. Icing of individual components of the turbine engine compressor can cause undesired blade vibration and deterioration of the fluid flow conditions in the interstitial space. The icing of the turbine engine's inlet can cause a power loss of up to 40% of the rated power and even lead to the motor compressor stall. In extreme cases it may turn off the flame in the combustion chamber or permanently damage the engine.

The location of the ice layer in the inlet system of the turbine engine depends on the size of the water molecules contained in the air and on the type of construction of the element located in the airflow zone. Small molecules of water are icing single protruding elements of the inlet system. These include temperature and air pressure measuring elements placed in the inlet for controlling the engine systems. Due to the operation of airplanes under the conditions of icing, the pressure measurement elements in the engine inlet and the hull of the aircraft used for determining the flight velocity are continuously heated to prevent the freezing of these components and thus losing basic flight parameters. On the other hand, during freezing rain or high-watered clouds, icing already occurs at the edge of the inlet of turbine engine, on the walls of intake duct, the cap, inlet wrenches of compressor steering wheel. This type of icing is intense and can destabilize the engine.

The accumulation of ice on the turbine engines inlet causes diameter reduction of the engine air inlet, aggravating the airflow conditions, which results in the deterioration of engine performance. The basic problem is the reduction of engine thrust value. String is a basic parameter characterizing a jet engine. By adopting a one-dimensional thermodynamic thermoform model, the engine sequence can be defined as the difference in the flow rate of the medium at the inlet and outlet of the jet engine. The icing of turbine engine air inlet leads to reduction in the air mass flow rate, by narrowing the inlet flow channel and thereby reducing the value of the thrust generated. A single-turbine engine sequence is represented by the dependence (1).

* Corresponding author: jaroslaw.markowski@put.poznan.pl

$$F_c = \dot{m}\,[(1 + \tau)\,C_5 - V_H] \qquad (1)$$

where:
F_c – engine thrust value [N],
$\dot{m}$ – mass flow rate of the medium [kg/s],
τ – percentage of fuel relative to airflow [%],
C_5 – speed at the engine outlet [m/s],
V_H – speed at the inlet to the engine [m/s].

The icing of inlet systems is a particular problem for the operation of engines without de-icing systems. Undoubtedly, aircrafts equipped with turbine engines without anti-icing installations include unmanned aerial vehicles. An unmanned aircraft equipped with a turbine engine, which for structural reasons is devoid of anti-icing installations, is exposed to icing of the intake system. Therefore, the problem of safe use of turbine-powered unmanned aerial vehicles in autumn-winter and winter-spring periods is emerging. Unmanned aerial vehicles, depending on their purpose, fly at low altitude during most of their operating hours, where during this period the air is characterized by high humidity and low temperatures. Therefore, in order to evaluate changes in micro engine turbine parameters, a stationary test was performed on selected turbine microengine under conditions conducive to freezing of the inlet.

2 Methodology

The object of the stationary test was the GTM-120 turbine engine (Fig. 1). It is a single-flow engine, with single-stage radial compressor, annular combustion chamber and single-stage axial turbine. The stationary test and measurement station consists of additional pneumatic starter system and electronic control and data recording system for the GTM-120 engine [6-8]. Fuel consumption was measured using a fuel gauge operating according to the gravimetric method. Measurement of the thrust value was made by the KMM20-200N force sensor. In order to assess the temperature distribution of the inlet of the turbine engine during the test, a thermal imaging camera was used. To simulate precipitation conditions that increase air humidity within the engine inlet, a spray nozzle for water supplied from the water supply was used. The air temperature at the time of the test was $t_{ot} = -2$ °C, atmospheric pressure $p_{ot} = 1025$ hPa.

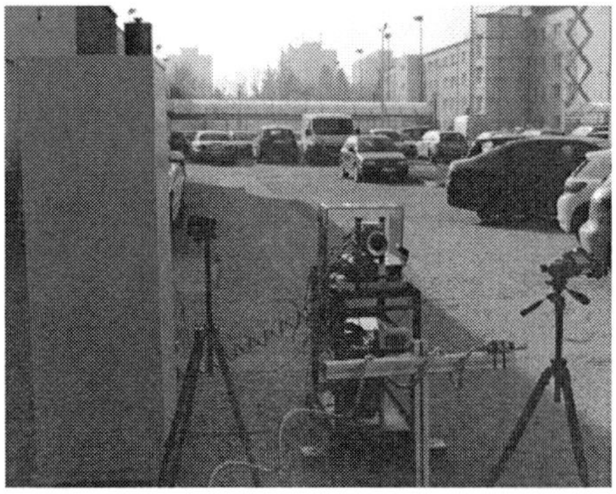

Fig. 1. The view of a test bench during a stationary test of the GTM-120.

The research test was divided into two phases. In the first stage the speed characteristics of the full-range engine operating at ambient conditions were recorded. Measurements of operating parameters were made for the set thrust in the minimum stroke range of approximately $F_c = 15$ N, then $F_c = 20$ N, subsequent set points with 10 N step incrementally increasing the engine sequence to a maximum value of $F_c = 130$ N.

The second stage of the study consisted of evaluating changes in engine operating parameters caused by changes in ambient conditions by spraying water in the vicinity of the engine inlet. These tests were performed for the selected operating range corresponding to the set point of the thrust force $F_c = 90–130$ N. This is a range of 70–100% of the maximum thrust. This value selection was based on the experience of studies and data available in the literature [6, 7], which indicate the value of the turbine engine load as the load most likely to occur during engine operation.

After setting the engine parameters for the resulting thrust value $F_c = 90$ N, the water was sprayed at a distance of about 700 mm from the engine inlet. The distance of the spray nozzle relative to the inlet was determined by the direction and speed of the wind so as to provide a mist of water sprayed by the engine before the inlet (Fig. 2). The test was performed until the dynamic pressure measuring tube in the engine inlet was frozen. The measuring instrument allowed registration the operating parameters of the engine with frequency of 1 Hz. Fuel consumption was recorded at fixed engine operating points as averaged over a 5 second period. Registered data was analyzed.

Fig. 2. The view of the GTM-120 stationary test during spraying water before engine inlet, $t_{ot} = -2$ °C, $p_{ot} = 1025$ hPa.

3 Analysis of the results

During the tests basic operating parameters were determined at ambient conditions at air temperature $t_{ot} = -2$ °C, atmospheric pressure $p_{ot} = 1025$ hPa. For these conditions the engine characteristics have been determined, showing the dependence of the measured parameters: T_1 – air temperature at the engine inlet [K], T_2 – air temperature behind the compressor [K], T_3 –

DOI: 10.1051/matecconf/201711800031

temperature behind the combustion chamber [K], T_4 – exhaust temperature in the exhaust nozzle [K]; p_c – pre-turbine pressure [kPa], G_e – fuel consumption [g/s], as a function of thrust F_c. The waveforms of individual parameters as a function of thrust are shown in graphical form (Fig. 3–5).

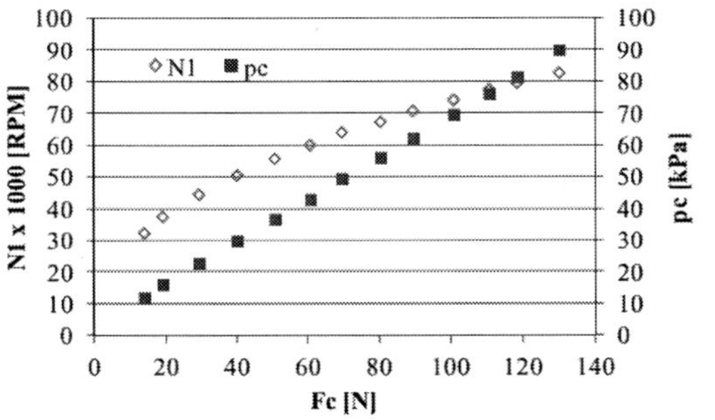

Fig. 3. GTM-120 engine shaft speed and pre-turbine pressure as a function of thrust under ambient conditions.

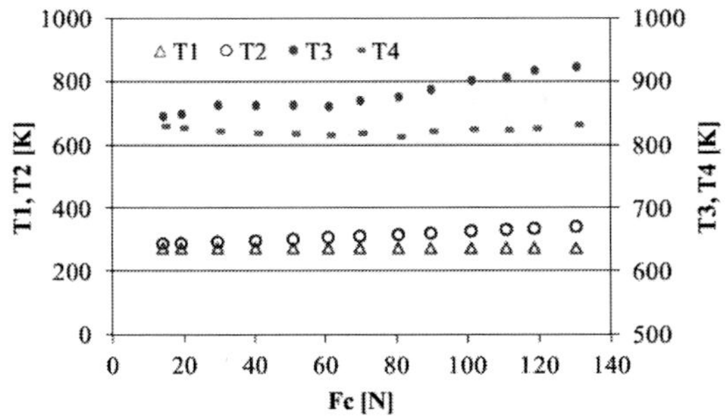

Fig. 4. The temperature values at the individual measuring points of the GTM-120 engine as a function of thrust under ambient conditions.

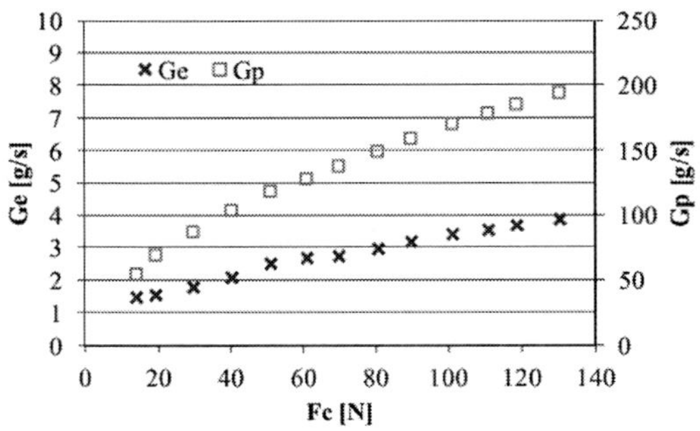

Fig. 5. Fuel consumption and air consumption as a function of thrust under ambient conditions.

Presented characteristics refer to changes in engine operating parameters provided in the second stage of the test, during which water was sprayed before the engine's inlet. The study was conducted for four thrust values. The first set of thrust was F_c = 90 N. During the operation of the engine with a set thrust value, the spray nozzle was positioned in the space before the engine inlet. Correction of the thrust set was made and measurements were conducted. In the following steps, thrust values were changed and operating parameters were measured. Registered changes of individual operating parameters were applied to the characteristics obtained during the first stage of the test (Fig. 6–8).

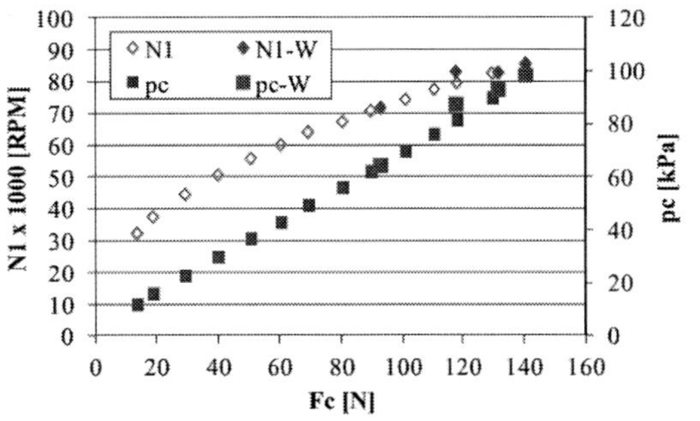

Fig. 6. GTM-120 engine shaft speed and pre-turbine pressure as a function of thrust with applied values obtained by spraying water, under ambient conditions.

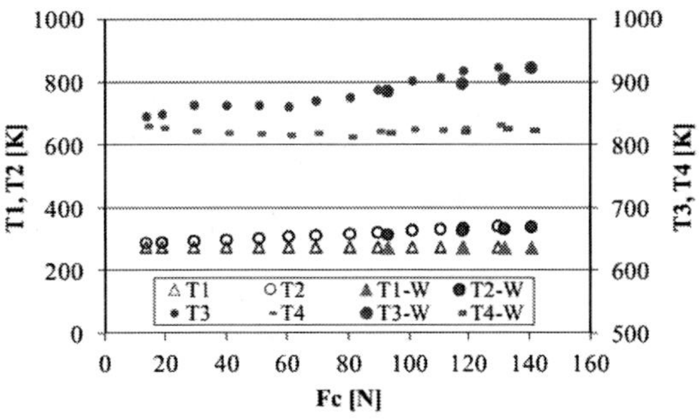

Fig. 7. The temperature values at the individual measuring points of the GTM-120 engine as a function of the thrust force with the applied values obtained when spraying water, under ambient conditions.

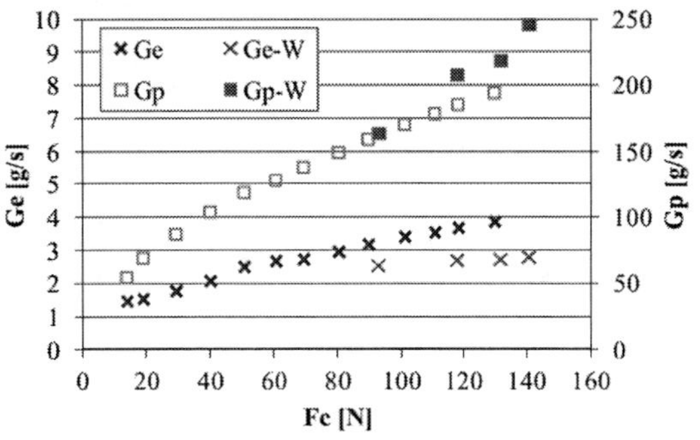

Fig. 8. Fuel consumption and air consumption as a function of thrust for GTM-120 engine with the values obtained when spraying water, under ambient conditions.

Comparative assessment of operating parameters obtained in the first and second stage of the test indicates that the spraying of water before the inlet of the turbine engine does not cause significant changes in the rotational speed of the motor shaft N1 and the pressure of the medium before the turbine p_c. It also does not significantly affect the temperature at the inlet of the engine T_1, the temperature behind the compressor T_2, the temperature before the turbine T_3. This results in a decrease in the flue gas temperature in the engine exhaust nozzle by approximately 20 K. The changes in air parameters caused by the spray of water resulted in an increase of air flow and engine thrust while fuel consumption G_p decreased. Airflow increased by approximately 10% and fuel consumption G_e reduced by approximately 28%.

During the tests temperature measurements were made on the inner surface of the engine inlet with the use of a thermal camera. The engine inlet image was recorded under conditions set for the thrust setting point of F_c = 90 N, without water spray (Fig. 9), and during engine operation with the same thrust setting with water spraying in space before engine air inlet. (Fig. 10).

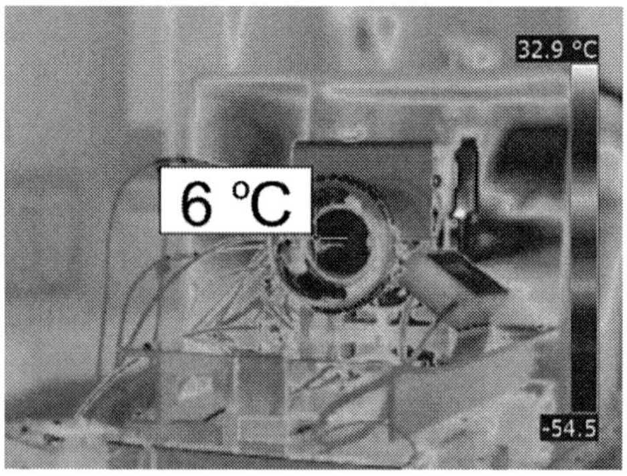

Fig. 9. Thermal distribution of the engine (inlet) operating in ambient conditions with thrust F_c = 90 N.

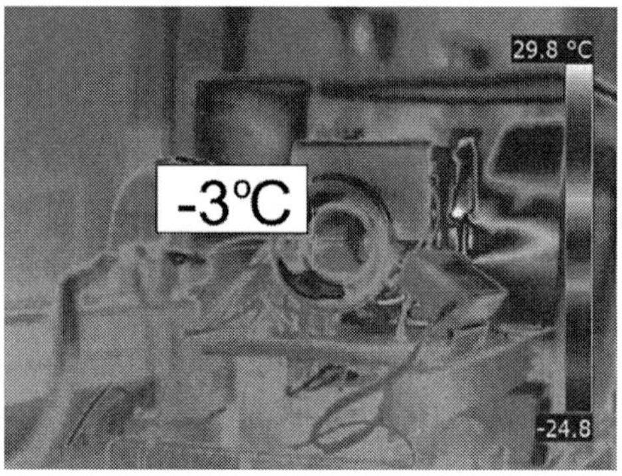

Fig. 10. Thermal distribution of the engine (inlet) with thrust of F_c = 90 N, while spraying water in space before inlet.

The analysis of the thermal distribution shows the difference in temperature at the inlet of the engine measured at the same point. The surface temperature at the inlet of the engine with a thrust of F_c = 90 N, without spray of water was 6 °C, and when the water was sprayed at the inlet area, the surface of the inlet at the measuring point was –3 °C. Consequently, the change in the air space before the air inlet at the ambient temperature t_{ot} = –2 °C and the ambient pressure p_{ot} = = 1025 hPa causes a drop by 9 °C. Such change in the surface temperature of the engine inlet caused water droplets freezing settling on the inlet edge and on the dynamic pressure measuring elements positioned inside of the inlet (Fig. 11). It should be emphasized that the interior surface of the engine air inlet during the test did not overlap with ice.

The study was stopped when a layer of ice appeared on the dynamic pressure probe. The possibility of engine damage by the ice layer, forced to stop the measurements.

Fig. 11. The view of the GTM-120 inlet after tests.

4 Conclusions

The micro turbine engine tests conducted at ambient conditions with an air temperature of t_{ot} = -2 °C and p_{ot} = 1025 hPa, and an analysis of the obtained results, allowed to make a few conclusions. The spraying of water in the space before the air inlet of the turbine engine under ambient conditions does not cause significant changes in the temperature of the refrigerant behind the compressor and before the engine turbine. Also, there were no significant change in the pre-turbine pressure and engine shaft speed. However, it increases the mass flowing factor by about 10%. The medium mass increase according to equation (1) results in an increase in thrust. In the case of tests conducted in relation to thrust level, the increase in the mass of the flowing medium contributes to the reduction of fuel consumption. Reduction in fuel consumption was about 28%. Reducing fuel consumption has also contributed to reducing exhaust gas temperature at the engine outlet. Changes in these parameters confirm the theoretical dependence of the turbine engines.

The comparative analysis of changes in inlet surface temperature during operation showed that the surface temperature of the inlet at the point of measurement could be reduced by 9 °C under ambient conditions due to spraying of water in the inlet area, which may cause the engine inlet to freeze.

The study additionally indicated that the inlet temperature of the engine just before the compressor is positive. The reason for this is the compact design of the engine, which provides the inlet heat from the compressor housing. This situation causes the turbine engine of such a design to have a high resistance to freezing without additional intake heating systems.

The study presented in this article was performed within the statutory research (contract No. 05/52/DSPB/0244).

References

1. R. Chachurski, P. Waślicki, *Instalacje przeciwoblodzeniowe i odladzające statków powietrznych* (WAT, 2011)
2. R. Chachurski, P. Waślicki, *Wykrywanie i sygnalizacja oblodzenia statków powietrznych* (WAT, 2011)
3. R. Chachurski, *Zagrożenia oblodzeniem lotniczych silników tłokowych* (WAT, 2011)
4. A. Gębura, K. Janusiak, M. Poradowski, *Journal of Konbin* (2014)
5. A. Milkiewicz, R. Stepaniuk, *Praktyczna aerodynamika i mechanika lotu samolotu odrzutowego, w tym wysokomanewrowego* (Warszawa, 2009)
6. J. Merkisz, J. Pielecha, S. Radzimirski, *Springer Tracts on Transportation and Traffic* **4**, 67-90 (2014), DOI: 10.1007/978-3-319-02705-0_4
7. J. Merkisz, J. Markowski, J. Pielecha, *Selected Issues in Exhaust Emissions from Aviation Engines* (Nova Science Publishers, 2014)
8. J. Merkisz, J. Markowski, J. Pielecha, D. Karpinski, M. Galant, *18 ETH-Conference on Combustion Generated Nanoparticles* (Zurich, 2014)

MATEC Web of Conferences **118**, 00032 (2017) DOI: 10.1051/matecconf/201711800032
VII International Congress on Combustion Engines

Innovative changes in the cylinder liners surface shaping methods

Jozef Gruszka[1,*]

[1]Poznan University of Technology, Faculty of Engineering Management, ul. Strzelecka 11, 60-965 Poznan, Poland

Abstract. The main directions of changes in new internal combustion engine designs are determined not only by legislation on the toxic components emission in the exhaust gases, but also by the changes resulting from technological development, which are the results of research and development activities. One of the basic systems that has undergone intensive development recently is the piston-rings-cylinder (PRC) node. This article contains an original analysis of the direction of changes in cylinder surface shaping in terms of the cylinder's main functional features in the PRC system (the casting material and the opening surface topography after the finishing process). The results of the research on cast iron materials for cylinder liner castings with strength of R_m > 300 MPa were analyzed based on the centrifugal casting method and their finishing stage in the finishing process meeting the criteria for reduction of oil consumption and particle emissions for new HDD type engine designs. The author also points to innovations in surface structure metrology based on new 3D optical measurement methods and the quality rating method by Mercedes company.

1 Introduction

In the construction of each combustion engine, the central position is occupied by a piston-cylinder system, often called the piston-ring-cylinder system (abbreviated as PRC, Fig. 1). In the operating range of this system, the basic processes of the of thermodynamic conversion of combustion energy of a gas into a reciprocating motion force take place, which are fundamental for the engine operation. In terms of working loads the PRC is the most complex and heavily loaded tribological system in the internal combustion engine [1, 2]. Its individual components: the piston, the piston rings and the cylinder sleeve form a movement pattern between each other which, in tribological terms, each constitute separate systems.. Analysis of individual tribological systems (cylinder liner-piston, piston-piston rings, piston rings-cylinder liner, cylinder-engine block) based on [3-5] as well as own investigations results [6-10] seem to indicate that the piston rings-cylinder liner system has a dominant functional and tribological significance in the PRC system. From the tribological properties of cylinder liners and piston rings, the material and its mechanical properties are first listed (strength, hardness), then its structure and chemical composition. For cylinder liners, errors in the macroscopic shape of the opening are formed (roundness, straightness or coaxiality) and micro geometry of the opening (the shape of grooves, angles of their orientation, surface roughness, deformation depth and surface cleanliness).Thus, it can be assumed that the cylinder liner material and the topography of the lining surface are the basic functional features of the sleeve and thus the technological parameters for these features can be determined (Fig. 1).

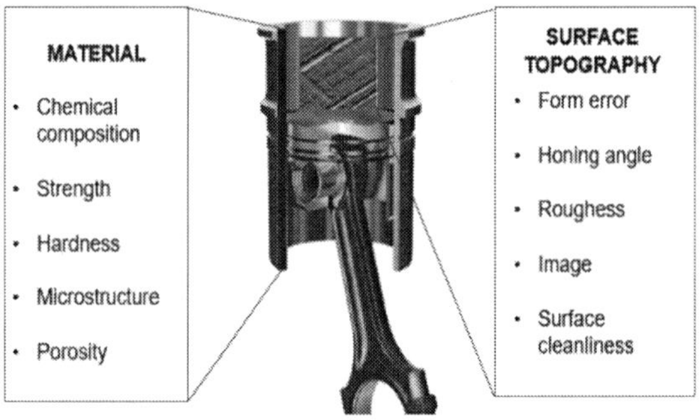

Fig. 1. Functional characteristics of the cylinder liners in the system PRC [15].

The current global scale development of propulsion systems is based on the conventional SI and CI piston engine technology, which have undergone extensive technological development in recent years. With regard to cylinder liners, the current development tendencies are primarily directed towards the sleeve material, mainly relating to the increase of the ad hoc tensile strength due to an increase in maximum combustion pressures exceeding P_{max} of 23-25 MPa (Heavy Duty Diesel Engines), which require materials with a strength rating of R_m > 300 MPa [11-15].

New materials, aside from the transfer of increased thermal-mechanical loads and good thermal conductivity, should meet the following requirements:

* Corresponding author: jozef.gruszka@put.poznan.pl

- cylinder top opening macro and micro geometry stability,
- ability to reduce friction loss,
- high friction, abrasion and corrosion resistance (EGR system) and no scratching tendency,
- good vibration damping ability and associated resistance to cavitation,
- overall weight reduction.

The whole solution should have a significant impact on meeting the environmental requirements for reducing oil consumption and particle emissions at economically justified manufacturing costs [1, 2, 14].

2 Directions of changes in cylinder liners casting shape

The basic construction material used for cylinder liners is still the gray cast iron alloy with flake graphite in a pearlite matrix, mainly due to its structural and tribological properties:

- good strength and fatigue resistance at room temperature and at elevated temperatures (such as EN-GJL-250 tensile strength R_m = 250–300 MPa, at 400 °C),
- modulus of elasticity E = 100–120 GPa,
- good thermal expansion α = 10–12 (20–200 °C) 10^{-6} K^{-1},
- high thermal conductivity λ = 47 (20 °C) $W \cdot m^{-1}$ K^{-1},
- wear resistance (abrasive, corrosive and cavitation),
- no scratching tendency,
- good vibration damping ability [1, 7, 15-17].

In addition, cast iron alloy with pearlite matrix flake graphite has good casting properties (mold filling ability), machinability, and the possibility of material recovery in the machining process (closed material circulation and relatively low manufacturing costs). It is also a structural material used for cylinder blocks similarly to aluminum alloys (Tab. 1).

Table 1. Comparison of basic properties of cast iron for cylinder blocks and cylinder liners [1, 14].

Properties	Unit	Material property					
		A 390	EN-GJL 250	EN-GJL 300	EN-GJV 450	EN-GJS 600	42CrMo
Tensile strength Rm	MPa	275	250	300-375	450–525	> 600	900-1150
Elastic modulus Eo	GPa	80	105	115	145	170	210
Elongation at break point Amin	%	1	1,5	1,5	1,0	0,5	(10)
Fatigue strength (20° – 225°C)	MPa	100-35	110-100	125-120	210-205	-	(280)
Thermal conductivity λ (20°C)	$W \cdot m^{-1} K^{-1}$	130	33-34		30-37	28-35	(50)
Thermal expansion α (20°-200°C)	$10^{-6} K^{-1}$	18	12,5		12,3	12	11-12
Density	g/cm³	2,6-2,7	7,1-7,3				7,7
Hardness	HB	110-150	260-310		260-310	280-330	240
Structure			Pearlitic with ferrite				Martensitic
Graphite form			Flake		Vermicular	Ball	
Technological process		Casting	Centrifugal casting				Forging
Material availability		Base materials in production		Developing materials partly in production			

Research and analysis of new cast iron groups based on available literature data [1, 2, 15-17], cylinder liner manufacturers findings (FM, GKN and MAHLE) as well as own research in [14] have shown that new construction materials in order to meet the increased requirements of tensile strength limit (R_m > 300 MPa) belong to the cast iron group:

- with flake graphite (EN-GJL 300),
- with vermicular graphite (EN-GJV 450),
- cast iron with ball graphite (EN-GJS 600),

whose basic properties are listed in Table 1 against the background materials used for cylinder blocks and cylinder liner castings?

As a result of the castings investigated by means of centrifugal casting from standard cast iron previously used for cylinder liners (EN-GJL-250) and cast iron castings with alloy elements (EN-GJL-300) and ductile cast iron (EN-GJV 450 and EN-GJS 600) the following results were obtained (Table 2):

- cast iron with addition of Mo and Cu alloys in relation to standard cast iron was mainly characterized by an increase in the Mo proportion from 0.16% to 0.4–0.5% and a Cu fraction of 0.75%, which allows for the tensile strength of R_m > 300 MPa. In the tested castings the minimum tensile strength was R_m = 320 MPa at hardness of 250 HBW. The arrangement of graphite and the matrix was similar to that of standard pearlite cast iron;
- cast iron with an addition of N (0.016%) was characterized by a reduced P (0.27%), S (0.01%), Mo (0.05%) and Ni (0.04%) but increased Cu (0, 80%). The minimum tensile strength was R_m = 347 MPa and hardness was 265 HBW. A significant change in the microstructure arrangement is the shape of the graphite, which has a changed size with the unmodified pearlite matrix compared to the standard cast iron matrix;
- cast iron with vermicular graphite is gaining in importance especially through the beneficial combination of mechanical, physical and technological properties. The properties of this cast iron lie between the properties of flake graphite and ball graphite, as illustrated in Tab. 1.

An important feature of this cast iron is that, with well-developed vermiculite grafts, material properties do not depend significantly on machining processes [16]. The problem is, the dependence of the structure and properties of the cast iron on the wall thickness and the cooling rate. In this relation, the behavior of this cast iron is similar to cast iron with flake graphite [17, 18].

The successful development of the cylindrical liner molding technology in the centrifugal casting process has led to the development of a series of new cast iron production programs involving Mo, Cu and N as well as Mg. Cast iron with Mo, Cu and N mainly as materials for cylinder liners for HDD motors (DAF, VOLVO, Daimler AG and Track). In turn, cast iron with vermicular graphite (Mg modifier) found use in Europe, Asia and the Americas mainly as a material for cylinder blocks of engines (reduction of wall thickness and weight reduction) with a displacement of 1.6–14.6 dm^3

HD type (Audi Ford, Hyundai, John Deere, MAN, Daimler Trucks, Renault) and HDD type (Caterpillar, VOVO, DAF) [16].

Table 2. Comparison of cast iron properties tested.

Type of material	Chemical composition [%]													Mechanical properties	
	C	Si	Mn	P	S	Cr	Mo	Ni	Cu	V	Ti	N	Mg	Rm [MPa]	HBW
	Image of metallic matrix formation														
Standard cast iron	3,28	2,31	0,5	0,57	0,06	0,49	0,16	0,11						255	244

Cast iron with Mo, Cu addition	2,95	1,95	0,6	0,32	0,04	0,3	0,3	0,09	0,75					320	264
	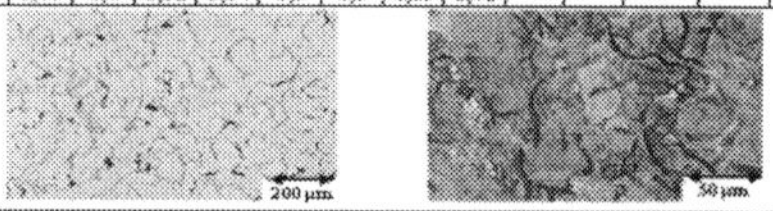														
Cast iron with N addition	2,9	2,04	0,61	0,26	0,01	0,49	0,05	0,04	0,84	0,01	0,01	0,016		347	265
	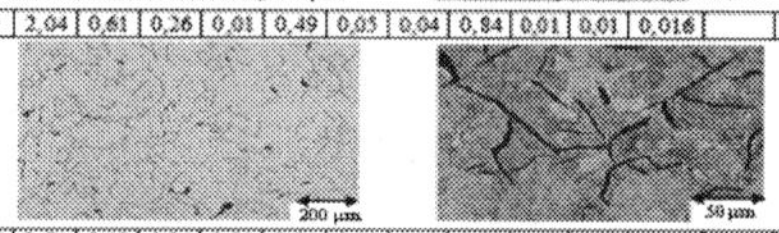														
Cast iron with Mg addition	3,38	2,54	0,48	0,04	0,01	0,01	0,06	0,02	0,03	0,73	0,01		0,06	660	266
	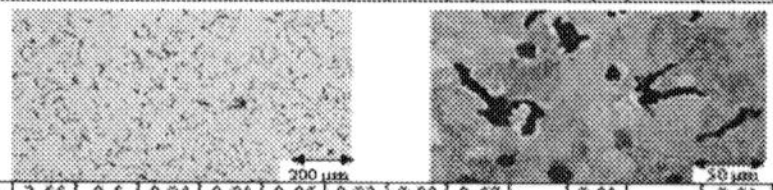														
	3,6	2,55	0,5	0,04	0,01	0,06	0,02	0,03	0,67		0,01		0,01	842	286
	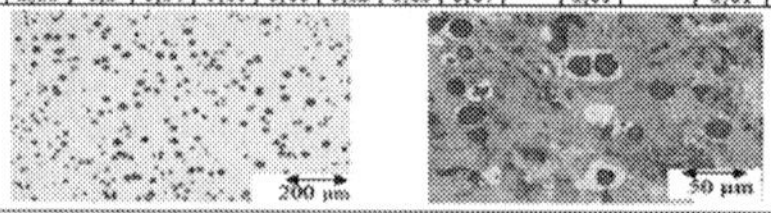														

In recent literature, many of the results of studies on flake, vermicular and ballistic graphite in cast iron, presented currently have been subjected to an isothermal transformation quenching process, which are referred to as cast iron types:

- ADI – Austempered Ductile Iron (spheroidal ductile iron tempered with isothermal conversion),
- DADI – Direct Austempered Ductile Iron (spheroidal cast iron subjected to direct tempering with isothermal conversion),
- AADI – AsformingAustempered Ductile Iron (spheroidal cast iron tempered with isothermal transformation under low temperature heat-plastic treatment),
- AVCI – Austempered Vermicular Cast Iron (cast iron with vermicular graphite tempered with isothermal conversion),
- AGI – Austempered Gray Iron (cast iron with flake graphite tempered with isothermal conversion) [17].

The use of isothermal tempering allows the cast iron to have exceptionally good strength and durability characteristics, with exceptional abrasion resistance and less than 10% density compared to steel or cast steel [2, 17]. Significant barriers to the serial application of this type of a solution are the manufacturing costs (significant costs of isothermal tempering followed by machining costs) [14].

3 The directions of changes in shaping the topology of the liner opening surface after the finishing process

The final shape of the cylinder opening surface geometry is formed in the finishing process. Over the last dozen or so years, this process has been the subject of intensive development mainly due to:

- development of the finishing machinery and moving from single-stage finishing machines to multi-stage finishing with the ability to automate the finish process and scanning for shape errors and its correction in the process along with the possibility of surface roughness and SPC analysis,
- ability to remove traces of previous conventional treatment (roughness, reinforcements and surface tension),
- cutting tools with high durability, homogeneous crystal structure, and self-sharpening diamond tips (ND, MD) and SiC carborundum cutting tips, cubitron (SGG) and Norton Quantum (NQD),
- possibility of using lasers, as a technological tool in the process of honing (cutting of cracks in the form of micro oil trays, laser ablation, thermal spray coating) ,
- development of techniques used in surface topography (contact measurements, optical, scanning microscopy), increase in instrument range (resolution, various filters and speeds, 2D and 3D measurements),
- the size and shape of the oil reservoirs in the laser-forming TDC zone,
- surfacecleanliness [11-15, 18, 19].

The availability of many technological variants of the finishing process has enabled cylinder liner manufacturers to carry out optimization work on new engine designs, in terms of:

- macro structure of the opening shape with shape error parameters (straightness and roundness or concentricity) at the level of < 6–10 μm,
- shaping the surface micro geometry of the surface with roughness parameters R_{pk} and $R_k \rightarrow 0$ μm, R_{vk} according to the engine design requirements (5–15 μm) the honing angle α = 50–140° and the shape and arrangement of the cracks,
- the size and shape of the oil reservoirs in the laser-guided TDC zone,
- surface cleanness after the finishing process [14].

In the new solutions, the formation of the cylindrical surface topography is a trend towards the method of multi-step finishing, taking into account the dependence of the R_k parameter on the exhaust emission standard, as illustrated in the data presented in Fig. 2. Examples of the shape study of crests, vertices, recesses and surfaces between the grooves are summarized in the form surface profilography (Fig. 3) and scanning microscopy images (Fig. 4, 5).

From the data in Figure 2, the decrease of the average values of the parameters R_{pk} (0.27 μm to 0.13 μm) R_k (from 1.08 μm to 0.41 μm) and the R_{vk} groove depth reduction (from 2.05 μm to 1.4 μm) can be observed with the increase of emissions requirements from Euro III to Euro V.

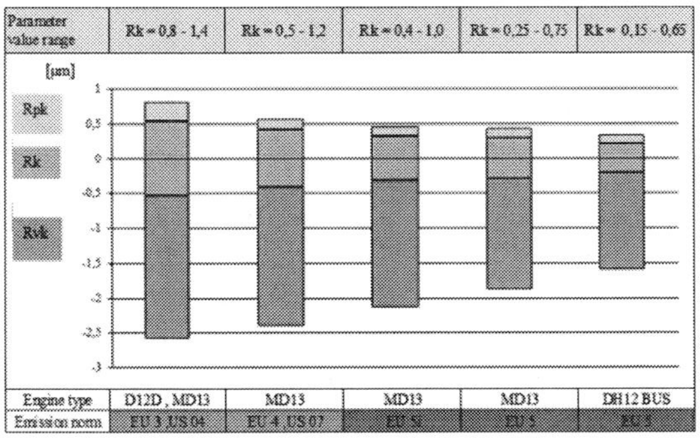

Fig. 2. Cylinder liners surface roughness parameters test results according to the purpose of the liner for the VOLVO HDEP type of engine [1, 14].

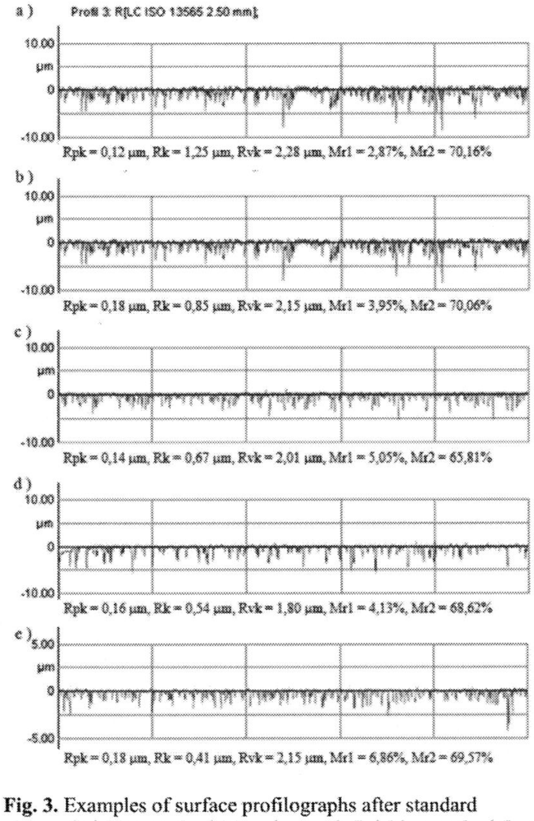

Fig. 3. Examples of surface profilographs after standard plateau finishing method (a) and smooth finishing method (b-e) depending on the range of parameter R_k.

The results of examining the images of the structure of vertices, recesses and surfaces between grooves by scanning electron microscopy (Fig. 4, 5) were the basis for determining the surface deformation levels as a result of the finishing process and determining levels of qualification for permissible and unacceptable surface images (VOLVO) [2, 14].

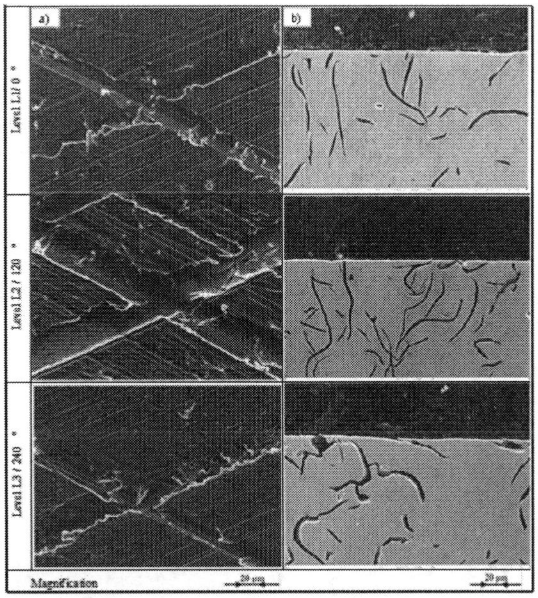

Fig. 4. Surface topography images (a) and surface layer deformation images (b) after standard finishing using SEM imaging.

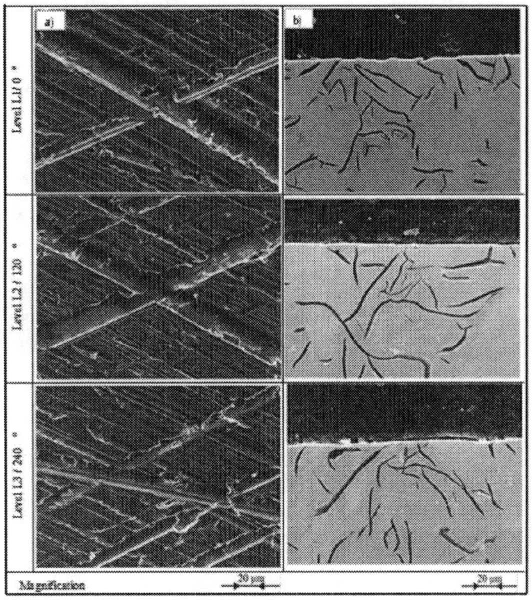

Fig. 5. Surface topography images (a) and surface layer deformation images (b) after smooth finishing using SEM imaging.

The use of white light interferometry (WLI) is new in surface metrology. Studies [19-21] have shown the possibility of wide applications for this method in spatial 3D surface roughness measurement including cylinder liner surface [1]. So far in industrial practice, the methodology of contact measurements in the 2D parameter system (R_{pk}, R_k and R_{vk}, M_{r1} and M_{r2}) is used to evaluate the topography of the liner surface. In industrial conditions, the method of contact measurement in 3D was very rare. Measurements require highly qualified personnel in the field of 3D parameters and related software (TalyMap-Form Talysurf Series 2 from Taylor Hobson) [20]. The problem of the lack of industrial use of 3D parameters is also due to the lack of formalized parameter ranges with respect to the topography of the finish and their influence on, for example, the functional properties of the surface. For 2D parameters, the R_{pk} parameter defines the bearing surface area for the piston rings), the R_k parameter defines the effective surface area and the R_{vk} parameter describes the surface area of the oil retention.

Figure 6 shows an example of an image analysis of a cylinder liner surface topography based on the MBN 37800-1 (Mercedes Benz Norm) using the WLICyl-S white light interferometer (WLI) (WLICyl-S measurement station for honing structures by the BMT company) [21].

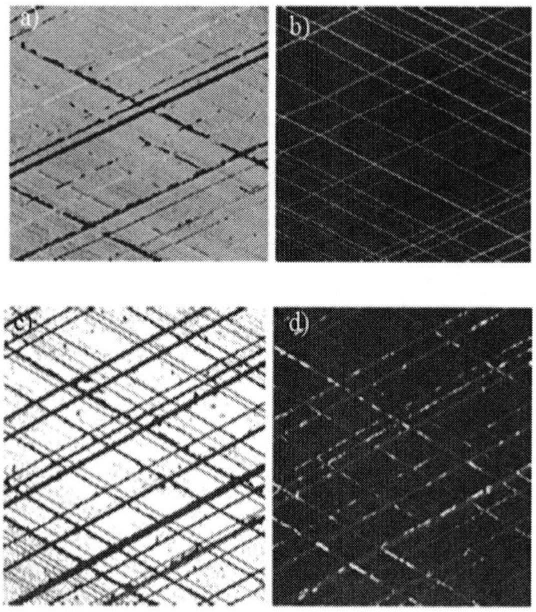

Fig. 6. Honing structure according to MBN 37800-1. From left (a) to right (d): original topography after the F-operator application, raster image, contact surface (white), undesired surface structures on the sheet metal part (white) [13].

The MBN 37800-1 specifies tolerance limits, measurement and evaluation conditions, and the distribution of measurement points in both 2D (R_{pk}, R_k and R_{vk}) and 3D layouts with a range of parameters and determines tolerable casting errors in the porosity range on the finish. When using the WLICyl measuring instrument, it is necessary to have a license to use MBN 37800-1. According to many publications [19-21], the use of scanning white light interferometry will require a lot of research into new test methods or an international comparative research initiative (round-robin), for example, when measuring Talysurf CCI 600 [20] and WLICyl-Swith the use of the MBN 37800-1 standard or the proposed standard ISO/DIS 25178.

4 Conclusions

Compliance with the new design requirements of the type HDD/HDEP internal combustion engines (increase of maximum combustion pressures exceeding P_{max} = = 23-25 MPa) requires the construction of cylinder liners that can meet the strict criteria of material strength (R_m > 300 MPa) and the surface topography to meet the requirements of standards related to reduction of oil consumption and particulate emissions.

The cylinder liner material and topography of the liner surface are the basic functional features of the cylinder in the PRC system of an internal combustion engine and thus the technological parameters for each feature can be determined, which in turn define the key criteria for selecting the manufacturing technology.

The current development of the cylinder liner molding technology in the centrifugal casting process allows for the production of series of new cast irons with Mo, Cu and N as well as Mg additions. The use of new cast iron grades requires a production cost analysis from the foundry which is due to the share of alloyed material in relation to the base materials.

In the process of forming the surface of the top opening in the size range (< 6–10 μm) and micro geometry of the cylinder liner surface the smooth multi-process finishing technique seems prevalent. Smooth finishing technology allows the topography of the surface to be adjusted in 2D parametric and 3D surface configurations to meet oil consumption and particulate emissions standards.

New methods of optical 3D measurements (WLI) using the MBN 37800-1 standard will requirethe implementation of research methodology and international benchmarking and comparative studies with the round-robin method based on the instrument used (Talysurf CCI 600 or WLICyl-S).

References

1. Technical information from meetings MAHLE – VOLVO, DAF, MAN, PSA-BMW, Unpublished material (Krotoszyn, 2005-2010)
2. Zylinderkomponenten, Eigenschaften, Anwendungen, Werkstoffe MAHLE (Stuttgart, 2009)
3. W. Karpiuk, T. Borowczyk, M. Idzior, R. Smolec, *2016 International Conference on Sustainable Energy, Environment and Information Engineering*, 16-22 (2016)
4. W. Karpiuk, R. Smolec, M. Idzior, *2016 International Conference on Sustainable Energy, Environment and Information Engineering*, 37-43 (2016)
5. G.M. Szymanski, F. Tomaszewski, Mechanical Systems and Signal Processing **68-69**, 479-490 (2016), DOI: 10.1016/j.ymssp.2015.07.015
6. J. Merkisz, M. Idzior, P. Lijewski, P. Fuc, W. Karpiuk, *Proceedings of the Ninth Asia-Pacific International Symposium on Combustion and Energy Utilization* 352-356 (2008)

7. P. Fuc, P. Kurczewski, A. Lewandowska, E. Nowak, J. Selech, A. Ziolkowski, International Journal of Life Cycle Assessment **21**, 1438-1451 (2016), DOI: 10.1007/s11367-016-1104-y
8. J. Merkisz, M. Idzior, J. Pielecha, W. Gis, *WIT Transactions on the Built Environment* **111**, 181-189 (2010), DOI: 10.2495/UT100171
9. Z. Stelmasiak, D. Pietras, *IOP Conference Series-Materials Science and Engineering* **148**, UNSP 012087 (2016), DOI: 10.1088/1757-899X/148/1/012087
10. S. Carras, J. Gruszka, T.G. Mathia, D. Mazuyer, F-P. Ninove, P. Pawlus, *10th International Symposium on Measurement and Quality* (Osaka, 2010)
11. J. Gruszka, A. Suchecki, Silniki Spalinowe **3**, 16-26 (2007)
12. J. Gruszka, Inżynieria Maszyn **12**, 42-53 (2009)
13. J. Gruszka, Archiwum Technologii Maszyn i Automatyzacji **30**, 73-78 (2010)
14. J. Gruszka, Silniki Spalinowe 2009-SC1-157, 350-357 (2009)
15. J. Gruszka, *Technologiczne kształtowanie cech funkcjonalnych warstwy powierzchniowej tulei cylindrowych* (Poznan 2012)
16. S. Dawson, *68th WFC – World Foundry Congress*, 93-99 (2008)
17. E. Guzik, Archiwum Odlewnictwa, 29-85 (2001)
18. W.L. Guesser, T. Fundições, P.V. Duran, W. Krause, *Congres Le diesel*, 11 (2004)
19. NAGEL: Honsymposium (2009)
20. C. Łukianowicz, PAK **56**, 1055-1058 (2010)
21. J. Seewig, C. Wiehr, S. Groger, Charakterisierung technischer Bauteiloberflachen – Stand der Oberflachenmesstechnikheute (ak-rauheit.de, 2017)

The experimental determination of the relationship between the energising time of the common rail injector and the set fuel quantity and rail pressure

Dariusz Pietras[1,*], *Tomasz* Knefel[1], and *Mateusz* Satława[1]

[1]University of Bielsko-Biala, Faculty of Mechanical Engineering and Computer Science, ul. Willowa 2, 43-309 Bielsko-Biala, Poland

Abstract. The article discusses the issue of experimentally determining the relationship between the energising time of the common rail electromagnetic injector and the set fuel quantity and rail pressure. Experimental studies according to the assumed methodology were made on a test bench enabling the dynamic flow rate measurement of the injector. The fuel system mounted on the test bench was controlled by the laboratory CI engine control unit based on the original concept of one of the authors of the article. The results of the experimental studies have made it possible to determine many of the characteristics of the fuel flow rate depending on the specified rail pressure and the energising time of the injector. An analysis was then performed followed by extrapolation of the obtained results. The data obtained from these analyses are the basis for the development of the energising time control algorithm based on a set fuel quantity and rail pressure.

1 Introduction

The authors of this paper have been working for many years in a research facility that specializes in the issues related to the selection and optimization of the engine control process parameters for both gasoline and diesel engines [1-3].

The current state of knowledge suggests that the number of injections per one engine work cycle, the amounts of particular fuel dose parts, and the delay or advance angles of the beginning of injection of particular parts, i.e. in general, the overall volume-time characteristics of the fuel injection process significantly affect the heat release and, consequently, the engine performance, as well as the contents of harmful substances in the exhaust gas, including solid particles (mainly soot), and the noise generated by the engine in operation [4-7].

Fuel dose division also applies to issues related to the mathematical modeling of the engine work cycle process and the mathematical modeling of fuel supply systems for gasoline and diesel engines [8-10].

Factory-setting controllers developed for specific engine models often do not allow for easy modification of the engine performance parameters. Due to its complicated construction, the number of registered and operational parameters is up to 3000 and interdependencies between them make it impossible to use them in scientific research.

Thus, it became necessary to design and execute, yet another, less complex system driver in terms of the interrelation of control parameters that are relevant when the engine is mounted into the vehicle. The main feature of the newly developed driver was the ability to enable dynamic engine operation on the engine dynamometer, because the driver structure is based on array algorithms [2].

The basic parameter in the control of modern diesel engines equipped with common rail fuel supply systems is the driver of the injector energising time (*ET*) (coil supply). Important in this issue is to know the relationship between the energising time and the real fuel injection process [8, 11, 12]

For the driver development design assumptions, it was assumed that the control of this parameter is to be carried out on the basis of the set fuel quantity (Q) and rail pressure (p_{rail}). The driver based on the programmed algorithm will determine the energising time.

Conducting research based on the developed methodology was meant to provide the basis for the development of an algorithm controlling the injector energising time based on the set fuel dose and the rail pressure.

2 Methodology

2.1 Test stand

Experimental research was carried out in the laboratory of the Department of Combustion Engines and Vehicles of the University of Bielsko-Biala. The test bench power supply is electronically controlled using a test stand based on the Star 8 test stand [8].

On the test stand, a common-rail power supply system was installed for the 1.3 SDE engine.

* Corresponding author: pietras@ath.bielsko.pl

Precise electronic control of the test stand was done using:

- Three-phase AC motor with an inverter, driving the test stand spindle. It enables high torque at low speeds and also stabilizes the rotational speed.
- Fuel heating system. It allows for setting the required temperature and stabilize it when it is reached.
- The control system of a set number of injection injections that works in conjunction with the electromagnetic displacement actuator and with a reluctant rotational speed sensor. The value of the set injection cycle number is used to determine the specific dosage of the fuel.

The test stand is shown in Fig. 1 [13]. The injection system was controlled by a laboratory universal diesel engine driver. The driver was created within the framework of the research project No. 5178/B/T02/2011/40 financed by the National Science Center according to the original concept of the project manager. In addition to the CI engine control function on research test stands, the driver also enables the testing of common rail systems on appropriate test stands.

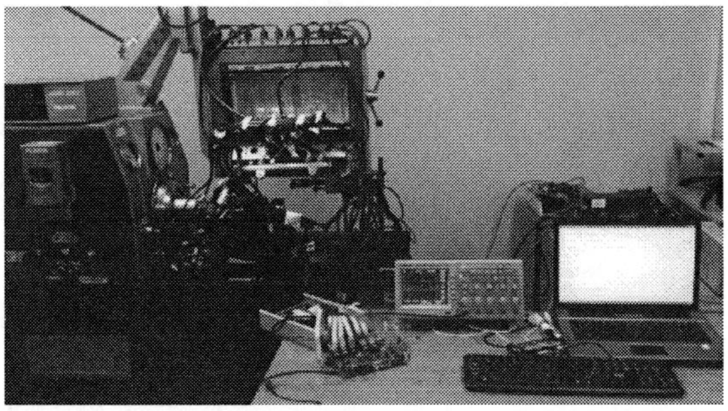

Fig. 1. View of test stand [13].

2.2 Research plan

The study consisted of measuring the fuel mass injected into the burette by each of the four injectors for set values of energising time, the rail pressure and the set number of injection. The range of injector energising time and rail pressure changes are shown in Tab. 1. The number of injection was set so that the minimum injection volume was not less than 3 cm^3. The number of injections was 250, 500, and 1000 in the test. The value of 1000 was related to the short time of the injector energising time and the low rail pressure, while 250 injections per measurement test were carried out for long periods of injector energising time and high rail pressure. In the research description, as well as in the rest of the article, the fuel rail pressure values were reported in bars, because this unit is most commonly used in the technical descriptions of the injector manufacturers and in other papers in this subject. On test stand to fueled the injection equipment has been used the diesel fuel. Increasingly, CI engines with common rail systems are fueled by bio-fuels [14-16]. The results obtained on diesel should be consistent for other fuels because of assumptions about the parameters of physical properties of bio-fuels, although in some engine states (fuel temperature) the results may be different [17].

The fuel weight was measured using a RADWAG WPE 600 precision scale.

Table 1. Range of changes in set parameters (ET [μs])

p_{rail} [bar]										
270	400	500	600	700	800	900	1000	1100	1200	1300
250	250	250	250	250	250	250	250	250	250	250
300	300	300	300	300	300	300	300	300	300	300
400	400	400	400	400	400	400	400	400	400	400
500	500	500	500	500	500	500	500	500	500	500
600	600	600	600	600	600	600	600	600	600	600
700	700	700	700	700	700	700	700	700	700	
	800	800	800	800	800	800	800	800	800	
			900	900	900	900	900	900		
					1000	1000	1000			

Each of the 77 research points was performed in accordance with the following plan:

- preparation of the test stand: maintaining the set fuel temperature, draining of the burettes,
- entering the required ET and p_{rail} in the driver software,
- setting the number of injections,
- measurement of the mass of empty burettes,
- performing the test,
- recording the diesel fuel temperature,
- measuring the weight of the burettes with fuel.

3 Research results and their analysis

Research was carried out based on the methodology outlined in Section 2.2. Taking into account the number of injections in each test, the mass of fuel per injection was determined. The methodology for presenting the results on the charts would be the similar to the research work of other authors [18, 19].

The graphical results are shown in Fig. 2 to 6 for selected pressures: 250, 500, 700, 900 and 1300 bar. For graphs with a green background, a maximum value for the fuel mass axis of 45 mg/cycle was used and a maximum value for the injector energising time axis was 1050 μs. In this way, it is easier to compare the values of the mass increase for all the pressures tested. Furthermore, two additional graphs for a cylinder pressure of 267 and 500 bar are also shown to better illustrate the difference in fuel mass for individual injectors.

For a storage rail pressure of 267 bar (Fig. 2), the mass values of fuel injected into the burettes are small. The first and second injectors did not inject the fuel into the burettes at all. The maximum injection value is obtained for injector 4. It should be added that for the tested injection system, the rail pressure of 270 bar corresponds only to the idling engine operation.

DOI: 10.1051/matecconf/201711800033

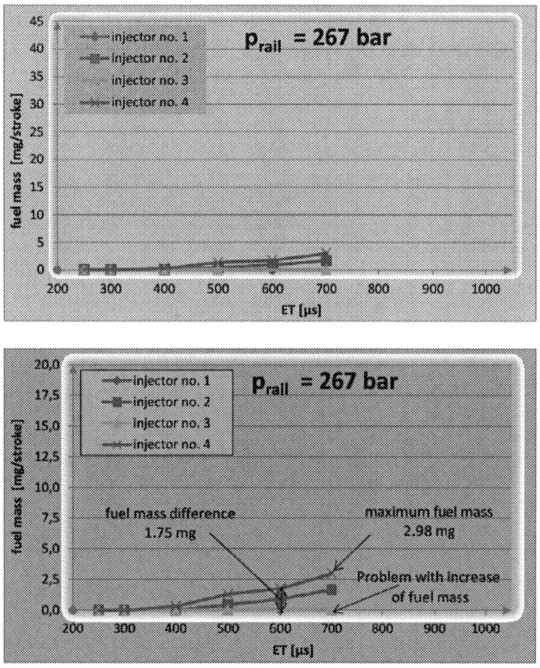

Fig. 2. Change of fuel mass as a function of energising time for rail pressure equal 267 bar.

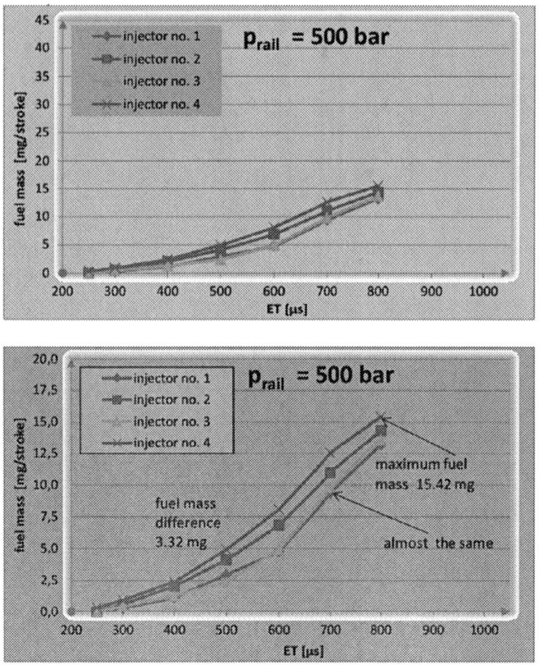

Fig. 3. Change of fuel mass as a function of energising time for rail pressure equal 500 bar.

Fig. 3 shows the change of the injected fuel mass as a function of the energising time and the rail pressure at 500 bar. Compared to the case discussed above, weight increase was observed for each of the investigated injectors. Just like before, the maximum injected mass of the fuel in the burettes occurred for injector 4. The functions for injectors 1 and 3 practically overlapped, but their values diverged from the values obtained for injectors 2 and 4. As before, the linear nature of the change in fuel mass can be observed.

Fig. 4, 5 and 6 show the change in weight of the injected fuel as a function of the energising time for the rail pressure of 700, 900 and 1300 bar, respectively.

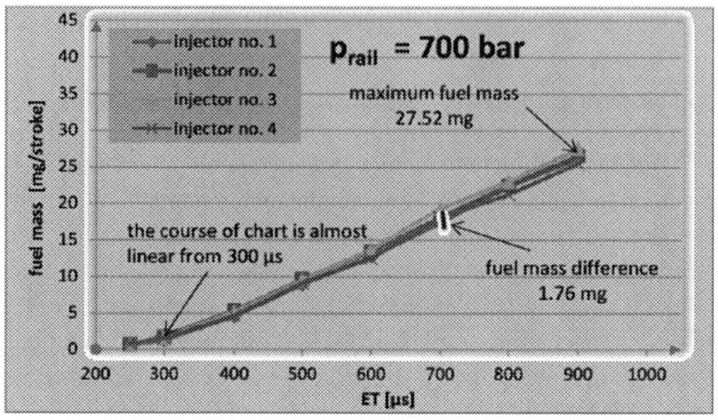

Fig. 4. Change of fuel mass as a function of energising time for rail pressure equal 700 bar.

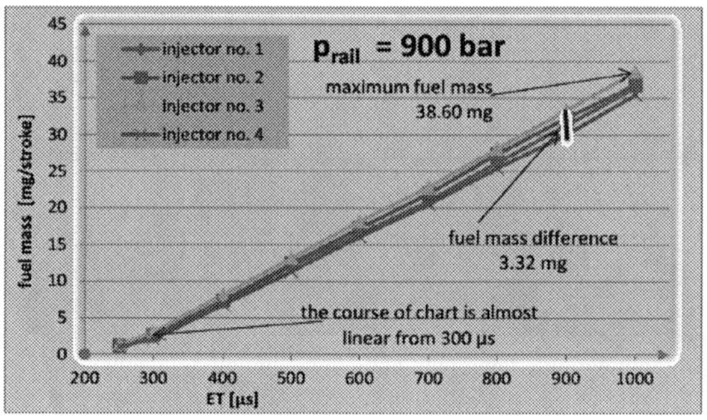

Fig. 5. Change of fuel mass as a function of energising time for rail pressure equal 900 bar.

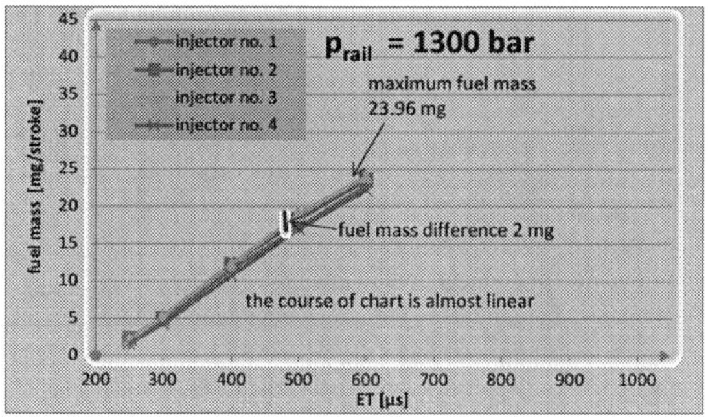

Fig. 6. Change of fuel mass as a function of energising time for rail pressure equal 1300 bar.

The injected fuel mass relation with the energising time shown in the three figures has a linear character even at the shortest energising times. At the same time, the differences in the injected fuel mass between different injectors are small or even negligible.

4 Determination of fuel quantity as a function of ET and p_{rail}

Based on the measurement of the fuel temperature at each measurement point and the relation of the test fuel density as a function of temperature, the volume of fuel injected was determined. The waveforms shown in Fig. 7-9 show the averaged fuel injection results for injectors 2 and 4. Due to the large discrepancies in the results for injectors 1 and 3, especially for low pressures, they were omitted in further analyzes. Figure 7 shows the actual fuel dosage as a function of the injector's energising time for the individual rail pressures. The functions are linear, which is directly tied to the results of the individual pressure tests discussed in Chapter 2. The nature of the variation in fuel dosage from energising times correlates with the literature data particularly for high fuel rail pressure values.

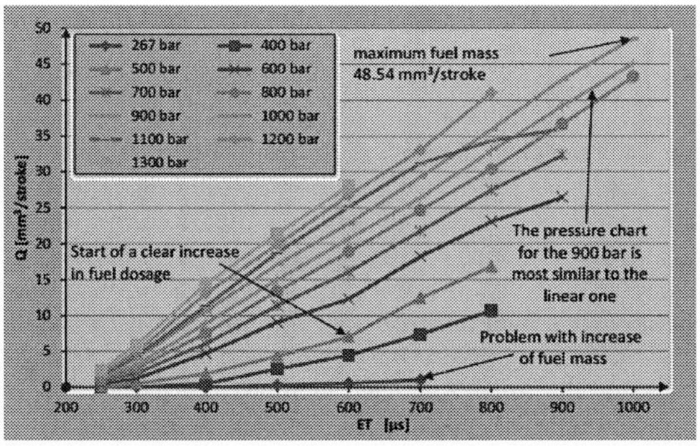

Fig. 7. Change of fuel quantity as a function of energising time for different rail pressure.

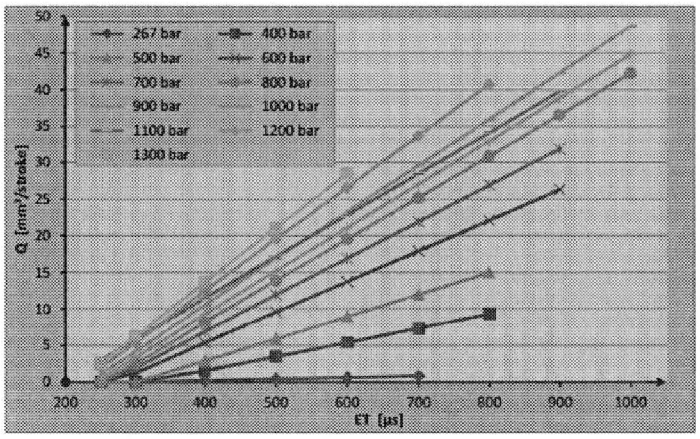

Fig. 8. The effect of the linear regression on fuel quantity as a function of energising time for different rail pressure.

For a rail pressure of 1100 bar, there is a divergence from the linear nature of the fuel dose increase starting at the energising time of 700 µs. Most likely this is due to technical problems that occurred during this measurement. These were related to the maintenance of the pressure in the supply system, which has been noted as significant after the last test at a pressure of 1100 bar.

The next step of the analysis was the linear regression of the obtained results shown in Fig. 7. The effect of the regression process is shown in Fig. 8. The linear results obtained with high accuracy corresponded to the real results.

The fuel level is clearly higher for higher fuel pressure values in the rail. The only deviation from this character of changes is the fuel dose at a rail pressure of 1100 bar, which is due to the reasons described above.

5 Development of data for the energising time control algorithm

As mentioned in the introduction, the purpose of the research and the analyses discussed above was to determine the energising time relation between the common rail injector control and the pre-set dose and the rail fuel pressure. To achieve the intended effect, after performing a linear regression of the results further analysis was required. This problem was realized through the extrapolation of linear regression results (Fig. 9). Straight lines marked in color correspond to values in the measuring range. The functions shown in black are the result of extrapolation. Through this method the values of the energising time dependence from the set fuel dose were determined in a wider range of fuel dose variation than those obtained in the performed experimental studies.

Taking the assumptions that the time ET is to be determined from the set fuel dose into consideration, the axes are further inverted relative to the graph 8 so that the graph represents the dependence of the injector energising time on the fuel dose for the different rail pressure values.

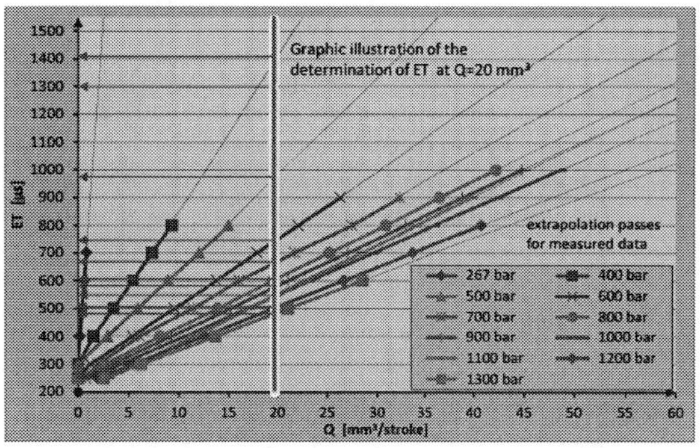

Fig. 9. Graphic illustration of the determination *ET* for different values of the rail pressure and the prescribed fuel quantity.

Figure 9 shows a graphical representation of the energising time control method for a test fuel dose of 20 mm^3 and different rail pressure values. It can be observed how the injector *ET* changes depending on the rail pressure. The higher the rail pressure, the shorter the opening time of the injector. However, the differences between the injector energising time are different for the various fuel pressures in the rail. For example, the difference between 500 and 600 bar is approximately 231 µs, whereas for 1200 and 1300 bar is only about 21 µs.

Given that the relationships shown in Figure 9 are linear, it is possible to precisely determine the energizing time for the individual fuel doses and the various rail pressures. The values of the determined energising times are shown in Table 2 in the form of an array algorithm.

Table 2. Parameters of the control algorithm for the energising time.

Q [mm^3/ stroke]	p_{rail} [bar]									
	400	500	600	700	800	900	1000	1100	1200	1300
0	0	0	0	0	0	0	0	0	0	0
1	346	319	288	282	271	260	249	215	237	229
2	402	354	312	302	289	277	265	232	251	243
3	458	389	337	322	307	294	281	250	265	256
4	514	424	361	342	324	311	297	268	280	270
5	570	459	385	362	342	328	312	285	294	283
6	626	494	410	382	360	345	328	303	308	297
7	682	528	434	402	377	362	344	321	322	310
8	738	563	458	422	395	379	360	338	336	324
9	794	598	483	442	413	395	375	356	350	337
10	850	633	507	462	431	412	391	374	365	351
15	1130	807	628	562	519	497	470	462	435	418
20	1409	981	750	662	608	581	549	550	506	485
25	1689	1156	872	761	696	666	627	639	577	553
30	1969	1330	993	861	784	750	706	727	648	620
35	2249	1504	1115	961	873	835	785	815	719	688
40	2528	1678	1236	1061	961	919	864	904	790	755

The algorithm specifies all injector energising times as a function of the set fuel dose and rail pressure. However, not all the values are present in engine control. For example, high fuel dose injection is not performed at very low pressures. To illustrate this case, a graph limiting the maximum injector energising time to 1100 μs was made. Also, measurements for the fuel supply rail pressure value of 267 bar was omitted intentionally because, as previously mentioned, it only occurs for the idle speed.

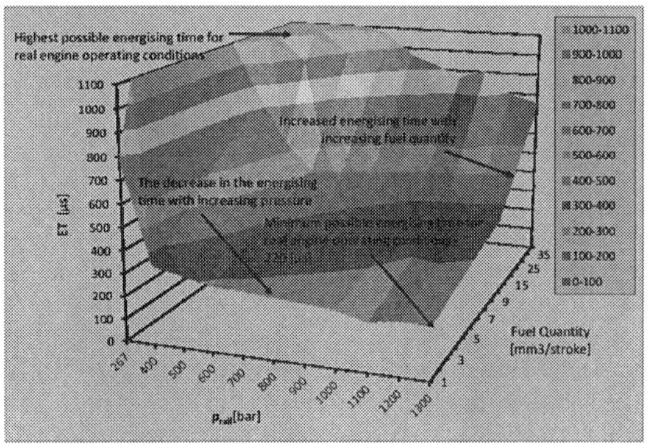

Fig. 10. Graphic interpretation of determining the energising time based on parameters contained in the algorithm shown in Table 2.

6 Conclusions

Planned and performed experimental studies as well as their further analysis have proven useful. One of the main goals set out in the development of the concept of laboratory operation of a universal diesel engine driver, i.e. determining the relationship between the set fuel dose, the fuel pressure in the cylinder, and the injector energising time.

By analyzing the research results (Fig. 7), a reduction in the fuel dose increase relative to the increase in injection pressure for the various injection energising times. This may be due to the reduction in the active cross section of the fuel stream that flows through the spray nozzle openings while increasing the injection pressure for the same time as the injector's energising time.

Divergence in the results for individual injectors, especially for short energising times and low fuel pressure values in the rail may be caused by three different reasons. The first is measurement errors that become very significant when measuring very small doses of fuel. The second reason is probably the lack of any adjustments in the timing of the individual injectors applied during the test. At the same time, this confirms the correctness of the factory-specified injector control corrections included in the original software codes. The injectors used in the experiment were not brand-new. There could therefore be variations in dosage resulting from the operational wear on the hydraulic and electrical parts of the injectors. This could be the third cause of injector dosage inaccuracies for small energising times and low rail fuel pressures.

At the same time, the authors acknowledge that these considerations concerned studies corresponding to a single injection dose. For engine tests, the fuel injection algorithms will also be programmed for two, three, or even five injection steps in the driver algorithm. There may be a discrepancy between the planned and the actual injected fuel dose resulting from the injectors not completely closing at short dwell times between the various injection steps [20]. The injector solenoid coil is an inertial element, so the current continues to flow for a time after the supply voltage has disappeared.

The publication has been produced as part of the research project no. 5178/B/T02/2011/40 financed from the resources of the National Science Centre.

References

1. T. Knefel, Maintenance and Reliability **14**, 42-53 (2012)
2. Z. Stelmasiak, T. Knefel, J. Larisch, Pomiary, Automatyka, Kontrola **56**, 217-220 (2010)
3. Z. Stelmasiak, T. Knefel, J. Larisch, Combustion Engines, **2007-SC2**, 173-180 (2007)
4. I. Pielecha, W. Cieslik, Journal of Thermal Analysis and Calorimetry **126**, 815-827 (2016), DOI: 10.1007/s10973-016-5544-1

5. L.J. Kapusta, I. Pielecha, K. Wisłocki, A. Teodorczyk, Journal of Thermal Analysis and Calorimetry **123**, 819-828 (2016), DOI: 10.1007/s10973-015-4927-z
6. I. Pielecha, K. Wislocki, P. Borowski, W. Cieslik, Journal of Thermal Analysis and Calorimetry **122**, 473-485 (2015), DOI: 10.1007/s10973-015-4735-5
7. D. Pietras, Polish Maritime Research **3**, 3-12 (2016), DOI: 10.1515/pomr-2016-0032
8. T. Knefel, Scientific dissertations (Scientific Publishing University of Bielsko-Biala, 2013)
9. I. Pielecha, International Journal of Automotive Technology **15**, 47-55 (2014), DOI: 10.1007/s12239-014-0005-y
10. I. Pielecha, P. Borowski, J. Czajka, K. Wislocki, J. Kazmierowski, Journal of Thermal Analysis and Calorimetry **119**, 695-703 (2015), DOI: 10.1007/s10973-014-4139-y
11. T. Knefel, J. Nowakowski, Combustion Engines **169**, 113-119 (2017), DOI: 10.19206/CE-2017-220
12. M. Skowron, I. Pielecha, K. Wislocki, *IOP Conference Series-Materials Science and Engineering* **148**, UNSP 012085 (2016), DOI: 10.1088/1757-899X/148/1/012085
13. M. Satlawa, Archive of diploma theses (University of Bielsko Biala, 2017)
14. W. Karpiuk, T. Borowczyk, M. Idzior, R. Smolec, *2016 International Conference on Sustainable Energy, Environment and Information Engineering* 16-22 (2016)
15. W. Karpiuk, R. Smolec, M. Idzior, *2016 International Conference on Sustainable Energy, Environment and Information Engineering* 37-43 (2016)
16. J. Merkisz, M. Idzior, P. Lijewski, P. Fuc, W. Karpiuk, *Proceedings of the Ninth Asia-Pacific International Symposium on Combustion and Energy Utilization* 352-356 (2008)
17. A. Rybak, J. Hunicz, P. Krzaczek, Combustion Engines **171**, 39-43 (2017), DOI: 10.19206/CE-2017-407
18. W. Busz, A. Walaszyk, Combustion Engines **162**, 978-981 (2015)
19. T. Osipowicz, K.F. Abramek, T. Stoeck, Combustion Engines **162**, 688-694 (2015)
20. M. Reza Herfatmanesh, Z. Peng, A. Ihracska, Y. Lin, L. Lu, C. Zhang, Advances in Mechanical Engineering **8**, 1-8 (2016), DOI: 10.1177/1687814016648246

MATEC Web of Conferences **118**, 00034 (2017) DOI: 10.1051/matecconf/201711800034
VII International Congress on Combustion Engines

Method of reduction of diagnostic parameters during observation on the example of a combustion engine

Malgorzata Orczyk[1,*], and *Franciszek* Tomaszewski[1]

[1]Poznan University of Technology, Faculty of Machines and Transport, ul. Piotrowo 3, 60-965 Poznan, Poland

Abstract. The article presents a method of selecting diagnostic parameters which map the process of damaging the object. This method consists in calculating, during the observation, the correlation coefficient between the intensity of damage and the individual diagnostic parameters; and discarding of those parameters whose correlation coefficient values are outside of the narrowest confidence interval of the correlation coefficient. The characteristic feature of this method is that the parameters are reduced during the diagnostic experiment. The essence of the proposed method is illustrated by the vibration diagnosis of an internal combustion engine.

1 Introduction

In vibroacoustic diagnostics, it is important to select the parameters of the vibration signal or noise that contain information about the characteristics of the examined object. While conducting diagnostic tests, many researchers measure arbitrarily selected or measurable diagnostic parameters [1-3]. After the studies are completed, scientists reduce the parameters which map selected or simulated object states. However, most of researchers measure only specific parameters of the vibroacoustic signal – usually speed and acceleration – and then they search for the relationships between them and the changed states of the object.

Methods for reducing the vector of diagnostic signal parameters are presented in works: [4-6]. In papers [4, 5], the author proposed selection of diagnostic parameters based on the passive diagnostic experiment performed in compliance with the algorithm of the BEDIND program (the name stands for "Bierny Eksperyment Detekcji Informacji Diagnostycznej" in Polish, which means "Passive Experiment of Detection of Diagnostic Information"). The program worked in two versions: non-standardized and standardized. In the non-standardized version, the covariance matrix is the basis for the selection of diagnostic parameters, so that the program selects the parameters that carry the maximum information about the state of the object. They are characterized by, among other things, a high dynamics of changes. In the standardized version, the basis for the selection is the correlation matrix, which enables the selection of parameters that are independent of each other and can carry information about two different damages. In the paper [7] the author proposed the choice of diagnostic parameters based on the correlation matrix (similarly to the BEDIND program) by selecting independent parameters.

The mentioned methods of reducing diagnostic parameters have one common feature: they select parameters without correlating their changes with the state of the object and they are realized after the experiment. The method of reducing diagnostic parameters proposed by the author of this work consists in choosing the parameters that measure the reliability of the object. What is important, this method enables reduction of the parameters during the diagnostic experiment. This approach enables the reduction of the parameters during the experiment on an ongoing basis. After the reduction is done, researcher obtains parameters that map the process of damaging the object, while the cost of the experiment is minimized. This method requires performing the passive diagnostic experiment and parallel registration of the damage done to the object.

2 Diagnostic experiments

The planning of every experiment is connected with the use of well-known experiment planning techniques [8-10]. However, these techniques are limited in case of planning of diagnostic experiments, which are often conditioned by the possibilities of their realization. Therefore, in diagnostic tests, aimed at defining the relationship state → signal, with the assumption of a cause-and-effect action of the diagnosed object, there are three types of diagnostic experiments distinguished [6, 7, 11, 12]:

- active,
- passive,
- active-passive.

Active experiment – consists in the observation of the signal after deliberately changing the value of state parameters. This type of experiment can, to a limited

* Corresponding author: malgorzata.orczyk@put.poznan.pl

extent, benefit from the latest techniques of experiment planning and techniques of developing results [8].

Passive experiment – involves observing signal parameters without knowing the state parameters. It gives the possibility of control and interference. In this type of experiment two variants are distinguished. The first is to observe the diagnostic signals of the object from the beginning of its exploitation to its end or until the object is damaged. The second variant consists in simultaneous observation of signals of a particular group of objects in different states of exploitation.

Active-passive experiments – consist in observing signal parameters with simultaneous measurement of state parameters for one or two object state values or exploitation time without interfering in state parameter values. In most cases, exploited objects (before repair) are used as a single value of the state of the object. The two state values are: first value – the state of new objects (before use), and second value – for exploited objects (before repair).

The types of experiments presented here are partial experiments. The active experiment is related to simulated states, as the author presented in work [2]. However, in many cases the results of disassembly of the object are not connected with the exploitation. On the other hand, in the passive diagnostic experiment, the obtained results of measurement of diagnostic parameters are not related to the states of the object. Active-passive experiment refers only to two states of objects: usable and exploited, without information about intermediate states.

In order to identify the causes of changes in diagnostic parameters in a passive diagnostic experiment and to investigate the process of damaging in diagnostic parameters, a new type of experiment was proposed, called passive-reliability experiment.

This experiment consists of periodic measurements of selected diagnostic (vibroacoustic) signal parameters with parallel recording of all exploitation events in form of faults in the intervals between subsequent diagnostic observations [13]. The essence of this experiment is shown in Figure 1.

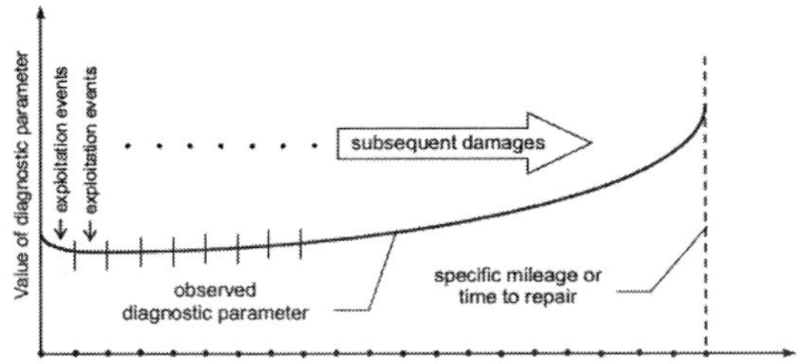

Fig. 1. Scheme of the passive-reliability experiment.

Periodic observations of the diagnostic parameters of the object result in obtaining the curves of the technical state of the object (life cycle curves), which are formed on the basis of the individual signal parameters.

Recorded damage information make it possible to describe the state of the object (reliability status, damage process) using reliability indicators. With diagnostic data in the form of the object's technical condition curve and with data of damages from the intervals between subsequent diagnostic observations, it is possible to select the parameters of the diagnostic signal that represent the reliability state or the object's damaging process.

3 Reliability characteristics for renewable objects

The obtained information about exploitation events (damages) of a diagnosed object can be used to determine the basic reliability characteristics describing the nature and intensity of the damaging process. Since mechanical objects are renewable, the assessment of their reliability may involve using reliability functions and indicators typical for renewable objects. In addition, the functions and indicators of reliability until the first damage can be used to assess the reliability of renewable objects in the initial period of their exploitation, i.e. until the first damage occurs [14, 15].

For the purposes of the sample, the research considers the following basic characteristics of the reliability of renewable objects [14]:

- renewal function

$$H(\Theta)=\frac{\sum_{i=1}^{r} n_i'(\Theta)}{r} \quad (1)$$

$n'_i(\Theta)$ – number of renewals in i-th sample until time Θ,
r – number of trials, covered by study, after n objects,

- availability ratio

$$k_g(\Theta)=\frac{n-n(\Theta)}{n}=1-\frac{n(\Theta)}{n} \quad (2)$$

$n(\Theta)$ – number of damaged objects in moment Θ,
n – sample size, number of usable objects
Θ – operating time (exploitation measure of aging),

- cumulative distribution function

$$F_k(\Theta)=\frac{m_k(\Theta)}{n} \quad (3)$$

$m_k(\Theta)$ – number of damaged objects in the n-element sample in period $[0,t]$, counting since the completion of $(k-1)$ repair of each of n objects,

- probability density function

$$f_k(\Theta)=\frac{m_k(\Theta,\Theta+\Delta t)}{n\cdot\Delta\Theta} \quad (4)$$

$m_k(\Theta,\Theta+\Delta\Theta)$ – number of damaged objects in the n-element sample in period $[\Theta, \Theta+\Delta\Theta]$, counting since the completion of $(k-1)$ repair of each of n objects,

- damage parameter

$$\omega(\Theta) = \frac{m(\Theta, \Theta + \Delta t)}{n \cdot \Delta\Theta} \tag{5}$$

$m(\Theta,\ \Theta + \Delta\Theta)$ – number of objects damaged in the time interval $(\Theta,\ \Theta + \Delta\Theta)$,

- average time of object usability between damages

$$\Theta_k = \frac{1}{n}\sum_{i=1}^{n}\Theta_{k_i} \tag{6}$$

Θ_{k_i} realizations of usability time between $(k-1)$ and k-th damage.

Estimation of the presented reliability features and indicators for renewable objects is possible only after the completion of tests on a particular sample of objects. The specified reliability functions and indicators (especially the damage parameter) can be used to study the relationships between them and the diagnostic signal parameters. Finding the relationship between the reliability status of an object (described by selected reliability indicators) and selected vibroacoustic signal parameters may be the basis for the selection (reduction) of signal parameters correlated to the object's damaging process.

4 Reliability characteristics for a single object

For selection of diagnostic parameters it is important to determine the reliability indicators for a single object rather than a specific sample. Apart from that, the importance lies in the ability to estimate reliability indicators in the subsequent steps of diagnostic measurements. Determination of reliability indicators for a single object after each diagnostic measure will help to reduce the diagnostic signal parameters during the diagnostic experiment, so that only those are left, which are correlated with the process of the object damage.

To evaluate the reliability of a single object at the end of the study and to present a rating after each diagnostic measure, reliability indicators have been proposed, basing on the damage parameter. For evaluation of objects after the observation, these indicators are:

- relative intensity of damages IU_w

$$IU_w(\Theta) = \frac{n_i(\Theta + \Delta\Theta)}{N \cdot \Delta\Theta_i} \tag{7}$$

$n_i(\Theta + \Delta\Theta)$ – number of damages in $\Theta + \Delta\Theta$,
N – total number of recorded damages,
$\Delta\Theta_i$ – interval length (interval between subsequent observations),

- cumulative relative intensity of damages SIU_w

$$SIU_w(\Theta) = \sum_{i=1}^{n}\frac{n_i(\Theta + \Delta\Theta)}{N \cdot \Delta\Theta_i} \tag{8}$$

For evaluation of a single object after each diagnostic observation (during the experiment), the proposed indicators are:

- current relative intensity of damages BIU_w

$$BIU_w(\Theta) = \frac{n_i(\Theta + \Delta\Theta)}{N_i(\Theta) \cdot \Delta\Theta_i} \tag{9}$$

$n_i(\Theta + \Delta\Theta)$ and $\Delta\Theta_i$ – as in dependency (7),
$N_i(t)$ – the total number of damages to the i-th interval (measurement),

- cumulative current relative intensity of damages $SBIU_w$

$$SBIU_w(\Theta) = \sum_{i=1}^{n}\frac{n_i(\Theta + \Delta\Theta)}{N_i(\Theta) \cdot \Delta\Theta_i} \tag{10}$$

$n_i(\Theta + \Delta\Theta)$ and $N_i(\Theta) \cdot \Delta\Theta_i$ – as in dependency (7).

Current intensities of damages are calculated beginning from the fourth interval (fourth diagnostic observation). Relative intensity of damages $IU_w(\Theta)$ characterizes in every moment of t (mileage) relative deterioration of the reliability of the object referred to the unit of mileage $\Delta\Theta = 1$.

The cumulative relative intensity of damages $SIU_w(\Theta)$ can be defined as a measure of the depletion of object's capabilities to perform a task.

5 Selection of diagnostic parameters mapping the damaging process

The method of reduction of parameters during the diagnostic experiment consists in determining the correlation coefficient between the values of the diagnostic parameters obtained from subsequent observations and the values of the damage intensity functions. It is done according to the following relation [16]:

$$r = \frac{\sum_{i=1}^{n}(x_i - \bar{x})(y_i - \bar{y})}{\sqrt{\sum_{i=1}^{n}(x_i - \bar{x})^2 \sum_{i=1}^{n}(y_i - \bar{y})^2}} \tag{11}$$

x_i – subsequent values of the given signal parameter,
$\bar{x}$ – average value of all realizations of a given parameter,
y_i – subsequent values of a given reliability function,
$\bar{y}$ – average value of all realizations of a given reliability function.

For the determined correlation coefficient a confidence interval is estimated using the Fisher transform [16]:

$$Z = 0.5\ln\frac{1+r}{1-r} \tag{12}$$

and reverse transformation, which results in:

DOI: 10.1051/matecconf/201711800034

$$r = \frac{e^{2z} - 1}{e^{2z} + 1} \tag{13}$$

If r is the empirical correlation coefficient of a random sample from a population of two-dimensional normal distribution, then the random variable:

$$Z = 0.5 \ln \frac{1+r}{1-r} \tag{14}$$

has a Fischer distribution, whose asymptotic distribution is the normal distribution:

$$N\left(z(\rho) \frac{1}{\sqrt{n-3}} \right) \tag{15}$$

Using the Laplace transform function table, the $z(\rho)$ parameter can be evaluated:

$$z_d(\rho) = z - \lambda_{\frac{\alpha}{2}} \frac{1}{\sqrt{n-3}} \tag{16}$$

$$z_g(\rho) = z + \lambda_{1-\frac{\alpha}{2}} \frac{1}{\sqrt{n-3}} \tag{17}$$

Thus, the lower bound of the confidence interval for the correlation coefficient will be:

$$R_d = \frac{e^{2 \cdot z_d(\rho)} - 1}{e^{2 \cdot z_d(\rho)} + 1} \tag{18}$$

and the upper bound of the confidence interval will be:

$$R_g = \frac{e^{2 \cdot z_g(\rho)} - 1}{e^{2 \cdot z_g(\rho)} + 1} \tag{19}$$

Having a defined correlation coefficient and its upper and lower confidence intervals, a researcher may proceed to select the signal parameters correlated to the engine's damaging process. This choice is based on rejection of these parameters, for which the correlation coefficient determined in the subsequent steps of observation goes beyond the boundaries of the calculated confidence interval. As further measurements are taken, the boundaries of the confidence interval become narrowed, thus making a more accurate representation of the damage process in those parameters that fall within the designated confidence interval.

6 Selection of diagnostic parameters on the example of an internal combustion engine

The basis for selection of diagnostic parameters correlated with the process of damaging the engine are: waveforms of diagnostic parameters (technical status curves) obtained from the diagnostic measurements of an engine [17] and relative and current intensities of engine damages, determined on the basis of the collected data about the damages [18]. The collected data on engine damage was grouped into sections, where the end of each section was mileage, during which diagnostic measurements were made. Basing on such grouped information about damages, researchers proposed reliability indicators for the evaluation of a single engine, dependencies 7-10. Figures 2 show examples of standardized diagnostic parameters waveforms in relation to the first observation in a function of vehicle mileage.

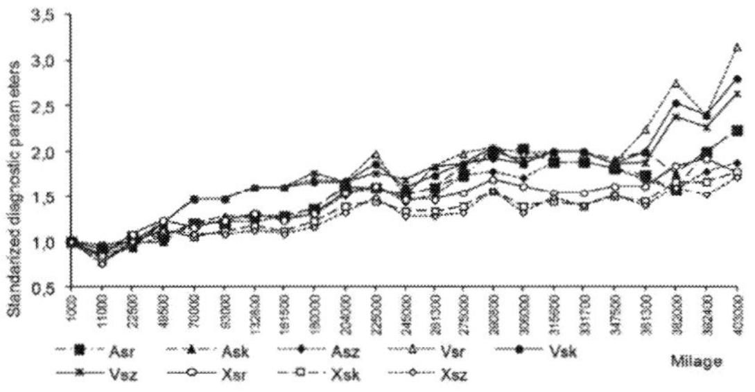

Fig. 2. Waveforms of standardized diagnostic parameters of basic vibration magnitudes in the function of the vehicle mileage.

Figures 4 and 5 illustrate the waveforms of cumulative intensity damage values determined in dependencies (8) and (10). The waveforms of cumulative relative intensity of damages SIU_w and cumulative current intensity of damages $SBIU_w$ illustrate process of damaging the engine in the function of vehicle's mileage. The characteristic feature of the presented intensities is that they show high engine damageability in the initial period of its operation. This period can be counted as the engine's break-in period.

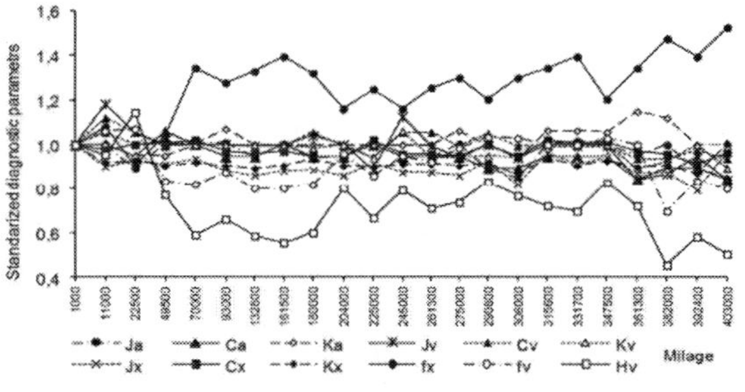

Fig. 3. Waveforms of standardized diagnostic parameters of coefficients of vibration magnitudes in the function of vehicle mileage.

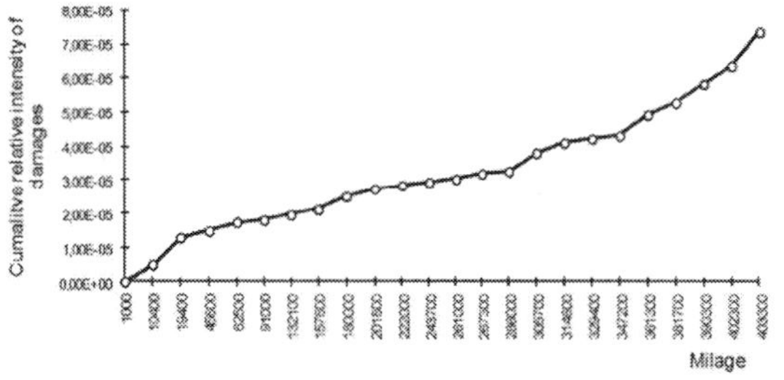

Fig. 4. Changes of the cumulative relative intensity of damages SIU_w of the engine in the function of vehicle mileage.

The cumulative intensity of damages is most useful for selecting the signal parameters at the end of the tests. This is due to the fact that as the wear process of the engine elements progresses, the damaging processes increase and their intensity grows, resulting in new damages due to interactions between the engine components. Because of these phenomena, the process of damaging the engine is cumulative. It is best shown by cumulative relative intensity of damages SIU_w and cumulative current relative intensity of damages $SBIU_w$.

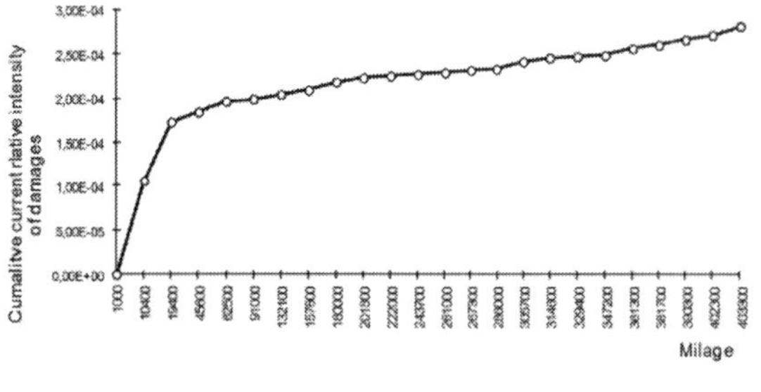

Fig. 5. Changes of the cumulative current relative intensity of damages $SBIU_w$ of the engine in the function of vehicle mileage.

The presented reliability characteristics for single engine evaluation after the completion of the SIU_w tests and for the assessment of engine reliability during the $SBIU_w$ experiment will be used to reduce the parameters of the diagnostic vibration signal of the internal combustion engine. The reduction of the signal parameters will be carried out for two options. The first will concern the selection of parameters based on the reliability indicator defined by the dependency (8) and will concern the selection of parameters after the completion of the test. The second option will concern the selection of parameters during the experiment, basing on the reliability indicator defined by the dependency (10).

7 Selection of signal parameters after the experiment

As previously noted, the reduction of the signal vector – to the signal parameters that best illustrate the process of damaging the engine – consists in determining the correlation coefficient between the various signal parameters and the proposed reliability indicators. For the determined correlation coefficients, the confidence interval boundary values were estimated in accordance with dependencies (18) and (19). To select the parameters correlated to the damaging process, researchers used the narrowest confidence interval for the parameter which best shows the damage process. For the analyzed engine this is *Ask* – the effective value of vibrations acceleration.

Figure 6 shows an example of the correlation coefficient values calculated – beginning from the fourth observation – between the signal parameters: *Ask*, *Vsk*, *Vsz*, *Xsk*, *Ja*, *Jv*, *Jx*, *Fx*, *Fv*, *Hv* and cumulative relative intensity of damages (SIU_w). The figure also shows upper and lower boundaries of the confidence interval (bold dashed line).

The correlation coefficients shown in Figure 6, determined after the subsequent diagnostic measurements, between the signal parameters and the cumulative relative intensity of damages SIU_w, enable – after the experiment – the selection of parameters correlated to the process of damaging the engine. The parameters best correlated (covariated) with the engine damaging process described by the cumulative relative intensity of damages are: the effective value of the vibration accelerations (*Ask*) and Rice frequency for the vibration velocity *Fv*.

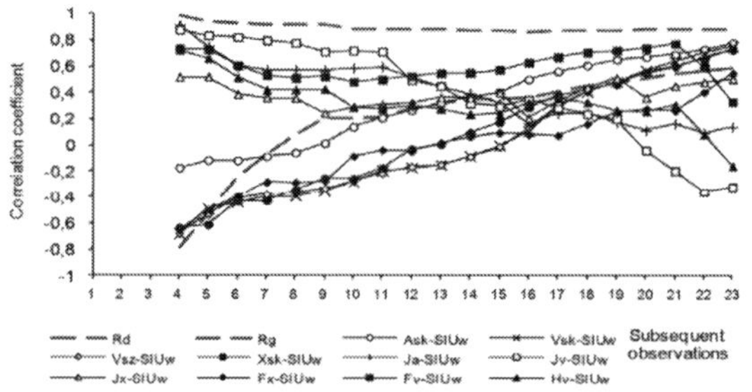

Fig. 6. Changes in correlation coefficient between SIU_w and signal parameters.

8 Selection of signal parameters during the experiment

Reduction of diagnostic parameters during the diagnostic-reliability experiment consists in:

- determining – starting from the fourth diagnostic observation – the cumulative current relative intensity of damages $SBIU_w$,
- determining the correlation coefficient between the cumulative current relative intensity of damages $SBIU_w$ and individual signal parameters; estimation of confidence intervals' boudaries of correlation coefficient,
- selecting the narrowest interval of boundary value of correlation coefficient,
- rejection of such a diagnostic parameter for which, in three consecutive observations, the correlation coefficient value does not fall within the narrowest confidence interval of the correlation coefficient.

A diagnostic parameter whose correlation coefficient goes beyond the boundaries of narrowest confidence interval of the correlation coefficient does not map the damaging process and therefore should be rejected. With this procedure, in the following steps of observation the vector of parameters is decreased (reduced) to those that are covariated with the object's damaging process. As a result, at the end of the diagnostic-reliability experiment, only the parameter describing the damaging process will be obtained.

Figures 7 to 10 show the essence of the proposed method of selecting parameters (which map the damaging process) during the experiment, with the example of several vibrational diagnostic parameters.

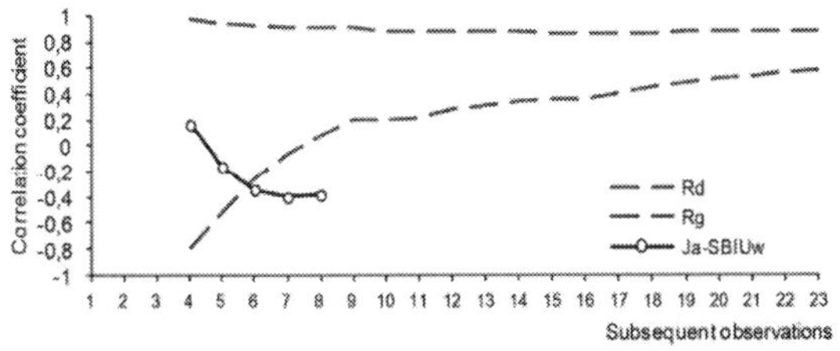

Fig. 7. Changes in correlation coefficient between $SBIU_w$ and the coefficient of impulsiveness of vibration accelerations *Ja*.

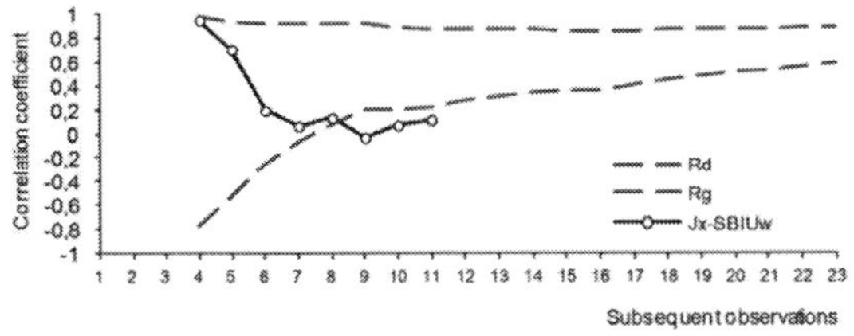

Fig. 8. Changes in correlation coefficient between $SBIU_w$ and the coefficient of impulsiveness of vibration relocations *Jx*.

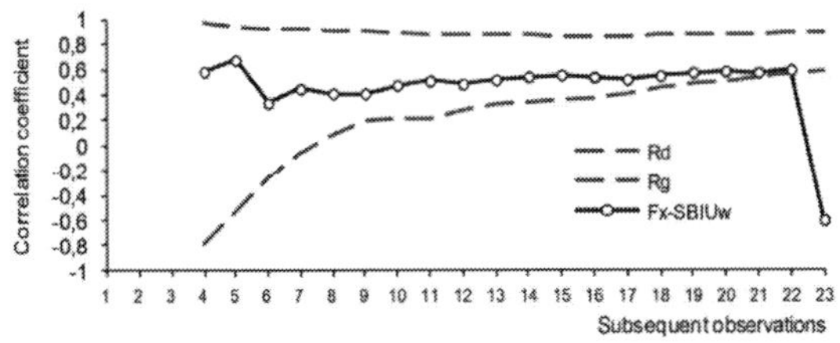

Fig. 9. Changes in correlation coefficient between $SBIU_w$ and the Rice frequency for vibration relocations *Fx*.

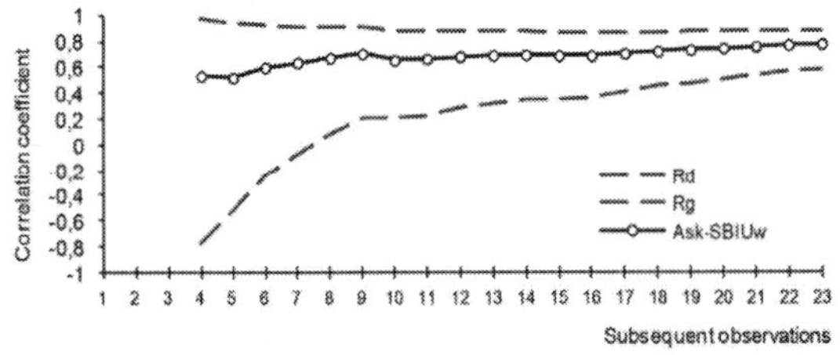

Fig. 10. Changes in correlation coefficient between $SBIU_w$ and effective values of vibration accelerations *Ask*.

The changes of correlation coefficient shown in Figures 7-10, for the sample four diagnostic parameters, show that the parameter that best maps the damaging process is the effective value of the vibration acceleration Ask (Fig. 10). The correlation coefficient for this parameter until the last observation falls within the confidence interval of the correlation coefficient. The coefficient of impulsiveness of vibration accelerations Ja can be discarded (not registered further) right after the eighth observation (Fig. 7). The coefficient of impulsiveness of vibration relocations *Jx* can be discarded after the eleventh observation (Fig. 8). The correlation coefficients of these parameters go beyond the confidence interval of the correlation coefficient and thus do not map the damaging process of the internal combustion engine. Rice's frequency correlation coefficient for vibrations' relocations *Fx* moves beyond the confidence interval of the correlation coefficient only for the last observation.

The conducted analysis of the reduction of all recorded diagnostic parameters of the internal combustion engine during the passive-reliability experiment, presented in Figures 2 and 3 showed that the parameters that represent the engine damaging process are the basic vibration values (acceleration, speed and relocation). The damaging of engine is best mapped by the effective value of vibration accelerations *Ask*.

Conclusion

The proposed method of reducing diagnostic parameters during the passive-reliability diagnostic experiment enables selection of a parameter that is covariated with the progressive process of damaging the object. However, this experiment is labor-intensive because it requires a long time of object observation and recording of diagnostic and reliability data. Recording of diagnostic data in the form of diagnostic signal parameters is performed periodically with a given interval of time or mileage. Logging of reliability data requires continuous observation of the object and recording of all damages and the time or course of their occurrence.

This method can be used for diagnostically unrecognised objects, i.e. it is not known what parameters to measure to obtain maximum information about the object's state. It can be used for single objects because the analysis of the selection of diagnostic parameters concerns a single object. With the use of the presented method of rejecting subsequent diagnostic parameters which do not map the damaging process, research costs are reduced by performing fewer measurements and analyses.

The limitation of the presented method is that it can only be applied to complex objects. In the case of simple objects with a small number of elements, this method does not enable determining of the proposed reliability indicators and diagnostic parameters describing the damage process.

Presented research and the paper is partly funded by Statutory Activities fund of the Institute of Combustion Engines and Transport, PUT (PL) 05/52/DSPB/0259.

References

1. J. Merkisz, M. Idzior, J. Pielecha, W. Gis, *WIT Transactions on the Built Environment* **111**, 181-189 (2010), DOI: 10.2495/UT100171
2. Z. Korczewski, M. Zacharewicz, Transactions of the Institute of Measurement and Control **34**, 937-946 (2012), DOI: 10.1177/0142331211426170
3. Z. Korczewski, J. Rudnicki, *Computational Methods in Marine Engineering VI*, 535-545 (2015)

4. C. Cempel, *Podstawy wibroakustycznej diagnostyki maszyn* (WNT, 1982)
5. C. Cempel, Zagadnienia Eksploatacji Maszyn, **44**, 571-585 (1980)
6. A. Gill, A. Kadzinski, Eksploatacja i Niezawodnosc – Maintenance and Reliability **17**, 282-287 (2015), DOI: 10.17531/ein.2015.2.16
7. B. Żółtowski, *Projektowanie eksperymentów w diagnostyce maszyn* (Torun, 1984)
8. K. Mańczak, *Technika planowania eksperymentu* (WNT, 1976)
9. H. Golas, A. Mazur, J. Gruszka, P. Szafer, *IOP Conference Series-Materials Science and Engineering* **145**, 062005 (2016), DOI: 10.1088/1757-899X/145/6/062005
10. H. Golas, A. Mazur, J. Gruszka, *ACSR-Advances in Computer Science Research* **57**, 257-264 (2016)
11. G.M. Szymanski, M. Josko, F. Tomaszewski, R. Filipiak, *Mechanical Systems and Signal Processing* **58-59**, 298-307 (2015)
12. G.M. Szymanski, F. Tomaszewski, Mechanical Systems and Signal Processing **68-69**, 479-490 (2016)
13. F. Tomaszewski, *Zagadnienia wyznaczania stanu technicznego złożonego obiektu mechanicznego za pomocą sygnału wibroakustycznego. Na przykładzie silnika spalinowego pojazdu szynowego* (Poznań 1998)
14. J.N. Migdalski (Wyd. Przemysłu Maszynowego WEMA, 1982)
15. W. Sawczuk, Maintenance and Reliability **18**, 565-571 (2016)
16. D. Bobrowski, *Probabilistyka w zastosowaniach technicznych* (WNT, 1980)
17. F. Tomaszewski, *Wyniki badań diagnostycznych silników spalinowych 2112 SSF lokomotyw serii SP45* (Poznan, 1985)
18. F. Tomaszewski, *Wyniki badań niezawodnościowych silników spalinowych 2112 SSF lokomotyw serii SP45* (Poznan, 1985)

The influence of engine warm up phase on nitrogen oxides emission for heavy-duty Euro VI diesel engine

Rafał Sala[1,*], *Jarosław* Krasowski[1], and *Jakub* Dzida[1]

[1]Automotive Research and Development Institute BOSMAL, ul Sarni Stok 93, 43-300, Bielsko-Biała, Poland

Abstract. This article presents test results achieved under World Harmonized Transient Driving Cycle (WHTC) run on heavy-duty diesel engine compliant with Euro VI standard. The emission cycle was performed both for cold and hot engine start up conditions. Modal analysis of emission compounds was carried out with the focus on nitro-gen oxides (NOx) emission. The scope of the work was to assess the influence of cold start effect on selective catalytic reduction (SCR) system operation efficiency. Due to the fact that the urea injection strategy in SCR system can by activated only after certain exhaust gas temperature threshold is achieved, the NOx emission on WHTC cold start cycle is significantly greater than for the hot start one. Result analysis included also determination of time period needed for achieving the readiness of SCR system to work efficiently. The occurrence of ammonia slip emissions downstream SCR was also monitored.

1 Introduction

Current legislative emission limits of passenger and heavy duty vehicles further tighten requirements of emitted NOx. Certification procedures of both vehicle groups include cold start condition as a part of a test cycle. Nevertheless there is a significant difference of cold start influence at test overall outcome between passenger cars and heavy duty engines. The chassis dyno cycle of light vehicles e.g. WLTC comprises cold start as one of the most crucial and influential phase of the test. On the other hand the World Harmonized Heavy Duty Certification Procedure minimalizes the meaning of cold start emission. This procedure incorporates world harmonized steady-state test cycle (WHSC) and transient cold and hot test cycles (WHTC). Regardless presence of cold cycle, its importance at final WHTC result is relatively low giving it only approx. 10% of total value. Consequently, the heavy duty engines after-treatment systems (ATS) do not focus on cold start emissions as much as of light vehicles. Nowadays the most common solution for reducing NOx emission of HD diesel engines is an introduction of Selective Catalytic Reduction to ATS. This, as all catalyst-based cleaning up devices, starts to work effectively only when proper level of exhaust temperature is achieved. This technology meets legislation limits of Euro VI ensuring low NOx emission for rare cold start engine operation. However when a cold start occurs the emission is significantly greater. The investigation of engine out and tailpipe NOx concertation traces clearly outlines engine operation periods practically without NOx after-treatment.

2 SCR operation under engine warm up conditions

Improving the vehicle fuel economy that basically means decreasing of CO_2 emission, is a real technical challenge when it comes to simultaneous reduction of NOx emissions to meet tight legislative limits. Introduction of advanced fuel combustion strategies leads to an improvement of engine efficiency and decreases CO_2 emission, but at the same time causes an increase of raw NOx emission [1]. This in turn requires an application of highly efficient NOx removal systems. Additionally, optimized combustion processes result in dropping exhaust temperature even further and that has an adverse effect on after-treatment system operation, especially in a warm up phase.

Considering the lean NOx after-treatment methods, the selective catalytic reduction (SCR) system provides the highest NOx reduction efficiency and it is the most reliable over entire engine operation range. This basic features make the SCR a dominant technology in the light of current and upcoming stringent NOx emission limits [2, 3]. Further advantages of SCR are durable performance, width of performance window, reasonable cost and also available infrastructure. In order to meet current Euro VI regulations, the average NOx conversion efficiency calculated from cold and hot cycles often needs to exceed the 95% [4-6].

The process of selective catalytic reduction relies on dosing the NOx reduction agent, which is water urea solution, upstream of the SCR catalyst. The atomized into the hot exhaust gas stream urea decomposes in three steps that are physically separated in time and space: evaporation, the thermal decomposition of finely sprayed

* Corresponding author: rafal.sala@bosmal.com.pl

urea into ammonia (NH_3) and isocyanic acid (HNCO), hydrolysis, which would occur along the exhaust pipe prior to the catalyst, and the hydrolysis of isocyanic acid, which would occur on the catalyst surface [7, 8]. When sprayed into the exhaust line, the evaporation of the water decreases the exhaust gas temperature by some 10-15 °C [9]. This cooling effect has an influence on activity, especially in the low temperature region where the conversion is limited. Moreover, there is solid deposit formation in the exhaust and difficulties in dosing at exhaust temperatures below 200 °C.

With urea solution dosed as a reductant, there exists an additional constraint in the lower temperature range, because the conversion of urea to NH3 in the exhaust line does not take place completely below 200 °C. The limited conversion of isocyanic acid by hydrolysis is perhaps the biggest obstacle facing the use of urea-SCR for achieving high NOx conversion in urban areas [10, 11] due to low exhaust temperatures. While the SCR catalyst still has activity down to at least 150°C, the urea dosing has to be stopped to avoid formation of deposits and undesired by-products.

A lot of effort is being made to reduce the cold-start emissions, which has become critical in the practical applications and also to improve deNOx performance in urban driving or other low load conditions. As a result, higher activity of SCR catalysts at low temperatures is desired to effectively reduce real-world NOx emission while fuel economy requirement has to be simultaneously met. One of the main obstacles is that during low-speed urban driving, the efficient combustion engines generate exhaust gas temperatures below 200 °C, which is insufficient for the conventional urea-SCR based systems [12, 13]

Further development focuses on high-efficiency SCR catalysts with broad temperature windows, with superb thermal durability, with precise control algorithm incorporating ammonia storage capacity, and with the entire after-treatment system optimized to meet the current and emerging NOx regulations [14-16].

There are already existing solutions and ongoing research works on solid materials delivering already volatile ammonia into exhaust gas. This kind of methods help to faster initialize the SCR system during engine warm up phase and to improve the mixing process in exhaust.

Currently, the-state-of-art deNOx technology is urea-SCR over primarily CuZ SCR, with relatively wide temperature window, better durability, and low NH_3 slip [17]. In many applications to reduce the NH_3 slip, the additional oxidation clean-up catalyst (CUC) is installed downstream of SCR.

3 Test methodology description

3.1 Test stand

The procedure chosen for testing was the World Harmonized Transient Cycle (WHTC), part of world-wide harmonized heavy-duty certification (WHDC) procedure for engine exhaust emissions. The WHTC is a transient test cycle with both cold and hot start requirements that lasts 1800 s, with several motoring segments. The execution of certification procedures obligate a test stand capable to fulfill all legislative requirements. BOSMAL's Engine Development Laboratory incorporates highly automated fully dynamic dyno with up to date emission measurement instruments.

Fig. 1. Aftertreatment system equipped with selective catalytic reduction system.

The undiluted exhaust gas is measured by two independent sets of analyzers allowing continues acquisition of engine out and tailpipe emissions. The direct concentration measurement of the following compounds is carried out: total hydrocarbons (THC), methane (CH_4), carbon monoxide (CO), nitrogen oxides (NOx), nitrogen oxide (NO), oxygen (O_2) and carbon dioxide (CO_2). The results obtained allow to calculate the concentration of non-methane hydrocarbons (NMHC) and nitrogen dioxide (NO_2). Since ammonia (NH_3) emission is expected solely downstream SCR, the NH_3 measurement by laser diode detection instrument is done only at the tailpipe. For the particulate mass (PM) measurement the partial dilution tunnel is applied. The diluted sample is transferred through fine fiber filter which is weighed after test execution. The secondary sample dilution is carried out in order to measure particulate number (PN). The acquired results of gas compounds concentration and particulates are calculated accordingly to legislative formulas and specific values of emission and fuel consumption are obtained [18-20].

3.2 Engine and aftertreatment system

The engine assigned to the testing was a Euro VI compliant heavy duty six cylinder engine of 6.7 L displacement and rated power of 235 kW. The basic features of the engine includes single stage turbocharging and common rail system with maximum fuel injection pressure of 1800 bar. The advanced aftertreatment system allows no-EGR architecture decreasing soot formation and consequently lowering the particulate filter load. The duty of NOx emission reduction is shifted to the aftertreatment system. The fulfilment of Euro VI limits requires the ATS of high complexity. The system incorporates DOC, DPF, SCR technologies and NH_3 clean-up reactor to minimalize

DOI: 10.1051/matecconf/201711800035

ammonia slip. The SCR functionality is based upon exhaust mass flow and readings of number of sensors: NOx upstream and downstream ATS, NH_3 at tailpipe and exhaust temperature.

4 Test results and discussion

The WHTC cycle lasts 1800 seconds and it is run from cold and hot engine start. Presented below are total emission results calculated from cold and hot WHTC cycles, compared to Euro VI limits (Table 1).

The profile of NOx conversion efficiency and basic exhaust and engine parameters of cold and hot WHTC cycles are presented on the Fig. 3 and Fig. 4. Considering the engine and ATS warm up phase during WHTC cold cycle (Fig. 3) three are three characteristic sub phases that can be distinguished:

Table 1. WHTC total results and EU Emission Standards for Heavy-Duty Diesel.

Stage	WHTC	CO	NMHC	CH4	NOx	PM	PN	NH_3
		g/kWh					1/kWh	ppm
Euro IV	Euro VI limits	4,0	0,16	0,5	0,46	0,01	$6{,}0\cdot10^{11}$	10
	WHTC result	0,01	0,0	0,02	0,33	0,0017	$2{,}53\cdot10^{11}$	0,0

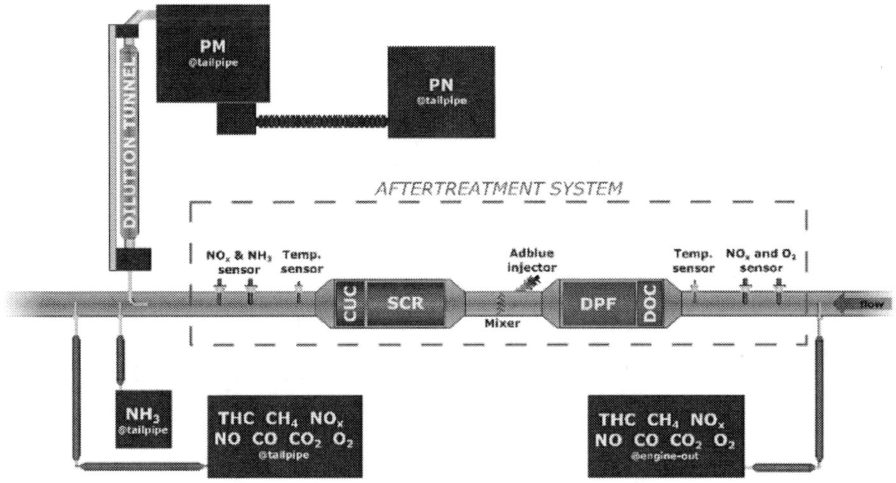

Fig. 2. Aftertreatment system layout with sampling points.

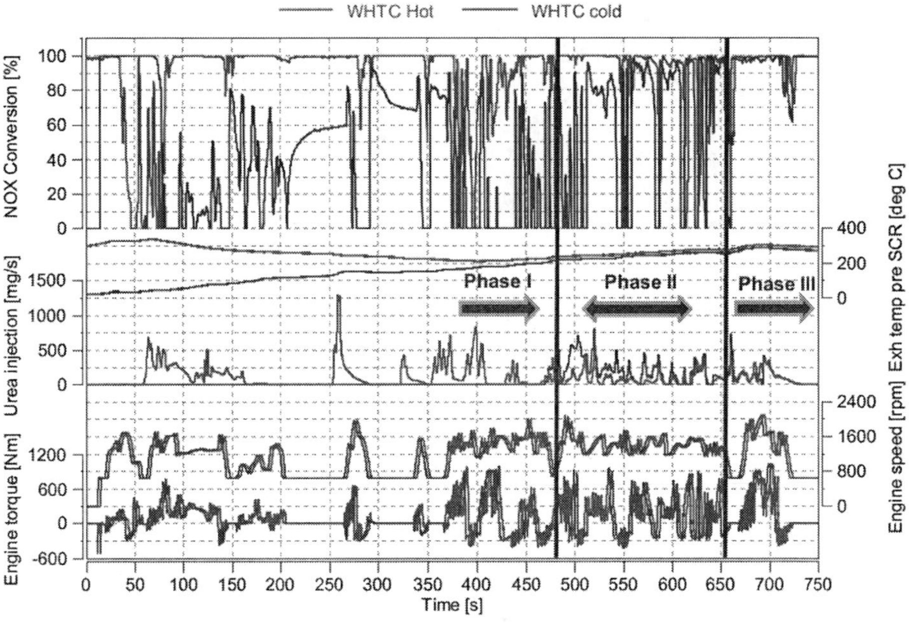

Fig. 3. NOx conversion efficiency and base operating parameters over WHTC cycle – warm up phases.

Phase I – starts just after engine cranking up and lasts until urea dosing is switched on. During that phase the exhaust gas temperature is too low to start the dispensing of the reducing agent. The catalyst light-off temperature was not yet reached therefore efficiency of catalytic reduction and oxidation processes are very poor. In this phase the lowering of NOx emission greatly relies on in-cylinder measures such as retarded fuel injection process or post injection. The last aims to increase the pace of catalyst warm up. The duration of phase I is about 470 seconds.

Phase II – extends from the point when desired exhaust temperature is reached and the urea injection starts. The phase lasts until the moment when the SCR system achieves its nominal NOx conversion efficiency. In this phase the exhaust gas temperatures measured pre SCR for cold and hot cycles converge (Fig. 3) but the NOx reduction efficiency for the cold cycle in still lower than for the hot one. The response time of the SCR system in NOx reduction from the start of urea dosing is 60 seconds. The reason for this phenomena can be explained by different thermal state of catalyst substrate and lower SCR saturation with ammonia. The duration time of phase II is approximately 200 seconds.

Phase III – is considered from the time when the traces of NOx conversion efficiency for both cold and hot cycles are inline and the SCR system efficiency reaches 98% or more. This phase lasts to the end of WHTC cycle with the duration time of 1130 seconds (Fig. 4).

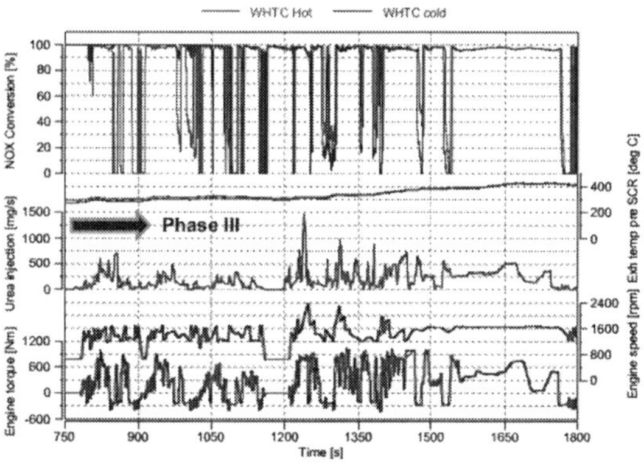

Fig. 4. NOx conversion efficiency and base operating parameters over WHTC cycle – after warm up.

The length of phases I and II determines the exhaust system warm up. The length of both phases is 670 seconds which is 30% of cold cycle duration. During this period the NOx emission is noticeably high. There are many technical efforts undertaken to decrease this critical warm up phase. Frequent usage of the vehicle with SCR system on a short distances with cold start events considerably increases the real NOx emission level.

The effect of engine thermal state on harmful emission components and CO_2 is presented on Fig. 5. It can be seen that the NOx emission on hot cycle is 65% lower comparing to the cold one. There is also a great impact of engine warm up on CO and PM emission, as well as on THC and CO_2 compounds.

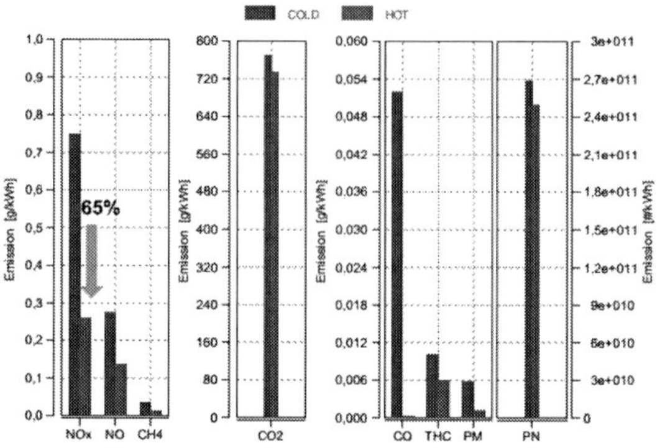

Fig. 5. Emission components of cold end hot WHTC cycle.

5 Conclusions

A major challenge in the development was the control of cold start NOx emissions, making thermal management strategies a key component of the project. The engine initial calibration for the cold start portion has been modified, including EGR modifications, multiple injections, intake throttling, and elevated idle speed. The engine reverted to the original, high fuel economy calibration when hot.

Passive NOx absorbers remain a promising technology for the control of cold start emissions in future low NOx engines. The formulation stores NOx below 200 °C and has a peak NOx release 250 °C. The NOx storage capacity becomes fully regenerated after exposure to temperatures above 300 °C.

The development of a high-efficiency SCR system for future engines targets at developing SCR formulation capable to achieve 90% of NOx reduction at 150°C with favourable range of NO/NOx in exhaust gas, provided by the DOC. Further the urea droplet sizes should be decreased below 10 micron in order to minimize deposits formation at low gas temperature level.

While NOx emission standards for heavy-duty engines have not changed since 2010, the real world SCR performance has been continuously improving since the launch of SCR equipped engines. In particular, newer SCR equipped engines achieve higher NOx conversions at low temperature conditions – an important issue identified with many early SCR implementations. Also, the control strategy changes have improved system response and tightened the range of dosing to achieve closer to stoichiometric NOx/NH_3 conditions.

Further design optimisation is taken into account to minimalize NOx emission at low exhaust temperature. It includes thermal insulation improvement, lowering dosing onset temperature, improved catalysts, and optimized fluid dynamics during urea injection.

References

1. K. Brzozowski, J. Nowakowski, Eksploatacja i Niezawodnosc – Maintenance and Reliability **4**, 56-62 (2011)
2. J. Han, E. Kim, T. Lee, J. Kim, N. Ahn, H.S. Han, SAE Technical Paper 2011-01-1315 (2011)
3. P. Blakeman, K. Arnby, P. Marsh, C. Newma, G. Smedler, SAE Technical Paper 2009-26-013 (2009)
4. M. Brzezanski, R. Sala, Materials Science and Engineering **148**, 012062 (2016), DOI:10.1088/1757-899X/148/1/012062
5. M. Bajerlein, L. Rymaniak, Experimental and Applied Mechanics **518**, 96-101 (2014), DOI: 10.4028/www.scientific.net/AMM.518.96
6. P. Lijewski, J. Merkisz, P. Fuc, A. Ziolkowski, L. Rymaniak, W. Kusiak, European Journal of Forest Research **136**, 153-160 (2017), DOI: 10.1007/s10342-016-1015-2
7. H.L. Fang, H.F.M. DaCosta, Appl. Catal. B: Environ. **46**, 14-34 (2003)
8. M. Koebel, M. Elsener, M. Kleemann, Catal. Today **56**, 335-345 (2000)
9. T. Johhannessen, H. Schmidt, J. Svagin, J. Johansen, J. Oechsle, R. Bradley, SAE Technical Paper 2008-01-1027 (2008)
10. T. Maunula, T. Kinnunen, K. Kanniainen, A. Viitanen, A. Savimaki, SAE Technical Paper 2013-01-1063 (2013)
11. Sala, R. Ph.D. Dissertation (Cracow, 2014)
12. T. Maunula, SAE Technical Paper 2013-24-0161 (2013)
13. J. Merkisz, S. Tarkowski, Eksploatacja i Niezawodnosc – Maintenance and Reliability **2**, 50-58 (2011)
14. L. Yang, V. Franco, A. Campestrini, J. German, P. Mock, ICCT, White Paper (2015)
15. European Environment Agency, CSI004 (2016)
16. J. Merkisz, J. Pielecha, P. Molik, M. Nowak, *2014 IEEE Vehicle Power and Propulsion Conference,* Coimbra (2014)
17. Z. Stepien, L. Ziemianski, G. Zak, M. Wojtasik, L. Jeczmionek, Z. Burnus, Fuel **161**, 278-286 (2015), DOI: 10.1016/j.fuel.2015.08.071
18. J. Merkisz, J. Pielecha, *Springer Tracts on Transportation and Traffic* **8**, 69-92 (2015), DOI: 10.1007/978-3-319-15928-7_7
19. J. Merkisz, J. Pielecha, S. Radzimirski, *Springer Tracts on Transportation and Traffic* **4**, 67-90 (2014), DOI: 10.1007/978-3-319-02705-0_4
20. W. Gis, J. Pielecha, J. Waskiewicz, M. Gis, M. Menes, *IOP Conference Series-Materials Science and Engineering* **148**, UNSP 012040 (2016), DOI: 10.1088/1757-899X/148/1/012040

Analysis of fuel consumption of a spark ignition engine in the conditions of a variable load

Andrzej Bieniek[1,*], *Mariusz* Graba[1], *Krystian* Hennek[1], and *Jarosław* Mamala[1]

[1] Opole University of Technology, ul. Proszkowska 76, 45-758 Opole, Poland

Abstract. This paper reports the results of a study into operating parameters of an system consisting of an SI engine and a drivertrain in a Fiat Panda passenger car in the conditions of a variable load. The analysis was primarily concerned with the variability of fuel consumption resulting of the changing load applied to the driving wheels in the conditions of a test performed on dynamometer rollers. The bench test included a dedicated driving cycle in which a constant linear speed of the car was changed every 10 km/h and subsequently maintained for registration and analysis purposes. The focus of the study involved the identification of the points characterized with the minimum specific fuel consumption. Throughout the test, the load applied to the transmission was alternated by simulating an increase of road inclination for which the variability of the engine efficiency was analyzed.

1 Introduction

Despite the considerable media noise focusing on the developments in electric powertrains, throughout the many years to come, we can expect further development of vehicle powertrains powered by traditional combustion engines. At present, engine designs are continuously improved with the purpose of fulfilling ever stricter emission standards, including CO_2 emissions, which are directly relative to the fuel consumption [1-3]. The developmental tests that are considered to be particularly interesting include studies into variable power output resulting from the application of Variable Compression Ratio (VCR) and Controlled Auto-Ignition (CAI). However, the complete exploitation of the operating parameters of such up-to-date engine designs is only possible by the application of powertrain systems with a considerable capability of torque transfer. Concurrently, we know that the factor representing the utilization of the installed power is relative to the integration of the Driver-Vehicle-Environment (DVE) model. Although the role of the driver is to decide on the demanded linear car velocity in a free or induced manner, this is already largely limited in modern powertrains. In this context, the adequately selected traction characteristic of the vehicle should provide a high efficiency of the overall system. However, the efficiency of the engine varies depending on the position of the acceleration pedal, and this is due to the fact that it is utilized in variable ambient conditions – including rain, wind, snow or variable terrain. Problems in these areas lead to the small efficiency of the use of engine power, in particular during drive on a low gear and extremely variable linear vehicle speeds during urban drive.

The research reported in this paper was undertaken with the purpose of a detailed analysis of the impact of variable engine load on the fuel consumption of a passenger car. The variable load in this paper refers to the increase of road inclination angle expressed in per cent for the particular gearbox ratios. As a result of the additional load of the driving wheels (and thus, a variable load of the powertrain of a passenger car), the fuel use is known is increase. The testing procedure was performed on a Maha MSR 500 chassis dynamometer by application of a test car.

2 Measurement methodology

The analysis of fuel consumption of the conditions of the variable engine load was performed on the basis of the research conducted on a MAHA MSR 500 chassis dynamometer installed at the Department of Road and Agricultural Vehicles, Opole University of Technology. The testing applied Fiat Panda passenger car with the engine capacity 1242 cm^3 and a maximum power of 44 kW (Fig. 1).

A dedicated driving cycle was defined for the purposes of the research with a section involving a constant linear car velocity for each of the gearbox ratios. The range of the linear vehicle velocity was selected individually for each of the gears so that the rotational speed can be changed in the range from 1000 rpm to 5500 rpm. The linear velocity was increased stepwise by 10 km/h and the velocity profile followed in the driving cycle is presented in Fig. 2.

Throughout the duration of the test, the car was in motion at a constant linear car velocity and the gearbox ratio was given in accordance with the cycle described above. In addition, the scope of the study involved the

* Corresponding author: a.bieniek@po.opole.pl

DOI: 10.1051/matecconf/201711800036

change of the additional load of the powertrain associated with the simulated variation in the road inclination. The basic study was conducted for the car rolling on a horizontal surface and, subsequently, simulation tests were repeated for the 3%, 5%, 7% and 10% road inclinations.

Fig. 1. Test bench.

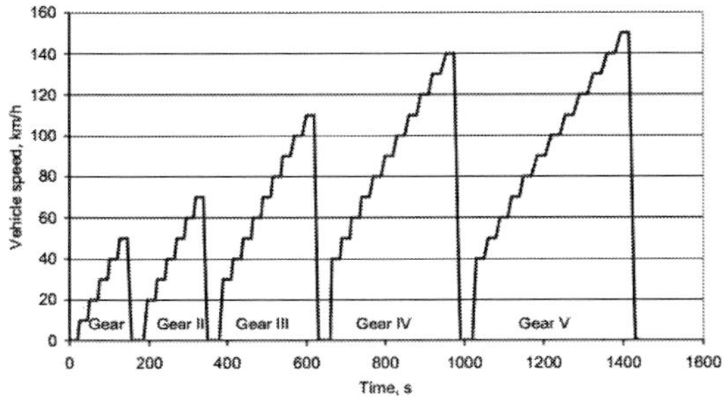

Fig. 2. Course of the driving cycle.

The testing procedure was accompanied by the measurements of fuel consumption undertaken with Flowtronic 215 fuel measuring system accompanied by pollutant emission measurement by means of MGT 5 analyzer. The analysis applied averaged data registered at the given points (for a constant car velocity) throughout the driving cycle. Exemplary results of actual registrations made by the measuring system are presented in Fig. 3.

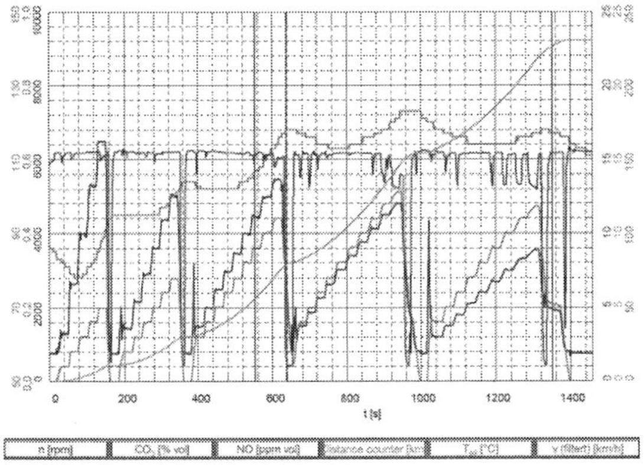

Fig. 3. Examples of data registered by the dynamometer system for 3% road inclination.

For each of the registered cycles, the measuring system of the dynamometer recorded 11 variables accompanied by fuel consumption measurement by the system indicated above.

3 Fuel consumption during constant car

The literature in the area of fundamentals of vehicle dynamics [4, 5] gives that driving force acting on a car in the conditions of its acceleration or rolling at a constant speed, we have to do with an equilibrium resulting from the resistance to motion, consisting of rolling resistance, drag forces, climbing resistance and inertia forces resulting from the car acceleration.

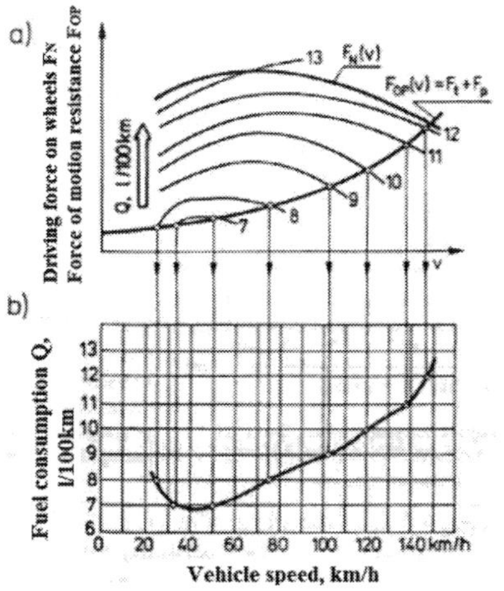

Fig. 4. Fuel consumption for a constant car velocity [5].

The figure above represents the theoretical description of the characteristic of fuel consumption for a constant car velocity. The points in Fig. 4a define the characteristics of the instantaneous equilibrium of the transmission system, where the driving force is balanced by the forces acting on the car. In such conditions, the total of values of the basic resistive forces, including rolling resistance F_t, drag force F_P, climbing resistance F_W have to compensate the driving force and the formula of motion states that at such an operating point the car acceleration is close to zero.

$$\frac{dv}{dt} = \frac{F_N - (F_{tp} + F_w)}{m\delta} \qquad (1)$$

where: m – mass of car, F_N – driving force on wheels, F_{TP} – total rolling and air resistance on the horizontal road, F_W – resistance of the road inclination, δ – spin mass coefficient.

Given the simplifications above, for a constant linear car velocity, we can perform an analysis of the course of the driving force in a car powertrain in an experimental manner by following adequate procedures during road test. However, we need to bear in mind that it is beneficial to perform all these activities during a single road test, whose course should be similar to the standard driving maneuvers performed by a driver throughout an actual road drive. However, since it is difficult to perform such a test during actual drive, there is a possibility of performing an analysis on a chassis dynamometer comprising electrodynamic brakes acting on the particular driven axles in a vehicle. In such conditions, it is possible to apply approximate forces representing resistance to motion to the vehicle powertrain and analyze the engine response in the form of variations in fuel consumption. Based on the insights from literature, we know that the increase of the engine load resulting from is climbing, in turn leads to an increased fuel consumption.

4 Variable load of the powertrain

The variable load of the transmission in a car is realized by application of an electrodynamic brake to the drive axle. The registered driving force acting on the dynamometer rollers is formed by the product of the driving force output and the linear car velocity. On the basis of the conducted studies, we have undertaken an analysis of the course of the power generated by the engine, and an exemplary power curve for a given constant load (for 3% road inclination) and variable linear car velocity corresponding to the particular gearbox ratios is presented in Fig. 5.

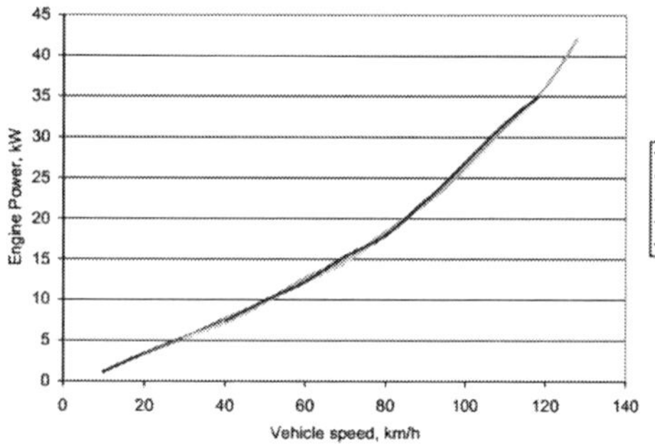

Fig. 5. Course of engine power for 3% road inclination.

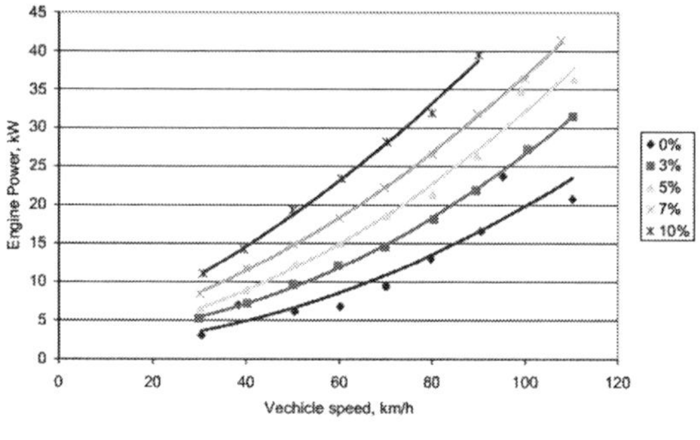

Fig. 6. Course of the engine power for 3rd gear corresponding to various road inclinations.

The engine power measured in the powertrain is in range of the forecasted results (Fig. 4). However, within the range of the constant ratio in the 3rd gear, the variability of the engine power is already considerable during the changing road inclination, as presented in Fig. 6. The change in the road inclination from 0% to 10% leads to almost 2.5 times increase of the engine power. Concurrently, in the conditions of 10% road inclination, the car is not capable of developing an output adequate to accelerate the car to the demanded constant velocity specified in the driving cycle. This condition was registered for 3rd gear (partly), as well as 4th and 5th gears. We also noted that a two times increase of the engine power is necessary to overcome a road inclination of 7% for the 3rd gear, for which fuel consumption was also registered.

An analysis of the fuel consumption was performed, whose level also increases two times for the car speeds of 40 and 60 km/h, however, for greater speeds, it decreases and the engine power output approaches the peak value.

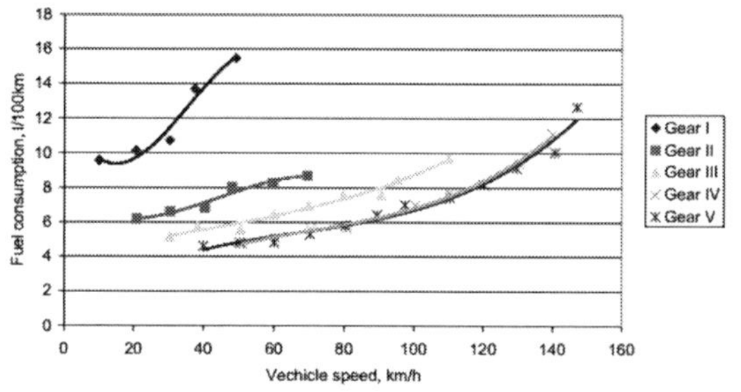

Fig. 7. Distance-based fuel consumption for a constant gearbox ratio and 0% road inclination.

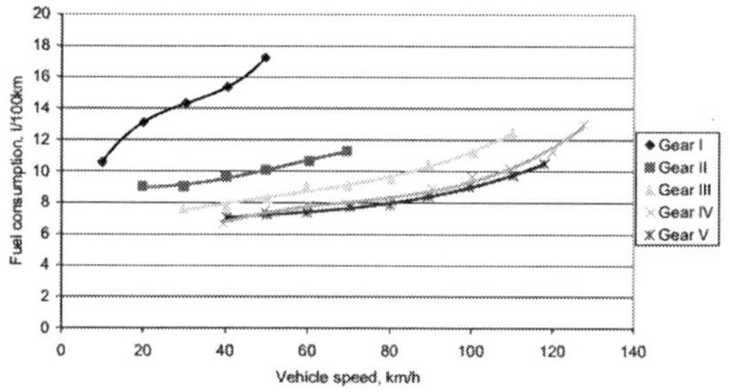

Fig. 8. Distance-based fuel consumption for a constant gearbox ratio and 3% road inclination.

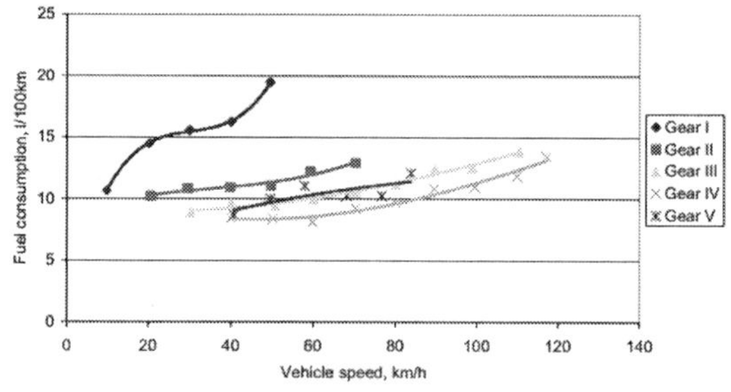

Fig. 9. Distance-based fuel consumption for a constant gearbox ratio and 5% road inclination.

DOI: 10.1051/matecconf/201711800036

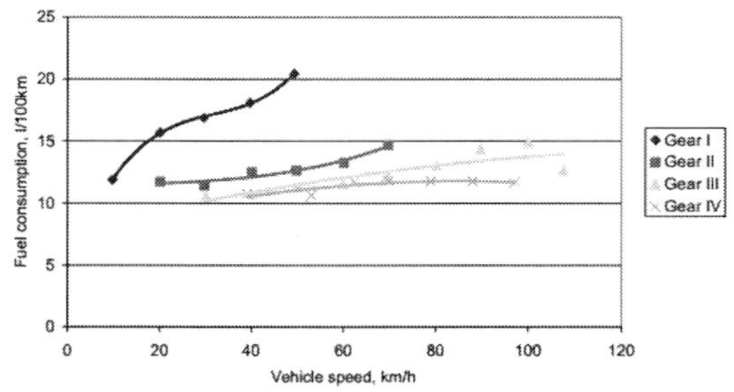

Fig. 10. Distance-based fuel consumption for a constant gearbox ratio and 7% road inclination.

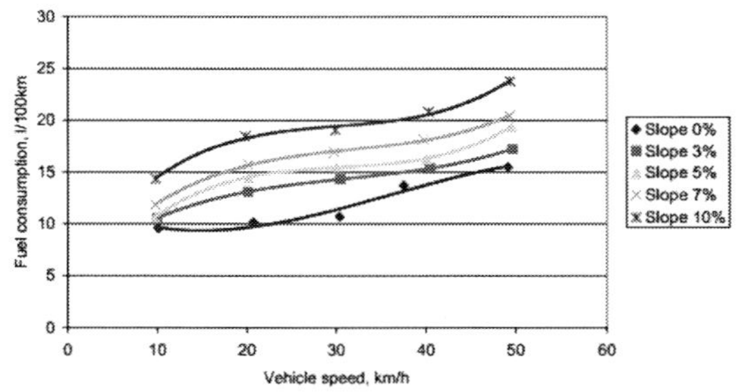

Fig. 11. Distance-based fuel consumption for gear I and variable road inclination.

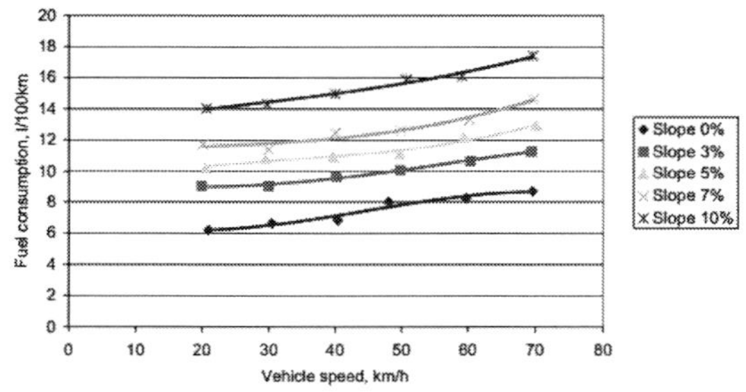

Fig. 12. Distance-based fuel consumption for gear II and variable road inclination.

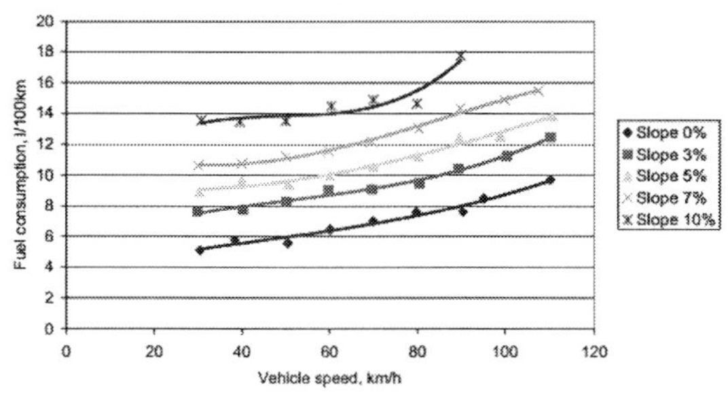

Fig. 13. Distance-based fuel consumption for gear III and variable road inclination.

The engine characteristic does not always include a minimum fuel consumption based on the criterion of the distance driven, as given in [6-8], and this fact was also observed with regard to the examined engine and confirmed on its characteristic. This is quite standard with regard to engines with a sequential multi-point fuel injection. Concurrently, very important differences can be noted for the case when the distance-based fuel consumption is converted into flow rate of fuel supply and the result is presented in the function of the rotational speed of the engine crankshaft, as presented on Fig. 15-18.

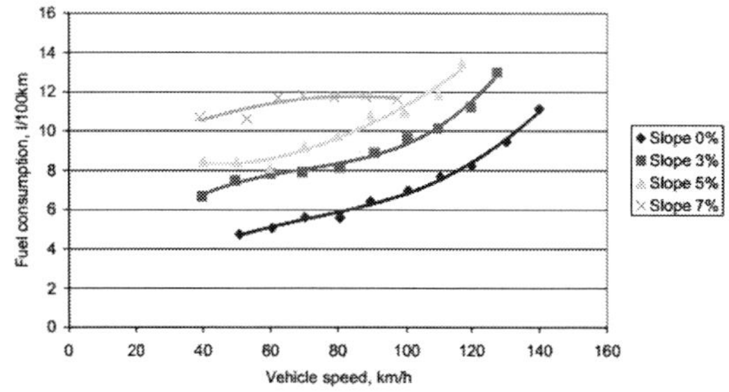

Fig. 14. Distance-based fuel consumption for gear IV and variable road inclination.

The summary of data regarding flow rate of fuel for a selected constant gearbox ratio, as visible on Fig. 15-18, demonstrates the considerable dependence of fuel consumption on the rotational engine speed and to a lesser degree relative to the increase of the road inclination. The greatest level of the change of the flow rate was noted for the 4th gear, where the consumption increased from around 3 l/h to 16 l/h. For the 3rd gear, we noted a flow rate of fuel of 1.5 l/h (for 0% road inclination corresponding to the rotational speed of 1500 rpm. However, an increase in fuel consumption is equal to as much as 6 l/h (i.e. from 6 l/h to 12 l/h) for the rotational engine speed of 5000 rpm.

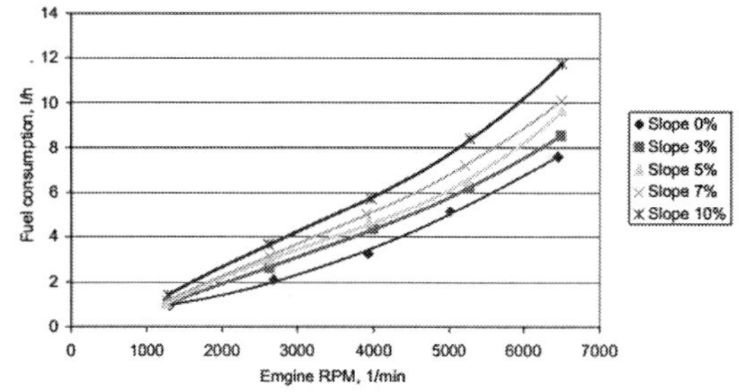

Fig. 15. Flow rate of fuel consumption of the engine for gear I and variable road inclination.

Hence, bearing in mind that the variable level of the engine load has such a considerable impact on fuel consumption, it is necessary to identify a gearbox ratio for which an operating point of the engine is maintained so as to achieve a maximum engine efficiency. For this reason, an important operating parameter applied to describe the characteristics of a transmission is associated with its overall efficiency.

DOI: 10.1051/matecconf/201711800036

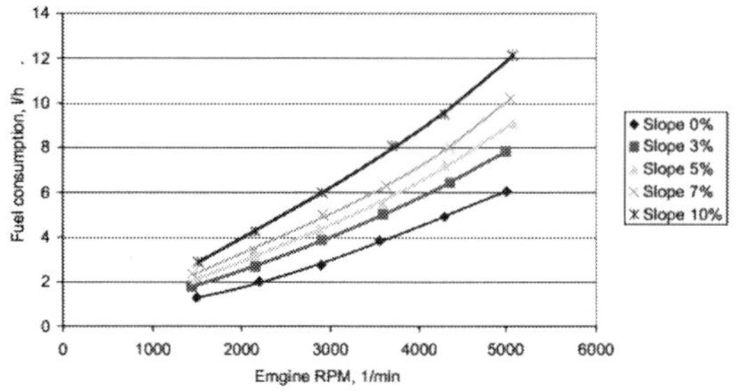

Fig. 16. Flow rate of fuel consumption of the engine for gear II and variable road inclination.

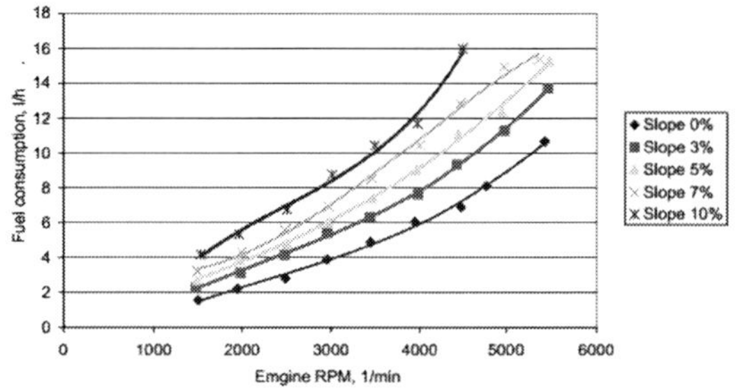

Fig. 17. Flow rate of fuel consumption of the engine for gear III and variable road inclination.

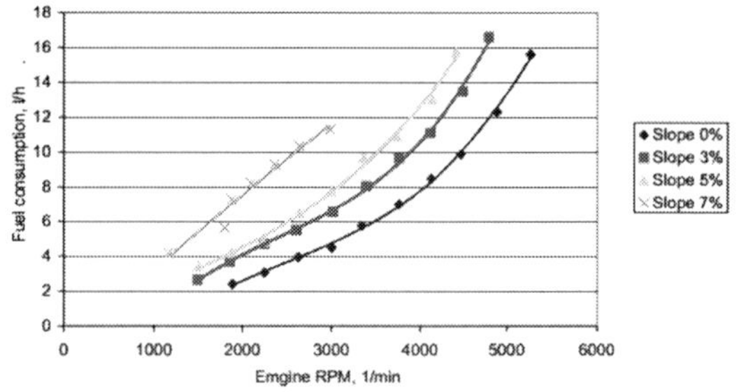

Fig. 18. Flow rate of fuel consumption of the engine for gear IV and variable road inclination.

On the basis of the analysis of the formula representing the overall engine efficiency (2), we can express it as the inverse of the product of the specific fuel consumption and its calorific value

$$\eta_o = \frac{N_e}{G_e W_d} = \frac{1}{g_e W_d} \qquad (2)$$

Hence, the up-to-date powertrains are designed in such a manner that the engine is suitably matched to the drivetrain and the car gives the driver the freedom to select the linear car velocity by depressing the accelerator pedal, whereas the selection of the engine load and gearbox ratio is performed automatically. For this reason, the car applied in the study includes a multi-speed *automatic transmission*.

Fig. 19, 20, 21, 22 presents the determined values of engine efficiency of the tested car, where the maximums of this parameter area are visible for each of the tested gearbox ratios and load levels. Thus, we can see that the maximum engine efficiency is in the range of 30% with a note that this level shifts from the 5th gear to the 4th one along with the increase in the load.

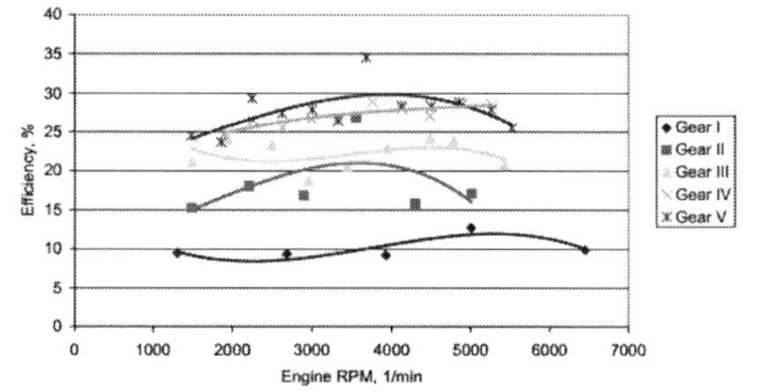

Fig. 19. Efficiency of the engine for 0% road inclination load and variable gearbox ratios.

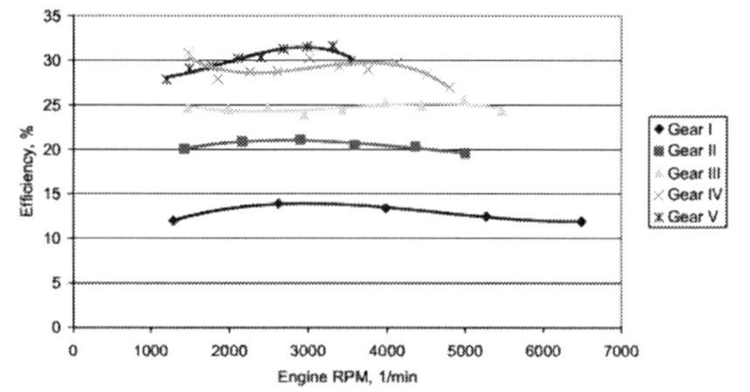

Fig. 20. Efficiency of the engine for 3% road inclination load and variable gearbox ratios.

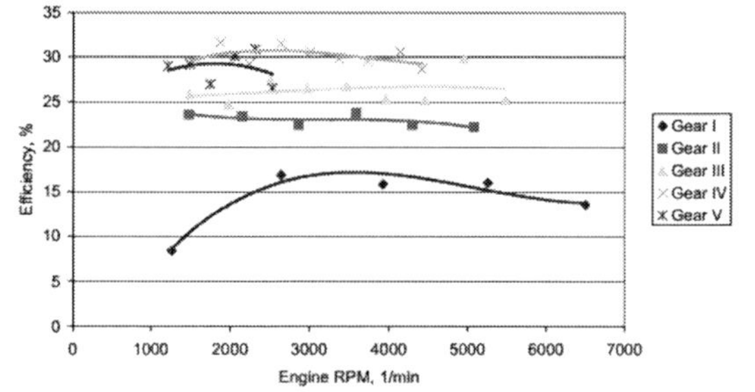

Fig. 21. Efficiency of the engine for 5% road inclination load and variable gearbox ratios.

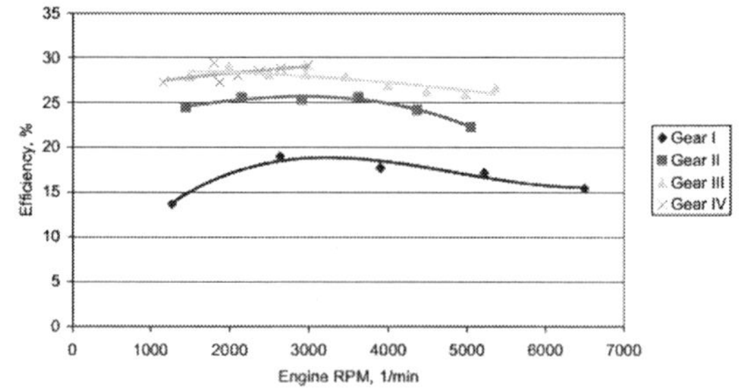

Fig. 22. Efficiency of the engine for 7% road inclination load and variable gearbox ratios.

Concurrently, the maximum engine efficiency can be noted for a horizontal road section in the conditions of a rotational speed of 4000 rpm, which shifts towards lower speeds for greater road inclinations, i.e. at 3% climb – for the rotational speed of 3000 rpm, at 5% climb – for the rotational speed of 2500 rpm (4th gear) and during 7% climb – the rotational speed is already equal to 1800 rpm and the gear is 3rd. These are the areas for the operation of the engine with the greatest efficiency.

Accounting for the data from literature worldwide, we have to consider the tendency towards the greater automation of powertrain systems as a justified and correct one. Only in such conditions should we be able to select an operating point of an engine in an adequate manner to meet the demand with respect to the speed and load so that its maximum efficiency can be achieved.

5 Conclusions

The study conducted into powertrain characteristics leads to a clear statement that the fuel consumption and efficiency of a combustion engine in a passenger car are affected equally by the engine load resulting from the road inclination as well as the adequate selection of the gearbox ratio. The effect of the road inclination has an impact on the driver in terms of their response in the form of a greater depression of the accelerator pedal with the purpose of maintaining the required car speed as well as in the form of the selection of an adequate transmission ratio. The considerations above lead to a conclusion that in the conditions of a constant progress on the way to the improvement of the overall engine performance, it is important to find a manner in which the selection of an instantaneous operating point on engine characteristic is made so as to obtain an increase of the efficiency by selecting an adequate utilization rate of the power generated by the engine.

References

1. A. Bieniek, J. Jantos, J. Mamala, Transport **22**, 247-251 (2007)
2. E. Horvath, A. Torok, Production Engineering Archives **7**, 6-10 (2015)
3. J. Mamala, *Kompensacja niedostatku siły napędowej w procesie rozpędzenia samochodu osobowego* (Opole, 2011)
4. Transport Research Journal of Vilnius Giediminas **XXII**, 247-251 (2007)
5. M. Mitschke, *Dynamika samochodu* (WKL, Warszawa, 1972)
6. N. Reza, Jazar, *Springer Science* (New York, 2008)
7. J. Merkisz, J. Pielecha, S. Radzimirski, *Springer Tracts on Transportation and Traffic* **4**, 1-170 (2014), DOI: 10.1007/978-3-319-02705-0
8. P. Lijewski, J. Merkisz, P. Fuc, A. Ziolkowski, L. Rymaniak, W. Kusiak, European Journal of Forest Research **136**, 153-160 (2017), DOI: 10.1007/s10342-016-1015-2

AUTHOR INDEX